MODULAR FORMS AND STRING THEORY

An indispensable resource for readers in physics and mathematics seeking a solid grasp of the mathematical tools shaping modern theoretical physics, this book comprises a practical introduction to the mathematical theory of modular forms and their application to the physics of string theory and supersymmetric Yang–Mills theory.

Suitable for adventurous undergraduates, motivated graduate students, and researchers wishing to navigate the intersection of cutting-edge research in physics and mathematics, it guides readers from the theory of elliptic functions to the fascinating mathematical world of modular forms, congruence subgroups, Hecke theory, and more. Having established a solid basis, the book proceeds to numerous applications in physics, with only minimal prior knowledge assumed. Appendixes review foundational topics, making the text accessible to a broad audience, along with exercises and detailed solutions that provide opportunities for practice. After working through the book, readers will be equipped to carry out research in the field.

ERIC D'HOKER obtained his PhD in physics from Princeton University and is currently Distinguished Professor of Theoretical and Mathematical Physics at the University of California, Los Angeles (UCLA) and a fellow of the American Physical Society. He was previously a Simons fellow, a Dyson Distinguished Visiting Professor at Princeton's Institute for Advanced Study, and has served as President of the Aspen Center for Physics.

JUSTIN KAIDI obtained his PhD in physics from UCLA. After two years as a research assistant professor at the Simons Center for Geometry and Physics at Stony Brook University, he joined the University of Washington as an assistant professor. He is currently an associate professor at the Institute for Advanced Study and Department of Physics at Kyushu University.

MODULAR FORMS AND STRING THEORY

ERIC D'HOKER
University of California, Los Angeles

JUSTIN KAIDI
Kyushu University

CAMBRIDGE
UNIVERSITY PRESS

Shaftesbury Road, Cambridge CB2 8EA, United Kingdom

One Liberty Plaza, 20th Floor, New York, NY 10006, USA

477 Williamstown Road, Port Melbourne, VIC 3207, Australia

314–321, 3rd Floor, Plot 3, Splendor Forum, Jasola District Centre,
New Delhi – 110025, India

103 Penang Road, #05–06/07, Visioncrest Commercial, Singapore 238467

Cambridge University Press is part of Cambridge University Press & Assessment,
a department of the University of Cambridge.

We share the University's mission to contribute to society through the pursuit of
education, learning and research at the highest international levels of excellence.

www.cambridge.org
Information on this title: www.cambridge.org/9781009457538

DOI: 10.1017/9781009457521

When citing this work, please include a reference to the DOI 10.1017/9781009457521

First published 2025

A catalogue record for this publication is available from the British Library

A Cataloging-in-Publication data record for this book is available from the Library of Congress

ISBN 978-1-009-45753-8 Hardback

To Jody

Contents

Organization   *page* xvi

Acknowledgments   xvii

1   **Introduction**   1

Part I   Modular forms and their variants   7

2   **Elliptic functions**   9

   2.1   Periodic functions of a real variable   9

       2.1.1   Unfolding and Poisson resummation formulas   10

       2.1.2   Application to analytic continuation   11

   2.2   Periodic functions of a complex variable   13

       2.2.1   Application to calculating $\zeta(s)$ at special values   15

   2.3   Elliptic functions   16

       2.3.1   Construction by the method of images   18

   2.4   The Weierstrass elliptic function   19

       2.4.1   The field of elliptic functions in terms of $\wp$ and $\wp'$ 19

       2.4.2   The discriminant and the roots of the cubic   21

       2.4.3   The Weierstrass ζ-function   22

       2.4.4   Rescaling the periods   23

   2.5   Jacobi elliptic functions   23

   2.6   Jacobi ϑ-functions   25

       2.6.1   The number of zeros of ϑ 26

       2.6.2 ϑ-functions with characteristics   27

       2.6.3   The Riemann relations for Jacobi ϑ-functions   28

       2.6.4   Elliptic functions from their zeros and poles   29

   2.7   Uniformization of cubics and quartics   31

	2.8	Elliptic curves	32
		2.8.1 Embedding into $\mathbb{CP}^2$	32
	2.9	Addition formulas and group structure of elliptic curves	34
	2.10	Meromorphic functions and Abelian differentials	35
		2.10.1 Perspective of the torus $\mathbb{C}/\Lambda$	36
		2.10.2 Perspective of the elliptic curve $\mathcal{E}$	37
	2.11	Abelian and elliptic integrals	39
		2.11.1 Abelian integrals of the first kind	39
		2.11.2 Abelian integrals of the second kind	40
		2.11.3 Abelian integrals of the third kind	40
	Exercises		42
3	**Modular forms for** $\mathrm{SL}(2,\mathbb{Z})$		**43**
	3.1	Automorphisms of lattices and the modular group	43
		3.1.1 Structure of the modular group	44
	3.2	The fundamental domain for $\mathrm{SL}(2,\mathbb{Z})$	45
		3.2.1 The Poincaré upper half plane	45
		3.2.2 The fundamental domain for $\mathrm{SL}(2,\mathbb{Z})$	46
		3.2.3 Elliptic points	48
		3.2.4 Cusps	49
	3.3	Modular functions, modular forms, and cusp forms	49
		3.3.1 Comments on terminology and differential forms	50
		3.3.2 The ring of modular forms	52
	3.4	Eisenstein series	52
		3.4.1 Fourier decomposition of Eisenstein series	53
		3.4.2 Poincaré series representation	55
		3.4.3 The polynomial ring of Eisenstein series	56
	3.5	Dimension and generators of the ring of modular forms	58
		3.5.1 Dimension and generators of $\mathcal{M}_k$ and $\mathcal{S}_k$	60
	3.6	Modular functions and the j-function	61
		3.6.1 The j-functions and moonshine	62
		3.6.2 Classification of modular functions	62
		3.6.3 Modular forms with prescribed zeros	63
	3.7	Modular transformations of Jacobi ϑ-functions	64
		3.7.1 The discriminant Δ in terms of ϑ-constants	65
		3.7.2 Sums and products of eighth powers	67
	3.8	The Dedekind η-function	68
	Exercises		70
4	**Variants of modular forms**		**71**
	4.1	Quasi-modular and almost-holomorphic modular forms	71

	4.1.1	E_2 and E_2^*	71
	4.1.2	Ring structure	74
	4.1.3	Differential equations for quasi-modular forms	74
4.2	Non-holomorphic Eisenstein series		76
	4.2.1	Analytic continuation of $E_s(\tau)$	77
	4.2.2	Fourier series of $E_s(\tau)$	78
4.3	Maass forms		80
	4.3.1	The space of Maass forms	82
	4.3.2	Cusp forms	83
	4.3.3	Physical interpretation of the spectrum	85
4.4	Spectral decomposition		85
4.5	Maass forms of arbitrary weight		87
4.6	Mock modular forms		88
	4.6.1	Examples of mock modular forms	89
4.7	Quantum modular forms		92
Exercises			94
5	**Quantum fields on a torus**		95
5.1	Quantum fields		95
	5.1.1	Conformal fields	96
	5.1.2	Free conformal fields	97
5.2	The bc system		98
	5.2.1	The bc system on an annulus	98
	5.2.2	The operator product expansion	99
5.3	Correlators of the bc system on the torus		101
	5.3.1	The 4-point correlator	102
	5.3.2	Arbitrary correlators	103
5.4	The scalar field		103
5.5	The scalar Green function on the torus		105
	5.5.1	The Laplacian on the torus	106
	5.5.2	The scalar Green function via Fourier series	107
	5.5.3	The scalar Green function via ϑ-functions	108
	5.5.4	Modular properties	109
5.6	Scalar determinant and Kronecker's first limit formula		110
	5.6.1	The first Kronecker limit formula	111
	5.6.2	The scalar partition function	111
5.7	Spinor fields		112
	5.7.1	Spinor fields on the annulus	113
	5.7.2	Spinor fields on the torus	113
5.8	Spinor Green functions on the torus		114

	5.8.1	Odd spin structure	115
	5.8.2	Even spin structure via Fourier analysis	115
	5.8.3	Even spin structure via ϑ-functions	116
5.9	Spinor determinant and Kronecker's second limit formula		117
Exercises			120

6   Congruence subgroups and modular curves   **121**
6.1	Definition of congruence subgroups of $\mathrm{SL}(2,\mathbb{Z})$	121
6.2	The classic congruence subgroups	122
6.3	Computing the order of $\mathrm{SL}(2,\mathbb{Z}_N)$	123
6.4	Modular curves	125
	6.4.1 Fundamental domain for a congruence subgroup Γ	125
	6.4.2 The example of $\Gamma(2)$	126
6.5	Compactification of the modular curves	127
6.6	The genus of $X(\Gamma)$	129
	6.6.1 The Hurwitz formula for branched coverings	129
	6.6.2 Calculating the genus of $X(\Gamma)$	129
6.7	Formulas for d, $\varepsilon_2, \varepsilon_3$, and ε_∞ for $\Gamma(N), \Gamma_1(N)$, and $\Gamma_0(N)$	130
	6.7.1 The degree	131
	6.7.2 The absence of elliptic points	131
	6.7.3 Counting elliptic points for $\Gamma_0(N)$	132
	6.7.4 Counting the number of cusps	133
6.8	Explicit expressions for the genus	134
Exercises		136

7   Modular forms for congruence subgroups   **137**
7.1	Modular forms and cusp forms with respect to Γ	137
7.2	Dimension formulas for modular forms	138
	7.2.1 Modular forms of even weight, $k \geq 2$	139
	7.2.2 Modular forms of odd weight, $k \geq 1$	140
7.3	The fields of modular functions on $X(\Gamma)$	142
7.4	Holomorphic Eisenstein series for $\Gamma(N)$	142
	7.4.1 Invariance under $\Gamma(N)$	144
	7.4.2 Asymptotics near the cusps	144
7.5	Fourier decomposition of Eisenstein series for $\Gamma(N)$	145
7.6	Holomorphic Eisenstein series for $\Gamma_0(N)$: a first look	146
7.7	Holomorphic Eisenstein series for $\Gamma_1(N)$ and $\Gamma_0(N)$	147
	7.7.1 The rings of modular forms for congruence subgroups	149

	7.7.2	SageMath	150
7.8		Jacobi's theorem on sums of four and eight squares	150
Exercises			153

8 Modular derivatives and vector-valued modular forms — 155
8.1		Modular covariant derivatives	155
8.2		Modular differential equations	156
	8.2.1	First-order MDE	156
	8.2.2	Second-order MDE	157
	8.2.3	Third-order MDE	158
	8.2.4	Example	159
	8.2.5	Modular invariance of solution spaces	160
8.3		Vector-valued modular forms	160
	8.3.1	Examples	160
	8.3.2	Integrality and $\Gamma(N)$	163
	8.3.3	Relation to modular differential equations	164
Exercises			166

9 Modular graph functions and forms — 168
9.1		Non-holomorphic modular forms of arbitrary weight	168
9.2		One-loop modular graph functions are Eisenstein series	169
9.3		Maass operators and Laplacians	170
9.4		Two-loop modular graph functions	172
9.5		General modular graph functions and forms	175
	9.5.1	Modular properties	176
	9.5.2	Examples: Eisenstein series and dihedral graphs	177
	9.5.3	Examples: holomorphic modular graph forms	177
	9.5.4	Algebraic relations	178
	9.5.5	Differential relations	178
	9.5.6	Higher loop algebraic relations	179
9.6		Relating modular graph functions to holomorphic forms	179
9.7		Modular graph functions and iterated modular integrals	182
Exercises			186

Part II Extensions and applications — 187

10 Hecke operators — 189
10.1	Definition of Hecke operators	189
10.2	Parametrization of equivalence classes $\mathrm{SL}(2,\mathbb{Z})\backslash M_n$	191
10.3	Hecke operators map $\mathcal{M}_k$ to $\mathcal{M}_k$	192
10.4	Fourier expansions	193

10.5   Example: the Ramanujan tau function   194
10.6   Multiplicative properties of Hecke operators   195
10.7   Hecke eigenforms   196
       10.7.1   Petersson inner product   197
10.8   Hecke operators acting on Maass forms   198
10.9   Hecke operators on vector-valued modular forms   200
10.10 Further physics applications   201
Exercises   207

11   **Singular moduli and complex multiplication**   208
11.1   Conditions for complex multiplication   209
11.2   Elliptic functions at complex multiplication points   209
11.3   Examples   210
       11.3.1   Heegner points   211
11.4 $j(\tau)$ as an algebraic integer   212
       11.4.1   The examples of $n = 2, 3, 4, 5$ 215
       11.4.2 $j(\tau)$ as a rational integer   216
11.5 ϑ-functions at complex multiplication points   217
       11.5.1   The point $\tau = i$ 217
       11.5.2   The point $\tau = 2i$ 218
       11.5.3   The point $\tau = \sqrt{2}\,i$ 219
       11.5.4   The points $\tau = in$ and $\tau = i/n$ with $n \in \mathbb{N}$ 219
11.6   The values of E_2, E_4, and E_6 at the points $\tau = i, \rho$ 220
Exercises   221

12   **String amplitudes**   223
12.1   Overview   223
12.2   The Polyakov formulation   225
12.3   String amplitudes as integrals over moduli space   228
12.4   Conformal invariance and decoupling negative norm
       states   230
12.5   String amplitudes in terms of vertex operators   233
12.6   Superstring amplitudes   235
       12.6.1   The Arakelov–Green function   237
       12.6.2   Physical singularity structure of amplitudes   238
12.7   Ultraviolet finiteness from modular invariance   239
12.8   Effective interactions from the four-graviton amplitude   240
       12.8.1   Low-energy expansion at tree level   241
       12.8.2   Transcendental weight   241
12.9   Genus one in terms of modular graph functions   242
12.10 Genus two in terms of modular graph functions   244

12.10.1	Contributions to low weight	245
12.11	Integration over the genus-one moduli space	246
12.11.1	Partitioning the genus-one moduli space	247
12.11.2	Integrals involving Eisenstein series	248
12.11.3	The nonanalytic contribution $\mathcal{A}_R^{(1)}$	249
12.11.4	The analytic contribution $\mathcal{A}_L^{(1)}$	251
12.11.5	Assembling analytic and nonanalytic parts	251
12.12	Integration over the genus-two moduli space	252
12.13	The Selberg trace formula and functional determinants	253
Exercises		258
13	**Toroidal compactification**	**259**
13.1	Conformal field theory on flat manifolds	259
13.2	Lattices and tori of dimension d	260
13.3	Fields taking values in a torus	261
13.4	T-duality on a circle	262
13.5	T-duality on a torus $\mathbb{T}^d$	265
13.5.1	A first look at T-duality	265
13.5.2	Holomorphic block decomposition	267
13.5.3	T-duality in terms of the lattice	267
13.6	Rationality and complex multiplication	270
13.6.1	The case $\mathbb{T}^1$	270
13.6.2	The case $\mathbb{T}^2$	271
Exercises		272
14	**S-duality of Type IIB superstrings**	**274**
14.1	Type IIB supergravity	274
14.1.1	Fields of Type IIB supergravity	275
14.1.2	Field equations of Type IIB supergravity	276
14.2	The $\mathrm{SL}(2,\mathbb{R})$ symmetry of classical Type IIB supergravity	277
14.3	$\mathrm{SL}(2,\mathbb{R})$ to $\mathrm{SL}(2,\mathbb{Z})$ via an anomaly mechanism	278
14.4	$\mathrm{SL}(2,\mathbb{R})$ to $\mathrm{SL}(2,\mathbb{Z})$ via the Dirac quantization condition	281
14.5	Low-energy effective interactions	282
14.6	$\mathrm{SL}(2,\mathbb{Z})$ duality in Type IIB superstring theory	284
14.7	Eisenstein series from supersymmetry and S-duality	285
14.7.1	Non-renormalization theorems	290
Exercises		292
15	**Dualities in $\mathcal{N}=2$ super Yang–Mills theories**	**294**
15.1	Super Yang–Mills: states and fields	294
15.1.1	States	295
15.1.2	Fields	296

15.2   Super Yang–Mills: Lagrangians   298
    15.2.1 $\mathcal{N} = 1$ Lagrangians   299
    15.2.2   Renormalizable $\mathcal{N} = 2$ Lagrangians   299
15.3   Low-energy Lagrangian on the Coulomb branch   301
15.4   BPS states, monopoles, and dyons   303
15.5   SL$(2, \mathbb{Z})$-duality of the $\mathcal{N} = 4$ theory 304
    15.5.1   The global structure of $\mathcal{N} = 4$ theories   305
    15.5.2   Example: $\mathfrak{su}(2)$ SYM   309
    15.5.3   SL$(2, \mathbb{Z})$ on line operators   310
15.6   The Seiberg–Witten solution   311
    15.6.1   The $N_f < 2N$ theory for gauge group SU(N) 312
15.7   The $\mathcal{N} = 2^*$ theory for gauge group SU(N) 313
    15.7.1   The Seiberg–Witten solution   313
    15.7.2   The Seiberg–Witten curve   314
    15.7.3   The vacuum expectation values   315
15.8   The $\mathcal{N} = 2^*$ theory for gauge group SU(2) 316
    15.8.1   Expansion in powers of the mass m 317
    15.8.2   Low orders and perturbative contribution   318
    15.8.3   Modular properties   319
15.9   Linear quiver chains from a limit of $\mathcal{N} = 2^*$ 320
15.10 $\mathcal{N} = 2$ dualities   321
    15.10.1 Conformal symmetry   322
    15.10.2 Superconformal symmetry   323
    15.10.3 Superconformal field theories   324
    15.10.4 SU(2) quivers and Riemann surfaces   326
Exercises   331

16   Basic Galois theory   332
16.1   Fields   332
16.2   Field extensions   333
    16.2.1   Simple extensions   334
    16.2.2   Splitting fields   334
    16.2.3   Normal extensions   335
16.3   Field automorphisms and the Galois group   335
    16.3.1   Intermediate fields   337
16.4   The fundamental theorem of Galois theory   338
16.5   Solvability by radicals   340
16.6   Cyclotomic fields and Abelian extensions   341
16.7   Galois theory in rational conformal field theory   342
Exercises   344

Part III   Appendix   345

Appendix A **Some arithmetic**   347
   A.1   Arithmetic mod N   347
   A.2   Chinese remainder theorem   348
   A.3   Solving polynomial equations   348
   A.4   Quadratic residues and quadratic residue symbols   350
   A.5   Gauss sums   354
   A.6   Quadratic reciprocity   355
   A.7   Characters   355
   A.8   Dirichlet characters   356
   A.9   Dirichlet *L*-functions   357

Appendix B **Riemann surfaces**   362
   B.1   Topology   362
   B.2   Metrics and complex structures   367
   B.3   Uniformization   370
   B.4   Fuchsian groups   373
   B.5   Construction of Riemann surfaces via Fuchsian groups   374

Appendix C **Line bundles on Riemann surfaces**   378
   C.1   Holomorphic line bundles on a Riemann surface   378
   C.2   Holomorphic sections and the Riemann–Roch theorem   382
   C.3   Vanishing theorem and dimension formulas   383
   C.4   Tensors and spinors on Σ 385
   C.5   Proof of the Riemann–Roch theorem   387
   C.6   Proof of the vanishing theorem   387
   C.7   The dimension of moduli space   388

Appendix D **Riemann ϑ-functions and meromorphic forms**   391
   D.1   The Siegel half-space   391
   D.2   The Riemann theta function   392
   D.3   Jacobian, Abel map, and Riemann vanishing theorem   394
   D.4   The prime form   396
   D.5   Holomorphic differentials   396
   D.6   Meromorphic differentials   397
   D.7   The *bc* system   399

Appendix E **Solutions to exercises**   403

References   455
Index   477

Organization

This book is organized into three parts. Each part is preceded by a one-page organizational summary.

Part I provides an introduction to elliptic functions and modular forms and to variants such as quasi-modular forms, almost-holomorphic modular forms, non-holomorphic modular forms, mock modular forms, and quantum modular forms. Full chapters are dedicated to modular forms for congruence subgroups, vector-valued modular forms, and modular graph functions.

Part II provides various mathematical extensions and physical applications of the material of Part I. The mathematical extensions include Hecke operators, complex multiplication, and Galois theory. The physical applications include string amplitudes, T-duality of toroidal compactifications of string theories, S-duality in Type IIB string theory, dualities in Yang–Mills theories with extended supersymmetry, Seiberg–Witten theory, and two-dimensional conformal field theory.

Part III contains four appendixes of material that is central to the core chapters but may be read independently thereof, including introductions to modular arithmetic, the topology and geometry of Riemann surfaces, line bundles on Riemann surfaces, and higher rank ϑ-functions on higher-genus Riemann surfaces. A fifth appendix provides solutions to the Exercises that are formulated at the end of each one of Chapters 2–16.

Bibliographical notes are provided at the end of each chapter and appendix. For mathematics references, many excellent textbooks are available on elliptic functions, modular forms, Riemann surfaces, and Galois theory. For physics references, we shall refer as much as possible to textbooks, review papers, and lecture notes that we find useful and to research papers whenever the material is not readily available otherwise.

Acknowledgments

Eric is indebted to Michael Green, Boris Pioline, Oliver Schlotterer, and especially to Duong Phong for productive and enjoyable collaborations on subjects closely related to this book over the span of many years. Four decades of financial support from the National Science Foundation, as well as a Simons Foundation Fellowship are gratefully acknowledged. Last, but by no means least, this book is dedicated to his wife Jody Enders, whose love and brilliance serve as a never-ending inspiration.

Justin thanks Jan Albert, Ying-Hsuan Lin, Kantaro Ohmori, Julio Parra-Martinez, Eric Perlmutter, Yuji Tachikawa, Gabi Zafrir, and Yunqin Zheng for fruitful collaborations on topics closely related to the content of this book, which in particular influenced several of the chapters in Part II. He also thanks the Simons Center for Geometry and Physics, the Kavli Institute for Physics and Mathematics of the Universe, the University of Washington, Kyushu University, the US Department of Energy, and the Inamori Foundation for support during the writing of this text.

Both authors wish to express their gratitude to their colleagues Zvi Bern, Thomas Dumitrescu, Michael Gutperle, Per Kraus, Julio Parra-Martinez, Mikhail Solon, and Terry Tomboulis for the stimulating intellectual atmosphere at the Mani L. Bhaumik Institute for Theoretical physics, where part of this book was written, and to Dr. Mani Bhaumik for making it all possible.

Finally, they are happy to thank Nicholas Gibbons, Stephanie Windows and Jane Chan of Cambridge University Press for their help during the publication process.

1

Introduction

Integers form a group under addition but not under multiplication. However, matrices with integer entries can form groups under multiplication. For example, 2×2 matrices of unit determinant and integer entries $a, b, c, d \in \mathbb{Z}$,

$$\begin{pmatrix} a & b \\ c & d \end{pmatrix} \qquad ad - bc = 1$$

form the group $\mathrm{SL}(2, \mathbb{Z})$ under multiplication, referred to as the *modular group*. Functions and differential forms that are invariant under $\mathrm{SL}(2, \mathbb{Z})$ are referred to as *modular functions* and *modular forms*, respectively. Modular functions generalize *periodic functions* and *elliptic functions*, to which they are intimately related. In turn, modular functions and modular forms are special cases of *automorphic functions* and *automorphic forms* which are invariant under more general infinite discrete subgroups of $\mathrm{SL}(2, \mathbb{R})$, or under more general multiplicative groups of matrices with integer entries such as $\mathrm{SL}(m, \mathbb{Z})$ and $\mathrm{Sp}(2m, \mathbb{Z})$, referred to as *arithmetic groups*.

In mathematics, the theory of elliptic functions was developed by Gauss, Jacobi, and Weierstrass, building on the study of elliptic integrals by Euler, Legendre, and Abel, and was motivated in part by questions ranging from number theory to the solvability of algebraic equations by radicals. Riemann generalized elliptic functions to Riemann surfaces of arbitrary genus.

The development of modular functions and forms dates back to Eisenstein, Kronecker, and Hecke. Automorphic functions and forms were studied by Fuchs, Fricke, Klein, and Poincaré. In modern times, among many other developments, the Taniyama–Shimura–Weil conjecture ultimately led to the proof of Fermat's Last Theorem by Wiles and Taylor and to a proof of the Modularity Theorem by Breuil, Conrad, Diamond, and Taylor.

A fundamental role was played by modular forms and quasi-modular forms in the solution to the sphere packing problem in eight dimensions by Viazovska, work for which she was awarded the Fields Medal in 2022.

In physics, elliptic functions provide solutions to various boundary problems in electrostatics and fluid mechanics and play a fundamental role in the theory of integrable systems. Elliptic integrals arise in the solution to even the simplest mechanical problems, such as the pendulum.

The modular group $\mathrm{SL}(2, \mathbb{Z})$ first made its appearance in string theory in 1972 when Shapiro identified it as a symmetry of the integrand for the one-loop closed bosonic string amplitude. Shapiro defined the amplitude as the integral over the quotient of the Poincaré upper half plane by $\mathrm{SL}(2, \mathbb{Z})$ and argued that the amplitude thus obtained is free of the short-distance divergences that arise in quantum field theory. This fundamental observation extends to the perturbative superstring theories, and to all loop orders, provided that the group $\mathrm{SL}(2, \mathbb{Z})$ is replaced by the modular group $\mathrm{Sp}(2g, \mathbb{Z})$ for genus g Riemann surfaces. The absence of short-distance divergences uniquely qualifies string theory for the task of unifying gravity with the strong and electro-weak interactions into a consistent quantum theory.

A second context in which $\mathrm{SL}(2, \mathbb{Z})$ arises in string theory is as follows. The five perturbative superstring theories in ten-dimensional Minkowski space-time with spacetime supersymmetry are the Type IIA and Type IIB theories with the maximal number of thirty-two supersymmetries, and the Type I and two Heterotic string theories with sixteen supersymmetries. The massless states of each one of these superstring theories are described by an associated supergravity theory, which is an extension of the Einstein–Hilbert theory of general relativity. For example, Type IIB supergravity contains, in addition to the space-time metric, a complex-valued axion-dilaton scalar field τ. The imaginary part of τ is related to the string coupling and must be positive on physical grounds. Thus, the field τ takes values in the Poincaré upper half plane $\mathrm{SL}(2, \mathbb{R})/\mathrm{SO}(2)$, where the group $\mathrm{SL}(2, \mathbb{R})$ acts on τ by Möbius transformations,

$$\tau \to \frac{a\tau + b}{c\tau + d} \qquad\qquad \begin{pmatrix} a & b \\ c & d \end{pmatrix} \in \mathrm{SL}(2, \mathbb{R})$$

While $\mathrm{SL}(2, \mathbb{R})$ is a symmetry of Type IIB supergravity, certain quantum effects in string theory reduce the $\mathrm{SL}(2, \mathbb{R})$ symmetry to its discrete subgroup $\mathrm{SL}(2, \mathbb{Z})$, also referred to as the part of the *S-duality group* of Type IIB string theory that acts on bosonic fields. Type IIB string solutions related by an $\mathrm{SL}(2,\mathbb{Z})$ transformation are physically identical, so that the space of inequivalent solutions is given by the double coset $\mathrm{SL}(2,\mathbb{Z})\backslash\mathrm{SL}(2,\mathbb{R})/\mathrm{SO}(2)$. The implications of this conjectured symmetry were fully appreciated only with the discovery of NS- and D-brane solutions in the 1990s.

A third context where $\mathrm{SL}(2,\mathbb{Z})$ and higher arithmetic groups emerge is when string theory is considered on a space-time of the form $\mathbb{R}^{10-d} \times \mathbb{T}^d$, where $\mathbb{T}^d$ is a d-dimensional flat torus. Such a setup is referred to as *toroidal compactification*. While the Fourier analysis of supergravity fields on a torus produces only momentum modes, the compactification of a string theory produces both momentum and winding modes. The winding modes are responsible for quintessentially string-theoretic discrete symmetries, referred to as *T-dualities*, that have no counterpart in quantum field theory. For example, Type IIB superstring theory on a circle of radius R is T-dual to Type IIA superstring theory on a circle of radius α'/R, where α' is a constant with dimensions of length squared, which is proportional to the inverse of the string tension. Toroidal compactification converts some of the components of the bosonic fields, such as the metric, into scalar fields, which combine with the axion–dilaton field τ of Type IIB to live on a larger coset space that enjoys a larger arithmetic symmetry group. As a function of the dimension d, the following arithmetic symmetry groups $G(\mathbb{Z})$, and corresponding coset spaces $G(\mathbb{R})/K(\mathbb{R})$, arise starting with the ten-dimensional Type IIB superstring theory for $d = 0$.

d	$G(\mathbb{Z})$	$G(\mathbb{R})$	$K(\mathbb{R})$
0	$\mathrm{SL}(2,\mathbb{Z})$	$\mathrm{SL}(2,\mathbb{R})$	$\mathrm{SO}(2)$
1	$\mathrm{SL}(2,\mathbb{Z})$	$\mathrm{SL}(2,\mathbb{R}) \times \mathbb{R}^{\times}$	$\mathrm{SO}(2)$
2	$\mathrm{SL}(2,\mathbb{Z}) \times \mathrm{SL}(3,\mathbb{Z})$	$\mathrm{SL}(2,\mathbb{R}) \times \mathrm{SL}(3,\mathbb{R})$	$\mathrm{SO}(2) \times \mathrm{SO}(3)$
3	$\mathrm{SL}(5,\mathbb{Z})$	$\mathrm{SL}(5,\mathbb{R})$	$\mathrm{SO}(5)$
4	$\mathrm{SO}(5,5,\mathbb{Z})$	$\mathrm{SO}(5,5,\mathbb{R})$	$(\mathrm{SO}(5) \times \mathrm{SO}(5))/\mathbb{Z}_2$
5	$E_{6,6}(\mathbb{Z})$	$E_{6,6}(\mathbb{R})$	$\mathrm{USp}(8)/\mathbb{Z}_2$
6	$E_{7,7}(\mathbb{Z})$	$E_{7,7}(\mathbb{R})$	$\mathrm{SU}(8)/\mathbb{Z}_2$
7	$E_{8,8}(\mathbb{Z})$	$E_{8,8}(\mathbb{R})$	$\mathrm{Spin}(16)/\mathbb{Z}_2$

On the last three lines, $E_{n,n}$ for $n = 6, 7, 8$ denotes a particular real form of the corresponding complex Lie groups E_6, E_7, and E_8. Compactification on Calabi–Yau manifolds or orbifolds exhibit similarly quintessential string theoretic relations that go under the name of *mirror symmetry*.

A fourth context where the modular group $\mathrm{SL}(2,\mathbb{Z})$ arose in physics is

Yang–Mills theory. The Standard Model of Particle Physics is a Yang–Mills theory with gauge group $SU(3) \times SU(2) \times U(1)$ in which the masses of quarks, leptons, and the gauge bosons $W^{\pm}, Z^0$ are generated via spontaneous symmetry breaking. In a *grand unified field theory*, the group $SU(3) \times SU(2) \times U(1)$ itself arises by spontaneous symmetry breaking of a *simple gauge group* such as $SU(5), SO(10), E_6, E_7$, or E_8. The spectrum of these theories contains a massless photon and various electrically charged particles, but also contains 't Hooft–Polyakov magnetic monopoles. In 1977, Goddard, Nuyts, and Olive conjectured that such theories may exhibit *electric–magnetic duality* which swaps electric particles and magnetic monopoles. Dyons, which carry both electric and magnetic charges, extend this duality to the full modular group $SL(2, \mathbb{Z})$, or a subgroup thereof. A concrete realization of electric–magnetic duality, referred to as Montonen–Olive duality, is provided by Yang–Mills theories with extended supersymmetry, and culminated in the Seiberg–Witten solution in 1994.

A fifth context where modular forms are of great importance is two-dimensional conformal field theory. Cardy linked the $SL(2, \mathbb{Z})$ symmetry of a conformal field theory on a torus $\mathbb{T}^2$ to its unitarity properties. The powerful constraints on the operator product expansion of conformal primary fields in terms of representations of $SL(2, \mathbb{Z})$, obtained by Erik Verlinde in 1988, may be implemented in the *modular bootstrap* program and have been used to great effect to advance the classification of conformal field theories. Thanks to the *gauge–gravity correspondence*, these constraints on conformal field theory give rise to constraints on theories of quantum gravity in three-dimensional anti-de Sitter space, which indeed was one of the original motivations for the modular bootstrap program. Additionally, different conformal field theories may be related to one another by Hecke operators.

Each of the contexts where modular forms play a role in physics, described in the previous paragraphs, continues to provide an active and fertile area of current research, with myriad open questions remaining. There are a number of further important and beautiful contexts where modular forms play a central role but that will not be addressed directly in this book. They include orbifold, Calabi–Yau, and F-theory compactifications; the counting of microstates of four-dimensional black holes; the various incarnations of moonshine; three-dimensional topological field theory; topological modular forms; and modular cosmology. No doubt, the theory of modular forms continues to teach us important conceptual and computational lessons about these well-weathered fields, to say nothing of its promise for up-and-coming topics such as generalized global symmetries.

The goal of this book is to exhibit the profound interrelations between modular forms and string theory, which are numerous. Our presentation is intended to be informal but mathematically precise, logically complete, and self-contained. We have made every effort to render the exposition as simple as possible and accessible to adventurous undergraduates, motivated graduate students, and dedicated professionals interested in the interface between theoretical physics and pure mathematics. To paraphrase Einstein, "Everything should be made as simple as possible, but not simpler."

Assuming little more than a knowledge of complex function theory, some planar differential geometry, and basic group theory, we introduce elliptic functions and elliptic curves as a lead-in to modular forms and modular curves for $SL(2, \mathbb{Z})$ and its congruence subgroups. A prior background in modular arithmetic, Riemann surfaces, or line bundles is not required as those subjects are presented in some detail in four separate appendixes. Free quantum fields on a torus provide an excellent illustration of how elliptic functions and modular forms can be used to solve problems in two-dimensional conformal field theory of relevance to string theory. A basic understanding of the operator formulation of quantum mechanics and some Lie algebra theory will prove useful here but is not absolutely required. As will be explained in the *organizational introductions*, further mathematical topics, ranging from quasi-modular forms and modular graph functions to Hecke operators and Galois theory, are included in order to broaden the spectrum of applications in conformal field theory and string theory.

Even the most economical introduction to string theory proper, attempted here, inevitably benefits from some familiarity with general relativity, classical fields, and some basic elements of scattering theory, though much may be picked up at an intuitive level during a first read-through. By contrast, the chapter on toroidal compactifications should be accessible without further physics prerequisites, while the chapter on S-duality will expose the reader to supergravity. Although the chapter on dualities in super Yang–Mills theory also appeals to several further physics concepts, such as Yang–Mills theory, supersymmetry, magnetic monopoles, and effective field theory, we have attempted to introduce each one of these vast subjects with the minimal amount of detail needed to exhibit their interplay with modular forms.

Part I
Modular forms and their variants

In Part I, we introduce the basic concepts and properties of elliptic functions, holomorphic and non-holomorphic modular forms, and a number of their variants that will be of interest to physics. Mostly mathematical aspects will be discussed here and little prior physics background is required.

In both mathematics and applications to physics, modular forms are intimately tied to elliptic functions. Thus, we start off Part I with a thorough exposition of elliptic functions, elliptic curves, Abelian differentials, and Abelian integrals in Chapter 2. Universal tools, such as Poisson resummation, the analytic continuation of parametric integrals, and the method of images are developed and illustrated along the way.

We introduce holomorphic and meromorphic modular forms under the modular group $SL(2, \mathbb{Z})$ in Chapter 3, account for their properties, and construct them explicitly in terms of holomorphic Eisenstein series. Special attention is devoted to the discriminant cusp form Δ, the j-function, and the Dedekind η-function, which play key roles throughout.

In Chapter 4, we introduce variants of modular forms, including quasimodular forms, almost-holomorphic modular forms, non-holomorphic Eisenstein series, Maass forms, mock modular forms, and quantum modular forms, and present the spectral decomposition theorem, all of which turn up in one physics application or another.

We illustrate the use of elliptic functions and modular forms in two-dimensional conformal field theory in Chapter 5 by constructing Green functions, correlation functions, and functional determinants for scalar and spinor fields on a torus and relating the results to the Kronecker limit formulas. The basics of conformal field theory are introduced here via various examples and no prior knowledge of the subject is required.

In Chapter 6, we review congruence subgroups of $SL(2, \mathbb{Z})$ and their associated modular curves, elliptic points, cusps, and genera. Modular forms for the classic congruence subgroups are introduced in Chapter 7 and applied to counting the number of representations of an integer by sums of squares. In Chapter 8, we further generalize to vector-valued modular forms that transform under nontrivial representations of $SL(2, \mathbb{Z})$, but whose individual components are modular forms under a congruence subgroup and systematically satisfy modular differential equations.

In Chapter 9, we introduce modular graph functions and forms, which provide nontrivial generalizations of non-holomorphic Eisenstein series, multiple zeta values, elliptic polylogarithms, and iterated modular integrals. Physically, they arise in the low energy expansion of superstring amplitudes, as will be shown in Chapter 12.

2

Elliptic functions

In this chapter, elliptic functions are introduced via the method of images following a review of periodic functions, Poisson resummation, the unfolding trick, and analytic continuation applied to the Riemann ζ-function. The differential equations and addition formulas obeyed by periodic and elliptic functions are deduced from their series representation. The classic constructions of elliptic functions, in terms of their zeros and poles, are presented in terms of the Weierstrass $\wp$-function, the Jacobi elliptic functions sn, cn, dn, and the Jacobi ϑ-functions. The elliptic function theory developed here is placed in the framework of elliptic curves, Abelian differentials, and Abelian integrals.

2.1 Periodic functions of a real variable

A function $f : \mathbb{R} \to \mathbb{C}$ is periodic with period 1 if for all $x \in \mathbb{R}$ it satisfies

$$f(x + 1) = f(x) \tag{2.1.1}$$

Equivalently, f is a function of the circle $S^1 = \mathbb{R}/\mathbb{Z} \to \mathbb{C}$. The function f is completely specified by the values it takes in the unit interval $[0, 1)$. There are various other useful systematic methods for constructing periodic functions, some of which we shall now review.

If a function $g : \mathbb{R} \to \mathbb{C}$ decays sufficiently rapidly at ∞, then a periodic function may be constructed by the *method of images*,

$$f(x) = \sum_{n \in \mathbb{Z}} g(x + n) \tag{2.1.2}$$

The functions $e^{2\pi i m x}$ for $m \in \mathbb{Z}$ are all periodic with period 1 and, in fact, form a basis for the square-integrable periodic functions or, equivalently, for $L^2(S^1)$. Their orthogonality and completeness are manifest from the following relations (in the sense of distributions),[1]

$$\int_0^1 dx\, e^{2\pi imx}\, e^{-2\pi im'x} = \delta_{m,m'}$$

$$\sum_{m\in\mathbb{Z}} e^{2\pi imx}\, e^{-2\pi imy} = \sum_{n\in\mathbb{Z}} \delta(x-y+n) \qquad (2.1.3)$$

Any periodic function f with period 1 may be decomposed into a Fourier series as follows,

$$f(x) = \sum_{m\in\mathbb{Z}} f_m\, e^{2\pi imx} \qquad\qquad f_m = \int_0^1 dx\, f(x)\, e^{-2\pi imx} \qquad (2.1.4)$$

The notation $e(x) = e^{2\pi ix}$ is often used in the mathematics literature, but we shall adhere to the physics notation and write the exponential explicitly.

2.1.1 Unfolding and Poisson resummation formulas

A very simple but extremely useful tool is the *unfolding trick*. If a function $g : \mathbb{R} \to \mathbb{C}$ decays sufficiently rapidly at ∞ to make its integral over $\mathbb{R}$ converge absolutely, then the unfolding trick gives,

$$\sum_{n\in\mathbb{Z}} \int_0^1 dx\, g(x+n) = \int_{-\infty}^{\infty} dx\, g(x) \qquad (2.1.5)$$

More generally, combining the method of images with Fourier decomposition, we construct a periodic function f from a nonperiodic function g using (2.1.2) and then calculate the Fourier coefficients f_m using the unfolding trick (2.1.5),

$$f_m = \int_0^1 dx\, e^{-2\pi imx} \sum_{n\in\mathbb{Z}} g(x+n) = \int_{-\infty}^{\infty} dx\, g(x)\, e^{-2\pi imx} \qquad (2.1.6)$$

The Fourier transform of g will be denoted by $\hat{g}$ and is given by,

$$\hat{g}(y) = \int_{-\infty}^{\infty} dx\, g(x)\, e^{-2\pi ixy} \qquad (2.1.7)$$

so that $f_m = \hat{g}(m)$. Therefore, the function f may be expressed in two different ways,

$$f(x) = \sum_{n\in\mathbb{Z}} g(x+n) = \sum_{m\in\mathbb{Z}} \hat{g}(m)\, e^{2\pi imx} \qquad (2.1.8)$$

---

[1] Throughout, the Kronecker symbol $\delta_{m,n}$ equals 1 when $m=n$ and 0 otherwise, while the Dirac δ-function is normalized by $\int_{\mathbb{R}} dy\, \delta(x-y)f(y) = f(x)$ for an arbitrary test function $f(x)$.

Setting $x = 0$ (or any integer) gives the Poisson resummation formula.

Theorem 2.1.1. *A function g and its Fourier transform $\hat{g}$ defined in (2.1.7) obey the Poisson resummation formula,*

$$\sum_{n \in \mathbb{Z}} g(n) = \sum_{m \in \mathbb{Z}} \hat{g}(m) \qquad (2.1.9)$$

assuming both sums are absolutely convergent.

An immediate application is to the case where g is a Gaussian. It will be convenient to normalize Gaussians as follows,

$$g(x) = e^{-\pi t x^2} \qquad\qquad \hat{g}(y) = \frac{1}{\sqrt{t}} e^{-\pi y^2 / t} \qquad (2.1.10)$$

The formulas in (2.1.10) are valid as long as $\mathrm{Re}\,(t) > 0$ and may be continued to $\mathrm{Re}\,(t) = 0$. The corresponding Poisson resummation formula then reads,

$$\sum_{n \in \mathbb{Z}} e^{-\pi t n^2} = \frac{1}{\sqrt{t}} \sum_{m \in \mathbb{Z}} e^{-\pi m^2 / t} \qquad (2.1.11)$$

We shall soon see that this relation admits an important generalization to Jacobi ϑ-functions, where it will correspond to a modular transformation.

2.1.2 Application to analytic continuation

The Riemann ζ-function, known already to Euler, is defined by the series,

$$\zeta(s) = \sum_{n=1}^{\infty} \frac{1}{n^s} \qquad (2.1.12)$$

which converges absolutely for $\mathrm{Re}\,(s) > 1$ and thus defines a holomorphic function of s in that region. Its arithmetic significance derives from the fact that it admits a product representation, due again to Euler,

$$\zeta(s) = \prod_{p \text{ prime}} \left(1 - \frac{1}{p^s} \right)^{-1} \qquad (2.1.13)$$

The Riemann ζ-function is just one example of the family of ζ-functions that may be associated with certain classes of self-adjoint operators whose spectrum is discrete, free of accumulation points, and bounded from below (we shall take them to be bounded from below by zero without loss of generality). Consider such a self-adjoint operator H and its associated discrete spectrum of real eigenvalues λ_n with $n \in \mathbb{N}$ and $\lambda_1 > 0$.[2] We may associate

---

[2] Throughout $\mathbb{N} = \{1, 2, 3, \ldots\}$ denotes the set of positive integers.

a ζ-function to the operator H as follows,

$$\zeta_H(s) = \sum_{n=1}^{\infty} \frac{1}{\lambda_n^s} = \operatorname{Tr}\left(H^{-s}\right) \tag{2.1.14}$$

provided λ_n grows with n sufficiently fast, such as $\lambda_n \sim n^{\alpha}$ for $\alpha > 0$ at large n. If the growth is $\lambda_n \sim n$ for large n, then the series is absolutely convergent for $\operatorname{Re}(s) > 1$ and defines a holomorphic function in s in that region. The Riemann ζ-function may be associated in this way with the Hamiltonian of, say, the harmonic oscillator, whose spectrum is linear $\lambda_n \sim n$, or the free nonrelativistic particle in an interval with periodic boundary conditions, whose spectrum is quadratic $\lambda_n \sim n^2$.

It is a famous result of Riemann that the function $\zeta(s)$ may be analytically continued to the entire complex plane, has a single pole at $s = 1$, and admits a functional relation. The result may be derived using the Poisson relation (2.1.11). Consider the function $\zeta^*(s)$ defined by the integral,

$$\zeta^*(s) = \frac{1}{2} \int_0^{\infty} \frac{dt}{t}\, t^{\frac{s}{2}} \left(\sum_{n \in \mathbb{Z}} e^{-\pi t n^2} - 1\right) \tag{2.1.15}$$

For large t, the integral is convergent for any s in the complex plane, but for small t, it is convergent only for $\operatorname{Re}(s) > 1$, as may be seen by using (2.1.11) for $t \to 0$. For $\operatorname{Re}(s) > 1$, the integral defining $\zeta^*(s)$ may be evaluated by integrating term by term. The contribution of the $n = 0$ term in the sum is cancelled by the subtraction of 1, and we find,

$$\zeta^*(s) = \pi^{-\frac{s}{2}} \Gamma\left(\frac{s}{2}\right) \zeta(s) \tag{2.1.16}$$

The function $\Gamma(z)$ is defined for $\operatorname{Re}(z) > 0$ by its integral representation,

$$\Gamma(z) = \int_0^{\infty} dt\, t^{z-1}\, e^{-t} \tag{2.1.17}$$

and elsewhere by analytic continuation in z. It satisfies the relation,

$$\Gamma(z)\Gamma(1-z)\sin(\pi z) = \pi \tag{2.1.18}$$

To construct an integral representation for $\zeta^*(s)$ that may be analytically continued, we partition the integration region into the intervals $[0, 1]$ and $[1, \infty]$. Leaving the integral over $[1, \infty]$ unchanged, we change variables $t \to 1/t$ in the integral over $[0, 1]$, thereby transforming it into an integral over $[1, \infty]$, and then perform a Poisson resummation in n to obtain,

$$\int_1^{\infty} \frac{dt}{t}\, t^{-\frac{s}{2}} \left(\sum_{n \in \mathbb{Z}} e^{-\pi n^2/t} - 1\right) = \int_1^{\infty} \frac{dt}{t}\, t^{\frac{1-s}{2}} \left(\sum_{n \in \mathbb{Z}} e^{-\pi t n^2} - \frac{1}{\sqrt{t}}\right) \tag{2.1.19}$$

The pure power integrals may be performed and admit an obvious analytic continuation with poles at $s = 0$ and $s = 1$. The final result is given by,

$$\zeta^*(s) = -\frac{1}{s(1-s)} + \int_1^\infty \frac{dt}{t} \left(t^{\frac{s}{2}} + t^{\frac{1-s}{2}} \right) \sum_{n=1}^\infty e^{-\pi t n^2} \qquad (2.1.20)$$

The integral produces a function that is holomorphic throughout $\mathbb{C}$, and the entire expression manifestly enjoys the functional relation,

$$\zeta^*(s) = \zeta^*(1-s) \qquad (2.1.21)$$

which in turn, using (2.1.16), gives the standard functional relation of the Riemann ζ-function. Immediate consequences are that the residue of $\zeta(s)$ at $s = 1$ is one, that $\zeta(s)$ is holomorphic elsewhere throughout $\mathbb{C}$ and, from the functional relation, that $\zeta(-2n) = 0$ for all $n \in \mathbb{N}$.

2.2 Periodic functions of a complex variable

Before turning to elliptic functions, we look at how complex analytic periodic functions and trigonometric functions arise naturally from applying the method of images. We start with a meromorphic function $g : \mathbb{C} \to \mathbb{C}$ that has a single simple pole $g(z) = 1/z$ and use the method of images to construct a periodic meromorphic function,

$$f(z) = \sum_{n \in \mathbb{Z}} \frac{1}{z+n} \qquad (2.2.1)$$

The sum is not absolutely convergent, but it is *conditionally convergent*, which means that its value depends on the order in which we choose to sum the terms. In quantum field theory, we refer to this process as *regularizing* the infinite sum. A natural way to do this is by requiring $f(-z) = -f(z)$ and taking a symmetric limit of a finite sum with this property. Equivalently, we may group terms with n and $-n$ together under the summation,

$$f(z) = \lim_{N \to \infty} \sum_{n=-N}^{N} \frac{1}{z+n} = \frac{1}{z} + \sum_{n=1}^\infty \left(\frac{1}{z+n} + \frac{1}{z-n} \right) \qquad (2.2.2)$$

which renders the sum on the right absolutely convergent for $z \notin \mathbb{Z}$. The meromorphic nature of f allows us to evaluate the sum by finding a function with identical poles, namely with unit residue at every integer, and identical asymptotic behavior at ∞. The difference between this function and $f(z)$ is then holomorphic throughout $\mathbb{C}$ and bounded, and therefore must be constant by Liouville's theorem (which states that a function that is holomorphic and bounded in $\mathbb{C}$ must be constant). The symmetry property $f(-z) = -f(z)$ sets this constant to zero, resulting in the following formula,

$$f(z) = \pi \frac{\cos \pi z}{\sin \pi z} = -i\pi \frac{1 + e^{2\pi i z}}{1 - e^{2\pi i z}} \qquad (2.2.3)$$

Integrating the expressions for f in (2.2.2) and (2.2.3) and determining the integration constant by matching the behavior at $z = 0$ give Euler's celebrated infinite product formula,

$$\sin \pi z = \pi z \prod_{n=1}^{\infty} \left(1 - \frac{z^2}{n^2}\right) \qquad (2.2.4)$$

Linearity of the method of images allows us to treat any function that admits a partial fraction decomposition into poles in z in an analogous manner.

We can also proceed by deriving a differential equation for f directly from its series representation in (2.2.2). For $|z| < 1$, we may expand the second representation in (2.2.2) in a convergent power series in z,

$$f(z) = \frac{1}{z} - 2 \sum_{m=1}^{\infty} z^{2m-1} \zeta(2m) \qquad (2.2.5)$$

Its derivative is given by,

$$f'(z) = -\frac{1}{z^2} - 2 \sum_{m=1}^{\infty} (2m - 1) z^{2m-2} \zeta(2m) \qquad (2.2.6)$$

The combination $f' + f^2$ is free of poles, bounded in $\mathbb{C}$, and therefore constant by Liouville's theorem. The constant is computed by evaluating $f'(z) + f(z)^2$ at $z = 0$, and we find the following differential equation,

$$f'(z) + f(z)^2 = -6\,\zeta(2) = -\pi^2 \qquad (2.2.7)$$

which may, in turn, be solved by quadrature,

$$\int \frac{df}{f - i\pi} - \int \frac{df}{f + i\pi} = -2\pi i \int dz \qquad (2.2.8)$$

Interpreting $z(f)$ as the *inverse function of* $f(z)$, we see that $z(f)$ must be multiple-valued as a function of f since the integrals in df are logarithms and thus multiple-valued around the poles at $f = \pm i\pi$. Since $z(f)$ is multiple-valued with period 1, its inverse function $f(z)$ must be periodic with period 1. The integral relation of (2.2.8), along with the symmetry condition $f(-z) = -f(z)$ used to fix the integration constant, reproduces (2.2.3).

One application is to the derivation of the addition formulas for trigonometric functions. Using only the fact that $f(z)$ of (2.2.3) has a simple pole
at $z = 0$ and is periodic with period 1, we can derive the addition formulas
for trigonometric functions. Indeed, for fixed w, the function $f(z + w)$ has
simple poles at $z = -w \pmod 1$ with unit residue, which match those poles
in $f'(z)/(f(z) + f(w))$. However, the latter also has simple poles at $z \in \mathbb{Z}$,
which may be matched by the function $f(z)$. Implementing the symmetry
under swapping z and w, we arrive at the following identity,

$$2f(z + w) - \frac{f'(z) + f'(w)}{f(z) + f(w)} - f(z) - f(w) = 0 \qquad (2.2.9)$$

from which every trigonometric addition theorem may be deduced.

2.2.1 Application to calculating $\zeta(s)$ at special values

As another application, we evaluate the Riemann ζ-function at even positive
integers by equating the representations for $f(z)$ given in (2.2.3) and (2.2.5),

$$-i\pi \frac{1 + e^{2\pi i z}}{1 - e^{2\pi i z}} = \frac{1}{z} - 2 \sum_{m=1}^{\infty} z^{2m-1}\zeta(2m) \qquad (2.2.10)$$

It follows that $\zeta(2m)$ is π^{2m} times a rational number. To see this, we express
the power series in z of the left side in terms of Bernoulli numbers $B_m \in \mathbb{Q}$,
which are defined as follows,

$$\frac{w}{e^w - 1} = \sum_{m=0}^{\infty} \frac{B_m}{m!} w^m \qquad (2.2.11)$$

Clearly, we have $B_0 = 1$, $B_1 = -1/2$, and $B_{2m+1} = 0$ for all $m \in \mathbb{N}$. The
next few nonzero values for the Bernoulli numbers are as follows,

$$
\begin{array}{llll}
B_2 = \frac{1}{6} & B_6 = \frac{1}{42} & B_{10} = \frac{5}{66} & B_{14} = \frac{7}{6} \\
B_4 - -\frac{1}{30} & B_8 = -\frac{1}{30} & B_{12} = -\frac{691}{2730} & B_{16} = -\frac{3617}{510}
\end{array}
\qquad (2.2.12)
$$

An equivalent formula for Bernoulli numbers of even index is as follows,

$$\frac{w}{2} \times \frac{e^w + 1}{e^w - 1} = \sum_{m=0}^{\infty} \frac{B_{2m}}{(2m)!} w^{2m} \qquad (2.2.13)$$

Setting $w = 2\pi i z$ and identifying term by term in the series (2.2.10) give,

$$\zeta(2m) = -(2\pi i)^{2m} \frac{B_{2m}}{2(2m)!} \qquad (2.2.14)$$

which, in turn, give the following low-order values,

$$\zeta(2) = \frac{\pi^2}{6} \quad \zeta(6) = \frac{\pi^6}{945} \quad \zeta(10) = \frac{\pi^{10}}{93555} \quad \zeta(14) = \frac{2\pi^{14}}{18243225} \quad (2.2.15)$$
$$\zeta(4) = \frac{\pi^4}{90} \quad \zeta(8) = \frac{\pi^8}{9450} \quad \zeta(12) = \frac{691\pi^{12}}{638512875} \quad \zeta(16) = \frac{3617\pi^{16}}{325641566250}$$

By using the functional relation, we may also compute the ζ-function at other values of interest in physics, such as,

$$\zeta(0) = -\frac{1}{2} \qquad \zeta'(0) = -\frac{1}{2}\ln(2\pi) \qquad \zeta(-1) = -\frac{1}{12} \quad (2.2.16)$$

The last formula here is key to the argument that superstrings live in 10 spacetime dimensions!

2.3 Elliptic functions

Periodic functions of one real variable may be generalized to functions of several real variables by considering functions that are periodic in each real variable separately. *Elliptic functions* are periodic functions in 2 real variables, or equivalently doubly periodic functions in a complex variable z, supplemented with the requirement of complex analyticity in z.

Denoting the two periods in the complex variable z by $\omega_1, \omega_2 \in \mathbb{C}$,[3] we require that the periods be linearly independent as vectors in $\mathbb{R}^2$ so that $\omega_2/\omega_1 \notin \mathbb{R}$. The periods then generate a two-dimensional lattice Λ defined by (see Figure 2.1),

$$\Lambda = \mathbb{Z}\omega_1 + \mathbb{Z}\omega_2 \qquad (2.3.1)$$

The lattice Λ is an Abelian group under addition. The quotient $\mathbb{C}/\Lambda$ of the complex plane by the lattice Λ is a torus of dimension 2 over $\mathbb{R}$, that is also an Abelian group. The torus $\mathbb{C}/\Lambda$ may be represented in $\mathbb{C}$ by a parallelogram P_p anchored at a point $p \in \mathbb{C}$ on which opposite sides are pairwise identified with one another (see Figure 2.1),

$$P_p = \{p + t_1\omega_1 + t_2\omega_2, \ 0 \le t_1, t_2 \le 1\} \qquad (2.3.2)$$

A function of z that is doubly periodic with periods ω_1, ω_2 is periodic under the lattice Λ or equivalently is a well-defined function on the torus $\mathbb{C}/\Lambda$. We note that it is often as single-valued functions on a torus that elliptic functions appear in physics.

Collecting these ingredients, we obtain the following definition.

---

[3] In Bateman [Erd81a], the periods are denoted by 2ω and $2\omega'$, respectively, and are related to the periods defined here by $\omega_1 = 2\omega$ and $\omega_2 = 2\omega'$. Our notation for the periods ω_1 and ω_2 is not to be confused with the half-periods ω_α used in equation (18) in Section 13.12 of [Erd81a].

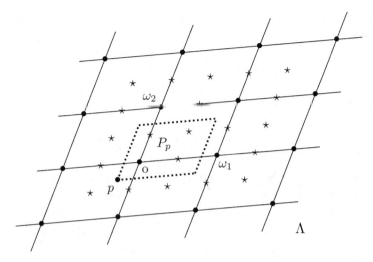

Figure 2.1 The points of the lattice Λ are represented in the complex plane by dots. The lattice Λ is generated by the periods ω_1, ω_2 so that $\Lambda = \mathbb{Z}\omega_1 \oplus \mathbb{Z}\omega_2$. The boundary ∂P_p of the fundamental parallelogram P_p is drawn in a dotted line, where p has been chosen so that ∂P_p contains neither periods, nor half-periods, which are represented by stars.

Definition 2.3.1. *A meromorphic function $f : \mathbb{C} \to \mathbb{C}$ is an elliptic function with periods $\omega_1, \omega_2 \in \mathbb{C}$ with $\omega_2/\omega_1 \notin \mathbb{R}$ if any one of the following equivalent conditions holds,*

- *f is periodic with periods ω_1, ω_2, namely $f(z+\omega_1) = f(z+\omega_2) = f(z)$ for all $z \in \mathbb{C}$;*
- *f is invariant under translations by the lattice Λ, namely $f(z+\omega) = f(z)$ for all $\omega \in \Lambda$ and for all $z \in \mathbb{C}$;*
- *f is single-valued on the torus $\mathbb{C}/\Lambda$.*

Elliptic functions for a given lattice Λ form a *field*, as sums, products, and inverses of elliptic functions for a lattice Λ are elliptic functions for the same lattice Λ, and the constant function 1 is the identity under multiplication.

As we shall establish in Section 2.6.4, elliptic functions are essentially determined by their zeros and poles, subject to certain conditions. For example, Liouville's theorem implies that an elliptic function without poles must be a constant function. To obtain the relations between zeros and poles of elliptic functions more generally, we proceed by defining the residue function $\operatorname{res}_f(w)$ to be the *residue of f* at the point w, and the order function $\operatorname{ord}_f(w)$ to be the *order* of f at the point w, namely the integer such that $f(z)(z-w)^{-\operatorname{ord}_f(w)}$ is a finite nonzero constant as $z \to w$.

Theorem 2.3.2. *The residues and orders of an arbitrary elliptic function f with lattice Λ and corresponding fundamental parallelogram P_p satisfy the following relations,*

$$\sum_{w \in P_p} \mathrm{res}_f(w) = 0$$

$$\sum_{w \in P_p} \mathrm{ord}_f(w) = 0$$

$$\sum_{w \in P_p} w \cdot \mathrm{ord}_f(w) = 0 \ (\mathrm{mod}\ \Lambda) \tag{2.3.3}$$

These relations are proven by integrating the functions $f(z)$, $f'(z)/f(z)$, and $z f'(z)/f(z)$ over the boundary ∂P_p, respectively. When considering a function f on a torus $\mathbb{C}/\Lambda$, it will often be convenient to choose p such that no zeros or poles of f lie on ∂P_p.

In two-dimensional electrostatics, the electric potential Φ may be viewed as a function of z and $\bar{z}$, and it is a harmonic function away from the support of the electric charges. The electric field $\mathbf{E} = -\nabla \Phi$ may be decomposed into derivatives with respect to z and $\bar{z}$, so that $E_z = -\partial_z \Phi$ is a holomorphic function away from the charges. Considering a distribution of point-like charges (without accumulation points) on a torus $\mathbb{C}/\Lambda$, the electric field is an elliptic function for the lattice Λ. The first condition in (2.3.3) requires the sum of all the charges to vanish and must hold on any compact space. The second and third conditions are specific to the torus. The conditions (2.3.3) imply that there exists no elliptic function with just one simple pole in P_p, corresponding to the fact that the torus does not support a single point charge, but there do exist elliptic functions with two simple poles at arbitrary positions and opposite residues corresponding in electrostatics to two opposite point charges. There also exist elliptic functions with a single double pole (which may be considered as a limit of the case with two simple poles) corresponding to the electric field of an electric dipole.

2.3.1 Construction by the method of images

Just as periodic functions may be constructed using the method of images in (2.1.2), an elliptic function may be constructed by using the method of images for the lattice Λ with periods ω_1, ω_2 and $\omega_2/\omega_1 \notin \mathbb{R}$. If a meromorphic function $g : \mathbb{C} \to \mathbb{C}$ decays sufficiently rapidly at ∞, then an elliptic function f may be constructed from g by the method of images,

$$f(z) = \sum_{\omega \in \Lambda} g(z + \omega) = \sum_{m,n \in \mathbb{Z}} g(z + m\omega_1 + n\omega_2) \qquad (2.3.4)$$

Any function $g(z)$ that is bounded by $|g(z)| \leq |z|^{-2-\epsilon}$ for $\epsilon > 0$ as $z \to \infty$ satisfies this criterion. The most famous example is given by the Weierstrass $\wp$-function, in terms of which all other elliptic functions may be expressed, as we will now explain.

2.4 The Weierstrass elliptic function

All elliptic functions for a lattice Λ may be built up from the Weierstrass elliptic function, which has one double pole (mod Λ) and no other poles. The Weierstrass elliptic function $\wp(z|\Lambda)$ for the lattice Λ may be constructed via the method of images. While it may be natural to consider the sum over $\omega \in \Lambda$ of translates $(z + \omega)^{-2}$ of the double pole z^{-2}, in actual fact this series converges only conditionally. We shall learn how to handle such series in Chapter 4. Here we shall define $\wp$ by the manifestly convergent series over the lattice $\Lambda' = \Lambda \setminus \{0\}$,

$$\wp(z|\Lambda) = \frac{1}{z^2} + \sum_{\omega \in \Lambda'} \left(\frac{1}{(z+\omega)^2} - \frac{1}{\omega^2} \right) \qquad (2.4.1)$$

The function $\wp$ is even in z since the lattice sum is invariant under $\Lambda' \to -\Lambda'$.

2.4.1 The field of elliptic functions in terms of $\wp$ and $\wp'$

We start by describing the field of elliptic functions for a fixed lattice Λ in terms of $\wp(z|\Lambda)$ for even functions in z. The elliptic function $\wp(z|\Lambda) - \wp(w|\Lambda)$, viewed as a function of z for fixed $w \notin \Lambda$, has a double pole at $z = 0$ and no other poles, and therefore must have two zeros in view of the second relation in (2.3.3). For generic w, the two zeros $z = \pm w$ are distinct and must be the only two zeros. However, for the nongeneric case where $w \equiv -w \pmod{\Lambda}$, the zero is double. There are exactly four points in P_p with $w \equiv -w \pmod{\Lambda}$, namely the half-periods 0, $\omega_1/2$, $\omega_2/2$, and $\omega_3/2 \pmod{\Lambda}$, where $\omega_3 = \omega_1 + \omega_2$. At the three nonzero half-periods w, the function $\wp(z|\Lambda) - \wp(w|\Lambda)$ has a double zero in z.

The first key result is that every *even* elliptic function f for the lattice Λ

is a rational function of $\wp(z|\Lambda)$, given by,

$$f(z) = \prod_{w \in P_p} \Big(\wp(z|\Lambda) - \wp(w|\Lambda) \Big)^{\mathrm{ord}_f(w)} \tag{2.4.2}$$

which may be established by matching poles and zeros. To represent elliptic functions which are odd under $z \to -z$, or to describe functions without definite parity, we differentiate $\wp$,

$$\wp'(z|\Lambda) = -2 \sum_{w \in \Lambda} \frac{1}{(z+w)^3} \tag{2.4.3}$$

This series is absolutely convergent for $z \notin \Lambda$, and we have $\wp'(-z|\Lambda) = -\wp'(z|\Lambda)$. The function $\wp'(z|\Lambda)^2$ is even and has a single pole of order 6 at $z = 0$ and is therefore a polynomial of degree 3 in $\wp(z|\Lambda)$. To determine this polynomial, we proceed by analogy with the trigonometric case of Section 2.2. We expand both $\wp$ and $\wp'$ in powers of z near $z = 0$ in convergent series,

$$\wp(z|\Lambda) = \frac{1}{z^2} + \sum_{k=1}^{\infty} (k+1)G_{k+2}(\Lambda)\, z^k$$

$$\wp'(z|\Lambda) = -\frac{2}{z^3} + \sum_{k=1}^{\infty} k(k+1)G_{k+2}(\Lambda)\, z^{k-1} \tag{2.4.4}$$

The coefficients G_k are given by the convergent series,

$$G_k(\Lambda) = \sum_{w \in \Lambda'} \frac{1}{w^k} \qquad\qquad k \geq 3 \tag{2.4.5}$$

The reflection symmetry $-\Lambda = \Lambda$ implies,

$$G_{2k+1}(\Lambda) = 0 \qquad\qquad k \in \mathbb{N} \tag{2.4.6}$$

so that the expansions to low order are as follows,

$$\wp(z|\Lambda) = \frac{1}{z^2} + 3\, G_4(\Lambda)z^2 + 5\, G_6(\Lambda)z^4 + \mathcal{O}(z^6)$$

$$\wp'(z|\Lambda) = -\frac{2}{z^3} + 6\, G_4(\Lambda)z + 20\, G_6(\Lambda)z^3 + \mathcal{O}(z^5) \tag{2.4.7}$$

The pole in z of order six, which is present in both $\wp'(z|\Lambda)^2$ and $\wp(z|\Lambda)^3$, is cancelled in the combination $\wp'(z|\Lambda)^2 - 4\wp(z|\Lambda)^3$, whose remaining pole is of order 2. By matching all terms of negative or zero power in z and then using Liouville's theorem, we find the relation,

$$\wp'(z|\Lambda)^2 = 4\wp(z|\Lambda)^3 - g_2(\Lambda)\wp(z|\Lambda) - g_3(\Lambda) \tag{2.4.8}$$

where it is conventional to introduce the notation,

$$g_2(\Lambda) = 60\, G_4(\Lambda) \qquad\qquad g_3(\Lambda) = 140\, G_6(\Lambda) \qquad (2.4.9)$$

We may assemble these results in the following theorem.

Theorem 2.4.1. *Every elliptic function $f(z)$ for a lattice Λ may be expressed as a rational function of the Weierstrass functions $\wp(z|\Lambda)$ and $\wp'(z|\Lambda)$ for lattice Λ, which equivalently may be reduced to a linear function in $\wp'(z)$ using the relation (2.4.8).*

To prove the theorem, we decompose $f(z)$ into its even and odd parts, $f(z) = f_+(z) + f_-(z)$ with $f_\pm(-z) = \pm f_\pm(z)$. Since $f_+(z)$ and $f_-(z)/\wp'(z|\Lambda)$ are both even elliptic functions, we use the result derived in (2.4.2) that every even elliptic function may be expressed as a rational function of $\wp(z|\Lambda)$. Illustrations of Theorem 2.4.1 are provided in Exercises 2.2 and 2.3.

2.4.2 The discriminant and the roots of the cubic

The roots of the cubic polynomial in (2.4.8) must all produce double zeros as a function of z, since the left side of (2.4.8) is a perfect square of a meromorphic function. But we had seen earlier that double zeros of $\wp(z|\Lambda) - \wp(w|\Lambda)$ occur if and only if $w \notin \Lambda$ and $w \equiv -w \pmod{\Lambda}$, namely at the three nonzero half-periods $\frac{1}{2}\omega_1, \frac{1}{2}\omega_2$, and $\frac{1}{2}\omega_3 = \frac{1}{2}\omega_1 + \frac{1}{2}\omega_2$. Denoting the values of $\wp$ at the half-period $\frac{1}{2}\omega_\alpha$ by e_α with $\alpha = 1, 2, 3$,

$$e_\alpha(\Lambda) = \wp\left(\tfrac{1}{2}\omega_\alpha|\Lambda\right) \qquad (2.4.10)$$

we obtain the following factorized form of the cubic polynomial,[4]

$$\wp'(z|\Lambda)^2 = 4\big(\wp(z|\Lambda) - e_1\big)\big(\wp(z|\Lambda) - e_2\big)\big(\wp(z|\Lambda) - e_3\big) \qquad (2.4.11)$$

where the three symmetric functions of e_α may be identified with the coefficients g_2 and g_3 of the cubic (2.4.8) as follows,

$$e_1 + e_2 + e_3 = 0$$
$$e_1 e_2 + e_2 e_3 + e_3 e_1 = -\tfrac{1}{4}g_2$$
$$e_1 e_2 e_3 = \tfrac{1}{4}g_3 \qquad (2.4.12)$$

The discriminant Δ of the cubic $4x^3 - g_2 x - g_3$ is defined as follows,

$$\Delta = 16(e_1 - e_2)^2(e_2 - e_3)^2(e_3 - e_1)^2 \qquad (2.4.13)$$

---

[4] Henceforth, we shall suppress the Λ-dependence of $g_2, g_3,$ and e_α.

In terms of the coefficients g_2 and g_3, it is given by,

$$\Delta = g_2^3 - 27 g_3^2 \qquad (2.4.14)$$

The discriminant is nonzero if and only if all roots are distinct. When two of the roots or all three roots coincide, the discriminant vanishes and the cubic is singular. To obtain e_α in terms of g_2 and g_3, it suffices to find the roots of the cubic polynomial $4x^3 - g_2 x - g_3$, which are given in Lagrange's presentation of the roots by,

$$e_\alpha = \frac{1}{2\sqrt{3}} \left(\rho^\alpha \delta + \rho^{2\alpha} g_2 \, \delta^{-1} \right) \qquad \delta^3 = \sqrt{27} \, g_3 + \sqrt{-\Delta} \quad (2.4.15)$$

for $\alpha = 1, 2, 3$ and $\rho = e^{2\pi i/3}$ a cube root of unity.

2.4.3 The Weierstrass ζ-function

Since $\wp(z|\Lambda)$, viewed as a doubly periodic function on $\mathbb{C}$, has only double poles, it is the derivative in z of a single-valued meromorphic function $\zeta(z|\Lambda)$ on $\mathbb{C}$, referred to as the Weierstrass ζ-functions (not to be confused with the Riemann ζ-function). More precisely, $\zeta(z|\Lambda)$ is defined uniquely by,

$$\wp(z|\Lambda) = -\zeta'(z|\Lambda) \qquad \qquad \zeta(-z|\Lambda) = -\zeta(z|\Lambda) \qquad (2.4.16)$$

and has simple poles in z with unit residue at every point in Λ, and no other poles. As a result, it has a single pole in any fundamental parallelogram P_p and therefore cannot be double periodic in the lattice Λ. Since its derivative is doubly periodic the monodromies of $\zeta(z|\Lambda)$ are constant, and denoted by,

$$\begin{aligned}
\zeta(z + \omega_1|\Lambda) &= \zeta(z|\Lambda) + \eta_1(\Lambda) & \eta_1(\Lambda) &= 2\zeta(\tfrac{1}{2}\omega_1|\Lambda) \\
\zeta(z + \omega_2|\Lambda) &= \zeta(z|\Lambda) + \eta_2(\Lambda) & \eta_2(\Lambda) &= 2\zeta(\tfrac{1}{2}\omega_2|\Lambda)
\end{aligned} \qquad (2.4.17)$$

The expressions for $\eta_1(\Lambda)$ and $\eta_2(\Lambda)$ on the right in (2.4.17) are obtained by setting z equal to $-\tfrac{1}{2}\omega_1$ and $-\tfrac{1}{2}\omega_2$ in the left equations of (2.4.17), respectively, and using the odd parity of $\zeta(z|\Lambda)$. By integrating $z\wp(z|\Lambda)$ along ∂P_p, we obtain the Legendre relation,

$$\eta_1(\Lambda)\omega_2 - \eta_2(\Lambda)\omega_1 = 2\pi i \qquad (2.4.18)$$

Since the monodromies of $\zeta(z|\Lambda)$ are independent of z, the difference $\zeta(z - a|\Lambda) - \zeta(z - b|\Lambda)$ is an elliptic function for the lattice Λ with simple poles at a and b with residues $+1$ and -1, respectively. One also defines the Weierstrass σ-function by $\zeta(z|\Lambda) = \partial_z \ln \sigma(z|\Lambda)$ normalized so that $\sigma(z|\Lambda) = z + \mathcal{O}(z^2)$.

2.4.4 Rescaling the periods

From the definitions in (2.4.1) and (2.4.16), it is manifest that the Weierstrass functions ζ, $\wp$, and $\wp'$ are homogeneous of degree -1, -2, and -3, respectively, under simultaneous rescaling of z and the periods ω_1, ω_2, which is equivalent to rescaling the lattice $\Lambda \to \alpha\Lambda$ and,

$$\zeta(\alpha z | \alpha \Lambda) = \alpha^{-1} \zeta(z | \Lambda)$$
$$\wp(\alpha z | \alpha \Lambda) = \alpha^{-2} \wp(z | \Lambda)$$
$$\wp'(\alpha z | \alpha \Lambda) = \alpha^{-3} \wp'(z | \Lambda) \tag{2.4.19}$$

As a result, the coefficients g_2, g_3, G_m, the roots e_α, and the discriminant Δ are homogeneous of degree $-4, -6, -m, -2$, and -12, respectively. These rescaling properties show that the overall scale of the periods is an "easy" degree of freedom which in many, but not all, applications in mathematics and physics is immaterial. The freedom to rescale allows us to choose a canonical value for one of the periods which is usually taken to be $\omega_1 = 1$ and $\omega_2/\omega_1 = \tau$. In this normalization, the lattice is given by,

$$\Lambda = \mathbb{Z} \oplus \mathbb{Z}\tau \tag{2.4.20}$$

and the torus $\mathbb{C}/\Lambda$ may be represented in $\mathbb{C}$ by a parallelogram with vertices $0, 1, \tau$, and $\tau+1$ and opposite sides identified pairwise, as illustrated in Figure 2.2. The ratio τ is referred to as the *modulus*. For this normalization of the lattice, the Weierstrass function will be denoted as follows,

$$\wp(z) = \wp(z|\tau) = \wp(z|\Lambda) \tag{2.4.21}$$

The leftmost form $\wp(z)$ is reserved for when the lattice Λ and the value of τ are fixed and clear from the context. Out of the coefficients g_2 and g_3, one may construct one combination that is invariant under scaling and, by construction, depends only on the modulus τ,

$$j = 1728\, g_2^3/\Delta \tag{2.4.22}$$

The normalization factor is chosen in such a way that $j(\tau) \approx e^{-2\pi i \tau}$ as $\tau \to i\infty$ with unit coefficient. This so-called j-function will play a fundamental role in the theory of modular forms, to be studied in Chapter 3.

2.5 Jacobi elliptic functions

For a given lattice Λ, the elliptic function $\wp(z|\Lambda) - e_\alpha$ associated with a root e_α has a unique double pole at $z = 0$ and a unique double zero at $z \equiv \frac{1}{2}\omega_\alpha$ (mod Λ) for $\alpha = 1, 2, 3$. The latter follows from the fact that its derivative

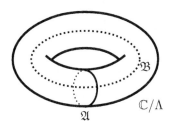

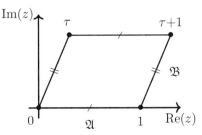

Figure 2.2 The torus $\mathbb{C}/\Lambda$ is represented in the plane by a parallelogram with complex coordinates $z, \bar{z}$ and opposite sides pairwise identified and is shown with a choice of canonical homology cycles $\mathfrak{A}$ and $\mathfrak{B}$ corresponding to the cycles $[0,1]$ and $[0,\tau]$ in $\mathbb{C}$.

$\wp'(z|\Lambda)$ vanishes at the half-periods. Hence $\sqrt{\wp(z|\Lambda) - e_\alpha}$ is a holomorphic function which is doubly periodic, up to sign factors, in z with periods ω_1 and ω_2. The Jacobi elliptic functions $\mathrm{sn}(u|k), \mathrm{cn}(u|k)$, and $\mathrm{dn}(u|k)$ may be defined in terms of these square roots as follows,

$$\mathrm{sn}(u|k) = \sqrt{\frac{e_1 - e_3}{\wp(z|\Lambda) - e_3}} \qquad\qquad \mathrm{dn}(u|k) = \sqrt{\frac{\wp(z|\Lambda) - e_2}{\wp(z|\Lambda) - e_3}}$$

$$\mathrm{cn}(u|k) = \sqrt{\frac{\wp(z|\Lambda) - e_1}{\wp(z|\Lambda) - e_3}} \qquad\qquad k^2 = \frac{e_2 - e_3}{e_1 - e_3} \qquad (2.5.1)$$

The variables u and z and the periods are related as follows,

$$u = \sqrt{e_1 - e_3}\, z \qquad\qquad K_\alpha = \sqrt{e_1 - e_3}\, \omega_\alpha \qquad (2.5.2)$$

The signs of the square roots are fixed by requiring $\mathrm{sn}(u|k) \sim u$, $\mathrm{cn}(u|k) \sim 1$, and $\mathrm{dn}(u|k) \sim 1$ as $u \to 0$. When the value of k is clear from context, one uses the abbreviated notations $\mathrm{sn}(u) = \mathrm{sn}(u|k)$, $\mathrm{cn}(u) = \mathrm{cn}(u|k)$, and $\mathrm{dn}(u) = \mathrm{dn}(u|k)$.

While the functions $\mathrm{sn}(u)^2$, $\mathrm{cn}(u)^2$, $\mathrm{dn}(u)^2$, and $\mathrm{sn}(u)\mathrm{cn}(u)\mathrm{dn}(u)$ are doubly periodic under translations of u by the lattice $\sqrt{e_1 - e_3}\,\Lambda$ generated by the periods K_1, K_2, the functions $\mathrm{sn}(u)$, $\mathrm{cn}(u)$, and $\mathrm{dn}(u)$ are doubly periodic only up to signs in view of the square roots in their definition. Under $u \to -u$, the function $\mathrm{sn}(u)$ is odd while $\mathrm{cn}(u), \mathrm{dn}(u)$ are even. Their zeros and poles are simple and may be read off from the definition.

The modulus k^2 plays a role for Jacobi elliptic functions analogous to the role played by the j-function for Weierstrass functions, and the two may be related to one another by expressing the roots e_α in the definition of k^2 in

(2.5.1) using (2.4.15), which gives the following relation,

$$j = 256 \frac{(k^4 - k^2 + 1)^3}{k^4(k^2 - 1)^2} \qquad (2.5.3)$$

The identities below result from the definitions of Jacobi elliptic functions,

$$
\begin{aligned}
\operatorname{sn}(u)^2 + \operatorname{cn}(u)^2 &= 1 & \operatorname{sn}'(u) &= \operatorname{cn}(u)\operatorname{dn}(u) \\
k^2\operatorname{sn}(u)^2 + \operatorname{dn}(u)^2 &= 1 & \operatorname{cn}'(u) &= -\operatorname{sn}(u)\operatorname{dn}(u) \\
\operatorname{dn}(u)^2 - k^2\operatorname{cn}(u)^2 &= 1 - k^2 & \operatorname{dn}'(u) &= -k^2\operatorname{sn}(u)\operatorname{cn}(u) \qquad (2.5.4)
\end{aligned}
$$

where the prime denotes the derivative with respect to u. The differential relations were obtained with the help of the following relation,

$$\wp'(z) = -2(e_1 - e_3)^{\frac{3}{2}} \frac{\operatorname{cn}(u)\operatorname{dn}(u)}{\operatorname{sn}(u)^3} \qquad (2.5.5)$$

where the prime on the left side denotes the derivative with respect to z. Upon taking the square of each differential relation and using the algebraic equations to express both sides in terms of the same function, we obtain the following differential equations,

$$
\begin{aligned}
\operatorname{sn}'(u)^2 &= \left(1 - \operatorname{sn}(u)^2\right)\left(1 - k^2\operatorname{sn}(u)^2\right) \\
\operatorname{cn}'(u)^2 &= \left(1 - \operatorname{cn}(u)^2\right)\left(1 - k^2 + k^2\operatorname{cn}(u)^2\right) \\
\operatorname{dn}'(u)^2 &= \left(1 - \operatorname{dn}(u)^2\right)\left(k^2 - 1 + \operatorname{dn}(u)^2\right) \qquad (2.5.6)
\end{aligned}
$$

Theorem 2.4.1 implies the following corollary for Jacobi elliptic functions.

Corollary 2.5.1. *An arbitrary elliptic function $f(u)$ with a lattice of periods $\sqrt{e_1 - e_3}\,\Lambda$ may be expressed as a rational function of either $\operatorname{sn}(u)^2$, $\operatorname{cn}(u)^2$, or $\operatorname{dn}(u)^2$ which is linear in the product $\operatorname{sn}(u)\operatorname{cn}(u)\operatorname{dn}(u)$.*

2.6 Jacobi ϑ-functions

In the Weierstrass approach to elliptic functions, the basic building blocks are the meromorphic $\wp$-function and its derivative $\wp'$, in terms of which every elliptic function is a rational function. An alternative approach is to produce elliptic functions in terms of Jacobi ϑ-functions, which are holomorphic at the cost of being multiple-valued on $\mathbb{C}/\Lambda$. Scaling the lattice Λ so that $\omega_1 = 1$ and $\omega_2 = \tau$ with $\operatorname{Im}(\tau) > 0$, the Jacobi ϑ-function is defined by,

$$\vartheta(z|\tau) = \sum_{n \in \mathbb{Z}} e^{i\pi\tau n^2 + 2\pi i n z} \qquad (2.6.1)$$

and is often denoted simply by $\vartheta(z)$ when the τ-dependence is clear. The series is absolutely convergent for $\text{Im}(\tau) > 0$ and defines a holomorphic function in $z \in \mathbb{C}$ and in τ for $\text{Im}(\tau) > 0$. The ϑ-function satisfies a complexified version of the heat diffusion equation,

$$\partial_z^2 \vartheta(z|\tau) = 4\pi i \partial_\tau \vartheta(z|\tau) \qquad (2.6.2)$$

The function $\vartheta(z|\tau)$ is even in $z \to -z$ and transforms under shifts in the lattice $\Lambda = \mathbb{Z} \oplus \mathbb{Z}\tau$ by,

$$\vartheta(z + 1|\tau) = \vartheta(z|\tau)$$
$$\vartheta(z + \tau|\tau) = \vartheta(z|\tau)\, e^{-i\pi\tau - 2\pi i z} \qquad (2.6.3)$$

Thus, $\vartheta(z|\tau)$ is not an elliptic function in z. Indeed, it could never be, as a holomorphic doubly periodic function on $\mathbb{C}$ must be constant by Liouville's theorem. We shall later see how ϑ can be naturally viewed as a holomorphic section of a holomorphic line bundle on $\mathbb{C}/\Lambda$, but for the time being we shall just work with ϑ as a multiple-valued function on $\mathbb{C}/\Lambda$.

2.6.1 The number of zeros of ϑ

To find the number of zeros of $\vartheta(z|\tau)$ as a function of z, we integrate its logarithmic derivative along the closed boundary of the fundamental parallelogram P_0 (with vertices $0, 1, \tau$, and $1 + \tau$), and decompose the integration as a sum of the line integrals along the four edges,

$$\oint_{\partial P_0} dz\, \partial_z \ln \vartheta(z|\tau) = \left(\int_0^1 + \int_1^{1+\tau} + \int_{1+\tau}^{\tau} + \int_\tau^0 \right) dz\, \partial_z \ln \vartheta(z|\tau) \quad (2.6.4)$$

By periodicity of $\vartheta(z|\tau)$ under $z \to z+1$, the contributions from the second and fourth integrals on the right side cancel one another. The contribution of the third integral is just a translate and opposite of the first integral, so that we have,

$$\oint_{\partial P_0} dz\, \partial_z \ln \vartheta(z|\tau) = \int_0^1 dz \Big(\partial_z \ln \vartheta(z|\tau) - \partial_z \ln \vartheta(z + \tau|\tau) \Big) \quad (2.6.5)$$

Using the second relation in (2.6.3), this integral is readily evaluated, and we find $2\pi i$. Since $\vartheta(z|\tau)$ is holomorphic it has no poles, and the result implies that $\vartheta(z|\tau)$ has exactly one zero in P_0. To determine the position of this zero, it will be useful to introduce characteristics.

2.6.2 ϑ-functions with characteristics

A convenient variant of the ϑ-function introduced in (2.6.1) is the ϑ-*function with characteristics* $\alpha, \beta \in \mathbb{C}$, defined by,

$$\vartheta \begin{bmatrix} \alpha \\ \beta \end{bmatrix} (z|\tau) = \sum_{n \in \mathbb{Z}} e^{i\pi\tau(n+\alpha)^2 + 2\pi i(n+\alpha)(z+\beta)} \qquad (2.6.6)$$

These functions are related to the original ϑ-function by translation of z and multiplication by an exponential factor,

$$\vartheta \begin{bmatrix} \alpha \\ \beta \end{bmatrix} (z|\tau) = \vartheta(z + \alpha\tau + \beta|\tau) \, e^{i\pi\tau\alpha^2 + 2\pi i\alpha(z+\beta)} \qquad (2.6.7)$$

The original $\vartheta(z|\tau)$-function corresponds to $\alpha = \beta = 0$. Under translations of z by the lattice Λ, and under a reflection $z \to -z$, we have,

$$\vartheta \begin{bmatrix} \alpha \\ \beta \end{bmatrix} (z+1|\tau) = \vartheta \begin{bmatrix} \alpha \\ \beta \end{bmatrix} (z|\tau) \, e^{2\pi i\alpha}$$

$$\vartheta \begin{bmatrix} \alpha \\ \beta \end{bmatrix} (z+\tau|\tau) = \vartheta \begin{bmatrix} \alpha \\ \beta \end{bmatrix} (z|\tau) \, e^{-2\pi i\beta - i\pi\tau - 2\pi iz}$$

$$\vartheta \begin{bmatrix} -\alpha \\ -\beta \end{bmatrix} (-z|\tau) = \vartheta \begin{bmatrix} \alpha \\ \beta \end{bmatrix} (z|\tau) \qquad (2.6.8)$$

Under integer shift in the characteristics, the transformations are as follows,

$$\vartheta \begin{bmatrix} \alpha+1 \\ \beta \end{bmatrix} (z|\tau) = \vartheta \begin{bmatrix} \alpha \\ \beta \end{bmatrix} (z|\tau)$$

$$\vartheta \begin{bmatrix} \alpha \\ \beta+1 \end{bmatrix} (z|\tau) = \vartheta \begin{bmatrix} \alpha \\ \beta \end{bmatrix} (z|\tau) \, e^{2\pi i\alpha} \qquad (2.6.9)$$

For a ϑ-function with characteristics to have definite parity under $z \to -z$, we must have $-\alpha \equiv \alpha \,(\text{mod } 1)$ and $-\beta \equiv \beta \,(\text{mod } 1)$, which requires $\alpha, \beta \in \{0, \frac{1}{2}\}$ (mod 1) and corresponds to the four half-periods. It is customary to give these four ϑ-functions special names,

$$\vartheta_1(z|\tau) = -\vartheta \begin{bmatrix} \frac{1}{2} \\ \frac{1}{2} \end{bmatrix} (z|\tau) \qquad\qquad \vartheta_2(z|\tau) = \vartheta \begin{bmatrix} \frac{1}{2} \\ 0 \end{bmatrix} (z|\tau)$$

$$\vartheta_3(z|\tau) = \vartheta \begin{bmatrix} 0 \\ 0 \end{bmatrix} (z|\tau) \qquad\qquad \vartheta_4(z|\tau) = \vartheta \begin{bmatrix} 0 \\ \frac{1}{2} \end{bmatrix} (z|\tau) \quad (2.6.10)$$

The minus sign for ϑ_1 is introduced for later convenience. From their definitions in (2.6.10), the parity properties in (2.6.8), and the monodromy properties in (2.6.9), it is manifest that ϑ_1 is odd under $z \to -z$ while $\vartheta_2, \vartheta_3, \vartheta_4$ are even. One designates the corresponding half-integer characteristics as *odd and even characteristics*, respectively.

The parity of ϑ_1 means that we know the location of its zero, namely at $z = 0$, and by periodicity we have zeros at every $z \in \Lambda$. Similarly, the zeros of ϑ_2, ϑ_3, and ϑ_4 are, respectively, at $\frac{1}{2} + \Lambda$, $\frac{1}{2} + \frac{\tau}{2} + \Lambda$, and $\frac{\tau}{2} + \Lambda$. For later use, we record the monodromy properties in z of the function $\vartheta_1(z|\tau)$,

$$\vartheta_1(z+1|\tau) = -\vartheta_1(z|\tau)$$
$$\vartheta_1(z+\tau|\tau) = -\vartheta_1(z|\tau)\, e^{-i\pi\tau - 2\pi i z} \tag{2.6.11}$$

which may be read off from (2.6.8).

2.6.3 The Riemann relations for Jacobi ϑ-functions

There are four basic Riemann relations on Jacobi ϑ-functions, given by,

$$\sum_\kappa \langle \kappa | \lambda \rangle \prod_{i=1}^{4} \vartheta[\kappa](\zeta_i) = 2 \prod_{i=1}^{4} \vartheta[\lambda](\zeta_i') \tag{2.6.12}$$

where κ and λ are half-integer characteristics,

$$\kappa = \begin{bmatrix} \kappa' \\ \kappa'' \end{bmatrix} \qquad\qquad \lambda = \begin{bmatrix} \lambda' \\ \lambda'' \end{bmatrix} \tag{2.6.13}$$

namely with $\kappa', \kappa'', \lambda', \lambda'' \in \{0, \frac{1}{2}\}$. The sign factor $\langle \kappa | \lambda \rangle$ is given by,

$$\langle \kappa | \lambda \rangle = \exp\{4\pi i (\kappa'\lambda'' - \kappa''\lambda')\} \tag{2.6.14}$$

and the relation between ζ_i and ζ_i' is given by,

$$\begin{pmatrix} \zeta_1' \\ \zeta_2' \\ \zeta_3' \\ \zeta_4' \end{pmatrix} = T \begin{pmatrix} \zeta_1 \\ \zeta_2 \\ \zeta_3 \\ \zeta_4 \end{pmatrix} \qquad\qquad T = \frac{1}{2} \begin{pmatrix} 1 & 1 & 1 & 1 \\ 1 & 1 & -1 & -1 \\ 1 & -1 & 1 & -1 \\ 1 & -1 & -1 & 1 \end{pmatrix} \tag{2.6.15}$$

The proof of the Riemann relations is left to the reader as Exercise 2.4.

An important special case is obtained by setting $\zeta_1 = \zeta_2 = z$ and $\zeta_3 = \zeta_4 = w$ so that $\zeta_1' = z + w$, $\zeta_2' = z - w$, and $\zeta_3' = \zeta_4' = 0$ for $i = 2, 3, 4$. The Riemann relation for $\lambda' = \lambda'' = 0$ then becomes an addition formula,

$$\sum_{i=1}^{4} \vartheta_i(z|\tau)^2 \vartheta_i(w|\tau)^2 = 2\vartheta_3(z+w|\tau)\vartheta_3(z-w|\tau)\vartheta_3(0|\tau)^2 \tag{2.6.16}$$

which, for $z = w$, becomes a duplication formula,

$$\vartheta_2(z|\tau)^4 + \vartheta_3(z|\tau)^4 + \vartheta_4(z|\tau)^4 = 2\vartheta_3(2z|\tau)\vartheta_3(0|\tau)^3 \tag{2.6.17}$$

Further setting $z = 0$, we obtain the famous Jacobi formula,

$$\vartheta_3(0|\tau)^4 - \vartheta_2(0|\tau)^4 - \vartheta_4(0|\tau)^4 = 0 \tag{2.6.18}$$

which guarantees the cancellation of the one-loop contribution to the cosmological constant in superstring theories in flat Minkowski space-time.

2.6.4 Elliptic functions from their zeros and poles

There are various classic constructions of elliptic functions in terms of Weierstrass elliptic functions, Jacobi elliptic functions, and Jacobi ϑ-functions, in addition to the construction given in Theorem 2.4.1. The most important two of those will be reviewed here. The relations between those different constructions have direct physical applications, for example, to Bose-Fermi equivalence in two-dimensional quantum field theory.

• The first construction is convenient when the zeros and poles of an elliptic function f are specified (subject to the conditions of Theorem 2.3.2) and proceeds by taking ratios of products of ϑ_1-functions. An elliptic function f must have the same number of zeros and poles in any fundamental parallelogram P_p by the second condition in Theorem 2.3.2. We then have the following theorem.

Theorem 2.6.1. *An elliptic function $f(z|\tau)$ for the lattice $\Lambda = \mathbb{Z} \oplus \mathbb{Z}\tau$ with N poles b_i and N zeros a_i in z modulo Λ and $i = 1, \cdots, N$ may be expressed as a ratio of Jacobi ϑ_1-functions in the following general form,*

$$f(z|\tau) = \prod_{i=1}^{N} \frac{\vartheta_1(z - a_i|\tau)}{\vartheta_1(z - b_i|\tau)} \tag{2.6.19}$$

provided $a_i, b_i \in \mathbb{C}$ satisfy,

$$\sum_{i=1}^{N} (a_i - b_i) \in \Lambda \tag{2.6.20}$$

for a_i, b_i not necessarily mutually distinct.

To prove the theorem, we observe that the poles and zeros of both sides clearly match, the right side is manifestly periodic under $z \to z + 1$, while periodicity in $z \to z + \tau$ may be verified to hold using (2.6.11) if and only if the condition on a_i, b_i of the theorem in (2.6.20) holds.

• The second construction is convenient when the poles of an elliptic function and their residues are given (subject to the conditions of Theorem 2.3.2) and proceeds by using the Weierstrass ζ-function and $\wp$-function, or equivalently by taking the logarithmic derivative of ϑ-functions. The equivalence is made clear by the following relations, presented here for the rescaled lattice $\Lambda = \mathbb{Z} \oplus \mathbb{Z}\tau$ with periods $\omega_1 = 1, \omega_2 = \tau$,

$$\wp(z|\tau) = -\eta_1(\tau) - \partial_z^2 \ln \vartheta_1(z|\tau)$$
$$\zeta(z|\tau) = \eta_1(\tau)z + \partial_z \ln \vartheta_1(z|\tau)$$
$$\ln \sigma(z|\tau) = \ln \vartheta_1(z|\tau) - \ln \vartheta_1'(0|\tau) + \tfrac{1}{2}\eta_1(\tau)z^2 \tag{2.6.21}$$

The first relation follows from the fact that both sides have a double pole $1/z^2$ and no other poles and are doubly periodic with respect to the lattice $\Lambda = \mathbb{Z} \oplus \mathbb{Z}\tau$. The constant $\eta_1(\tau)$ was already encountered in (2.4.17) and may be determined in terms of ϑ-functions by using the fact that $\vartheta_1(z|\tau)$ is odd in z and that $\wp(z|\tau) = z^{-2} + \mathcal{O}(z^2)$, and we find,

$$\eta_1(\tau) = -\frac{1}{3}\frac{\vartheta_1'''(0|\tau)}{\vartheta_1'(0|\tau)} = 2\zeta(\tfrac{1}{2}|\tau) \qquad (2.6.22)$$

The second relation results from integrating the first and requiring $\zeta(z|\tau)$ to be odd in z. Using (2.4.17), (2.4.18), and (2.6.3), one verifies that both sides have the same monodromy. The last equation follows from the second by integration and normalizing $\sigma(z) = z + \mathcal{O}(z^3)$.

In terms of these pieces, the second construction is as follows.

Theorem 2.6.2. *An elliptic function $f(z|\tau)$ with N_s poles at $z = b_{i,s}$ of order s and generalized residue $r_{i,s}$ for $s = 1, \cdots, S$ and $i = 1, \cdots N_s$ may be represented as a sum of Weierstrass ζ-functions and its derivatives, by identifying poles and generalized residues,*

$$f(z|\tau) = r_0 + \sum_{s=1}^{S}\sum_{i=1}^{N_s} r_{i,s}\, \zeta^{(s-1)}(z - b_{i,s}|\tau) \qquad \sum_{i=1}^{N_1} r_{i,1} = 0 \qquad (2.6.23)$$

where r_0 and $r_{i,s}$ may depend on τ but are independent of z, and we have $\zeta'(z|\tau) = -\wp(z|\tau)$.

We note that expressing $\wp(z|\tau)$ as an infinite sum with the help of (2.4.1) for the periods $\omega_1 = 1$ and $\omega_2 = \tau$, integrating the first equation of (2.6.21) twice in z, and matching integration constants gives a product formula for the ϑ_1-function and, by translations by half-periods, for the remaining ϑ_α-functions as well,

$$\vartheta_1(z|\tau) = 2q^{\frac{1}{8}}\sin \pi z \prod_{n=1}^{\infty}(1 - q^n e^{2\pi iz})(1 - q^n e^{-2\pi iz})(1 - q^n)$$

$$\vartheta_2(z|\tau) = 2q^{\frac{1}{8}}\cos \pi z \prod_{n=1}^{\infty}(1 + q^n e^{2\pi iz})(1 + q^n e^{-2\pi iz})(1 - q^n)$$

$$\vartheta_3(z|\tau) = \prod_{n=1}^{\infty}(1 + q^{n-\frac{1}{2}}e^{2\pi iz})(1 + q^{n-\frac{1}{2}}e^{-2\pi iz})(1 - q^n)$$

$$\vartheta_4(z|\tau) = \prod_{n=1}^{\infty}(1 - q^{n-\frac{1}{2}}e^{2\pi iz})(1 - q^{n-\frac{1}{2}}e^{-2\pi iz})(1 - q^n) \qquad (2.6.24)$$

with $q = e^{2\pi i\tau}$. A partial explicit derivation is provided in Exercise 2.5.

2.7 Uniformization of cubics and quartics

Trigonometric functions may be viewed as parametrizing all solutions to the quadratic equation $x^2 + y^2 = 1$ in terms of the functions $x = \sin t$ and $y = \sin' t = \cos t$. This process is referred to as *uniformization*.

Elliptic functions analogously parametrize all solutions to cubic and quartic equations. For example, the Weierstrass functions $\wp(z|\Lambda)$ and $\wp'(z|\Lambda)$ uniformize the following cubic equation,

$$y^2 = 4x^3 - g_2(\Lambda)x - g_3(\Lambda) = 4(x - e_1)(x - e_2)(x - e_3) \qquad (2.7.1)$$

by $x = \wp(z|\Lambda)$ and $y = \wp'(z|\Lambda)$ in view of (2.4.8), where $g_2(\Lambda)$ and $g_3(\Lambda)$ are associated with the lattice Λ via (2.4.9) and (2.4.5). Analogously, the Jacobi elliptic function $\mathrm{sn}(u|k)$ uniformizes the following quartic equation,

$$y^2 = (1 - x^2)(1 - k^2 x^2) \qquad (2.7.2)$$

by $x = \mathrm{sn}(u|k)$ and $y = \mathrm{sn}'(u|k)$ in view of the first equation in (2.5.6). The Jacobi elliptic functions $\mathrm{cn}(u|k)$ and $\mathrm{dn}(u|k)$ similarly uniformize different, but equivalent quartics given by (2.5.6). More generally, elliptic functions may be used to uniformize an arbitrary quartic of the form,

$$w^2 = (v - v_1)(v - v_2)(v - v_3)(v - v_4) \qquad (2.7.3)$$

for arbitrary points $v_1, \cdots, v_4 \in \mathbb{C}$. To show this, we use the fact that the group $\mathrm{SL}(2, \mathbb{C})$ acts conformally and bijectively by Möbius transformations on the compactified complex plane $\mathbb{C} \cup \{i\infty\}$. If the action of an arbitrary $\mathrm{SL}(2, \mathbb{C})$ transformation on the variables v and v_i is given by,

$$v = \frac{ax + b}{cx + d} \qquad v_i = \frac{ax_i + b}{cx_i + d} \qquad \begin{pmatrix} a & b \\ c & d \end{pmatrix} \in \mathrm{SL}(2, \mathbb{C}) \quad (2.7.4)$$

then w and the differences $v - v_i$ obey simple transformation rules,

$$w = \frac{w_0\, y}{(cx + d)^2} \qquad v - v_i = \frac{x - x_i}{(cx + d)(cx_i + d)} \qquad (2.7.5)$$

which map the general quartic (2.7.3) in (v, w) into a quartic in (x, y),

$$w_1^2\, y^2 = (x - x_1)(x - x_2)(x - x_3)(x - x_4)$$
$$w_1^2 = w_0^2(cx_1 + d)(cx_2 + d)(cx_3 + d)(cx_4 + d) \qquad (2.7.6)$$

where w_1 is independent of x, y and otherwise arbitrary. For $w_1 = 1$, the choice $(x_1, x_2, x_3, x_4) = (1, -1, k, -k)$ reproduces the quartic (2.7.2). The limit $x_4 \to \infty$ for $(x_1, x_2, x_3) = (e_1, e_2, e_3)$ and $4cw_0^2 \prod_{i=1}^{3}(ce_i + d) = -1$ reproduces the cubic (2.7.1). This completes the proof that an arbitrary quartic may be uniformized by Weierstrass or Jacobi elliptic functions.

2.8 Elliptic curves

Since an arbitrary lattice Λ is an Abelian subgroup of $\mathbb{C}$ the lattice Λ is a normal subgroup of $\mathbb{C}$ and the torus $\mathbb{C}/\Lambda$ is also an Abelian group. Since Λ acts on $\mathbb{C}$ without fixed points, the torus $\mathbb{C}/\Lambda$ is a compact Riemann surface. Representing the torus in $\mathbb{C}$ by a fundamental parallelogram P_p, the group law may be realized by addition in $\mathbb{C}$ modulo the lattice Λ. The group structure requires an identity element that is just $0 \in \mathbb{C}$ modulo Λ, and an additive inverse of z which is just $-z$ modulo Λ.

The procedure of uniformization, discussed in Section 2.7, shows that elliptic functions naturally parametrize arbitrary cubics and quartics. In particular, the pair of coordinates $(x, y) = (\wp(z|\Lambda), \wp'(z|\Lambda))$ parametrizes the cubic curve $y^2 = 4x^3 - g_2(\Lambda)x - g_3(\Lambda)$ in terms of the points in the torus $z \in \mathbb{C}/\Lambda$. The group structure of $\mathbb{C}/\Lambda$ maps to an Abelian group structure on the corresponding elliptic curve, as we shall now explain.

An elliptic curve $\mathcal{E}$ for given $g_2, g_3 \in \mathbb{C}$ is defined as follows,

$$\mathcal{E} = \left\{ (x, y) \in \mathbb{C} \text{ such that } y^2 = 4x^3 - g_2 x - g_3 \right\} \cup \{P_\infty\} \quad (2.8.1)$$

The point at infinity $P_\infty = (\infty, \infty)$ is added to obtain a compact space, just as the torus $\mathbb{C}/\Lambda$ is compact. While y^2 is determined in terms of x by the equation, the sign of y is not, which is why the second entry y is required. For every generic value of x where $y \neq 0, \infty$, there are two points in $\mathcal{E}$ corresponding to $\pm y$. For the points $x = e_1, e_2, e_3, e_4 = \infty$, however, we have $y = 0, \infty$ and there is only a single point in $\mathcal{E}$. These points correspond to branch points of the function $y = \sqrt{4x^3 - g_2 x - g_3}$ where the two copies of $\mathbb{C}$ intersect (see the left panel of Figure 2.3).

Alternatively, an elliptic curve may be defined in terms of a quartic,

$$\mathcal{E} = \left\{ (v, w) \in \mathbb{C} \text{ such that } w^2 = \prod_{i=1}^{4} (v - v_i) \right\} \cup \{P_{+\infty}, P_{-\infty}\} \quad (2.8.2)$$

The points $v = v_1, v_2, v_3, v_4$ correspond to the branch points of the function w where the two copies of $\mathbb{C}$ intersect (see the right panel of Figure 2.3). In this presentation two points at infinity $P_{\pm\infty}$ are required, one on each sheet.

2.8.1 Embedding into $\mathbb{CP}^2$

The projective space $\mathbb{CP}^2$ is compact, has complex dimension two, and may be defined by an equivalence relation $\approx$ under rescaling coordinates

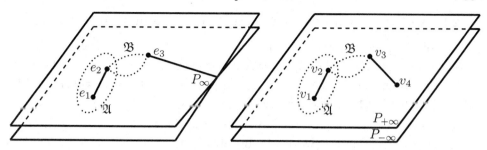

Figure 2.3 The elliptic curve $\mathcal{E}$ is represented in terms of a double cover of the Riemann sphere: with one branch point P_∞ at ∞ in the left panel and with all four branch points finite in the right panel and two distinct points $P_{\pm\infty}$ added at ∞ to compactify the elliptic curve. A choice of canonical homology cycles $\mathfrak{A}, \mathfrak{B}$ is also indicated.

in $(\mathbb{C}^3)^* = \mathbb{C}^3 \setminus \{(0,0,0)\}$ by $\lambda \in \mathbb{C}^* = \mathbb{C} \setminus \{0\}$,

$$\mathbb{CP}^2 = \left\{ (x,y,z) \in (\mathbb{C}^3)^* \, ; (\lambda x, \lambda y, \lambda z) \approx (x,y,z) \text{ for } \lambda \in \mathbb{C}^* \right\} \quad (2.8.3)$$

Actually, $\mathbb{CP}^2$ is a Kähler manifold isomorphic to $SU(3)/S(U(2) \times U(1))$. Every compact Riemann surface may be embedded in $\mathbb{CP}^n$ for some n. In particular, the elliptic curve $\mathcal{E}$ may be embedded in $\mathbb{CP}^2$ by rendering its defining equation homogeneous in three variables, namely the original x, y as well as the additional variable z, to obtain,

$$\mathcal{E} = \left\{ (x,y,z) \in \mathbb{CP}^2 \text{ such that } y^2 z = 4x^3 - g_2 x z^2 - g_3 z^3 \right\} \quad (2.8.4)$$

Since $\mathcal{E}$ is hereby defined as a closed subset of the compact space $\mathbb{CP}^2$, it is compact without the need to add points at infinity. The presentation of the elliptic curve $\mathcal{E}$ in each canonical coordinate patch of $\mathbb{CP}^2$ is given by,

$$
\begin{aligned}
x &= 1 & y^2 z &= 4 - g_2 z^2 - g_3 z^3 \\
y &= 1 & z &= 4x^3 - g_2 x z^2 - g_3 z^3 \\
z &= 1 & y^2 &= 4x^3 - g_2 x - g_3
\end{aligned}
\quad (2.8.5)
$$

The last entry gives the original equation for the elliptic curve. The second equation shows that the point $(0, 1, 0) \in \mathbb{CP}^2$ belongs to every elliptic curve $\mathcal{E}$. In fact, the point $(0, 1, 0)$ corresponds to the point $(x, y) = P_\infty$ in (2.8.1). This may be seen by considering the large x, y behavior in the $z = 1$ patch and allowing for a suitable rescaling $(x, y, 1) \approx (xy^{-1}, 1, y^{-1}) \to (0, 1, 0)$ as $x, y \to \infty$ subject to $y^2 = 4x^3 - g_2 x - g_3$.

2.9 Addition formulas and group structure of elliptic curves

The additive group structure of the torus $\mathbb{C}/\Lambda$ allows one to associate with two arbitrary points $P, Q \in \mathbb{C}/\Lambda$ a third point $R \in \mathbb{C}/\Lambda$ by requiring the sum of their complex coordinates z_P, z_Q, z_R to vanish $z_P + z_Q + z_R = 0 \pmod{\Lambda}$. The "addition" of the points P and Q gives $-R$.

To obtain the corresponding operation on the elliptic curve $\mathcal{E}$ in the Weierstrass presentation, we begin by obtaining the addition formula for the Weierstrass function, which generalizes the addition theorem of (2.2.9) for trigonometric functions,

$$\wp(z_P + z_Q) + \wp(z_P) + \wp(z_Q) - \frac{1}{4}\left(\frac{\wp'(z_P) - \wp'(z_Q)}{\wp(z_P) - \wp(z_Q)}\right)^2 = 0 \quad (2.9.1)$$

To prove it, one begins by showing that the left side has no poles, so that it must be constant by Liouville's theorem. One then shows that the constant vanishes by evaluating the expression at special points. Specifically, for fixed generic z_Q, the function $\wp(z_P + z_Q)$ is an elliptic function in z_P with one double pole at $z_P = -z_Q \pmod{\Lambda}$. Thus it is a rational function of $\wp(z_P)$ and $\wp'(z_P)$. Now $\wp(z_P) - \wp(z_Q)$ has a simple zero at $z_P = -z_Q$, so it should occur to the power -2, but it also has a simple zero at $z_P = z_Q$, which is cancelled by multiplying by the square of $\wp'(z_P) - \wp'(z_Q)$. The combination has a double pole in z_P at $z_P = 0$ which is cancelled by the addition of $\wp(z_P)$ and, by symmetry under the interchange of z_P and z_Q, also $\wp(z_Q)$. Thus the left side has no poles and is therefore constant. To evaluate the constant, we set $z_P = \frac{1}{2}\omega_1$, $z_Q = \frac{1}{2}\omega_2$, and $z_R = -\frac{1}{2}\omega_3$, use the fact that $\wp'$ vanishes at the nonzero half-periods, and then use the first formula in (2.4.12).

On the elliptic curve $\mathcal{E}$, we represent the points P, Q, R by the coordinates (x_P, y_P), (x_Q, y_Q), and (x_R, y_R), respectively and, for simplicity, assume that the points P and Q are distinct. A straight line $y = \alpha x + \beta$ through the points P, Q on $\mathcal{E}$ intersects the cubic $\mathcal{E}$ at exactly one further point R; see, for example, Figure 2.4. Therefore, its coordinates (x_R, y_R) satisfy $y_R^2 = 4x_R^3 - g_2 x_R - g_3$ and,

$$y_R = \frac{y_P - y_Q}{x_P - x_Q} x_R + \frac{x_P y_Q - x_Q y_P}{x_P - x_Q} \quad (2.9.2)$$

This system of equations may be solved for x_R as follows,

$$x_R + x_P + x_Q - \frac{1}{4}\left(\frac{y_P - y_Q}{x_P - x_Q}\right)^2 = 0 \quad (2.9.3)$$

with y_R given by (2.9.2). Parametrizing the coordinates by $x = \wp(z)$ and $y = \wp'(z)$ evaluated at $z = z_P, z_Q, z_R$, we recover the addition law (2.9.1) for

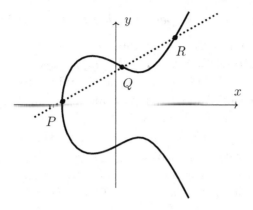

Figure 2.4 The real section of the elliptic curve $y^2 = x^3 - x + 1$, the dotted straight line $9y = 5x + 7.36$, and their intersection points $P, Q,$ and R.

the Weierstrass function. The point P_∞ plays the role of the unit element of the additive group $\mathcal{E}$. This may be seen by letting $x_Q, y_Q \to \infty$ which implies $(x_R, y_R) = (x_P, y_P)$.

2.10 Meromorphic functions and Abelian differentials

Much of the work on elliptic functions in the nineteenth century was motivated by the study of elliptic integrals, namely definite or indefinite integrals involving the square root of a cubic or quartic polynomial. In this section, we shall consider meromorphic functions and Abelian differentials before addressing their integrals in the subsequent section. For a discussion of differentials and their properties on arbitrary Riemann surfaces, see Appendix B.

The space of meromorphic functions on the Riemann sphere $\hat{\mathbb{C}} = \mathbb{C} \cup \{i\infty\}$ is the field $\mathbb{C}(x)$ of rational functions of a holomorphic coordinate $x \in \mathbb{C}$. Meromorphic functions on an elliptic curve $\mathcal{E}$ correspond to elliptic functions on the torus $\mathbb{C}/\Lambda$. Since elliptic functions on $\mathbb{C}/\Lambda$ are rational functions of $x = \wp(z|\Lambda)$ and $y = \wp'(z|\Lambda)$, the space of meromorphic functions on an elliptic curve $\mathcal{E}$ is the field $\mathbb{C}(x, y)$ of rational functions of x and y. The field $\mathbb{C}(x, y)$ may be viewed as a quadratic extension of $\mathbb{C}(x)$ by y, since $y^2 \in \mathbb{C}(x)$ on an elliptic curve $\mathcal{E}$. The Galois group of this extension is $\mathbb{Z}_2$ (see Chapter 16), the group of involutions of $\mathcal{E}$ that maps $(x, y) \to (x, -y)$ and swaps the two sheets in the elliptic representation of Figure 2.3. On the torus $\mathbb{C}/\Lambda$, the $\mathbb{Z}_2$ involution simply corresponds to $z \to -z$ (mod Λ).

2.10.1 Perspective of the torus $\mathbb{C}/\Lambda$

To investigate meromorphic differentials, we first consider the perspective of the torus. On a torus $\mathbb{C}/\Lambda$ with local complex coordinate z, there exists a translation invariant $(1,0)$-form dz, which is unique up to a multiplicative constant. The differential dz is holomorphic and nowhere vanishing. As a result, the vector spaces of meromorphic $(n,0)$-forms for $n \in \mathbb{Z}$ are all isomorphic to the space of meromorphic $(0,0)$-forms, that is, elliptic functions. Thus, the most general meromorphic differential $(n,0)$-form on $\mathbb{C}/\Lambda$ is given by $f(z)(dz)^n$, where $f(z)$ is an elliptic function on $\mathbb{C}/\Lambda$. We shall denote the vector space of meromorphic $(n,0)$-forms $\Omega^{n,0} = \Omega^{n,0}(\mathbb{C}/\Lambda)$.

The construction of elliptic functions in terms of the Weierstrass $\wp$-function, given in Theorem 2.4.1, allows us to express an arbitrary meromorphic $(n,0)$-form ϖ as a rational function in $\wp$ and $\wp'$ which is linear in $\wp'$,

$$\varpi = A\big(\wp(z)\big)(dz)^n + B\big(\wp(z)\big)\,\wp'(z)(dz)^n \tag{2.10.1}$$

where A and B are rational functions. Equivalently, Theorem 2.6.2 allows us to express ϖ as a sum of derivatives of the Weierstrass ζ-function,

$$\varpi = r_0\,dz^n + \sum_{s=1}^{S}\sum_{i=1}^{N_s} r_{i,s}\,\zeta^{(s-1)}(z - b_{i,s})\,dz^n \qquad \sum_{i=1}^{N_1} r_{i,1} = 0 \tag{2.10.2}$$

where the poles $b_{i,s}$ and the coefficients $r_0, r_{i,s}$ are arbitrary aside from the vanishing of the sum of $r_{i,1}$. In the remainder of this section, we shall concentrate on $(1,0)$-forms in $\Omega^{1,0}$ as they are the ones that constitute the integrands of Abelian integrals.

The space $d\Omega^{0,0}$ of $(1,0)$-forms obtained by taking the differential of elliptic functions is a subspace $d\Omega^{0,0} \subset \Omega^{1,0}$. This may be seen in the representation (2.10.1) by taking the differential of an arbitrary elliptic function $f \in \Omega^{0,0}$ and using the differential equation (2.4.8) and its derivative to recast the result in the form (2.10.1). In the representation of (2.10.2), the differential kills the first term and simply increases the order of the derivatives in the terms under the double sum, thereby producing a $(1,0)$-form of the type (2.10.2).

However, not every meromorphic $(1,0)$-form is the differential of an elliptic function. Constructing the quotient space $\Omega^{1,0}/d\Omega^{0,0}$ of meromorphic $(1,0)$ differentials modulo differentials of elliptic functions is a cohomology problem whose solution is fundamental in the theory of Riemann surfaces and in string theory. To obtain the generators of the quotient $\Omega^{1,0}/d\Omega^{0,0}$, we express an arbitrary $f \in \Omega^{0,0}$ and an arbitrary $\varpi \in \Omega^{1,0}$ in terms of the representation (2.10.2), and then take the differential of f,

$$df = \sum_{s=1}^{\tilde{S}} \sum_{i=1}^{\tilde{N}_s} \tilde{r}_{i,s}\, \zeta^{(s)}(z - \tilde{b}_{i,s})\, dz \qquad\qquad \sum_{i=1}^{\tilde{N}_1} \tilde{r}_{i,1} = 0$$

$$\varpi = r_0\, dz + \sum_{s=1}^{S} \sum_{i=1}^{N_s} r_{i,s}\, \zeta^{(s-1)}(z - b_{i,s})\, dz \qquad\qquad \sum_{i=1}^{N_1} r_{i,1} = 0 \qquad (2.10.3)$$

All terms in ϖ involving $\zeta^{(s-1)}(z - b_{i,s})$ for $s \geq 3$ may be matched by terms in df provided we choose the poles $\tilde{b}_{i,s-1} = b_{i,s}$ and coefficients $\tilde{r}_{i,s-1} = r_{i,s}$. For $s = 2$, this would also be the case if it were not for the constraint on the sum of the $\tilde{r}_{i,1}$, which means that a single $\zeta^{(1)}(z - b) = -\wp(z - b)$ is, manifestly, not the differential of an elliptic function. The term $r_0 dz$ and the terms proportional to $\zeta(z - b_{i,1})$ are not differentials of elliptic functions but, in view of the vanishing sum of residues, always occur as differences $\zeta(z - b_{i,1}) - \zeta(z - b_{j,1})$. In summary, $\Omega^{1,0}/d\Omega^{0,0}$ is generated by three type of $(1, 0)$-forms, referred to as Abelian differentials.

1. *Abelian differential of the first kind*: are holomorphic and given by a constant multiple of the coordinate differential dz;
2. *Abelian differentials of the second kind*: have a double pole at an arbitrary point $z_i \in \mathbb{C}/\Lambda$, no other poles, and are given by a constant multiple of,

$$\wp(z - z_i)dz = -\zeta^{(1)}(z - z_i)dz \qquad (2.10.4)$$

3. *Abelian differentials of the third kind*: have two simple poles with residues ± 1 at arbitrary points $z_i \neq z_j \in \mathbb{C}/\Lambda$, no other poles, and are given in terms of the Weierstrass ζ-function by,

$$\big(\zeta(z - z_i) - \zeta(z - z_j) + \zeta(z_i - z_j)\big)dz \qquad (2.10.5)$$

The third term inside the parentheses is included to make the differential invariant under Λ not only in z but also in the points z_i and z_j.

The characterization of Abelian differentials of the second and third kind in terms of their poles does not specify them uniquely as one may add an arbitrary constant multiple of dz to either.

2.10.2 Perspective of the elliptic curve $\mathcal{E}$

For an elliptic curve $\mathcal{E}$ with local coordinates (x, y) given by,

$$y^2 = 4x^3 - g_2 x - g_3 \qquad (2.10.6)$$

with $g_2, g_3 \in \mathbb{C}$, the $(1,0)$ form dz on $\mathbb{C}/\Lambda$ may be expressed in terms of $(x, y) = (\wp(z), \wp'(z))$ by using the relation $(2.4.8)$ and we find,

$$dz = \frac{dx}{y} = \frac{dx}{\sqrt{4x^3 - g_2 x - g_3}} \qquad (2.10.7)$$

The differential dx/y is single-valued and holomorphic on $\mathcal{E}$ since the square root at one of the branch points e_α is well defined in a suitable local coordinate ξ with $\xi^2 = x - e_\alpha$, in which the differential is given by $dx/y = c_\alpha^{-1} d\xi + \mathcal{O}(\xi)$ where $c_\alpha^2 = (e_\alpha - e_\beta)(e_\alpha - e_\gamma)$ for $\beta, \gamma \neq \alpha$. The coordinate differential dx vanishes at the branch points e_α and has a triple pole at ∞.

Mapping an arbitrary cubic $(2.10.6)$ to a quartic of the form $(2.7.3)$,

$$w^2 = (v - v_1)(v - v_2)(v - v_3)(v - v_4) \qquad (2.10.8)$$

the differential dx/y is mapped to dv/w by the $SL(2, \mathbb{C})$ Möbius transformation of $(2.7.4)$ which maps (x, y) to (v, w), x_i to v_i, and the differential dx to $dv = dx/(cx + d)^2$.

An arbitrary meromorphic differential ϖ on the torus $\mathbb{C}/\Lambda$, expressed in terms of $\wp$ and $\wp'$ in $(2.10.1)$, translates to the following expression on the corresponding elliptic curve $\mathcal{E}$,

$$\varpi = A(x)\frac{dx}{y} + B(x)dx \qquad (2.10.9)$$

where A and B are rational functions of x. Under the $\mathbb{Z}_2$ involution of $\mathcal{E}$ which maps $(x, y) \to (x, -y)$, the first term in ϖ is odd while the second term is even. We leave it as an exercise to show that the Abelian differentials on the torus $\mathbb{C}/\Lambda$ map to the following differentials on the elliptic curve $\mathcal{E}$,

1. *Abelian differential of the first kind*: a constant multiple of,

$$\frac{dx}{y} \qquad (2.10.10)$$

2. *Abelian differentials of the second kind*: have one double pole at an arbitrary point $(x_i, y_i) \in \mathcal{E}$, no other poles and, using the addition formula of $(2.9.1)$, are given by,

$$\left(-x - x_i + \frac{1}{4}\frac{(y + y_i)^2}{(x - x_i)^2} \right) \frac{dx}{y} \qquad (2.10.11)$$

3. *Abelian differentials of the third kind*: have two simple poles with opposite residues ± 1 at arbitrary points $(x_i, y_i) \neq (x_j, y_j)$, no other poles, and are

given by,

$$\frac{1}{2}\left(\frac{y+y_i}{x-x_i}+\frac{y_j+y}{x_j-x}+\frac{y_i+y_j}{x_i-x_j}\right)\frac{dx}{y} \tag{2.10.12}$$

The last term inside the parentheses arises from the $\zeta(z_i - z_j)dz$ contribution in the torus representation (2.10.5) and is proportional to the Abelian differential of the first kind.

2.11 Abelian and elliptic integrals

The decomposition of the space of meromorphic $(1,0)$-forms $\Omega^{1,0}$ into a sum of the space of exact differentials of elliptic functions $d\Omega^{0,0}$ plus the three kinds of basic Abelian differentials greatly simplifies the problem of integrating $(1,0)$-forms. Integration of exact differentials $\varpi \in d\Omega^{0,0}$ is straightforward, and we shall now concentrate on the integration of the three kinds of Abelian differentials. A canonical basis of first homology generators of the elliptic curve $\mathcal{E}$ will be denoted by $\mathfrak{A}, \mathfrak{B}$ and was shown already in Figures 2.2 and 2.3. Parametrizing the lattice $\Lambda = \mathbb{Z}\omega_1 \oplus \mathbb{Z}\omega_2$ by the periods ω_1 and ω_2 satisfying $\tau = \omega_2/\omega_1$ with $\tau \notin \mathbb{R}$, the homology basis may be represented on the torus $\mathbb{C}/\Lambda$ by the following simple cycles,

$$\begin{aligned}\mathfrak{A}: \quad & z \to z + \omega_1 \\ \mathfrak{B}: \quad & z \to z + \omega_2\end{aligned} \tag{2.11.1}$$

We shall study integrals of Abelian differentials along open and closed paths in $\mathcal{E}$ and $\mathbb{C}/\Lambda$. We consider open paths between arbitrary points $z_1, z_2 \in \mathbb{C}/\Lambda$ corresponding to points (x_1, y_1) and $(x_2, y_2) \in \mathcal{E}$, related by the Weierstrass function $x_i = \wp(z_i), y_i = \wp'(z_i)$ for $i = 1, 2$.

2.11.1 Abelian integrals of the first kind

The integral of the Abelian differential $dz = dx/y$ of the first kind is given as follows,

$$\int_{(x_1,y_1)}^{(x_2,y_2)} \frac{dx}{\sqrt{4x^3 - g_2 x - g_3}} = \int_{(x_1,y_1)}^{(x_2,y_2)} \frac{dx}{y} = \int_{z_1}^{z_2} dz = z_2 - z_1 \tag{2.11.2}$$

Viewed as taking values in $\mathbb{C}$, the integral depends on the path of integration. Indeed, while holomorphicity of $dz = dx/y$ guarantees independence under small deformations of the path of integration, the integral will depend on how many times the path circles the cycles $\mathfrak{A}$ and $\mathfrak{B}$. Combining (2.11.2) and (2.11.1), the integrals of $dz = dx/y$ around $\mathfrak{A}, \mathfrak{B}$ give the periods of the lattice Λ, with $\tau = \omega_2/\omega_1$ as follows,

$$\oint_{\mathfrak{A}} \frac{dx}{y} = \oint_{\mathfrak{A}} dz = \omega_1 \qquad \qquad \oint_{\mathfrak{B}} \frac{dx}{y} = \oint_{\mathfrak{B}} dz = \omega_2 \qquad (2.11.3)$$

While the value of the integral of (2.11.2) depends on the path of integration when its range is $\mathbb{C}$, the integral becomes single-valued as a map to $\mathbb{C}/\Lambda$. Abelian integrals of the first kind on Riemann surfaces of higher genus and the related Abel map are discussed in Appendix D.

2.11.2 Abelian integrals of the second kind

The integral of the Abelian differential ϖ of the second kind,

$$\varpi = \wp(z - z_i)dz = \left(-x - x_i + \frac{1}{4}\frac{(y+y_i)^2}{(x-x_i)^2}\right)\frac{dx}{y} \qquad (2.11.4)$$

is given by the Weierstrass ζ-function in view of $\zeta'(z) = -\wp(z)$,

$$\int_{z_1}^{z_2} \varpi = \zeta(z_1 - z_i) - \zeta(z_2 - z_i) \qquad (2.11.5)$$

The integral depends on the path of integration through the number of times the path winds around the cycles $\mathfrak{A}$ and $\mathfrak{B}$ and may be evaluated using (2.4.17),

$$\oint_{\mathfrak{A}} \varpi = -\eta_1 \qquad \qquad \oint_{\mathfrak{B}} \varpi = -\eta_2 \qquad (2.11.6)$$

The periods of ϖ do not belong to the lattice Λ, as is clear from the Legendre relation (2.4.18). In fact, it is not possible to form a linear combination of ϖ and the Abelian differential of the first kind dz to obtain a single-valued integral taking values in $\mathbb{C}/\Lambda$. But it is possible to form combinations with either vanishing $\mathfrak{A}$ period or vanishing $\mathfrak{B}$ period. For example, $\varpi + \eta_1 dz/\omega_1$ has vanishing $\mathfrak{A}$ period while its $\mathfrak{B}$ period is $2\pi/\omega_1$.

2.11.3 Abelian integrals of the third kind

The Abelian differential ϖ of the third kind, encountered already in (2.10.5) and (2.10.12), may be expressed as follows,

$$\varpi = \Big(\zeta(z - z_i) + \zeta(z_i - z_j) + \zeta(z_j - z)\Big)dz$$
$$= \frac{1}{2}\left(\frac{y+y_i}{x-x_i} + \frac{y_i+y_j}{x_i-x_j} + \frac{y_j+y}{x_j-x}\right)\frac{dx}{y} \qquad (2.11.7)$$

The equality between the two expressions may be verified directly by using the addition formula for the Weierstrass ζ-function,

$$\zeta(u - v) = \zeta(u) - \zeta(v) + \frac{1}{2}\frac{\wp'(u) + \wp'(v)}{\wp(u) - \wp(v)} \tag{2.11.8}$$

and then expressing $\wp$ and $\wp'$ in terms of x and y. The integral of ϖ is given in terms of the Weierstrass σ-function, which is related to $\zeta(z)$ by the last two equations in (2.6.21) and,

$$\frac{\sigma(z)'}{\sigma(z)} = \zeta(z) \tag{2.11.9}$$

via the logarithm,

$$\int_{z_1}^{z_2} \varpi = \ln \frac{\sigma(z_2 - z_i)\sigma(z_1 - z_j)}{\sigma(z_1 - z_i)\sigma(z_2 - z_j)} + (z_2 - z_1)\zeta(z_i - z_j) \tag{2.11.10}$$

For the canonical periods $\omega_1 = 1$ and $\omega_2 = \tau$, we may convert this expression to one in terms of Jacobi ϑ-functions using (2.6.21), and we obtain,

$$\int_{z_1}^{z_2} \varpi = \ln \frac{\vartheta_1(z_2 - z_i)\vartheta_1(z_1 - z_j)}{\vartheta_1(z_1 - z_i)\vartheta_1(z_2 - z_j)}$$
$$+(z_2 - z_1)\Big\{\zeta(z_i - z_j) - \eta_1(z_i - z_j)\Big\} \tag{2.11.11}$$

Its period integrals are obtained by using (2.6.11) and are given by,

$$\oint_{\mathfrak{A}} \varpi = \zeta(z_i - z_j) - \eta_1(z_i - z_j)$$
$$\oint_{\mathfrak{B}} \varpi = \tau\zeta(z_i - z_j) - \eta_2(z_i - z_j) \tag{2.11.12}$$

We conclude that all elliptic integrals can be evaluated in terms of Weierstrass functions. We leave it as an exercise to the reader to obtain the corresponding Abelian integrals in terms of Jacobi ϑ-functions.

• Bibliographical notes

The early history of the development of elliptic functions and modular forms, with emphasis on the contributions of Eisenstein and Kronecker, is discussed in the delightful book by Weil [Wei76]. Classic treatises on Weierstrass elliptic functions, Jacobi elliptic functions, and Jacobi ϑ-functions may be found in Whittaker and Watson [WW69]. The three volumes of the Bateman manuscript project [Erd81a, Erd81c, Erd81b] offer a treasure trove of practical definitions and useful formulas for a wealth of special functions, including Weierstrass and Jacobi elliptic functions as well as Jacobi ϑ-functions.

Mumford [Mum07] emphasizes Jacobi ϑ-functions, while Lang [Lan87] deals with elliptic functions as well as modular forms. The perspective of the geometry of algebraic curves is presented in detail in the book by Arbarello, Cornalba, Griffiths, and Harris [ACGH85]. Rational points on elliptic curves are pedagogically presented in the book by Silverman and Tate [ST92], while applications of elliptic curves to cryptography may be found in [CF06] and [Was03]. Applications of number theory in cryptography are discussed in detail in [Kob94]. An account of applications of elliptic functions to physics and technology may be found in a classic book by Oberhettinger and Magnus [OM13].

Exercises

2.1 Using the basic properties of the function $\Gamma(1+z)$, including its integral representation in (2.1.17) and the reflection relation (2.1.18), prove the following formula,

$$\frac{d^2}{dz^2} \ln \Gamma(1+z) = \sum_{n=1}^{\infty} \frac{1}{(z+n)^2} \qquad (2.11.13)$$

and use it to obtain the Taylor series expansion of $\Gamma(1+z)$ at $z = 0$.

2.2 Express $\wp(2z|\tau)$, $\wp(z|\tau/2)$, and $\wp(2z|\tau/2)$ as rational functions of $\wp(z|\tau)$.

2.3 Can one express $\wp(z|2\tau)$ as a rational function of $\wp(z|\tau)$? Explain and provide the relation between $\wp(z|2\tau)$ and $\wp(z|\tau)$.

2.4 Prove the Riemann relations with half-integer characteristics (2.6.12).

2.5 Establish the product formula for the Jacobi ϑ-function $\vartheta_1(z|\tau)$,

$$\frac{\vartheta_1(z|\tau)}{\vartheta_1'(0|\tau)} = \frac{1}{\pi} \sin \pi z \prod_{n=1}^{\infty} \frac{(1 - q^n e^{2\pi i z})(1 - q^n e^{-2\pi i z})}{(1 - q^n)^2} \qquad (2.11.14)$$

with $q = e^{2\pi i \tau}$ by combining (2.6.21) and (2.6.22) and integrating the series expression for $\wp(z|\tau)$ of (2.4.1) term-by-term in z.

3

Modular forms for $\mathrm{SL}(2, \mathbb{Z})$

In the Chapter 2, the dependence of elliptic functions on points in the torus $\mathbb{C}/\Lambda$ was studied for a fixed lattice Λ. In this chapter, it is the dependence on the lattice Λ that will be investigated. The modular group $\mathrm{SL}(2, \mathbb{Z})$ is introduced as the group of automorphisms of the lattice Λ, and its generators, elliptic points, and cusps are identified. The hyperbolic geometry of the Poincaré upper half plane is reviewed and the fundamental domain for $\mathrm{SL}(2, \mathbb{Z})$ is constructed. Modular forms and cusp forms are defined and shown to form a polynomial ring. They are related to holomorphic Eisenstein series, the discriminant function, the Dedekind η-function, and the j-function and are expressed in terms of Jacobi ϑ-functions. The Fourier and Poincaré series representations of Eisenstein series are analyzed as well.

3.1 Automorphisms of lattices and the modular group

A lattice Λ in the complex plane $\mathbb{C}$ may be generated by a pair of complex numbers $\omega_1, \omega_2 \in \mathbb{C}$ (also referred to as *periods*), which are linearly independent over $\mathbb{R}$. In terms of the modulus $\tau = \omega_2/\omega_1$, this condition translates to the requirement $\tau \notin \mathbb{R}$. We shall be interested in lattices Λ and corresponding tori $\mathbb{C}/\Lambda$ with a definite orientation so that the pair of periods (ω_1, ω_2) is chosen to be ordered by $\mathrm{Im}\,(\tau) > 0$.

The lattice Λ, and thus the torus $\mathbb{C}/\Lambda$, may be generated equivalently by other choices of periods, such as by ω_1 and $\omega_2 + n\omega_1$ for any $n \in \mathbb{Z}$. More generally, Λ is equivalently generated by any pair of periods $\tilde{\omega}_1, \tilde{\omega}_2$ given by a linear combination of ω_1, ω_2 with integer coefficients,

$$\begin{pmatrix} \tilde{\omega}_2 \\ \tilde{\omega}_1 \end{pmatrix} = \gamma \begin{pmatrix} \omega_2 \\ \omega_1 \end{pmatrix} \qquad \gamma = \begin{pmatrix} a & b \\ c & d \end{pmatrix} \qquad a, b, c, d \in \mathbb{Z} \qquad (3.1.1)$$

provided γ has an inverse with integer entries. This condition is equivalent to

$\det \gamma = \pm 1$, so that $\gamma \in \mathrm{GL}(2, \mathbb{Z})$. The corresponding action of $\gamma \in \mathrm{GL}(2, \mathbb{Z})$ on the modulus τ is by Möbius transformation,

$$\tau = \frac{\omega_2}{\omega_1} \qquad \gamma\tau = \tilde{\tau} = \frac{\tilde{\omega}_2}{\tilde{\omega}_1} = \frac{a\tau + b}{c\tau + d} \qquad (3.1.2)$$

The orientation $\mathrm{Im}\,(\tau) > 0$ of the lattice is preserved provided the new periods enjoy the same orientation $\mathrm{Im}\,(\gamma\tau) > 0$. Using the above transformation rule, the imaginary parts are found to transform as follows,

$$\mathrm{Im}\,(\gamma\tau) = \mathrm{Im}\,(\tilde{\tau}) = \frac{\det \gamma}{|c\tau + d|^2} \mathrm{Im}\,(\tau) \qquad (3.1.3)$$

The orientation of the lattice Λ will thus be preserved provided γ is restricted to the $\mathrm{SL}(2, \mathbb{Z})$ subgroup of $\mathrm{GL}(2, \mathbb{Z})$ for which $\det \gamma = 1$, identifying $\mathrm{SL}(2, \mathbb{Z})$ as the group of orientation-preserving automorphisms of Λ. The transformation $-I \in \mathrm{SL}(2, \mathbb{Z})$ maps the periods into their opposites but leaves the modulus τ invariant. Thus, the faithful action on the lattice Λ is by $\mathrm{SL}(2, \mathbb{Z})$, but on τ it is by the group $\mathrm{PSL}(2, \mathbb{Z}) = \mathrm{SL}(2, \mathbb{Z})/\mathbb{Z}_2$ where $\mathbb{Z}_2 = \{\pm I\}$.

3.1.1 Structure of the modular group

The group $\mathrm{SL}(2, \mathbb{Z})$ is an infinite discrete non-Abelian group, referred to as the *modular group*. It is generated by two of its elements,

$$S = \begin{pmatrix} 0 & -1 \\ 1 & 0 \end{pmatrix} \qquad T = \begin{pmatrix} 1 & 1 \\ 0 & 1 \end{pmatrix} \qquad (3.1.4)$$

These generators satisfy three fundamental relations,

$$S^2 = (ST)^3 = (TS)^3 = -I \qquad (3.1.5)$$

Any $\gamma \in \mathrm{SL}(2, \mathbb{Z})$ can be written as a finite word in the letters S, T. By using the first of the above relations, words with two consecutive S letters may be simplified and omitted when listing independent elements. Thus, any γ has the following product decomposition,

$$\gamma = \pm T^{\alpha_1} S T^{\alpha_2} S \cdots S T^{\alpha_n} \qquad (3.1.6)$$

for $\alpha_i \in \mathbb{Z}$ and a positive integer n, or $\gamma = \pm I, \pm S$ when $n = 0$. Furthermore, any word with three consecutive combinations of ST or TS may also be omitted by the last two relations.

To prove the validity of the decomposition into words in S and T, we consider an arbitrary element $\gamma \in \mathrm{SL}(2, \mathbb{Z})$,

$$\gamma = \begin{pmatrix} a & b \\ c & d \end{pmatrix} \qquad a, b, c, d \in \mathbb{Z}, \quad ad - bc = 1 \qquad (3.1.7)$$

If $c = 0$, then $ad = 1$, and $\gamma = aT^{ab}$. If $c < 0$ we multiply γ by $-I$ to reduce this case to the case $c > 0$. For $c \geq 1$ we proceed by induction on c. If $c = 1$, then $ad - b = 1$ and we have $\gamma = T^a S T^d$. Thus, the proposition holds true for $c = 0, 1$. If $c \geq 2$ we shall assume that the proposition holds true for all matrices whose lower left entry is positive and less than c, and the proposition remains to be proven for all matrices whose left lower entry equals c. Since $c \geq 2$, c cannot divide a, leaving the following two possibilities:

- If $|a| > c$, then there exists a non-zero integer n such that $a = nc + r$ with $0 < r < c$, and the matrix $\gamma' = ST^{-n}\gamma$ has lower left entry equal to $r < c$.
- If $|a| < c$, then the matrix $\gamma' = \mathrm{sign}(a)S\gamma$ has lower left entry $|a| < c$.

In either case, by applying a product of S and T generators we have constructed from γ a new matrix with strictly smaller value of the lower left entry. The validity of the proposition follows by induction on the value of c.

3.2 The fundamental domain for SL(2, ℤ)

In this section, we begin with a brief review of the geometry of the Poincaré upper half plane and then discuss the construction of the fundamental domain for SL(2, ℤ) as well as its fixed points and cusps.

3.2.1 The Poincaré upper half plane

We denote the Poincaré upper half plane by,

$$\mathcal{H} = \{\tau \in \mathbb{C}, \ \mathrm{Im}\,(\tau) > 0\} \qquad (3.2.1)$$

Under the action of SL(2, ℝ) by Möbius transformations on τ, the upper half plane $\mathcal{H}$ is mapped to $\mathcal{H}$ and its boundary $\partial\mathcal{H} = \mathbb{R}$ is mapped to $\partial\mathcal{H}$. The isotropy group of an arbitrary point in $\mathcal{H}$ is isomorphic to SO(2) and hence $\mathcal{H}$ is isomorphic to the symmetric space,

$$\mathcal{H} = \mathrm{SL}(2, \mathbb{R})/\mathrm{SO}(2) \qquad (3.2.2)$$

For example, the elements that leave $\tau = i$ invariant obey $i = (ai + b)/(ci + d)$ and $ad - bc = 1$ with $a, b, c, d \in \mathbb{R}$ and may be parametrized by

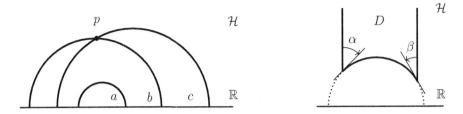

Figure 3.1 In the left panel the geodesics of the Poincaré upper half plane $\mathcal{H}$ are shown as the semi-circles centered on $\mathbb{R}$ with arbitrary radius; the geometry is non-Euclidean as several geodesics through a given point p, such as b and c, are parallel to the geodesic a. In the right panel, the area enclosed by the semi-infinite triangle D is area$(D) = \pi - \alpha - \beta$.

$b = -c, d = a$, and $a^2 + b^2 = 1$ to form a group isomorphic to SO(2). The Poincaré metric on $\mathcal{H}$, which is defined by the line element ds given by,

$$ds^2 = \frac{|d\tau|^2}{(\operatorname{Im}\tau)^2} \qquad (3.2.3)$$

is an SL(2, ℝ)-invariant Riemannian metric of constant negative curvature. The Poincaré upper half plane equipped with the Poincaré metric is a model of non-Euclidean geometry. Its geodesics (or "straight lines") are the half-circles of arbitrary radius centered on $\mathbb{R}$. Given a geodesic a and a point p not on this geodesic, there are an infinite number of geodesics through p that do not intersect a and are thus "parallel" to a, as shown in Figure 3.1.

The area of a polygon whose sides are geodesics may be obtained by decomposing the area into sums and differences of semi-infinite triangles with one vertex at the cusp and two corners with opening angles α and β, as illustrated in the right panel of Figure 3.1,

$$\operatorname{area}(D) = \int_D \frac{d^2\tau}{(\operatorname{Im}\tau)^2} = \pi - \alpha - \beta \qquad d^2\tau = \frac{i}{2}d\tau \wedge d\bar{\tau} \quad (3.2.4)$$

The area of a triangle with opening angles α, β, γ is then $\pi - \alpha - \beta - \gamma$ and is clearly always less than π: a well-known property of hyperbolic geometry. The derivation of the geodesics in the Poincaré upper half plane and the calculation of hyperbolic lengths and areas is the subject of Exercise 3.1.

3.2.2 The fundamental domain for SL(2, ℤ)

Since the modular group SL(2, ℤ) is a subgroup of SL(2, ℝ), it also maps $\mathcal{H}$ to $\mathcal{H}$ and $\partial\mathcal{H}$ to $\partial\mathcal{H}$. Every oriented lattice Λ, or equivalently every oriented torus $\mathbb{C}/\Lambda$, may be specified by a point $\tau \in \mathcal{H}$ up to overall rescaling of the lattice. Since the lattices corresponding to two points τ and τ' are equivalent to one another if and only if $\tau' = \gamma\tau$ for some $\gamma \in$ SL(2, ℤ), the space of inequivalent lattices or inequivalent tori is given by the quotients,

$$\text{SL}(2, \mathbb{Z})\backslash\mathcal{H} = \text{SL}(2, \mathbb{Z})\backslash\text{SL}(2, \mathbb{R})/\text{SO}(2) \tag{3.2.5}$$

Just as we represented the torus $\mathbb{C}/\Lambda$ by a fundamental parallelogram in $\mathbb{C}$, we can represent the coset $\text{SL}(2, \mathbb{Z})\backslash\mathcal{H}$ by a *fundamental domain* F in $\mathcal{H}$, which is a connected subset of $\mathcal{H}$. The fundamental domain is not unique since, for example, if F is a fundamental domain then so if γF for any $\gamma \in$ SL(2, ℤ). The standard choice is given by the following theorem.

Theorem 3.2.1. *The following F is a fundamental domain for* SL(2, ℤ),

$$F = \left\{ \tau \in \mathcal{H},\ |\tau| \geq 1,\ |\text{Re}(\tau)| \leq \tfrac{1}{2} \right\} \tag{3.2.6}$$

where it is understood that the boundary components at $\text{Re}(\tau) = \pm\frac{1}{2}$ *are identified with one another under* $\tau \to \tau + 1$, *while the boundary components at* $|\tau| = 1$ *for* $\text{Re}(\tau)$ *positive and negative are identified with one another under* $\tau \to -1/\tau$ *(see Figure 3.2). The area of F in the Poincaré metric is given by* $\text{area}(F) = \frac{\pi}{3}$ *in view of (3.2.4).*

To prove the theorem we show that, for an arbitrary point $\tau \in \mathcal{H}$, there exists a transformation $\gamma \in$ SL(2, ℤ) such that $\gamma\tau \in F$. To do so, we fix τ and make use of the relation (3.1.3) for $\det\gamma = 1$,

$$\text{Im}(\gamma\tau) = \text{Im}(\tilde{\tau}) = \frac{\text{Im}(\tau)}{|c\tau + d|^2} \tag{3.2.7}$$

to construct a $\gamma \in$ SL(2, ℤ) with the largest possible value of $\text{Im}(\gamma\tau)$. This is done by minimizing $|c\tau + d|$ for fixed τ as a, b, c, and d run over the integers satisfying $ad - bc = 1$. Taking γ to be the identity gives the bound $|c\tau + d| \leq 1$, which can be satisfied by only a finite number of integer pairs (c, d). We construct γ as a product of translations by T and inversions by S using an iterative process that terminates. Starting from an arbitrary $\tau \in \mathcal{H}$ which is not in F, apply a translation T^α for $\alpha \in \mathbb{Z}$ to map $\tau \to \tilde{\tau} = T^\alpha\tau$ into the vertical strip $\tilde{\tau} \in \{|\text{Re}(\tau)| \leq 1/2\}$. Under this transformation, the value of the imaginary part is unchanged. If $\tilde{\tau} \in F$, we are done. If $\tilde{\tau} \notin F$ then it must satisfy $|\tilde{\tau}| < 1$. Its image under S given by $S\tilde{\tau}$ has $|S\tilde{\tau}| > 1$ as well as $\text{Im}(S\tilde{\tau}) > \text{Im}(\tilde{\tau}) = \text{Im}(\tau)$ in view of the fact that $c = 1$ and $d = 0$ for S and that $|\tilde{\tau}| < 1$. If this image is in F we are done, and if not, we repeat the process. Since only a finite number of pairs (c, d) satisfy the bound $|c\tau + d| \leq 1$, and the iterative process strictly increases the value of $\text{Im}(\gamma\tau)$ at each inversion S, the process must terminate, which completes the proof.

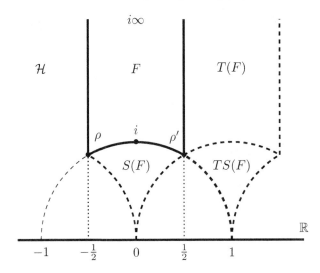

Figure 3.2 The standard fundamental domain F for SL$(2,\mathbb{Z})$; its images $T(F), S(F), TS(F)$; the elliptic points i, ρ, ρ' in F; and the cusp $i\infty$.

An immediate application of Theorem 3.2.1 to evaluating the class numbers of positive binary quadratic forms may be found in Exercise 3.2.

3.2.3 Elliptic points

Every point $\tau \in \mathcal{H}$ is invariant under the $\mathbb{Z}_2 = \{\pm I\}$ subgroup of SL$(2,\mathbb{Z})$. An elliptic point is a point in $\mathcal{H}$ that is invariant under a subgroup of SL$(2,\mathbb{Z})$ that is larger than $\mathbb{Z}_2$. We shall now show that the only elliptic points in F are i, ρ, and $\rho' = \rho + 1$, where $\rho^3 = 1$ with Im $(\rho) > 0$. Specifically, the point i is invariant under the subgroup $\{\pm I, \pm S\}$ of order 4, which is isomorphic to $\mathbb{Z}_4$. The point ρ is invariant under the subgroup $\{\pm I, \pm(ST), \pm(ST)^2\}$ of order 6, which is isomorphic to $\mathbb{Z}_6$, while the point ρ' is invariant under the subgroup $\{\pm I, \pm(TS), \pm(TS)^2\}$ of order 6.

To prove the assertion we consider a point $\tau_\gamma \in \mathcal{H}$ that is invariant under a given $\gamma \in$ SL$(2,\mathbb{Z})$,

$$\gamma\tau_\gamma = \tau_\gamma \qquad\qquad \gamma = \begin{pmatrix} a & b \\ c & d \end{pmatrix} \qquad\qquad (3.2.8)$$

and therefore satisfies the quadratic equation $c\tau_\gamma^2 - (a-d)\tau_\gamma - b = 0$ subject to Im $(\tau_\gamma) > 0$. For $c = 0$ there are no solutions that satisfy both conditions, while for $c \neq 0$ we may assume $c > 0$ without loss of generality. The point

τ_γ lies in $\mathcal{H}$ provided $|a+d| < 2$ in which case τ_γ is given by the solution to the quadratic equation that has positive imaginary part,

$$\tau_\gamma = \frac{a-d}{2c} + i\frac{\sqrt{4-(a+d)^2}}{2c} \qquad (3.2.9)$$

A transformation $\gamma \to \alpha\gamma\alpha^{-1}$ with $\alpha \in SL(2,\mathbb{Z})$ maps $\tau_\gamma \to \alpha\tau_\gamma$ and a suitable choice of α maps τ_γ to the standard fundamental domain F for $SL(2,\mathbb{Z})$. The choice $\tau_\gamma \in F$ imposes the condition $\mathrm{Im}\,(\tau_\gamma) \geq \mathrm{Im}\,(\rho) = \sqrt{3}/2$ which requires $(a+d)^2 + 3c^2 \leq 4$. Its solutions are $c = 1$ and either $a+d = 0$ or $a+d = \pm 1$. By a suitable translation of τ_γ to the fundamental domain F, we may set $a, d \in \{0, 1\}$ corresponding to the following matrices,

$$S = \begin{pmatrix} 0 & -1 \\ 1 & 0 \end{pmatrix} \qquad ST = \begin{pmatrix} 0 & -1 \\ 1 & 1 \end{pmatrix} \qquad TS = \begin{pmatrix} 1 & -1 \\ 1 & 0 \end{pmatrix} \quad (3.2.10)$$

for which $\tau_S = i$, $\tau_{ST} = \rho$, and $\tau_{TS} = \rho'$ of orders 4, 6, and 6, respectively. These are all the elliptic points of $SL(2,\mathbb{Z})$ in the fundamental domain F.

3.2.4 Cusps

Topologically, the fundamental domain F is a sphere with one puncture at $\tau = i\infty$, which is referred to as the *cusp* of the fundamental domain F of $SL(2,\mathbb{Z})$. The cusp does not belong to $\mathcal{H}$. It is invariant under the infinite Borel subgroup $\Gamma_\infty \subset SL(2,\mathbb{Z})$ of translations and sign reversal,

$$\Gamma_\infty = \{\pm T^n, \ n \in \mathbb{Z}\} \qquad (3.2.11)$$

Under $\gamma \in \Gamma_\infty$, the cusp in F is mapped to the cusp of the corresponding fundamental domain γF. Under an arbitrary $\gamma \in SL(2,\mathbb{Z})$ with $c \neq 0$, the cusp is mapped to a rational number, and more precisely,

$$SL(2,\mathbb{Z})\,(i\infty) = \mathbb{Q} \cup \{i\infty\} \qquad (3.2.12)$$

so that each rational number is a cusp of $SL(2,\mathbb{Z})$ in the corresponding fundamental domain.

3.3 Modular functions, modular forms, and cusp forms

We shall now define modular functions, modular forms, and cusp forms. Let $f(\tau)$ be a meromorphic function on the upper half plane $\mathcal{H}$, which obeys the following transformation law for some integer $k \in \mathbb{Z}$, and for all $\gamma \in SL(2,\mathbb{Z})$,

$$f(\gamma\tau) = (c\tau + d)^k f(\tau) \qquad\qquad \gamma = \begin{pmatrix} a & b \\ c & d \end{pmatrix} \qquad (3.3.1)$$

Since f is, in particular, a periodic function under $\tau \to \tau+1$, it has a Fourier expansion given by,

$$f(\tau) = \sum_{n \in \mathbb{Z}} a_n \, q^n \qquad\qquad q = e^{2\pi i \tau} \qquad\qquad (3.3.2)$$

Definition 3.3.1. *The definitions we shall adopt are as follows.*

1. f is a *meromorphic modular form of weight* k if f is meromorphic at $i\infty$, namely has only a finite number of nonvanishing Fourier coefficients a_n with negative n. In the special case of $k = 0$, we shall often use the terminology *meromorphic modular function*, or simply *modular function*;

2. f is a *holomorphic modular form of weight* k, or simply *modular form of weight* k if it is holomorphic in $\mathcal{H}$ and at $i\infty$, namely its Fourier coefficients a_n vanish for all negative n, and thus f is finite at $i\infty$;

3. f *is a cusp form of weight* k if it is a holomorphic modular form of weight k and vanishes at $i\infty$.

Next, we shall present a brief comparison of frequently used terminology, which is essentially universal for items 2 and 3 but not for item 1.

3.3.1 Comments on terminology and differential forms

The terminology for automorphic functions and automorphic forms and its relation with the terminology for modular functions and modular forms is not entirely standard in the mathematics literature, and may lead to some confusion, which we shall eliminate here with the help of precise definitions.

Automorphic functions and forms may be defined generally in terms of functions with specific transformation properties under a discrete group acting on a complex manifold. Here, we shall limit attention to the case where the complex manifold is the Poincaré upper half plane $\mathcal{H}$ and the discrete group Γ is a Fuchsian subgroup of $\mathrm{SL}(2,\mathbb{R})$, acting on $\mathcal{H}$ by Möbius transformations (see Appendix B for further discussion of Fuchsian groups).

In the complex analytic category, following Shimura [Shi71], an *automorphic function* f with respect to a Fuchsian group Γ is a meromorphic function on $\mathcal{H}$ that is invariant under all $\gamma \in \Gamma$,

$$f(\gamma \tau) = f(\tau) \qquad\qquad \gamma = \begin{pmatrix} a & b \\ c & d \end{pmatrix} \qquad\qquad (3.3.3)$$

and is meromorphic at the cusps of Γ in $\mathcal{H}$. Following Shimura [Shi71] and

Diamond and Shurman [DS05], an *automorphic form of weight k with respect to Γ* is a meromorphic function on $\mathcal{H}$ that instead transforms under all $\gamma \in \Gamma$ with a nontrivial automorphy factor $(c\tau + d)^k$,

$$f(\gamma\tau) = (c\tau + d)^k f(\tau) \qquad (3.3.4)$$

and is also meromorphic at all cusps. Equivalently, we may view an automorphic form f of weight k as a *differential $(\frac{k}{2}, 0)$-form*,

$$\mathfrak{f} = f(\tau)(d\tau)^{\frac{k}{2}} \qquad (3.3.5)$$

which is strictly invariant under Γ. The terminology of *forms* is particularly well suited here, and likely more familiar to a physics audience.

In the category of non-holomorphic functions, following Terras [Ter85], we may define an *automorphic function* as a complex-valued function on $\mathcal{H}$ that is invariant under Γ, thereby generalizing the notion from the meromorphic case. From the point of view of a physicist, it is natural to generalize the notion of an automorphic form to the non-holomorphic category by defining an automorphic form of weight (k, ℓ) to transform under all $\gamma \in \Gamma$ by,

$$f(\gamma\tau) = (c\tau + d)^k (c\bar{\tau} + d)^\ell f(\tau) \qquad (3.3.6)$$

corresponding to a differential $(\frac{k}{2}, \frac{\ell}{2})$-form $\mathfrak{f}$,

$$\mathfrak{f} = f(\tau)(d\tau)^{\frac{k}{2}}(d\bar{\tau})^{\frac{\ell}{2}} \qquad (3.3.7)$$

which is strictly invariant under Γ.

Following Shimura [Shi71] one refers to *modular functions* and *modular forms* in the special case, where $\Gamma = \mathrm{SL}(2, \mathbb{Z})$ or a congruence subgroup thereof, instead of automorphic functions or automorphic forms, respectively. Thus, in the complex category, meromorphic and holomorphic modular forms f of weight k transform under $\Gamma \subset \mathrm{SL}(2, \mathbb{Z})$ by the rule (3.3.1), and naturally map to $(\frac{k}{2}, 0)$ differentials (3.3.5), while meromorphic modular functions correspond to the special case $k = 0$, which is indeed the terminology we are adopting in Section 3.3.

However, while the terminology of items 2 and 3 in Definition 3.3.1 is essentially universal, the terminology for item 1 is not universal. The terminology *modular function of weight k* for item 1 is used in the books by Apostol [Apo76b], Iwaniec [Iwa97], Koblitz [Kob12], and Silverman [Sil85], while Diamond and Shurman [DS05] use the terminology *weakly modular of weight k* and Lang [Lan76] uses simply *modular of weight k*. The terminology *unrestricted modular form of weight k* is used in the books by Knapp [Kna92] and Rankin [Ran77].

In this book, we shall use the terminology of Section 3.3, which is close to the one adopted in the books by Köhler [Köh11], Ono [Ono04], and Zagier [BVdGHZ08]. Unless the meaning is crystal clear from the context, we shall add suitable qualifiers *meromorphic* versus *holomorphic*, or *non-holomorphic* whenever appropriate.

3.3.2 The ring of modular forms

The space $\mathcal{M}_k$ of holomorphic modular forms of weight k for the group SL(2, ℤ) is a vector space over ℂ. Since the product of a modular form of weight k and a modular form of weight ℓ is a modular form of weight $k + \ell$, the space of holomorphic modular forms of arbitrary weight forms a ring $\mathcal{M}$ graded by the weight,

$$\mathcal{M} = \bigoplus_k \mathcal{M}_k \qquad\qquad \mathcal{M}_k \mathcal{M}_\ell \subset \mathcal{M}_{k+\ell} \qquad (3.3.8)$$

The ratio of two holomorphic modular forms, in general, will not be a holomorphic modular form, as dividing by a cusp form will produce an object that is not holomorphic at $i\infty$. In Sections 3.4 and 3.5, we shall show that there are no holomorphic modular forms of negative weight and that $\mathcal{M}$ is a polynomial ring generated by the modular forms G_4 and G_6.

3.4 Eisenstein series

Eisenstein series G_k were already encountered when studying elliptic functions in (2.4.5). In this section, we shall show that they are holomorphic modular forms of weight k and obtain their Fourier series decomposition. In Section 3.4.3, we shall show that they actually generate the complete polynomial ring of all holomorphic modular forms.

We begin by considering the Eisenstein series defined in (2.4.5),[1]

$$G_k(\omega_1, \omega_2) = G_k(\Lambda) = \sum_{\omega \in \Lambda'} \frac{1}{\omega^k} = {\sum_{m,n \in \mathbb{Z}}}' \frac{1}{(m\omega_1 + n\omega_2)^k} \qquad (3.4.1)$$

The sums that define G_k are absolutely convergent for $k \geq 3$ and vanish for odd k in view of the lattice symmetry $\Lambda = -\Lambda$. The resulting $G_k(\omega_1, \omega_2)$ is invariant under SL(2, ℤ) transformations of the periods ω_1, ω_2 since G_k depends only on the lattice Λ and not on the specific periods chosen to represent Λ. The functions $G_k(\omega_1, \omega_2)$ are also manifestly homogeneous in

---

[1] Throughout, the prime superscript on the sum instructs us to omit the term with $m = n = 0$.

the periods of degree $-k$. In summary, under an arbitrary $\gamma \in \mathrm{SL}(2,\mathbb{Z})$, which maps Λ to Λ and the periods ω_1, ω_2 into periods $\tilde{\omega}_1, \tilde{\omega}_2$ given in terms of γ in (3.1.1), and an arbitrary complex scaling factor $\lambda \in \mathbb{C}^*$, we have,

$$G_k(\tilde{\omega}_1, \tilde{\omega}_2) = G_k(\omega_1, \omega_2)$$
$$G_k(\lambda\omega_1, \lambda\omega_2) = \lambda^{-k} G_k(\omega_1, \omega_2) \qquad (3.4.2)$$

Scaling allows us to define a function of $\tau = \omega_2/\omega_1$ on the upper half plane $\mathcal{H}$ by factoring out a power of ω_1 or, equivalently, by setting $\omega_1 = 1$,

$$G_k(\tau) = G_k(1, \tau) = \omega_1^k\, G_k(\omega_1, \omega_2) = \sum_{m,n \in \mathbb{Z}}{}' \frac{1}{(m + n\tau)^k} \qquad (3.4.3)$$

Combining the transformations (3.4.2) with the definition of $G_k(\tau)$ in (3.4.3) gives the modular transformation law for $G_k(\tau)$,

$$G_k(\gamma\tau) = \left(\frac{\tilde{\omega}_1}{\omega_1}\right)^k G_k(\tau) = (c\tau + d)^k G_k(\tau) \qquad (3.4.4)$$

which may also be inferred directly from (3.4.3). Thus, $G_k(\tau)$ vanishes for odd k and is a holomorphic modular form of weight k for even $k \geq 4$.

3.4.1 Fourier decomposition of Eisenstein series

Since the Eisenstein series $G_k(\tau)$ is a holomorphic function for $\tau \in \mathcal{H}$ and a periodic function of τ with period 1, it admits a Fourier expansion,

$$G_k(\tau) = \sum_{\nu \in \mathbb{Z}} a_\nu(k)\, q^\nu \qquad q = e^{2\pi i \tau} \qquad (3.4.5)$$

To evaluate the Fourier series, we decompose the double sum of (3.4.3) that defines $G_k(\tau)$ by isolating the contributions from $n = 0$,

$$G_k(\tau) = \sum_{m \neq 0} \frac{1}{m^k} + 2 \sum_{n=1}^{\infty} \sum_{m \in \mathbb{Z}} \frac{1}{(m + n\tau)^k} \qquad (3.4.6)$$

As τ approaches the cusp $\tau \to i\infty$, the double sum in (3.4.6) tends to zero. As a result, all expansion coefficients $a_\nu(k)$ vanish for $\nu < 0$ and the first sum in (3.4.6) gives $a_0(k)$,

$$a_0(k) = 2\zeta(k) \qquad 4 \leq k \in 2\mathbb{N} \qquad (3.4.7)$$

Thus, $G_k(\tau)$ has a finite limit at the cusp and is a holomorphic modular form of weight k. To evaluate the double sum in (3.4.6), we use the expression for the sum (2.2.1) in terms of (2.2.3) already proven in Chapter 2,

$$\sum_{m\in\mathbb{Z}}\frac{1}{z+m}=-i\pi\frac{1+e^{2\pi iz}}{1-e^{2\pi iz}}=-i\pi-2\pi i\sum_{\ell=1}^{\infty}e^{2\pi i\ell z} \qquad (3.4.8)$$

Actually, we shall need the derivative of order $k-1$ of this formula,

$$\sum_{m\in\mathbb{Z}}\frac{1}{(z+m)^k}=-i\pi\delta_{k,1}+\frac{(-2\pi i)^k}{\Gamma(k)}\sum_{\ell=1}^{\infty}\ell^{k-1}\,e^{2\pi i\ell z} \qquad (3.4.9)$$

valid for $k\geq 1$. Setting $z=n\tau$ and including the sum over n, we obtain the expression for the second sum in (3.4.6). Putting all together gives the following Fourier series decomposition for $G_k(\tau)$,

$$G_k(\tau)=2\zeta(k)+2\frac{(-2\pi i)^k}{\Gamma(k)}\sum_{n=1}^{\infty}\sum_{\ell=1}^{\infty}\ell^{k-1}q^{n\ell} \qquad (3.4.10)$$

Changing summation variables from (n,ℓ) to (N,ℓ) with $N=n\ell$, the sum over ℓ for given N may be rearranged in terms of the *sums of divisors functions*, defined for any $\alpha\in\mathbb{C}$ by,

$$\sigma_\alpha(N)=\sum_{\ell|N}\ell^\alpha \qquad (3.4.11)$$

or in terms of a certain generating function, as discussed in Exercise 3.3. The sum is over all positive divisors ℓ of N, including the divisors 1 and N. The function $\sigma_\alpha(N)$ is a *multiplicative arithmetic function*, but it is not completely multiplicative, as will be established in Exercise 3.3. In terms of $\sigma_\alpha(N)$, we obtain our final expression for the Fourier series of $G_k(\tau)$,

$$G_k(\tau)=2\zeta(k)+2\frac{(-2\pi i)^k}{\Gamma(k)}\sum_{N=1}^{\infty}\sigma_{k-1}(N)q^N \qquad (3.4.12)$$

Factoring out $2\zeta(k)$, and using the relation between Bernoulli numbers and $\zeta(k)$ for k even established in (2.2.14), that is, $(2\pi i)^k B_k=-2\,k!\,\zeta(k)$, we obtain the *normalized Eisenstein series* E_k, defined by,

$$\mathsf{E}_k(\tau)=\frac{G_k(\tau)}{2\zeta(k)}=1+\nu_k\sum_{N=1}^{\infty}\sigma_{k-1}(N)\,q^N \qquad \nu_k=-\frac{2k}{B_k} \qquad (3.4.13)$$

With this normalization, we manifestly have $\mathsf{E}_k(\tau)\to 1$ as $\tau\to i\infty$ for any even $k\geq 4$. The first few values for ν_k are given as follows,

$$\nu_4=240 \qquad \nu_6=-504 \qquad \nu_8=480 \qquad \nu_{10}=-264 \qquad \nu_{12}=\tfrac{65520}{691} \qquad (3.4.14)$$

In particular, the modular forms $g_2(\tau), g_3(\tau), G_4(\tau)$, and $G_6(\tau)$ and the discriminant $\Delta(\tau)$ of (2.4.14) for the lattice $\Lambda = \mathbb{Z} \oplus \mathbb{Z}\tau$ are given in terms of the normalized Eisenstein series by,

$$g_2(\tau) = \ 60\,G_4(\tau) = \frac{4\pi^4}{3}\mathsf{E}_4(\tau)$$

$$g_3(\tau) = 140\,G_6(\tau) = \frac{8\pi^6}{27}\mathsf{E}_6(\tau)$$

$$\Delta(\tau) = g_2^3 - 27g_3^2 = \frac{(2\pi)^{12}}{12^3}\left(\mathsf{E}_4(\tau)^3 - \mathsf{E}_6(\tau)^2\right) \qquad (3.4.15)$$

Since $\mathsf{E}_k(i\infty) = 1$ the discriminant vanishes at the cusp, and we have $\Delta(\tau) = (2\pi)^{12}q + \mathcal{O}(q^2)$. While $G_k(\tau)$ is a holomorphic modular form of weight k for even $k \geq 4$, the discriminant $\Delta(\tau)$ is a cusp form of weight 12.

3.4.2 Poincaré series representation

One construction of elliptic functions for a lattice Λ is through the method of images, exhibited in all generality in (2.3.4) and shown in detail for the case of the Weierstrass function in (2.4.1). One may view the construction by the method of images as performing a sum over the Abelian group Λ under which every elliptic function is invariant.

To obtain holomorphic and meromorphic modular forms of weight k under the group $\mathrm{SL}(2, \mathbb{Z})$, an analogous method is available through the so-called *Poincaré series*. This method may be applied to obtain meromorphic modular forms, as will be the case here, but also non-holomorphic modular functions and forms, as will be the case in Section 4.2 and Chapter 9.

To construct a modular form by a Poincaré series, one starts with a meromorphic function g or a meromorphic differential form $\mathfrak{g}$ on $\mathcal{H}$ and aims to sum over its images under the modular group. In many practical cases, however, the function g or the form $\mathfrak{g}$ is invariant under a *stability subgroup* Γ_{stab} of $\mathrm{SL}(2, \mathbb{Z})$. When Γ_{stab} has an infinite number of elements, as will often be the case, the sum over all images under $\mathrm{SL}(2, \mathbb{Z})$ would diverge. The Poincaré series is instead defined by summing over the images under the cosets $\Gamma_{\mathrm{stab}}\backslash\mathrm{SL}(2, \mathbb{Z})$. From a meromorphic function g, we obtain a meromorphic modular function f given by the Poincaré series,

$$f(\tau) = \sum_{\gamma \in \Gamma_{\mathrm{stab}}\backslash\mathrm{SL}(2,\mathbb{Z})} g(\gamma\tau) \qquad (3.4.16)$$

The function g is referred to as the *seed of the Poincaré series*. The corresponding formula defines the Poincaré series for a modular differential form f from a seed g. As noted originally by Poincaré, the seed function for a given modular function is not unique. In particular, the Poincaré sum of certain non-zero seed functions may vanish.

By way of example, we may take the seed to be simply $\mathfrak{g}_k = (d\tau)^k$ for some positive integer k. This seed is invariant under $-I \in \mathrm{SL}(2, \mathbb{Z})$ because τ is invariant, as well as under the following Borel subgroup of $\mathrm{SL}(2, \mathbb{Z})$ since $d\tau$ is invariant under shifts $\tau \to \tau + b$,

$$\Gamma_\infty = \left\{ \pm \begin{pmatrix} 1 & b \\ 0 & 1 \end{pmatrix}, b \in \mathbb{Z} \right\} \tag{3.4.17}$$

so that $\Gamma_{\mathrm{stab}} = \Gamma_\infty$. The remaining Poincaré series for $k \geq 2$ is given by,

$$\mathfrak{f}_k = \sum_{\gamma \in \Gamma_\infty \backslash \mathrm{SL}(2,\mathbb{Z})} \frac{(d\tau)^k}{(c\tau + d)^{2k}} = \frac{1}{2} \sum_{\substack{c,d \in \mathbb{Z}, \\ \gcd(c,d)=1}} \frac{(d\tau)^k}{(c\tau + d)^{2k}} \tag{3.4.18}$$

The restriction $\gcd(c, d) = 1$ implicitly assumes that $(c, d) \neq (0, 0)$. The form $\mathfrak{f}_k$ is closely related to the Eisenstein series G_{2k} defined in (3.4.3), as may be seen by factoring out the greatest common divisor $p = \gcd(m, n) \in \mathbb{N}$ from the summation variables in (3.4.3) by parametrizing them as follows $(m, n) = (pd, pc)$ with $\gcd(c, d) = 1$,

$$G_{2k}(\tau) = \sideset{}{'}\sum_{m,n\in\mathbb{Z}} \frac{1}{(m + n\tau)^{2k}} = \zeta(2k) \sum_{\substack{c,d \in \mathbb{Z}, \\ \gcd(c,d)=1}} \frac{1}{(c\tau + d)^{2k}} \tag{3.4.19}$$

so that $2\zeta(2k)\mathfrak{f}_k = G_{2k}(\tau)(d\tau)^k$ for $k \geq 2$.

3.4.3 The polynomial ring of Eisenstein series

The Eisenstein series $G_{2k} = G_{2k}(\tau)$ for the lattice $\Lambda = \mathbb{Z} + \mathbb{Z}\tau$ is a holomorphic modular form of weight $2k$ for any integer $k \geq 2$. We shall now show that G_{2k} may be expressed as a polynomial in G_4 and G_6 with rational coefficients, so that the vector space generated by the Eisenstein series G_{2k} is a polynomial ring generated by G_4 and G_6.

To show this, we differentiate the differential equation (2.4.8) for the Weierstrass $\wp$-function $\wp'(z)^2 = 4\wp(z)^3 - 60G_4\wp(z) - 140G_6$ with respect to z and obtain the following second-order differential equation,

$$\wp''(z) = 6\wp(z)^2 - 30G_4 \tag{3.4.20}$$

which is even in z. The Laurent expansion for $\wp$, obtained earlier in (2.4.4) for an arbitrary lattice Λ, may be expressed as follows,

$$\wp(z) = \frac{1}{z^2} + \sum_{k=1}^{\infty} (2k+1) G_{2k+2} \, z^{2k} \tag{3.4.21}$$

Substituting this expression and its double derivative into (3.4.20), the terms in z^{-4}, z^0, and z^2 automatically match. For the remaining terms, after some simplifications, we obtain the following recursion relation for $k \geq 4$,

$$G_{2k} = \sum_{\ell=1}^{k-3} \frac{3(2\ell+1)(2k-2\ell-3)}{(k-3)(2k-1)(2k+1)} G_{2\ell+2} G_{2k-2\ell-2} \tag{3.4.22}$$

which shows that G_{2k} for $k \geq 4$ is a polynomial in G_{2m+4} with $0 \leq m \leq k-4$, of homogeneous weight $2k$. Therefore, G_{2k} for $k \geq 4$ is a polynomial in G_4 and G_6. For example, to the lowest few orders we have,

$$G_8 = \tfrac{3}{7} G_4^2 \qquad\qquad G_{14} = \tfrac{30}{143} G_4^2 G_6$$
$$G_{10} = \tfrac{5}{11} G_4 G_6 \qquad\qquad G_{16} = \tfrac{9}{221} G_4^4 + \tfrac{300}{2431} G_4 G_6^2$$
$$G_{12} = \tfrac{18}{143} G_4^3 + \tfrac{25}{143} G_6^2 \qquad\qquad G_{18} = \tfrac{3915}{46189} G_4^3 G_6 + \tfrac{125}{4199} G_6^3 \tag{3.4.23}$$

Thus, the subspace of the graded ring of all holomorphic modular forms $\mathcal{M}$ that is generated by the Eisenstein series G_{2k} for all $k \geq 2$ is a polynomial ring generated by just two elements G_4 and G_6. While the modular forms G_4, G_6, G_8, G_{10}, and G_{14} are the only Eisenstein series of weight $4, 6, 8, 10$, and 14, respectively, we see that at weight 12, there are two independent modular forms, namely, G_4^3 and G_6^2, whose difference involves the discriminant Δ.

Equivalently, we may express G_{2k} in terms of the normalized Eisenstein series E_{2k}, to obtain the following formula,

$$E_{2k} = \sum_{\ell=1}^{k-3} \frac{6(2\ell+1)(2k-2\ell-3)\zeta(2\ell+2)\zeta(2k-2\ell-2)}{(k-3)(2k-1)(2k+1)\zeta(2k)} E_{2\ell+2} E_{2k-2\ell-2}$$

$$\tag{3.4.24}$$

in terms of which the relations (3.4.23) become,

$$E_8 = E_4^2 \qquad\qquad E_{14} = E_4^2 E_6$$
$$E_{10} = E_4 E_6 \qquad\qquad E_{16} = \tfrac{1617}{3617} E_4^4 + \tfrac{2000}{3617} E_4 E_6^2$$
$$E_{12} = \tfrac{441}{691} E_4^3 + \tfrac{250}{691} E_6^2 \qquad\qquad E_{18} = \tfrac{38367}{43867} E_4^3 E_6 + \tfrac{5500}{43867} E_6^3 \tag{3.4.25}$$

These relations, together with the Fourier series expansions of the E_{2k} in (3.4.13), imply an infinite number of identities between sums of divisor functions. For example, the identities corresponding to the relations $E_8 = E_4^2$ and

$E_{10} = E_4 E_6$ are given as follows (see Exercise 3.4),

$$\sigma_7(N) = \sigma_3(N) + 120 \sum_{M=1}^{N-1} \sigma_3(N-M)\sigma_3(M) \qquad (3.4.26)$$

$$11\,\sigma_9(N) = 21\sigma_5(N) - 10\sigma_3(N) + 5040 \sum_{M=1}^{N-1} \sigma_3(N-M)\sigma_5(M)$$

More generally, the divisor functions σ_{2k-1} may be expressed as polynomials in σ_3 and σ_5 with rational coefficients.

3.5 Dimension and generators of the ring of modular forms

To obtain the dimension of the space $\mathcal{M}_k$ of all holomorphic modular forms of weight k, we begin by proving the following theorem on meromorphic modular forms of weight k. Adapting the definitions used for Theorem 2.3.2, we shall denote by $\mathrm{ord}_f(p)$ the order of a zero (counted positively) or pole (counted negatively) of f at a point p, and by $\mathrm{ord}_f(i\infty)$ the exponent of the leading term in the q-expansion of $f(\tau)$. Meromorphicity of modular forms guarantees that $\mathrm{ord}_f(p)$ are integers, including for the cusp $p = i\infty$ and for the elliptic points $p = i$ and $p = \rho = e^{2\pi i/3}$.

Theorem 3.5.1. *A non-zero meromorphic modular form f of weight k satisfies the **valence formula**,*

$$\mathrm{ord}_f(i\infty) + \frac{1}{2}\mathrm{ord}_f(i) + \frac{1}{3}\mathrm{ord}_f(\rho) + \sum_{p \in F \backslash \{i\infty, i, \rho\}} \mathrm{ord}_f(p) = \frac{k}{12} \qquad (3.5.1)$$

where F is the standard fundamental domain in the Poincaré upper half plane $\mathcal{H}$ for the group $SL(2, \mathbb{Z})$ given in (3.2.6).

To prove the theorem, we count the number of zeros and poles of f by integrating the logarithmic derivative f'/f over a suitable contour. Doing so presents complications at the points $i\infty, i, \rho$, and ρ', so we regularize the domain F to a domain F_{reg} in which we cut off F near those points, as shown in Figure 3.3. The small circular arcs near i, ρ, and ρ' have radius $1/L$. For simplicity, we shall assume that $f(\tau)$ has neither zeros nor poles on $\partial F_{\mathrm{reg}}$ and refer to standard textbooks when this is not the case.

On the one hand, the residue theorem gives,

$$\lim_{L \to \infty} \frac{1}{2\pi i} \oint_{\partial F_{\mathrm{reg}}} d\tau\, \frac{f'(\tau)}{f(\tau)} = \sum_{p \in F_{\mathrm{reg}}} \mathrm{ord}_f(p) \qquad (3.5.2)$$

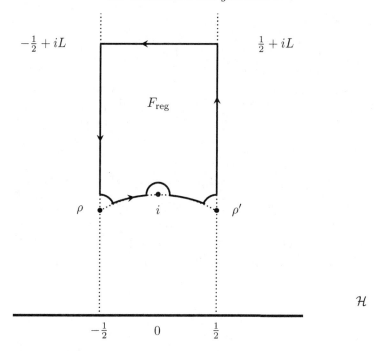

Figure 3.3 Integration contour of the regularized fundamental domain F_{reg}.

On the other hand, the integral along ∂F_{reg} may be evaluated by using the relations under modular transformations that identify the different line segments of ∂F_{reg} with one another. First, the contributions from the two vertical edges cancel one another, as their integrals are opposite. Second, we evaluate the contributions at the point $i\infty$ by using the dominant behavior at $\text{Im}\,(\tau) = L \gg 1$ given by $f(\tau) \sim q^{\text{ord}_f(i\infty)} = e^{2\pi i \tau\,\text{ord}_f(i\infty)}$ to obtain the contribution $-\text{ord}_f(i\infty)$. The minus sign arises from the orientation of the contour. Analogous contributions are collected at i but are counted with a factor of $\frac{1}{2}$ since the integration contour is only half a circle. Similarly, at ρ, the contribution is only $\frac{1}{6}$ of a full circle, but there are identical contributions from ρ and ρ' which accounts for the final factor of $\frac{1}{3}$. The above contributions account for the left side of (3.5.1). Finally, we collect the contribution along the $|\tau| = 1$ segments of the contour. The contributions to the left and to the right of i are related to one another by the transformation $S : \tau \to -1/\tau$, which acts on $f(\tau)$ by $f(S\tau) = \tau^k f(\tau)$, so that taking the differential of the logarithm, we have,

$$d(S\tau)\frac{f'(S\tau)}{f(S\tau)} = d\tau\frac{f'(\tau)}{f(\tau)} + k\frac{d\tau}{\tau} \qquad (3.5.3)$$

The integration arc goes from $\tau = \rho$ to $\tau = i$ and spans $\frac{1}{12}$ of a full circle, giving the right side of (3.5.1) and thereby completing the proof.

3.5.1 Dimension and generators of $\mathcal{M}_k$ and $\mathcal{S}_k$

While Theorem 3.5.1 applies to any modular form of weight k, whether holomorphic or meromorphic, we shall now specialize to holomorphic modular forms, and use Theorem 3.5.1 to compute the dimensions of the spaces $\mathcal{M}_k$ and $\mathcal{S}_k$. Holomorphic modular forms $f(\tau)$ can have no poles, so that $\mathrm{ord}_f(p) \geq 0$, including at the points $p = i\infty, i, \rho$, and ρ'.

Theorem 3.5.2. *Theorem 3.5.1 implies the following properties of $\mathcal{M}_k$:*

1. *For k odd, $k < 0$, and $k = 2$ there are no holomorphic modular forms, that is, $\dim \mathcal{M}_k = 0$;*

2. *For k even and $k \geq 4$ the space $\mathcal{M}_k$ is finite-dimensional; $\dim \mathcal{M}_0 = 1$;*

3. *For $k = 4, 6, 8, 10, 14$, we have $\dim \mathcal{M}_k = 1$ and $\mathcal{M}_k = \mathbb{C}G_k$;*

4. *The space $\mathcal{S}_k$ of weight k cusp forms is empty for $k < 12$, and $\mathcal{S}_{12} = \mathbb{C}\Delta$;*

5. *For $k \geq 16$, we have $\mathcal{S}_k = \Delta\mathcal{M}_{k-12}$ and $\mathcal{M}_k = \mathbb{C}G_k \oplus \mathcal{S}_k$.*

For low weight, these results are consistent with the results of (3.4.23). Moreover, the theorem confirms that $\mathcal{M}$ is precisely the polynomial ring freely generated by G_4 and G_6.

To prove Theorem 3.5.2, we use the valence formula (3.5.1) of Theorem 3.5.1. Considering the relation (3.5.1) multiplied by 12, and then mod 2, we see that there can be no solutions for k odd. Since modular forms have $\mathrm{ord}_f(p) \geq 0$, including at the points $p = i, \rho$, and $i\infty$, the left side of (3.5.1) is nonnegative so that there can be no solutions to (3.5.1) for $k < 0$. There is also no solution when $k = 2$, which concludes the proof of point 1. For k even and positive, there can only be a finite number of solutions since the sum on the left of (3.5.1) is over positive or zero numbers only, proving point 2. To proceed further, it is helpful first to prove two auxiliary statements,

$$G_4(\rho) = G_6(i) = 0 \qquad (3.5.4)$$

They follows from inspection of the sums,

$$G_4(\rho) = \sum_{(m,n)}' \frac{1}{(m\,\rho^3 + n\,\rho)^4} = \frac{1}{\rho^4} \sum_{(m,n)}' \frac{1}{(m\,\rho^2 + n)^4} = \rho^2 G_4(\rho)$$

$$G_6(i) = \sum_{(m,n)}' \frac{1}{(m\,i^4 + n\,i)^6} = \frac{1}{i^6} \sum_{(m,n)}' \frac{1}{(m\,i^3 + n)^6} = -G_6(i) \quad (3.5.5)$$

To recast the second sum on the first line, we use $m\rho^2 + n = m(-1-\rho) + n = n - m - m\rho$ so that the sum over (m,n) gives $G_4(\rho)$. To recast the second sum on the second line, we use $m\,i^3 + n = -im + n$, so that the sum over (m,n) gives $G_6(i)$.

By inspection of (3.5.1), we must have $\mathrm{ord}_f(i\infty) = 0$ for $k = 4, 6, 8, 10$ since the right side of (3.5.1) is less than 1, as well as for $k = 14$ since the remaining $1/6$ cannot be accounted for by nonvanishing $\mathrm{ord}_f(i)$ or $\mathrm{ord}_f(\rho)$. Therefore, the solutions to (3.5.1) for these values of k are unique and given by $(\mathrm{ord}_f(i), \mathrm{ord}_f(\rho)) = (0,1), (1,0), (0,2), (1,1), (1,2)$, respectively. They are generated by G_4, G_6, $G_8 \sim G_4^2$, $G_{10} \sim G_4 G_6$, and $G_{14} \sim G_4^2 G_6$, which proves point 3. To prove point 4, we note that a cusp form has $\mathrm{ord}_f(i\infty) \geq 1$, which requires $k \geq 12$, and uniqueness for $k = 12$. Finally for $k \geq 16$, the solutions manifestly satisfy the recursion relation $\mathcal{S}_k = \Delta \mathcal{M}_{k-12}$ and $\mathcal{M}_k = \mathbb{C}G_k \oplus \mathcal{S}_k$ by inspecting (3.5.1), proving point 5.

3.6 Modular functions and the j-function

Having classified all holomorphic modular forms under $\mathrm{SL}(2,\mathbb{Z})$ of weight k, we now classify all meromorphic modular forms of weight k. Given an arbitrary meromorphic modular form of weight k, we may always divide by a holomorphic modular form of weight k, such as G_k, to obtain a meromorphic modular function, namely a form of weight 0. Thus, it will suffice to classify the space of meromorphic modular functions. Sums, products, and ratios of modular functions produce again modular functions, so that the corresponding space is actually the field of meromorphic modular functions. We seek to obtain a complete description of this space.

Clearly, the only holomorphic modular function of weight 0 is constant in τ. We begin by constructing a modular function with a single pole at the cusp and no other poles. Normalizing the residue of the pole to unity, we define the famous j-function by,

$$j(\tau) = \frac{12^3 g_2^3}{g_2^3 - 27 g_3^2} = \frac{12^3 \, \mathsf{E}_4^3}{\mathsf{E}_4^3 - \mathsf{E}_6^2} = \frac{1}{q} + \mathcal{O}(q^0) \qquad (3.6.1)$$

where $q = e^{2\pi i \tau}$. Inspecting (3.5.1), we have $k = 0$ since we are dealing with a modular function of weight zero; $\mathrm{ord}_j(i\infty) = -1$ since j has a simple pole at $i\infty$; $\mathrm{ord}_j(i) = 0$ since $\mathsf{E}_4(i) \neq 0$; and $\mathrm{ord}_j(\rho) = 3$ since j has a triple zero at ρ in view of the identity $G_4(\rho) = \mathsf{E}_4(\rho) = 0$ established in (3.5.4). These values indeed satisfy (3.5.1), and we have,

$$j(i\infty) = \infty \qquad\qquad j(i) = 1 \qquad\qquad j(\rho) = 0 \qquad (3.6.2)$$

The q-expansion of the j-function may be derived from the q-expansions of $E_4(\tau)$ and $E_6(\tau)$ given in (3.4.13) with the help of (3.6.1). The first 13 orders in the expansion are given as follows,

$$
\begin{aligned}
j(\tau) = q^{-1} &+ 744 + 196884\,q + 21493760\,q^2 + 864299970\,q^3 & (3.6.3)\\
&+ 20245856256\,q^4 + 333202640600\,q^5 + 4252023300096\,q^6\\
&+ 44656994071935\,q^7 + 401490886656000\,q^8 + 3176440229784420\,q^9\\
&+ 22567393309593600\,q^{10} + 146211911499519294\,q^{11} + \mathcal{O}(q^{12})
\end{aligned}
$$

3.6.1 The j-functions and moonshine

It is a curious fact, first observed by McKay, that the Fourier coefficients of $j(\tau) - 744$ can be expanded in terms of dimensions of irreducible representations of the Monster group, which is the largest sporadic finite group. More concretely, the dimensions of the smallest irreducible representations of the Monster group are $1, 196883, 21296876, 842609326, \ldots$, and it is easy to witness the following relations,

$$
\begin{aligned}
196884 &= 1 + 196883\\
21493760 &= 1 + 196883 + 21296876\\
864299970 &= 2 \times 1 + 2 \times 196883 + 21296876 + 842609326 \qquad (3.6.4)
\end{aligned}
$$

and so on. This surprising connection between modular functions and sporadic groups has come to be known as *Monstrous moonshine* and is the first of a series of similar moonshine phenomena.

3.6.2 Classification of modular functions

The classification of meromorphic modular functions, of weight zero, is given by the following theorem.

Theorem 3.6.1. *Every rational function of $j(\tau)$ is a modular function of $SL(2, \mathbb{Z})$. Conversely, every meromorphic modular function $f(\tau)$ is a rational function of j. The number of poles of $f(\tau)$ in the fundamental domain F*

equals the number of zeros of $f(\tau)$ in F (both of which are to be counted with multiplicities, including at i and ρ). In terms of the zeros z_n and poles p_n for $n = 1, \cdots, N$, we have the explicit formula for $f(\tau)$,

$$f(\tau) = f_\infty \prod_{n=1}^{N} \frac{j(\tau) - j(z_n)}{j(\tau) - j(p_n)} \tag{3.6.5}$$

where f_∞ is a constant given by $f(i\infty)$. If $f(\tau)$ has a zero or pole at the cusp, a suitable limit of the above formula should be taken.

To prove the theorem, we use the valence formula (3.5.1) for $k = 0$, given that $f(\tau)$ has weight 0. Since $f(\tau)$ is meromorphic, the number of poles inside the compactified fundamental domain $\bar{F}$ must be finite. The vanishing of the sum of ord_f, counted with multiplicities, on the left side of (3.5.1) then implies that the number of zeros must equal the number of poles, thereby proving the first part of the converse statement of the theorem. Formula (3.6.5) matches all the zeros and poles of $f(\tau)$ by construction, which implies identity for some constant f_∞.

This result completes the construction of the field of meromorphic modular functions and, combining it with our earlier general construction of modular forms of arbitrary weight, also gives the construction for all meromorphic modular forms of $\mathrm{SL}(2, \mathbb{Z})$ of arbitrary weight.

3.6.3 Modular forms with prescribed zeros

A final comment in this subsection applies to holomorphic modular forms with prescribed zeros. From (3.5.1) we see that a zero at a regular point in F requires $k \geq 12$, and the addition of one regular zero requires shifting $k \to k + 12$. To construct, for example, the modular forms of weight 12 with one zero at an arbitrary point $p \in F \setminus \{i\infty, i, \rho\}$, we take a linear combination of the two modular forms available, namely E_4^3 and E_6^2, and arrange the coefficients so that a zero emerges at p,

$$f_p(\tau) = f_0(p) \left(\mathsf{E}_4(\tau)^3 \mathsf{E}_6(p)^2 - \mathsf{E}_4(p)^3 \mathsf{E}_6(\tau)^2 \right) \tag{3.6.6}$$

The normalization factor $f_0(p)$ may depend on p and can be fixed by evaluating $f_p(\tau)$ at any point $\tau \neq p$, including at $i\infty$. One proceeds analogously for modular forms with prescribed zeros at multiple points.

3.7 Modular transformations of Jacobi ϑ-functions

The Jacobi ϑ-functions obey more delicate modular transformation laws, which we shall now derive. Recall the definition (2.6.1) of the ϑ-function,

$$\vartheta(z|\tau) = \sum_{n \in \mathbb{Z}} e^{i\pi\tau n^2 + 2\pi i n z} \tag{3.7.1}$$

Setting $z = 0$ and $\tau = it$ for t real and positive, we may apply the Poisson formula (2.1.11) to obtain the transformation rule for $S \in$ SL$(2, \mathbb{Z})$,

$$\vartheta(0| - \tfrac{1}{\tau}) = (-i\tau)^{\frac{1}{2}} \, \vartheta(0|\tau) \tag{3.7.2}$$

Although originally derived for t real, the formula remains valid for $\tau \in \mathcal{H}$ provided we choose the proper branch cut for the square root in the prefactor of (3.7.2). Under the modular transformation T, however, the ϑ-function does not transform into itself, but instead we have,

$$\vartheta(z|\tau + 1) = \vartheta_4(z|\tau) \tag{3.7.3}$$

where ϑ_4 is a ϑ-function with half-integer characteristics defined in (2.6.6) and (2.6.10). The set of all four ϑ-functions with half-characteristics of (2.6.10) transforms into itself, and we have,

$$
\begin{aligned}
\vartheta_1(z|\tau + 1) &= \varepsilon\,\vartheta_1(z|\tau) & \vartheta_1(\tfrac{z}{\tau}| - \tfrac{1}{\tau}) &= -i(-i\tau)^{\frac{1}{2}} e^{i\pi z^2/\tau} \vartheta_1(z|\tau) \\
\vartheta_2(z|\tau + 1) &= \varepsilon\,\vartheta_2(z|\tau) & \vartheta_2(\tfrac{z}{\tau}| - \tfrac{1}{\tau}) &= (-i\tau)^{\frac{1}{2}} e^{i\pi z^2/\tau} \vartheta_4(z|\tau) \\
\vartheta_3(z|\tau + 1) &= \vartheta_4(z|\tau) & \vartheta_3(\tfrac{z}{\tau}| - \tfrac{1}{\tau}) &= (-i\tau)^{\frac{1}{2}} e^{i\pi z^2/\tau} \vartheta_3(z|\tau) \\
\vartheta_4(z|\tau + 1) &= \vartheta_3(z|\tau) & \vartheta_4(\tfrac{z}{\tau}| - \tfrac{1}{\tau}) &= (-i\tau)^{\frac{1}{2}} e^{i\pi z^2/\tau} \vartheta_2(z|\tau)
\end{aligned} \tag{3.7.4}
$$

where $\varepsilon = e^{2\pi i/8}$. One may collect these transformation laws in a single expression for each transformation, valid for all half-characteristics,

$$\vartheta \begin{bmatrix} \alpha \\ \beta \end{bmatrix} (z|\tau + 1) = e^{-i\pi\alpha(1+\alpha)}\, \vartheta \begin{bmatrix} \alpha \\ \beta + \alpha + \frac{1}{2} \end{bmatrix} (z|\tau)$$

$$\vartheta \begin{bmatrix} \alpha \\ \beta \end{bmatrix} \left(\tfrac{z}{\tau}| - \tfrac{1}{\tau}\right) = (-i\tau)^{\frac{1}{2}}\, e^{2\pi i\alpha\beta + i\pi z^2/\tau}\, \vartheta \begin{bmatrix} \beta \\ -\alpha \end{bmatrix} (z|\tau) \tag{3.7.5}$$

Combining the two, one finds that under a generic element $\gamma \in$ SL$(2, \mathbb{Z})$, the ϑ_1 function transforms as follows,

$$\vartheta_1(\tilde{z}|\tilde{\tau}) = \varepsilon_\vartheta(\gamma)(c\tau + d)^{\frac{1}{2}}\, e^{\pi i c z^2/(c\tau + d)}\, \vartheta_1(z|\tau) \qquad \gamma = \begin{pmatrix} a & b \\ c & d \end{pmatrix} \tag{3.7.6}$$

The transformation of the ϑ-function with arbitrary characteristics may be deduced from (3.7.6) using (2.6.7), and we find,

$$\vartheta \begin{bmatrix} \alpha + \frac{1}{2} \\ \beta + \frac{1}{2} \end{bmatrix} (\tilde{z}|\tilde{\tau}) = \varepsilon(\alpha, \beta; \gamma)(c\tau + d)^{\frac{1}{2}} e^{\pi i c z^2/(c\tau + d)} \, \vartheta \begin{bmatrix} \tilde{\alpha} + \frac{1}{2} \\ \tilde{\beta} + \frac{1}{2} \end{bmatrix} (z|\tau) \quad (3.7.7)$$

where z, τ, and the characteristics α, β transform as follows,

$$\tilde{z} = \frac{z}{c\tau + d} \qquad \tilde{\tau} = \frac{a\tau + b}{c\tau + d} \qquad \begin{bmatrix} \tilde{\alpha} \\ \tilde{\beta} \end{bmatrix} = \begin{bmatrix} a\alpha + c\beta \\ b\alpha + d\beta \end{bmatrix} \quad (3.7.8)$$

and the multiplier $\varepsilon(\alpha, \beta; \gamma)$ is given by,

$$\varepsilon(\alpha, \beta; \gamma) = \varepsilon_\vartheta(\gamma) \, e^{\pi i \{\alpha(\beta+1) - \alpha'(\beta'+1)\}} \quad (3.7.9)$$

Finally, $\varepsilon_\vartheta(\gamma)$ and $\varepsilon(\alpha, \beta; \gamma)$ are both eight roots of unity and $\varepsilon_\vartheta(\gamma)$ is independent of the characteristics α, β. The evaluation of $\varepsilon_\vartheta(\gamma)$ is in terms of the generalized Legendre symbol and will be given in Section 3.8.

Setting $z = 0$ in the ϑ-functions gives the so-called ϑ-*constants*, whose modular transformation laws are special cases of the general transformation laws presented above. None of these are modular forms under $\mathrm{SL}(2, \mathbb{Z})$, though we shall see in Chapter 7 that they are holomorphic modular forms under certain congruence subgroups of $\mathrm{SL}(2, \mathbb{Z})$. Furthermore, the derivative $\vartheta_1'(0|\tau)$ is closely related to the discriminant Δ while various sums of eight powers of ϑ_2, ϑ_3, and ϑ_4 are modular forms as we shall now discuss.

3.7.1 The discriminant Δ in terms of ϑ-constants

The modular transformation law for $\vartheta_1'(0|\tau)$ is readily obtained from (3.7.4) and is given below along with its eighth power,

$$\vartheta_1'(0|\tau + 1) = \varepsilon \, \vartheta_1'(0|\tau) \qquad \vartheta_1'(0| - \tfrac{1}{\tau}) = (-i\tau)^{\frac{3}{2}} \vartheta_1'(0|\tau)$$
$$\vartheta_1'(0|\tau + 1)^8 = \vartheta_1'(0|\tau)^8 \qquad \vartheta_1'(0| - \tfrac{1}{\tau})^8 = (-i\tau)^{12} \vartheta_1'(0|\tau)^8 \quad (3.7.10)$$

While $\vartheta_1'(0|\tau)$ itself is not a modular form, its eight power is a holomorphic modular form of weight 12 and therefore must be a linear combination of E_4^3 and E_6^2. To obtain a precise relation, we appeal to the infinite product formula for $\vartheta_1(z|\tau)$ given in (2.6.24) and find the following relations,

$$\vartheta_1'(0|\tau) = \pi \vartheta_2(0|\tau) \vartheta_3(0|\tau) \vartheta_4(0|\tau) = 2\pi q^{\frac{1}{8}} \prod_{n=1}^{\infty} (1 - q^n)^3 \quad (3.7.11)$$

This expression indeed fails to be a modular form in view of the $q^{\frac{1}{8}}$ prefactor, which is not invariant under $T : \tau \to \tau + 1$. However, its eight power produces a factor of q which is not only consistent with being a genuine modular form but also shows that it is a cusp form proportional to the discriminant Δ. The constant of proportionality is fixed by matching the term of order q in the Fourier expansion of both. As a result, we find a relation between $\vartheta_1'(0|\tau)^8$ and the discriminant Δ as well as a product formula for Δ,

$$\Delta(\tau) = (2\pi)^4 \vartheta_1'(0|\tau)^8 = (2\pi)^{12} q \prod_{n=1}^{\infty} (1 - q^n)^{24} \qquad (3.7.12)$$

The product formula for the discriminant shows that $\Delta(\tau)$ is nowhere vanishing for $\tau \in \mathcal{H}$ and has a simple zero in q at the cusp.

The Fourier coefficients $\tau(n)$ of Δ (not to be confused with the modulus of the torus!) are defined by,

$$\Delta(\tau) = (2\pi)^{12} \sum_{n=1}^{\infty} \tau(n) q^n \qquad (3.7.13)$$

They are referred to as the Ramanujan τ-function and have special arithmetic significance. The values of $\tau(n)$ for $n \leq 12$ are given as follows,

$$
\begin{aligned}
&\tau(1) = 1 && \tau(4) = -1472 && \tau(7) = -16744 && \tau(10) = -115920 \\
&\tau(2) = -24 && \tau(5) = 4830 && \tau(8) = 84480 && \tau(11) = 534612 \\
&\tau(3) = 252 && \tau(6) = -6048 && \tau(9) = -113643 && \tau(12) = -370944 \quad (3.7.14)
\end{aligned}
$$

The relation $1728\Delta = (2\pi)^{12}(\mathsf{E}_4^3 - \mathsf{E}_6^2)$ along with $\mathsf{E}_4^2 = \mathsf{E}_8$ allows us to express the τ-function in terms of sums of divisor functions as follows,

$$\tau(n) = \sum_{m=1}^{n-1} \left(\tfrac{200}{3} \sigma_3(m)\sigma_7(n-m) - 147\, \sigma_5(m)\sigma_5(N-m) \right)$$
$$+ \tfrac{5}{36} \sigma_3(n) + \tfrac{7}{12} \sigma_5(n) + \tfrac{5}{18} \sigma_7(n) \qquad (3.7.15)$$

Eliminating σ_7 in terms of σ_3 using the first formula in (3.4.26) confirms that $\tau(n)$ may be expressed solely in terms of σ_3 and σ_5. An alternative relation giving $\tau(n)$ in terms of σ_5 and σ_{11} will be obtained in Exercise 3.5.

Ramanujan conjectured that $\tau(n)$ is a *multiplicative arithmetic function*, which satisfies,

$$\tau(mn) = \tau(m)\tau(n) \qquad\qquad \gcd(m,n) = 1 \qquad (3.7.16)$$

where $\gcd(m,n)$ stands for the greatest common divisor of m and n. Ramanujan's function is, however, not *completely multiplicative* (which would mean that the first equation in (3.7.16) holds for arbitrary m, n, not necessarily relatively prime) and instead satisfies,

$$\tau(m)\tau(n) = \sum_{d\,|\,\gcd(m,n)} d^{11}\,\tau\left(\frac{mn}{d^2}\right) \tag{3.7.17}$$

Both conjectures on $\tau(n)$ were proven by Mordell in 1917 using the theory of Hecke operators applied to the cusp form $\Delta(\tau)$, and we shall provide this proof in Chapter 10.

3.7.2 Sums and products of eighth powers

Another way to construct a modular form using the ϑ-constants is by sums of eighth powers. The modular transformation properties show right away that the following sums,

$$T_{4k}(\tau) = \vartheta_2(0|\tau)^{8k} + \vartheta_3(0|\tau)^{8k} + \vartheta_4(0|\tau)^{8k} \tag{3.7.18}$$

are modular forms of weight $4k$. For $k = 1, 2$, the corresponding spaces of modular forms are one dimensional, so that we must have,

$$\mathsf{E}_4(\tau) = \tfrac{1}{2}T_4(\tau) = \tfrac{1}{2}\left(\vartheta_2(0|\tau)^8 + \vartheta_3(0|\tau)^8 + \vartheta_4(0|\tau)^8\right) \tag{3.7.19}$$

and $T_8 = 2\mathsf{E}_8 = 2\mathsf{E}_4^2$ by matching asymptotics at the cusp. The discriminant, on the other hand, may be expressed as a product of eight powers, by combining (3.7.11) and (3.7.12),

$$\Delta(\tau) = 16\pi^{12}\vartheta_2(0|\tau)^8\vartheta_3(0|\tau)^8\vartheta_4(0|\tau)^8 \tag{3.7.20}$$

so that the modular j-function takes the form,

$$j(\tau) = 1728\frac{g_2(\tau)^3}{\Delta(\tau)} = 32\frac{\left(\vartheta_2(0|\tau)^8 + \vartheta_3(0|\tau)^8 + \vartheta_4(0|\tau)^8\right)^3}{\vartheta_2(0|\tau)^8\,\vartheta_3(0|\tau)^8\,\vartheta_4(0|\tau)^8} \tag{3.7.21}$$

The asymptotics of the ϑ-constants at the cusp are given by (2.6.24) and confirm that $j(q)$ has a simple pole at the cusp with unit residue $j(q) \sim q^{-1}$. Finally, the modular form E_6 clearly cannot be written in terms of a polynomial in eight powers of ϑ-constants, but it does have a useful expression in terms of fourth powers, given by,

$$\mathsf{E}_6(\tau) = \left(\vartheta_3(0|\tau)^4 + \vartheta_4(0|\tau)^4\right) \tag{3.7.22}$$

$$\times\left(\frac{5}{2}\vartheta_3(0|\tau)^4\vartheta_4(0|\tau)^4 - \vartheta_3(0|\tau)^8 - \vartheta_4(0|\tau)^8\right)$$

The combination on the right side is a holomorphic modular form of weight 6, even if this property is not manifest.

3.8 The Dedekind η-function

The Dedekind η-function is defined as follows,

$$\eta(\tau) = q^{\frac{1}{24}} \prod_{n=1}^{\infty} (1 - q^n) \tag{3.8.1}$$

It is not a modular form of $\mathrm{SL}(2,\mathbb{Z})$, but its 24-th power is a modular form of weight 12 and, by (3.7.12), is related to the discriminant as follows,

$$\Delta(\tau) = (2\pi)^{12} \eta(\tau)^{24} \tag{3.8.2}$$

Under the generators T and S of the modular group, η transforms as follows,

$$\eta(\tau + 1) = e^{2\pi i/24} \, \eta(\tau)$$
$$\eta(-\tfrac{1}{\tau}) = (-i\tau)^{\frac{1}{2}} \, \eta(\tau) \tag{3.8.3}$$

The first relation may be read off from the definition of η, while the second relation may be derived first for $\tau \in i\mathbb{R}^+$ using Poisson resummation and then be extended to $\tau \in \mathcal{H}$.[2]

The transformation of $\eta(\tau)$ under an arbitrary $\gamma \in \mathrm{SL}(2,\mathbb{Z})$ may be expressed as follows,

$$\eta(\gamma\tau) = \varepsilon_\eta(\gamma)\,(c\tau+d)^{\frac{1}{2}}\eta(\tau) \qquad\qquad \gamma = \begin{pmatrix} a & b \\ c & d \end{pmatrix} \tag{3.8.4}$$

with $\varepsilon_\eta(\gamma)^{24} = 1$ or $\varepsilon_\eta : \mathrm{SL}(2,\mathbb{Z}) \to \mathbb{Z}_{24}$. The function $\varepsilon_\eta(\gamma)$ depends on γ, but not on τ. Its values may be determined from the transformation rules given in (3.8.3). Note that the actions of the transformations γ and $-\gamma$ coincide on τ but differ on the square root and on ε_η giving, in particular, the value $\varepsilon_\eta(-I) = (-1)^{\frac{1}{2}}$. In view of this property, $\varepsilon_\eta(\gamma)$ is not a proper homomorphism of $\mathrm{SL}(2,\mathbb{Z}) \to \mathbb{Z}_{24}$ but instead is referred to as a *multiplier system*. Special values may be deduced from the transformation laws (3.8.3),

$$\varepsilon_\eta(T^n) = e^{2\pi i n/24}$$
$$\varepsilon_\eta(S) = e^{-2\pi i/8}$$
$$\varepsilon_\eta(-I) = e^{2\pi i/4} = i \tag{3.8.5}$$

In particular, if γ is such that either $c > 0$, or $c = 0$ with $d > 0$, then $\varepsilon_\eta(-\gamma) = i\varepsilon_\eta(\gamma)$. The Petersson formula for $\varepsilon_\eta(\gamma)$ for a γ with either $c > 0$,

---

[2] The square root is defined via the argument function $\arg(\tau)$ for $\tau \in \mathbb{C}^*$, in terms of which $\tau = |\tau| \exp(i\arg(\tau))$ and $-\pi \leq \arg(\tau) < \pi$ by $\tau^{\frac{1}{2}} = |\tau|^{\frac{1}{2}} \exp(i\arg(\tau)/2)$ (see [Köh11]). This definition is compatible with the fact that for $\tau = it$ with $t \in \mathbb{R}^+$ we have $\eta(it), \eta(i/t) > 0$.

or $c = 0$ and $d > 0$ is given as follows (see [Köh11]),

$$c \text{ odd:} \qquad \varepsilon_\eta(\gamma) = (d|c) \exp\left\{ \tfrac{2\pi i}{24}\left((a+d)c + bd(1-c^2) - 3c\right)\right\}$$

$$c \text{ even:} \qquad \varepsilon_\eta(\gamma) = \text{sign}(d)(c|d) \exp\left\{ \tfrac{2\pi i}{24}\left((a+d)c + bd(1-c^2)\right.\right.$$

$$\left.\left. + 3d + 3 - 3cd\right)\right\} \qquad (3.8.6)$$

We recall from Appendix A, and in particular (A.4.3), that the Jacobi symbol $(n|N)$ for odd denominator N (which is always the case here) with decomposition into distinct primes p_i given by $N = \prod_i p_i^{\alpha_i}$ is defined by $(n|N) = \prod_i (n|p_i)$ where $(n|p_i)$ is the Legendre symbol for an odd prime p_i. The definition is extended to include $n = 0$ by $(0|1) = (0|-1) = 1$ and negative N by $(n|N) = (n|-N)$.

Knowledge of $\varepsilon_\eta(\gamma)$ allows us to produce a formula for the multiplier $\varepsilon_\vartheta(\gamma)$ encountered in the modular transformation law of the ϑ-functions in (3.7.6) by using the relation,

$$\vartheta_1'(0|\tau) = 2\pi\eta(\tau)^3 \qquad (3.8.7)$$

which is obtained by combining (3.7.11) with the definition of η in (3.8.1). The modular transformation of the left side may be derived from (3.7.6) while the transformation of the right side is given by (3.8.4), and we find,

$$\varepsilon_\vartheta(\gamma) = \varepsilon_\eta(\gamma)^3 \qquad (3.8.8)$$

consistent with $\varepsilon_\vartheta^8 = \varepsilon_\eta^{24} = 1$.

Finally, we note that the fourth power $\varepsilon_\eta^4(\gamma)$ no longer suffers from the subtleties of sign reversal of γ and the square root and is a genuine homomorphism $\varepsilon_\eta^4 : \text{SL}(2,\mathbb{Z}) \to \mathbb{Z}_6$ with $\varepsilon_\eta^4(-I) = 1$ and $\varepsilon_\eta^4(\gamma_1\gamma_2) = \varepsilon_\eta^4(\gamma_1)\varepsilon_\eta^4(\gamma_2)$.

• Bibliographical notes

The books by Apostol [Apo76b] and Lang [Lan76] and the lecture notes by Zagier [Zag92, Zag08] offer classic introductions to the subject of modular forms. The books by Shimura [Shi71] and Koblitz [Kob12] provide somewhat advanced presentations aimed more at arithmetic. A very clear and detailed presentation, to which we shall refer often, is given in the treatise by Diamond and Shurman [DS05]. An automorphic function perspective is given in the Bateman Manuscript Project [Erd81c] and the book by Iwaniec [Iwa97]. A detailed exposition of identities involving the Dedekind η-function and their role in the theory of modular forms may be found in the book by Köhler [Köh11]. Presentations with an emphasis on L-functions and q-series expansions may be found in the books by Bump [Bum98] and Ono [Ono04], respectively.

Exercises

3.1 Prove that the geodesics of the Poincaré metric $ds^2 = |dz|^2/y^2$ on the upper half plane $\mathcal{H} = \{z = x + iy \,|\, x, y \in \mathbb{R}, \, y > 0\}$ are half circles centered on the real line. Obtain an explicit formula for the hyperbolic distance $\ell(z_1, z_2)$ between two points $z_1, z_2 \in \mathcal{H}$. Prove the formula for the area of a hyperbolic triangle given in (3.2.4).

3.2 We consider integral binary quadratic forms $Q(x, y) = \alpha x^2 + \beta xy + \gamma y^2$ with $\alpha, \beta, \gamma \in \mathbb{Z}$ for $x, y \in \mathbb{Z}$. Clearly, the discriminant $D = \beta^2 - 4\alpha\gamma$ can only take the values $D \equiv 0, 1 \pmod 4$. We restrict to positive definite forms for which $\alpha, \gamma \geq 1$ and $D < 0$, which are primitive so that $\gcd(\alpha, \beta, \gamma) = 1$. A quadratic form $Q'(x, y)$ is equivalent to the quadratic form $Q(x, y)$ provided it is related by an invertible integer-valued linear map,

$$Q'(x, y) = Q(x', y') \qquad \begin{cases} x = ax' + by' \\ y = cx' + dy' \end{cases} \qquad (3.8.9)$$

with $a, b, c, d \in \mathbb{Z}$ and $ad - bc = 1$. The number of inequivalent primitive quadratic forms for a given discriminant $D < 0$ is referred to as the class number and is denoted $h(D)$.

(a) Recasting the quadratic form as $Q(x, y) = \alpha|x - \tau y|^2$, use Theorem 3.2.1 to characterize a single representative of each equivalence class.

(b) Compute the class numbers $h(D)$ for $-30 < D < 0$ (with the help of Maple or Mathematica, say).

3.3 In this problem, we establish two basic properties of the divisor sums.

(a) Prove the following identity for the generating function for $\sigma_s(m)$,

$$\sum_{n=1}^{\infty} \frac{n^s x^n}{1 - x^n} = \sum_{m=1}^{\infty} \sigma_s(m) x^m \qquad (3.8.10)$$

(b) Prove that the sum of divisor function is multiplicative. Provide an example that shows that it is not completely multiplicative.

3.4 Prove the identities (3.4.26) using (3.4.25) and (3.4.13).

3.5 In (3.7.15), the Ramanujan τ function was expressed in terms of sums of divisor functions σ_3, σ_5, and σ_7. Express $\tau(n)$ in terms of the divisor sums $\sigma_{11}(n)$ and $\sigma_5(n)$ and a sum of products of $\sigma_5(n)$. Using this result, prove Ramanujan's congruence relation $\tau(n) \equiv \sigma_{11}(n) \pmod{691}$.

4

Variants of modular forms

Building on the basics of modular forms developed in Chapter 3, we now discuss variants such as quasi-modular forms, almost-holomorphic modular forms, Maass forms, non-holomorphic Eisenstein series, mock modular forms, and quantum modular forms. Other generalizations, including modular forms for congruence subgroups, vector-valued modular forms, and modular graph forms, will be introduced in Chapters 7–9, respectively.

4.1 Quasi-modular and almost-holomorphic modular forms

Quasi-modular forms and almost-holomorphic modular forms are generalizations of the modular forms introduced in Chapter 3, which are obtained by relaxing certain parts of the definition of modular forms. We begin by reviewing the most famous of these objects, referred to as $\mathsf{E}_2(\tau)$ and $\mathsf{E}_2^*(\tau)$ respectively, before proceeding to the definition and presentation of the general properties of quasi-modular forms and almost-holomorphic modular forms.

4.1.1 E_2 *and* E_2^*

We have already introduced the holomorphic Eisenstein series G_k in (2.4.5) and (3.4.1), or equivalently E_k in (3.4.13). They are given for $k \geq 3$ in terms of absolutely convergent sums over a lattice $\Lambda = \mathbb{Z} \oplus \mathbb{Z}\tau$ and vanish for odd values of $k \geq 3$. For $k = 2$, the corresponding sum,

$$\sum_{\omega \in \Lambda'} \frac{1}{\omega^2} = \sum_{\substack{m,n \in \mathbb{Z} \\ (m,n) \neq (0,0)}} \frac{1}{(m+n\tau)^2} \tag{4.1.1}$$

fails to be absolutely convergent. Instead, it is conditionally convergent which means that its value will depend on how the infinite sums are arranged. Physicists would say that the sum needs to be *regularized*. There may be one or several choices for regularization that are deemed "natural" because they preserve one property or another of modular forms.

One choice for the regularization is to adopt the *Eisenstein summation convention*, which consists in cutting off the infinite sums in a symmetrical way, as was already used in (2.2.2) for a similar sum,

$$\mathsf{E}_2(\tau) = \lim_{N\to\infty} \lim_{M\to\infty} \frac{1}{2\zeta(2)} \sum_{n=-N}^{N} \sum_{\substack{m=-M \\ (m,n)\neq(0,0)}}^{M} \frac{1}{(m+n\tau)^2} \qquad (4.1.2)$$

For finite M, N, the sum is holomorphic in $\tau \in \mathcal{H}$ and the limit $M, N \to \infty$ converges to a finite holomorphic function $\mathsf{E}_2(\tau)$. However, $\mathsf{E}_2(\tau)$ fails to transform as a modular form under $\mathrm{SL}(2,\mathbb{Z})$. This is not surprising as Theorem 3.5.2 tells us there are no holomorphic modular forms of weight 2. Instead the transformation law of $\mathsf{E}_2(\tau)$ under $\gamma \in \mathrm{SL}(2,\mathbb{Z})$ is as follows,

$$\mathsf{E}_2(\gamma\tau) = (c\tau + d)^2 \mathsf{E}_2(\tau) + \frac{12}{2\pi i} c(c\tau + d) \qquad \gamma = \begin{pmatrix} a & b \\ c & d \end{pmatrix} \quad (4.1.3)$$

To show this, we decompose the sum of (4.1.2) into contributions from terms with $n = 0$ and $m \neq 0$, and from terms with $n \neq 0$ and arbitrary m,

$$\mathsf{E}_2(\tau) = 1 + \lim_{N\to\infty} \lim_{M\to\infty} \frac{1}{\zeta(2)} \sum_{n=1}^{N} \sum_{m=-M}^{M} \frac{1}{(m+n\tau)^2} \qquad (4.1.4)$$

and perform the sum over m using (3.4.9) in the convergent limit $M \to \infty$. The remaining sum over n admits a finite limit as $N \to \infty$ and evaluates to,

$$\mathsf{E}_2(\tau) = 1 - 24 \sum_{n=1}^{\infty} \sum_{\ell=1}^{\infty} \ell\, e^{2\pi i \ell n \tau} = 1 - 24 \sum_{\ell=1}^{\infty} \sigma_1(\ell) q^\ell \qquad (4.1.5)$$

where $q = e^{2\pi i \tau}$ and $\sigma_1(\ell)$ is a sum of divisors function defined in (3.4.11). Alternatively, the sum over n in (4.1.5) may be performed first and the remaining sum over ℓ may be rearranged in terms of the discriminant $\Delta(\tau)$,

$$\mathsf{E}_2(\tau) = \frac{1}{2\pi i} \partial_\tau \ln \Delta(\tau) \qquad (4.1.6)$$

Using the modular transformation law of the weight 12 modular form $\Delta(\tau)$ under $\mathrm{SL}(2,\mathbb{Z})$, we deduce the transformation law of $\mathsf{E}_2(\tau)$ by differentiating $\ln \Delta(\tau)$ to obtain (4.1.3). Remarkable relations with the Weierstrass ζ and $\wp$ functions were already encountered in (2.6.21) and (2.6.22) and include the following identities,

$$\zeta(\tfrac{1}{2}|\tau) = \frac{\pi^2}{6}\, \mathsf{E}_2(\tau)$$

$$\wp(z|\tau) = -\partial_z^2 \ln \vartheta_1(z|\tau) - \frac{\pi^2}{3}\, \mathsf{E}_2(\tau) \tag{4.1.7}$$

The latter demonstrates how $\mathsf{E}_2(\tau)$ acts as a *modular connection* to guarantee the proper modular transformation of $\wp$ given the modular transformation law of the ϑ_1-function in (3.7.4) and (3.7.6).

Another choice for the regularization is as follows,

$$\mathsf{E}_2^*(\tau) = \lim_{\varepsilon \to 0^+} \frac{1}{2\zeta(2)} \sum_{\omega \in \Lambda'} \frac{1}{\omega^2\, |\omega|^\varepsilon} \tag{4.1.8}$$

Finite $\varepsilon > 0$ renders the sum over Λ' absolutely convergent and the limit $\varepsilon \to 0$ turns out to be finite (consistent with the fact that the original sum in (4.1.1) was conditionally convergent). By construction, the resulting $\mathsf{E}_2^*(\tau)$ is a weight 2 modular form under $\mathrm{SL}(2,\mathbb{Z})$,

$$\mathsf{E}_2^*(\gamma\tau) = (c\tau + d)^2 \mathsf{E}_2^*(\tau) \tag{4.1.9}$$

However, $\mathsf{E}_2^*(\tau)$ fails to be holomorphic in τ. The regularization factor $|\omega|^\varepsilon$ introduces non-holomorphic behavior for $\varepsilon > 0$, which leaves a non-holomorphic remnant as $\varepsilon \to 0$,[1]

$$\mathsf{E}_2^*(\tau) = \mathsf{E}_2(\tau) - \frac{3}{\pi\tau_2} \tag{4.1.10}$$

where the holomorphic function $\mathsf{E}_2(\tau)$ was constructed earlier in (4.1.6). One verifies in Exercise 4.1 that the modular transformation laws of E_2 and E_2^* map into one another under the relation (4.1.10).

In summary, while the Eisenstein series E_k for even $k \geq 4$ are holomorphic modular forms of weight k, the regularization of the infinite sum for $k = 2$ allows for two "natural" choices: one is holomorphic but not modular, while the other is modular but not holomorphic. One cannot enforce both modularity and holomorphicity simultaneously. The incompatibility between holomorphicity and modular invariance may be viewed as perhaps the first example of what physicists refer to as an *anomaly*. Eisenstein's result dates back to before 1850!

---

[1] Throughout, we shall use the notation $\tau = \tau_1 + i\tau_2$ with $\tau_1, \tau_2 \in \mathbb{R}$.

4.1.2 Ring structure

A function $f : \mathcal{H} \to \mathbb{C}$ is an *almost-holomorphic modular form of weight k* for the modular group $\mathrm{SL}(2, \mathbb{Z})$ if f transforms as follows under $\gamma \in \mathrm{SL}(2, \mathbb{Z})$,

$$f(\gamma\tau) = (c\tau + d)^k f(\tau) \qquad\qquad (4.1.11)$$

and is a polynomial in τ_2^{-1} whose coefficients $f_n(q)$ are holomorphic in q,

$$f(\tau) = \sum_{n=0}^{N} f_n(q)\tau_2^{-n} \qquad\qquad q = e^{2\pi i \tau} \qquad (4.1.12)$$

By this definition, every holomorphic modular form of weight k is an almost-holomorphic modular form of weight k, while $\mathsf{E}_2^*(\tau)$ is an almost-holomorphic modular form of weight 2 that is not a holomorphic modular form. Clearly, any linear combination of almost-holomorphic modular forms of weight k is an almost-holomorphic modular form of weight k, and the product of almost-holomorphic modular forms of weights k and ℓ is an almost-holomorphic modular form of weight $k + \ell$, so that the space of almost-holomorphic is a ring graded by the weight.

A function $g : \mathcal{H} \to \mathbb{C}$ is a *quasi-modular form of weight k* if it is the holomorphic part $f_0(q)$ of some almost-holomorphic modular form f of weight k, as defined in (4.1.12). By this definition, every holomorphic modular form of weight k is a quasi-modular form of weight k, and $\mathsf{E}_2(\tau)$ is a quasi-modular form of weight 2 that is not a holomorphic modular form. The space of quasi-modular forms is a ring graded by the weight. More precisely, we have the following theorem.

Theorem 4.1.1. *The space of almost-holomorphic modular forms is a polynomial ring, graded by the weight k, and generated by $\mathsf{E}_2^*, \mathsf{E}_4$, and E_6. The space of quasi-modular forms is a polynomial ring, graded by the weight k, and generated by $\mathsf{E}_2, \mathsf{E}_4$, and E_6.*

The proof of the theorem is relegated to Exercise 4.3. Further relations between almost-holomorphic modular forms and holomorphic modular forms via differential operators are the subject of Exercise 4.2.

4.1.3 Differential equations for quasi-modular forms

The inhomogeneous nature of the transformation law of $\mathsf{E}_2(\tau)$ under $\mathrm{SL}(2, \mathbb{Z})$ suggests that $\mathsf{E}_2(\tau)$ may be viewed geometrically as a connection in the space of modular forms $\mathcal{M}$. The connection E_2 produces a covariant derivative D_k, referred to as the *Serre derivative*, which acts on $\mathcal{M}_k$ by,

$$D_k : \mathcal{M}_k \to \mathcal{M}_{k+2} \qquad\qquad D_k = \frac{1}{2\pi i}\frac{d}{d\tau} - \frac{k}{12}E_2(\tau) \qquad (4.1.13)$$

To show this, we use the fact that a modular form f of weight k, namely $f \in \mathcal{M}_k$, and its τ derivative f' satisfy the following transformation rules,

$$f(\gamma\tau) = (c\tau + d)^k f(\tau)$$
$$f'(\gamma\tau) = (c\tau + d)^{k+2} f'(\tau) + kc(c\tau + d)^{k+1} f(\tau) \qquad (4.1.14)$$

Combining (4.1.3) with the above relations and the definition of D_k gives,

$$D_k f(\gamma\tau) = (c\tau + d)^{k+2} D_k f(\tau) \qquad (4.1.15)$$

so that $D_k f$ is a modular form of weight $k + 2$, as announced. The Leibniz formula for the product of two modular forms f_1 and f_2 of respective weights k_1 and k_2 is given as follows,

$$D_{k_1+k_2}(f_1 f_2) = (D_{k_1} f_1)\, f_2 + f_1\, (D_{k_2} f_2) \qquad (4.1.16)$$

When the space $\mathcal{M}_{k+2}$ is one dimensional, one easily derives a differential equation for modular forms of weight k. The precise normalizations may be deduced from the asymptotic behavior at the cusp. For example, we have,

$$D_4\,E_4 = -\tfrac{1}{3}E_6 \qquad D_8\,E_8 = -\tfrac{2}{3}E_{10} \qquad D_{12}\,\Delta = 0$$
$$D_6\,E_6 = -\tfrac{1}{2}E_8 \qquad\qquad\qquad D_{12}\,E_{12} = -E_{14} \qquad (4.1.17)$$

Equivalent expressions may be obtained by using $E_8 = E_4^2$, $E_{10} = E_4 E_6$, and $E_{14} = E_4^2 E_6$ together with Leibniz's rule. The derivative of E_{10} is, however, slightly more involved since the space $\mathcal{M}_{12}$ is two dimensional.

Given that E_2 may naturally be interpreted as a *connection* in $\mathcal{M}_k$ there should be a natural notion of *curvature*. The associated curvature should transform tensorially, namely, it should be a modular form and should be obtained from the derivative and the square of the connection. Given the transformation law (4.1.3) for E_2, we obtain the transformations for its derivative with respect to τ and its square,

$$E_2'(\gamma\tau) = (c\tau + d)^4 E_2'(\tau) + 2c(c\tau + d)^3 E_2(\tau) + \tfrac{12c^2}{2\pi i}(c\tau + d)^2 \qquad (4.1.18)$$
$$E_2(\gamma\tau)^2 = (c\tau + d)^4 E_2(\tau)^2 + \tfrac{24}{2\pi i}\,c(c\tau + d)^3 E_2(\tau) + \left(\tfrac{12c}{2\pi i}\right)^2 (c\tau + d)^2$$

As a result, the following combination,

$$\frac{1}{2\pi i}E_2'(\gamma\tau) - \frac{1}{12}E_2(\gamma\tau)^2 = (c\tau + d)^4 \left(\frac{1}{2\pi i}E_2'(\tau) - \frac{1}{12}E_2(\tau)^2\right) \qquad (4.1.19)$$

transforms as a modular form of weight 4. Since this combination is both holomorphic in τ and a modular form of weight 4, it must belong to $\mathcal{M}_4$. But by Theorem 3.5.2, the space $\mathcal{M}_4$ has dimension one and is generated by E_4. Matching the value at $q = 0$, we obtain the relation,

$$\frac{1}{2\pi i}\frac{d}{d\tau}\mathsf{E}_2(\tau) - \frac{1}{12}\mathsf{E}_2(\tau)^2 = -\frac{1}{12}\mathsf{E}_4(\tau) \qquad (4.1.20)$$

It is natural to interpret $\mathsf{E}_4(\tau)$ as the curvature of the connection $\mathsf{E}_2(\tau)$, given indeed by a linear combination of the derivative and the square of $\mathsf{E}_2(\tau)$. The coefficient multiplying $\mathsf{E}_2(\tau)^2$ is half of what one would have expected if the derivative had been applied to a modular form of weight 2. This is as usual when obtaining the curvature from the connection. Implications of the relation (4.1.20) for identities between the divisor functions $\sigma_1(N)$ and $\sigma_3(N)$ are established in Exercise 4.4.

4.2 Non-holomorphic Eisenstein series

Next, we introduce non-holomorphic Eisenstein series, which may be viewed alternatively as generalizing the Riemann ζ-function, modular forms, or almost-holomorphic modular forms. Given a lattice $\Lambda = \mathbb{Z} + \mathbb{Z}\tau$, a non-holomorphic Eisenstein series may be defined by the following lattice sum,[2] or so-called *Kronecker–Eisenstein sum*,

$$E_s(\tau) = \sum_{\omega \in \Lambda'} \frac{\tau_2^s}{\pi^s |\omega|^{2s}} = {\sum_{m,n\in\mathbb{Z}}}' \frac{\tau_2^s}{\pi^s |m + n\tau|^{2s}} \qquad (4.2.1)$$

where $\tau_2 = \mathrm{Im}\,(\tau)$ and the prime on the sum indicates that the contribution $m = n = 0$ is to be omitted from the sum. The sum is absolutely convergent for $\mathrm{Re}\,(s) > 1$. The lattice sum definition of $E_s(\tau)$ may be used to show that E_s is modular invariant,

$$E_s(\gamma\tau) = E_s(\tau) \qquad \gamma = \begin{pmatrix} a & b \\ c & d \end{pmatrix} \in \mathrm{SL}(2,\mathbb{Z}) \qquad (4.2.2)$$

The lattice sum (4.2.1) may also be used to factor out the nontrivial common divisors of m and n in the double sum,

$$E_s(\tau) = \zeta(2s) {\sum_{\substack{c,d\in\mathbb{Z} \\ \gcd(c,d)=1}}}' \frac{\tau_2^s}{\pi^s |c\tau + d|^{2s}} \qquad (4.2.3)$$

---

[2] The notation $E_s(\tau)$ used for the non-holomorphic Eisenstein series should not be confused with the notation $\mathsf{E}_s(\tau)$ used for the holomorphic Eisenstein series. The alternative notation $E(s,\tau) = E_s(\tau)/2\zeta(2s)$ is also frequently used.

where $\zeta(2s)$ is the Riemann ζ-function and the remaining sum is over pairs $(c,d) \in \mathbb{Z}^2$ that have no common divisors strictly greater than 1. This expression in turn allows us to write $E_s(\tau)$ as a Poincaré series over the left coset $\Gamma_\infty \backslash \mathrm{SL}(2,\mathbb{Z})$ as follows,

$$E_s(\tau) = \frac{\zeta(2s)}{\pi^s} \sum_{\gamma \in \Gamma_\infty \backslash \mathrm{SL}(2,\mathbb{Z})} (\mathrm{Im}\,\gamma\tau)^s \qquad (4.2.4)$$

where the Borel subgroup Γ_∞ was defined in (3.4.17). This relation implies that $E_s(\tau)$ vanishes for all τ whenever $2s$ is a zero of the Riemann ζ-function, including at its zeros on the negative real axis, as well as at its nontrivial zeros when $\mathrm{Re}\,(4s) = 1$. Finally, the non-holomorphic Eisenstein series $E_s(\tau)$ is an eigenfunction,

$$\Delta E_s(\tau) = s(s-1)E_s(\tau) \qquad (4.2.5)$$

of the Laplace–Beltrami operator, or simply the *Laplacian*, on $\mathcal{H}$,

$$\Delta = 4\tau_2^2 \partial_\tau \partial_{\bar\tau} = \tau_2^2(\partial_{\tau_1}^2 + \partial_{\tau_2}^2) \qquad (4.2.6)$$

as may be verified by applying Δ term-wise in the sum of (4.2.1). Since the Laplacian Δ is invariant under the $\mathrm{SL}(2,\mathbb{R})$ isometry group of $\mathcal{H}$ it is automatically invariant under the $\mathrm{SL}(2,\mathbb{Z})$ subgroup of $\mathrm{SL}(2,\mathbb{R})$.

4.2.1 Analytic continuation of $E_s(\tau)$

While $E_s(\tau)$ is defined by the series (4.2.1) only for $\mathrm{Re}\,(s) > 1$, it may be analytically continued in s throughout the complex s-plane to produce a meromorphic function in s, just as was possible for the Riemann ζ-function. To exhibit the analytic structure of $E_s(\tau)$ in s, we use the following integral representation for the summands,

$$\Gamma(s)E_s(\tau) = \int_0^\infty \frac{dt}{t}\, t^s \left(\sum_{m,n\in\mathbb{Z}} e^{-\pi t|m+n\tau|^2/\tau_2} - 1 \right) \qquad (4.2.7)$$

The subtraction of 1 eliminates the zero mode $(m,n) = (0,0)$ so that the contribution from $t > 1$ to the integral converges absolutely for all $s \in \mathbb{C}$. The non-analyticity in s arises from the small t region. As we did in the case of the Riemann ζ-function, we split the integral at $t = 1$ and Poisson resum the integrand for $0 < t < 1$ in both m and n to obtain,

$$\Gamma(s)E_s(\tau) = \int_1^\infty \frac{dt}{t} t^s \left(\sum_{m,n\in\mathbb{Z}} e^{-\pi t|m+n\tau|^2/\tau_2} - 1 \right)$$

$$+ \int_0^1 \frac{dt}{t} t^s \left(\frac{1}{t} \sum_{m,n\in\mathbb{Z}} e^{-\pi|m+n\tau|^2/(t\tau_2)} - 1 \right) \qquad (4.2.8)$$

Next, we change variable in the second integral $t \to 1/t$, collect the resulting integrals for $t \in [1,\infty)$, and analytically continue the remaining integral of powers of t. The result is as follows,

$$\Gamma(s)E_s(\tau) = \frac{1}{s(s-1)} + \int_1^\infty \frac{dt}{t} \left(t^s + t^{1-s} \right) \left(\sum_{m,n\in\mathbb{Z}} e^{-\pi t|m+n\tau|^2/\tau_2} - 1 \right)$$

$$(4.2.9)$$

The first term on the right side is meromorphic in $s \in \mathbb{C}$, while the integral is manifestly holomorphic throughout $\mathbb{C}$. As a result, $E_s(\tau)$ admits a meromorphic analytic continuation to $s \in \mathbb{C}$ with simple poles at $s = 0$ and $s = 1$ with residue -1 and $+1$, respectively. Moreover, the expression is manifestly invariant under $s \to 1 - s$ which implies the functional equation,

$$\Gamma(s)E_s(\tau) = \Gamma(1-s)E_{1-s}(\tau) \qquad (4.2.10)$$

In particular, the functional relation requires $E_s(\tau) = 0$ whenever $s = -1, -2, \cdots$ in view of the fact that the right side is finite for these values while $\Gamma(s)$ on the left side diverges. This result agrees with our earlier observation that $E_s(\tau) = 0$ whenever $2s$ is a zero of the Riemann ζ-function.

4.2.2 Fourier series of $E_s(\tau)$

The Fourier series of $E_s(\tau)$ for all $\mathrm{Re}\,(s) > 1$ may be derived from the integral representation in (4.2.7), this time by Poisson resummation in the variable m only, using the formula,

$$\sum_{m,n\in\mathbb{Z}} e^{-\pi t|m+n\tau|^2/\tau_2} = \sqrt{\frac{\tau_2}{t}} \sum_{m,n\in\mathbb{Z}} e^{2\pi imn\tau_1} e^{-\pi m^2\tau_2/t - \pi n^2 t\tau_2} \quad (4.2.11)$$

In the mathematics literature, this procedure is referred to as the Chowla–Selberg method. Substituting this result into (4.2.7) and splitting the sum over m, n into contributions for $mn \neq 0$ and $mn = 0$, we obtain,

$$\Gamma(s)E_s(\tau) = A_s(\tau) + B_s(\tau) \qquad (4.2.12)$$

where,

$$A_s(\tau) = \sqrt{\tau_2} \sum_{m\neq 0} \sum_{n\neq 0} e^{2\pi i m n \tau_1} \int_0^\infty \frac{dt}{t} \, t^{s-\frac{1}{2}} \, e^{-\pi\tau_2(m^2/t+n^2 t)} \qquad (4.2.13)$$

$$B_s(\tau) = \int_0^\infty \frac{dt}{t} \, t^s \left(-1 + 2\sum_{n=1}^\infty \sqrt{\frac{\tau_2}{t}} \, e^{-\pi n^2 t \tau_2} + \sum_{m\in\mathbb{Z}} \sqrt{\frac{\tau_2}{t}} \, e^{-\pi m^2 \tau_2/t} \right)$$

The integral $A_s(\tau)$ in the summand is absolutely convergent and holomorphic in s throughout $\mathbb{C}$ and may be computed by performing the t-integral in terms of a modified Bessel function,

$$A_s(\tau) = 2\sqrt{\tau_2} \sum_{m,n\neq 0} \left|\frac{m}{n}\right|^{s-\frac{1}{2}} e^{2\pi i m n \tau_1} K_{s-\frac{1}{2}}(2\pi\tau_2|mn|) \qquad (4.2.14)$$

The integral $B_s(\tau)$ may be computed by Poisson resumming the m-sum in the integrand,

$$B_s(\tau) = \sum_{n\neq 0} \int_0^\infty \frac{dt}{t} \, t^s \left(\sqrt{\frac{\tau_2}{t}} \, e^{-\pi n^2 t \tau_2} + e^{-\pi n^2 t/\tau_2} \right) \qquad (4.2.15)$$

which is readily evaluated in terms of the Riemann ζ-function or ζ^*-function,

$$
\begin{aligned}
B_s(\tau) &= 2\Gamma(s)\pi^{-s}\zeta(2s)\tau_2^s + 2\Gamma(s-\tfrac{1}{2})\pi^{\frac{1}{2}-s}\zeta(2s-1)\tau_2^{1-s} \\
&= 2\zeta^*(2s)\tau_2^s + 2\zeta^*(2s-1)\tau_2^{1-s} \qquad (4.2.16)
\end{aligned}
$$

where $\zeta^*(2s) = \Gamma(s)\zeta(2s)/\pi^s$ is the function introduced already in (2.1.16). We may verify the functional relation (4.2.10) directly in terms of the properties of $A_s(\tau)$ and $B_s(\tau)$. For $A_s(\tau)$ this is manifest upon using the symmetry under interchange of m and n, while for $B_s(\tau)$ it is a consequence of the relation $\zeta^*(1-s) = \zeta^*(s)$ of (2.1.16). Assembling both contributions, and recasting $A_s(\tau)$ in terms of a summation over a single variable $N = mn$, we obtain the celebrated Fourier series decomposition for $E_s(\tau)$,

$$
\begin{aligned}
\Gamma(s)E_s(\tau) = &\, 2\,\zeta^*(2s)\,\tau_2^s + 2\,\zeta^*(2s-1)\,\tau_2^{1-s} \qquad (4.2.17) \\
&+ 4\sqrt{\tau_2}\sum_{N\neq 0} |N|^{\frac{1}{2}-s}\sigma_{2s-1}(|N|)\, e^{2\pi i N \tau_1} K_{s-\frac{1}{2}}(2\pi\tau_2|N|)
\end{aligned}
$$

where $\sigma_\alpha(n)$ is the divisors sum $\sigma_\alpha(n) = \sum_{d|n} d^\alpha$. One verifies that each Fourier mode, including the power-behaved terms, individually satisfies the Laplace eigenvalue equation (4.2.5), as required by the fact that the Laplacian Δ commutes with the translation operator ∂_{τ_1}.

For integer $s = k$, the Bessel function in the Fourier series becomes a spherical Bessel function $K_{k-\frac{1}{2}}(x)$, which is a combination of an exponential in x and a polynomial in $1/x$. As a result, the Fourier series of $E_k(\tau)$ simplifies and may be recast as follows,

$$E_k(\tau) = (-)^{k+1} \frac{B_{2k}}{(2k)!} (4\pi\tau_2)^k + \frac{4\,(2k-3)!\,\zeta(2k-1)}{(k-2)!\,(k-1)!\,(4\pi\tau_2)^{k-1}} \tag{4.2.18}$$

$$+ \frac{2}{(k-1)!} \sum_{N=1}^{\infty} N^{k-1}\sigma_{1-2k}(N) P_k(4\pi N\tau_2) \left(q^N + \bar{q}^N\right)$$

where $q = e^{2\pi i\tau}$. Furthermore, B_{2k} are the Bernoulli numbers and $P_k(x)$ is a polynomial in $1/x$ given by,

$$P_k(x) = \sum_{m=0}^{k-1} \frac{(k+m-1)!}{m!\,(k-m-1)!\,x^m} \tag{4.2.19}$$

Non-holomorphic Eisenstein series $E_k(\tau)$ will play a crucial role in the theory of modular graph functions and forms, to be developed in Chapter 9.

4.3   Maass forms

The non-holomorphic Eisenstein series $E_s(\tau)$ discussed in Section 4.2 are special examples of a more general class of non-holomorphic objects known as Maass forms. A non-holomorphic function $f(\tau)$ is called a Maass form if it is $\mathrm{SL}(2,\mathbb{Z})$-invariant, satisfies a Laplace eigenvalue equation,

$$\Delta f(\tau) = \lambda f(\tau) \qquad\qquad \Delta = 4\tau_2^2 \partial_\tau \partial_{\bar{\tau}} \tag{4.3.1}$$

and has at most polynomial growth at the cusps. Without loss of generality, we will set $\lambda = s(s-1)$ and denote the space of such Maass forms by $\mathcal{N}(s) = \mathcal{N}(1-s)$. From (4.2.5), we see that the non-holomorphic Eisenstein series $E_s(\tau)$ are elements of $\mathcal{N}(s)$. Indeed, below we will see that for *generic* $s \in \mathbb{C}$ the non-holomorphic Eisenstein series are the *only* elements of $\mathcal{N}(s)$.

The Fourier expansion of a Maass form $f(\tau)$ is given as follows,

$$f(\tau) = \sum_{N\in\mathbb{Z}} f_N(\tau_2)\, e^{2\pi i N\tau_1} \tag{4.3.2}$$

The coefficients $f_N(\tau_2)$ can be fixed uniquely, up to overall τ_2-independent factors. Indeed, substituting the expansion into the eigenvalue equation (4.3.1) and identifying Fourier modes gives a second-order differential equation for the coefficients $f_N(\tau_2)$,

$$\tau_2^2 \partial_{\tau_2}^2 f_N(\tau_2) = \left(s(s-1) + (2\pi N\tau_2)^2\right) f_N(\tau_2) \tag{4.3.3}$$

For $N = 0$, the general solution is of the form,

$$f_0(\tau_2) = a\,\tau_2^s + b\,\tau_2^{1-s} \qquad (4.3.4)$$

For $N \neq 0$, we obtain a modified Bessel equation with general solutions $I_{s-1/2}$ and $K_{s-1/2}$. Discarding the solutions $I_{s-1/2}$ since they grow exponentially at the cusp, we obtain,

$$f_N(\tau_2) = \alpha_N\,\sqrt{\tau_2}\,K_{s-1/2}\,(2\pi\tau_2|N|) \qquad N \neq 0 \qquad (4.3.5)$$

Therefore, a general element of $\mathcal{N}(s)$ has the following Fourier expansion,

$$f(\tau) = a\,\tau_2^s + b\,\tau_2^{1-s} + \sum_{N \neq 0} \alpha_N\,\sqrt{\tau_2}\,e^{2\pi i N \tau_1}\,K_{s-1/2}\,(2\pi\tau_2|N|) \qquad (4.3.6)$$

The Fourier expansions of the non-holomorphic Eisenstein series were obtained already in (4.2.17) and match with the above result upon setting,

$$a = \frac{2\zeta(2s)}{\pi^s} \qquad b = \frac{2\Gamma(s - \frac{1}{2})\pi^{\frac{1}{2}-s}\zeta(2s-1)}{\Gamma(s)} \qquad (4.3.7)$$

and identifying the nonconstant Fourier coefficients as follows,

$$\alpha_N = \frac{4|N|^{\frac{1}{2}-s}\sigma_{2s-1}(|N|)}{\Gamma(s)} \qquad (4.3.8)$$

The constant part of the Fourier expansion, that is, the term $a\,\tau_2^s + b\,\tau_2^{1-s}$, is known as the *Laurent polynomial* of the Maass form. Though it may at first sight seem that the space of Laurent polynomials is two dimensional and spanned by a and b, in fact this space is only one dimensional. This may be shown by considering Maass forms on the truncated fundamental domain,

$$F_L = \left\{ \tau \in \mathcal{H},\ |\tau| \geq 1,\ |\mathrm{Re}(\tau)| \leq \frac{1}{2},\ \tau_2 \leq L \right\} \qquad (4.3.9)$$

Indeed, consider two Maass forms $f, g \in \mathcal{N}(s)$ with Fourier expansions

$$f(\tau) = \sum_{N \in \mathbb{Z}} f_N(\tau_2)e^{2\pi i N \tau_1} \qquad g(\tau) = \sum_{N \in \mathbb{Z}} g_N(\tau_2)e^{2\pi i N \tau_1} \qquad (4.3.10)$$

Green's theorem on the region F_L states that if $\partial/\partial\tau_2$ is the normal derivative on ∂F_L and $d\tau_1$ is the differential arc length, then we have,

$$\int_{\partial F_L} d\tau_1 \left(f\frac{\partial g}{\partial\tau_2} - g\frac{\partial f}{\partial\tau_2} \right) = \int_{F_L} \frac{d\tau_1 d\tau_2}{\tau_2^2}(f\Delta g - g\Delta f) = 0 \qquad (4.3.11)$$

where the last equality follows from the eigenvalue equation (4.3.1) satisfied by f and g. The left side may be evaluated in terms of the Fourier coefficients. Noticing that modular invariance reduces the boundary integral to an integral over $\tau_2 = L$ (since the remaining boundaries of the fundamental domain give equal and opposite contributions), we have,

$$\int_{\partial F_L} d\tau_1 \left(f \frac{\partial g}{\partial \tau_2} - g \frac{\partial f}{\partial \tau_2} \right) = \sum_{N \in \mathbb{Z}} \left(f_N(L) g'_{-N}(L) - f'_N(L) g_{-N}(L) \right) \quad (4.3.12)$$

where the derivative is defined by,

$$f'_N(L) = \frac{\partial f_N(\tau_2)}{\partial \tau_2} \bigg|_{\tau_2 = L} \quad (4.3.13)$$

and similarly for the derivative $g'_{-N}(L)$. In the limit where $L \gg 1$, the components $f_N(L)$ and $g_N(L)$ decay exponentially to zero for $N \neq 0$ so that the only remaining relation is $f_0 g'_0 - f'_0 g_0 = 0$. Parametrizing the Laurent polynomial as follows,

$$\begin{aligned} f_0(\tau_2) &= a\,\tau_2^s + c\,\tau_2^{1-s} \\ g_0(\tau_2) &= b\,\tau_2^s + d\,\tau_2^{1-s} \end{aligned} \quad (4.3.14)$$

implies that, for $s \neq \frac{1}{2}$, the coefficients satisfy $ad - bc = 0$, so that the space of Laurent polynomials has dimension 1. For $s = \frac{1}{2}$ the space obviously has dimension 1, since then the two factors of τ_2 have the same exponent.

4.3.1 The space of Maass forms

As was mentioned earlier, at generic values of $s \in \mathbb{C}$ the space of Maass forms $\mathcal{N}(s)$ consists only of non-holomorphic Eisenstein series $E_s(\tau)$. On the other hand, at nongeneric values of s, the space of Maass forms can include cusp forms. In the current context, cusp forms are defined to be the subset of elements of $\mathcal{N}(s)$ such that the Laurent polynomial vanishes, that is to say $a = b = 0$ in (4.3.6). We will denote the space of cusp forms by $\mathcal{S}(s)$. The non-holomorphic Eisenstein series are clearly not cusp forms for any s. For generic values of s, the set $\mathcal{S}(s)$ will be empty.

The structure of $\mathcal{N}(s) = \mathcal{N}(1-s)$ can be summarized as follows:

Theorem 4.3.1. *The space of Maass forms* $\mathcal{N}(s)$ *is given by*

$$\mathcal{N}(s) = \begin{cases} \mathbb{C}E_s(\tau) & \operatorname{Re}s > \frac{1}{2} \text{ and } s \notin [0,1] \\ \mathbb{C}E_s(\tau) \oplus \mathcal{S}(s) & \operatorname{Re}s = \frac{1}{2} \text{ or } s \in [0,1) \\ \mathbb{C} & s = 1 \end{cases} \quad (4.3.15)$$

The proof of this theorem follows from the fact, proven earlier, that the space of Laurent polynomials has dimension 1. This means that for any $f(\tau) \in \mathcal{N}(s)$, there exists a constant c such that the combination $f(\tau) - cE(\tau, s)$ has zero Laurent polynomial, that is, is a cusp form.

To complete the proof, it remains to prove that the space of cusp forms is empty for all values of s other than $\operatorname{Re} s = \frac{1}{2}$ or $s \in [0, 1)$. The nonexistence for $s = 1$ is immediate: In this case, the cusp form would be harmonic, that is, satisfy a Laplace eigenvalue equation with zero eigenvalue and hence would have to vanish by the maximum modulus principle. More generally we may argue as follows. Above we have seen that it is possible to find a constant c such that for any $f(\tau) \in \mathcal{N}(s)$, the combination $h(\tau) = f(\tau) - cE_s(\tau)$ is a cusp form. Since cusp forms are square integrable over the fundamental domain $F = \mathrm{SL}(2, \mathbb{Z}) \backslash \mathcal{H}$ equipped with the Poincaré measure $d^2\tau / \tau_2^2$,

$$\int_F \frac{d^2\tau}{\tau_2^2} |h(\tau)|^2 < \infty \qquad\qquad d^2\tau = \frac{i}{2} d\tau \wedge d\bar\tau \qquad (4.3.16)$$

we may manipulate the following absolutely convergent integral by using the differential equation $\Delta h = s(s-1)h$ satisfied by h, and Green's theorem,

$$s(s-1) \int_F \frac{d^2\tau}{\tau_2^2} |h(\tau)|^2 = \int_F \frac{d^2\tau}{\tau_2^2} \Delta h(\tau) \overline{h(\tau)} = -4 \int_F d^2\tau \, |\partial_\tau h|^2 \quad (4.3.17)$$

Since the right side is real and nonpositive, cusp forms must have real $s(s-1) \leq 0$, which restricts us to $s \in [0, 1]$ or $\operatorname{Re} s = \frac{1}{2}$.

In closing, let us note that the triviality of the space of cusp forms for $\operatorname{Re} s \neq \frac{1}{2}$ and $s \notin \mathbb{R}$ implies that for these values the space $\mathcal{N}(s)$ is one-dimensional and hence there must be a functional relation relating $E_s(\tau)$ and $E_{1-s}(\tau)$, since both have Laplace eigenvalue $s(s-1)$. Indeed, precisely such a relation was identified in (4.2.10).

4.3.2 Cusp forms

We have just shown that the space $\mathcal{S}(s)$ of cusp forms is empty unless $\operatorname{Re} s = \frac{1}{2}$ or $s \in [0, 1)$. Actually, one can further show that $\mathcal{S}(s)$ vanishes for $s \in [0, 1)$, and hence that all nontrivial cusp forms must exist at $\operatorname{Re} s = \frac{1}{2}$ and away from the real axis: We refer to [Lan75] for the proof. Denoting $s = \frac{1}{2} + it$ for $t \in \mathbb{R}$, the corresponding eigenvalue is $\lambda = s(s-1) = -\left(\frac{1}{4} + t^2\right)$.

For any $t \in \mathbb{R}$, the space $\mathcal{S}\left(\frac{1}{2} + it\right)$ is finite dimensional. To show this, begin with M elements $v_1(\tau), \ldots, v_M(\tau) \in \mathcal{S}\left(\frac{1}{2} + it\right)$. Given an additional element $\tilde{v}(\tau)$, it is possible to choose coefficients $c_i \in \mathbb{C}$ such that the linear combination,

$$v(\tau) = \widetilde{v}(\tau) - \sum_{i=1}^{M} c_i v_i(\tau) \tag{4.3.18}$$

has vanishing Fourier coefficients up to order M. We will now show that for M sufficiently large but finite, the vanishing of Fourier coefficients up to order M implies vanishing to all orders. In other words, we will have $v(\tau) = 0$ exactly and hence all elements $\widetilde{v}(\tau) \in \mathcal{S}\left(\frac{1}{2} + it\right)$ can be written as a linear combination of a finite number M of other elements.

To prove the desired result, begin by recalling that a cusp form $v(\tau)$ is bounded on the upper half plane, and hence that $|v(\tau)|$ must have a maximum at some location $\tau^{(0)} = \tau_1^{(0)} + i\tau_2^{(0)}$. From the form of the Fourier expansion (4.3.6) and orthogonality of exponentials, we have,

$$\alpha_N \sqrt{\tau_2^{(0)}} \, K_{it}(2\pi\tau_2^{(0)}|N|) \tag{4.3.19}$$

$$= \int_{-1/2}^{1/2} d\tau_1 \, v(\tau_1 + i\tau_2^{(0)}/2) \, e^{-2\pi i N \tau_1} \, \frac{\sqrt{2}K_{it}(2\pi\tau_2^{(0)}|N|)}{K_{it}(\pi\tau_2^{(0)}|N|)}$$

It then follows that,

$$\left|v(\tau^{(0)})\right| \leq \sum_{|N|>M} \left|\alpha_N \sqrt{\tau_2^{(0)}} \, K_{it}(2\pi\tau_2^{(0)}|N|)\right|$$

$$\leq 2\sqrt{2} \left|v(\tau^{(0)})\right| \sum_{N>M} \frac{K_{it}(2\pi\tau_2^{(0)}N)}{K_{it}(\pi\tau_2^{(0)}N)} \tag{4.3.20}$$

If M is sufficiently large (depending on t), we may estimate the Bessel functions by exponentials, giving the inequality,

$$\left|v(\tau^{(0)})\right| \leq \left|v(\tau^{(0)})\right| C e^{-\pi M \tau_2^{(0)}} \tag{4.3.21}$$

for some M-independent constant C. Taking $M \geq (\ln C)/(\pi\tau_2^{(0)})$ we conclude that $|v(\tau^{(0)})| = 0$ and hence that $v(\tau)$ is identically zero.

We have just shown that $\mathcal{S}\left(\frac{1}{2} + it\right)$ is finite dimensional for any $t \in \mathbb{R}$. Furthermore, though we will not reproduce the proof here, it has been shown that there exist an infinite number of values of t for which the space of cusp forms is non-empty. However, there is no exact value of t for which this is known to be the case. The smallest approximate values of t for which cusp forms are believed to exist are (see, e.g., [Ter85]),

$$t \approx 9.53369, \ 12.17300, \ 13.77975, \ 14.35850,$$
$$16.13807, \ 16.64425, \ 17.73856, \ \ldots \tag{4.3.22}$$

In particular, we see that there is only a discretuum of such values.

4.3.3 Physical interpretation of the spectrum

The fundamental domain $F = \{\tau \in \mathcal{H}; \ |\mathrm{Re}\,(\tau)| \leq \frac{1}{2}; \ |\tau| \geq 1\}$ for $\mathrm{SL}(2, \mathbb{Z})$ is non-compact as it has a puncture at the cusp. From the point of view of scattering theory, one may view the cusp as an asymptotic region from which *a free wave* in the continuous spectrum can be sent in. This wave is indeed oscillatory and of the form $\tau_2^{\frac{1}{2}+it}$ for real t, from the first constant term in the Fourier expansion. Since the fundamental domain is otherwise bounded, the wave hits a wall at $|\tau| = 1$ and must be reflected out through the puncture and is asymptotically of the form $\tau_2^{\frac{1}{2}-it}$. This phenomenon is analogous to one-dimensional scattering on a semi-infinite line and explains in physical terms why the Eisenstein solutions have $E_{\frac{1}{2}+it}(\tau) = E_{\frac{1}{2}-it}(\tau)$.

The cusp forms are wave functions of bound states, as is confirmed by the fact that they decay exponentially toward the cusp. In most quantum mechanics problems, the number of bound states in the presence of a wall would be finite. But there are exceptions, such as the Coulomb problem. Here, it is the special properties of the hyperbolic metric on the fundamental domain of $\mathrm{SL}(2, \mathbb{Z})$ that produce an infinite number of cusp forms, or bound states, with a spectrum that rises all the way to infinity.

4.4 Spectral decomposition

We have now discussed various aspects of the spectrum of the Laplacian on the fundamental domain $\mathrm{SL}(2, \mathbb{Z})\backslash\mathcal{H}$. It is possible to extract from this data the following orthonormal basis of eigenfunctions of the Laplacian:

- **Discrete series:** The constant function $v_0 = 1/\sqrt{\mathrm{area}(F)} = \sqrt{3/\pi}$ (which is trivially an eigenfunction of the Laplacian) together with an orthonormal basis of cusp forms $v_n(\tau)$ with corresponding eigenvalues $s_n(s_n - 1) = -\left(\frac{1}{4} + t_n^2\right)$ for $t_n \in \mathbb{R}$. The first few values of t_n are given in (4.3.22).
- **Continuous series:** The non-holomorphic Eisenstein series $E_s(\tau)$ with $s = \frac{1}{2} + it$ and $t \in \mathbb{R}$ taking a continuum of values.

Orthogonality here is defined relative to the Petersson inner product with the Poincaré measure $d^2\tau/\tau_2^2$ on the fundamental domain,

$$\langle f|g\rangle = \int_F \frac{d^2\tau}{\tau_2^2}\,\overline{f(\tau)}\,g(\tau) \qquad\qquad d^2\tau = \frac{i}{2}d\tau\wedge d\bar\tau \qquad (4.4.1)$$

Let us use this inner product to verify the orthogonality claimed above. Consider the inner product between a non-holomorphic Eisenstein series $E_s(\tau)$ and a cusp form $v_n(\tau)$,

$$\langle v_n(\tau)|E_s(\tau)\rangle = \int_F \frac{d^2\tau}{\tau_2^2}\,\overline{v_n(\tau)}\,E_s(\tau) \qquad (4.4.2)$$

This inner product converges since cusp forms have exponential decay at the cusp, whereas non-holomorphic Eisenstein series have at most polynomial growth at the cusp. For $\operatorname{Re} s > 1$, we may proceed by using the Poincaré series for $E_s(\tau)$ given in (4.2.4), which gives,

$$\langle v_n(\tau)|E_s(\tau)\rangle = \frac{\zeta(2s)}{\pi^s}\int_F \frac{d^2\tau}{\tau_2^2}\,\overline{v_n(\tau)}\sum_{\gamma\in\Gamma_\infty\backslash\mathrm{SL}(2,\mathbb{Z})}(\operatorname{Im}\gamma\tau)^s$$

$$= \frac{\zeta(2s)}{\pi^s}\int_{-\frac12}^{\frac12}d\tau_1\int_0^\infty \frac{d\tau_2}{\tau_2^2}\,\tau_2^s\,\overline{v_n(\tau)} \qquad (4.4.3)$$

In the second equality, we have used modular invariance of $v_n(\tau)$ to unfold the integral over the fundamental domain F, that is, we have replaced the integral over $\mathrm{SL}(2,\mathbb{Z})$ orbits in F with an integral over the strip $\Gamma_\infty\backslash\mathcal{H}$, according to the general formula,

$$\int_{\mathrm{SL}(2,\mathbb{Z})\backslash\mathcal{H}}\frac{d^2\tau}{\tau_2^2}\sum_{\Gamma_\infty\backslash\mathrm{SL}(2,\mathbb{Z})}f(\tau) = \int_{\Gamma_\infty\backslash\mathcal{H}}\frac{d^2\tau}{\tau_2^2}f(\tau) \qquad (4.4.4)$$

for an arbitrary function $f(\tau)$ that is invariant under Γ_∞. This is the analog of the unfolding trick introduced for periodic functions in Section 2.1.1. The integral over τ_1 on the second line of (4.4.3) may now be performed and isolates the Laurent polynomial of $v_n(\tau)$, which is vanishing by definition. This gives the desired orthogonality for $\operatorname{Re} s > 1$. The analogous result for $\operatorname{Re} s < 1$ may be obtained by analytic continuation.

An arbitrary square integrable modular function on $\mathcal{H}$ can now be decomposed into this orthogonal basis. We state without proof the following spectral decomposition theorem.

Theorem 4.4.1 (Roelcke–Selberg). *Let $v_0 = \sqrt{3/\pi}$ be the normalized constant function and $\{v_n(\tau)\}_{n\geq 1}$ be an orthonormal basis of cusps forms. Then* *any square integrable modular function $f(\tau)$ admits the decomposition,*

$$f(\tau) = \frac{1}{4\pi i} \int_{\mathrm{Re}\,(s)=\frac{1}{2}} ds \, \langle E_s(\tau) | f(\tau) \rangle \, E_s(\tau)$$

$$+ \sum_{n \geq 0} \langle v_n(\tau) | f(\tau) \rangle \, v_n(\tau) \tag{4.4.5}$$

The reader may worry about the convergence properties of the continuous portion of this decomposition. Indeed, $E_s(\tau)$ is not square integrable because of its Laurent polynomial. However, when $\mathrm{Re}\,s = \frac{1}{2}$, one can show that $|E_s(\tau)| \ll \sqrt{\tau_2}$ when $\tau_2 \to \infty$, and the extra integration over t in the decomposition formula saves a factor of $\ln \tau_2$, which is sufficient to turn the right side into a square-integrable function.

An application of the spectral theorem to integrated correlators in supersymmetric Yang–Mills theory, suitably adapted to strip it from its physics prerequisites, is the subject of Exercise 4.5.

4.5 Maass forms of arbitrary weight

Thus far we have focused on Maass forms of modular weight zero. We now discuss Maass forms of arbitrary (integer) weight k, that is, those satisfying,

$$f(\gamma\tau) = (c\tau + d)^k f(\tau) \qquad \gamma = \begin{pmatrix} a & b \\ c & d \end{pmatrix} \in \mathrm{SL}(2, \mathbb{Z}) \tag{4.5.1}$$

The Maass property in this case must be recast in terms of a weight-k Laplace–Beltrami operator,

$$\Delta_k = 4\tau_2^{2-k} \partial_\tau \tau_2^k \partial_{\bar\tau} = \Delta - 2ik\tau_2 \partial_{\bar\tau} \tag{4.5.2}$$

in order to get a modular covariant eigenvalue equation,

$$\Delta_k f - \lambda f \tag{4.5.3}$$

Of particular interest are the *harmonic* Maass forms of weight k, that is, those satisfying an eigenvalue equation $\Delta_k f = 0$. Unlike in the case of weight 0, for nonzero weight the space of harmonic Maass forms is nontrivial.

A simple example of a weight-2 harmonic Maass form is the almost-holomorphic Eisenstein series $\mathsf{E}_2^*(\tau)$ introduced in Section 4.1. Indeed, from (4.1.10) and the fact that $\mathsf{E}_2(\tau)$ is holomorphic, we have,

$$\Delta_2 \mathsf{E}_2^*(\tau) = -\frac{3}{\pi}(\Delta - 4i\tau_2 \partial_{\bar\tau})\tau_2^{-1} = 0 \tag{4.5.4}$$

We close this section by mentioning two relations between the space $\mathcal{I}_k$ of weight-k harmonic Maass forms and the space of holomorphic modular forms. First, the space $\mathcal{M}_k$ of weight-k modular forms is a *subset* of the space $\mathcal{I}_k$, that is, $\mathcal{M}_k \subset \mathcal{I}_k$. This follows trivially from the fact that every weight-k modular form is annihilated by Δ_k. Second, the space $\mathcal{M}_{2-k}$ of weight-$(2-k)$ modular forms is a *quotient* of $\mathcal{I}_k$ by $\mathcal{M}_k$, that is, $\mathcal{M}_{2-k} = \mathcal{I}_k/\mathcal{M}_k$. To show this, first note that given any element $f(\tau) \in \mathcal{I}_k$, the function $g(\tau) = \tau_2^k \partial_\tau \overline{f(\tau)}$ is holomorphic (since by the Maass property of $f(\tau)$ it is annihilated by $\partial_{\overline{\tau}}$) and of weight $2 - k$ and is thus an element of $\mathcal{M}_{2-k}$. It furthermore vanishes if and only if $f(\tau)$ is in $\mathcal{M}_k$. There is thus an injective map from $\mathcal{I}_k/\mathcal{M}_k$ to $\mathcal{M}_{2-k}$. The proof of surjectivity can be found in [BF04, Theorem 3.7].

4.6 Mock modular forms

In Section 4.1, we saw that there exist two variants of the weight-2 holomorphic Eisenstein series, namely, the function $\mathsf{E}_2(\tau)$ defined in (4.1.2) which is holomorphic but not modular, and the function $\mathsf{E}_2^*(\tau)$ defined in (4.1.8) which is modular but not holomorphic. The former is an example of a quasi-modular form, whereas the latter is an example of an almost-holomorphic modular form. Section 4.1.2 described an isomorphism between the sets of quasi-modular forms and almost-holomorphic modular forms.

Having introduced non-holomorphic modular forms in Sections 4.2–4.5, it is natural to ask if their holomorphic parts lead to interesting analogs of quasi-modular forms. In particular, we will consider the holomorphic part of harmonic "weak" Maass forms of weight k,[3] which are known as *mock modular forms*.

Mock modular forms made their first appearance in a letter of Ramanujan to Hardy, in which they were referred to as *mock theta functions*. In the modern definition, a mock theta function is a q-series,

$$H(q) = \sum_{n \geq 0} a_n q^n \qquad (4.6.1)$$

such that for some $\lambda \in \mathbb{Q}$, the series $q^\lambda H(q)$ gives a mock modular form of weight $\frac{1}{2}$, with the shadow (defined below) being a weight-$\frac{3}{2}$ function of the form $\sum_{n \in \mathbb{Z}} \epsilon(n) n q^{|\kappa| n^2}$ with $\kappa \in \mathbb{Q}$ and ϵ an odd periodic function.

Given a harmonic Maass form $F^*(\tau)$ of weight k, we may extract its

---

[3] The adjective "weak" indicates the potential for exponential growth at the cusps; by contrast, a "Maass form" is required to have at most polynomial growth at the cusp.

holomorphic part as follows. We begin by defining,

$$g(\tau) = \tau_2^k \, \partial_\tau \overline{F^*(\tau)} \tag{4.6.2}$$

The condition that $F^*(\tau)$ satisfies $\Delta_k F^* = 0$ implies that $g(\tau)$ is holomorphic since $\Delta_k = \tau_2^{2-k} \partial_\tau \tau_2^k \partial_{\bar\tau}$. If one can identify a new function $g^*(\tau)$ such that $\overline{g(\tau)} = \tau_2^k \partial_{\bar\tau} g^*(\tau)$, then it is clear that the function $F(\tau)$ defined by,

$$F(\tau) = F^*(\tau) - g^*(\tau) \tag{4.6.3}$$

will be holomorphic. Such a function $g^*(\tau)$ can be obtained by simply inverting the differential operator $\tau_2^k \partial_{\bar\tau}$, giving the expression,

$$g^*(\tau) = (2i)^k \int_{-\bar\tau}^{i\infty} dz \, \frac{\overline{g(-\bar{z})}}{(z+\tau)^k} \tag{4.6.4}$$

known as a non-holomorphic Eichler integral. This integral is independent of the path from $-\bar\tau$ to $i\infty$ since the integrand is holomorphic in z.

The function $F(\tau)$ is holomorphic, but will in general fail to be modular, since $g^*(\tau)$ is not modular. The function $g(\tau)$ in terms of which $F(\tau)$ is defined is referred to as the *shadow* and is a modular function of weight $(2 - k)$. The pair of $F(\tau)$ and $g(\tau)$ are often together referred to as a mock modular form, though in some contexts the function $F(\tau)$ alone is given that name.

Given a harmonic Maass form $F^*(\tau)$, it is always possible to extract from it a unique mock modular form by taking the holomorphic part in the way described above. Conversely, given a mock modular form, it is always possible to complete it to a harmonic Maass form by taking the sum of $F(\tau)$ with the function $g^*(\tau)$ obtained from the shadow. This establishes an isomorphism between the two spaces, exactly analogous to the isomorphism between the spaces of almost-holomorphic modular forms and quasi-modular forms given in Section 4.1.

4.6.1 Examples of mock modular forms

We now give some concrete examples of mock modular forms.

A first example is provided by $\mathsf{E}_2(\tau)$. In Section 4.5, we showed that $\mathsf{E}_2^*(\tau)$ is a weight-2 harmonic Maass form, and hence by definition, the holomorphic part of $\mathsf{E}_2^*(\tau)$, namely $\mathsf{E}_2(\tau)$, is a weight-2 mock modular form. The shadow $g(\tau)$ in this case is a constant, namely $g(\tau) = -\frac{12}{\pi}$. One verifies that this constant gives $g^*(\tau) = -\frac{3}{\pi\tau_2}$, reproducing the result in (4.1.10). Thus, $\mathsf{E}_2(\tau)$ is both quasi-modular and mock modular. This is not in general the case: quasi-modular forms are generally not mock modular. For example, $\mathsf{E}_2(\tau)^2$ is a quasi-modular form but fails to be mock modular since its modular completion is not a harmonic Maass form.

A second example is provided by the following function,

$$F^*(\tau) = F(\tau) + \frac{6}{\sqrt{i}} \int_{-\bar{\tau}}^{i\infty} dz \, \frac{\overline{\eta(-\bar{z})^3}}{(z + \tau)^{1/2}} \qquad (4.6.5)$$

with $F(\tau)$ defined by,

$$F(\tau) = \frac{1}{\eta(\tau)^3} \left(- \mathsf{E}_2(\tau) + 24 \sum_{\substack{r > s > 0 \\ r - s = 1 \bmod 2}} (-1)^r \, s \, q^{rs/2} \right) \qquad (4.6.6)$$

which appears rather unexpectedly in physics via the elliptic genus of strings propagating on a K3 manifold. It is known that $F^*(\tau)$ is modular with weight $\frac{1}{2}$. Hence its holomorphic part $F(\tau)$ must be a mock modular function. The shadow in this case is $g(\tau) = 3\sqrt{2}\,\eta(\tau)^3$. A curious fact is that the first few terms in the Fourier expansion of $F(\tau)$,

$$F(\tau) = q^{-\frac{1}{8}} \left(-1 + 45q + 231q^2 + 770q^3 + 2277q^4 + 5796q^5 + \dots \right) \ (4.6.7)$$

give dimensions of irreducible representations of the Mathieu group M_{24}, a sporadic simple group of order 244,823,040. This correspondence goes under the name of *Mathieu moonshine*, a cousin of the Monstrous moonshine mentioned briefly in Section 3.6.1.

As a third and final example, we mention a classic result of Zagier, which states that the following two functions are harmonic of weight $k = \frac{3}{2}$,

$$F_0^*(\tau) = 3 \sum_{n \geq 0} H(4n) \, q^n + 6\tau_2^{-1/2} \sum_{n \in \mathbb{Z}} \beta(4\pi n^2 \tau_2) \, q^{-n^2} \qquad (4.6.8)$$

$$F_1^*(\tau) = 3 \sum_{n > 0} H(4n-1) \, q^{n-\frac{1}{4}} + 6\tau_2^{-1/2} \sum_{n \in \mathbb{Z}} \beta \left(4\pi(n+1/2)^2 \tau_2 \right) q^{-\left(n+\frac{1}{2}\right)^2}$$

where $\beta(t)$ is an incomplete gamma function

$$\beta(t) = \frac{1}{16\pi} \int_1^\infty du \, u^{-3/2} e^{-ut} \qquad (4.6.9)$$

and $H(n)$ are the Hurwitz class number of imaginary quadratic fields. Recall that the class numbers $h(D)$ of integral binary quadratic forms were the subject of Exercise 3.2. The Hurwitz class numbers are defined for $n > 0$ to be the number of $\mathrm{SL}(2, \mathbb{Z})$ equivalence classes of integral binary quadratic forms of discriminant $-n$, weighted by the reciprocal of the number of their automorphisms. The first few nonzero entries are given by

$$H(3) = \frac{1}{3} \qquad H(4) = \frac{1}{2} \qquad H(7) = H(8) = H(11) = 1 \quad (4.6.10)$$

For $n = 0$ one sets $H(0) = -\frac{1}{12}$ and for $n < 0$ we have $H(n) = 0$.

The holomorphic parts $F_{0,1}(\tau)$ of $F_{0,1}^*(\tau)$ are easy to read off,

$$F_0(\tau) = 3 \sum_{n \geq 0} H(4n)q^n$$

$$F_1(\tau) = 3 \sum_{n > 0} H(4n - 1)q^{n-\frac{1}{4}} \qquad (4.6.11)$$

and have corresponding shadows,

$$g_0(\tau) = -\frac{3}{4\pi}\vartheta_3(0|2\tau) \qquad g_1(\tau) = -\frac{3}{4\pi}\vartheta_2(0|2\tau) \qquad (4.6.12)$$

It is important to note that these functions are not mock modular forms in the strict sense discussed earlier, since under $\tau \to -1/\tau$ the original harmonic functions $F_{0,1}^*(\tau)$ are not individually invariant, but rather transform into one another as a doublet,

$$\begin{pmatrix} F_0^*(-1/\tau) \\ F_1^*(-1/\tau) \end{pmatrix} = -\frac{(-i\tau)^{3/2}}{\sqrt{2}} \begin{pmatrix} 1 & 1 \\ 1 & -1 \end{pmatrix} \begin{pmatrix} F_0^*(\tau) \\ F_1^*(\tau) \end{pmatrix} \qquad (4.6.13)$$

There are two ways of interpreting this fact. The first is to think of $F_{0,1}^*(\tau)$ as forming a *vector-valued* harmonic Maass form. In this interpretation, the functions $F_{0,1}(\tau)$ likewise form a vector-valued mock modular form. Vector-valued modular forms will be discussed in more detail in Chapter 8.

The second approach is to notice that $F_0^*(\tau)$ is invariant under $T : \tau \to \tau + 1$ and $F_1^*(\tau)$ is invariant under $T^4 : \tau \to \tau + 4$. Hence we can say that $F_0^*(\tau)$ individually is modular invariant under the subgroup of SL$(2, \mathbb{Z})$ generated by T and ST^4S. This subgroup is denoted by $\Gamma_0(4)$ and is one of the so-called *congruence subgroups* of SL$(2, \mathbb{Z})$, to be discussed in detail in Chapter 6. In this interpretation, one would say that $F_0^*(\tau)$ is a harmonic Maass form for the congruence subgroup $\Gamma_0(4)$ and that $F_0(\tau)$ is likewise a mock modular form for this subgroup. Modular forms for congruence subgroups will be discussed in detail in Chapter 7.

4.7 Quantum modular forms

Finally, let us briefly discuss the notion of a *quantum modular form*.[4] Unlike for standard modular forms for which the domain is the upper half plane, quantum modular forms are defined on the space of cusps $\mathbb{Q} \cup \{i\infty\}$ (or some subset thereof). Since this domain is a discrete space, quantum modular forms do not obey the usual analyticity properties of modular forms. Furthermore, they are not modular in the standard sense.

Instead, a weight-k quantum modular form $f(\tau)$ is defined such that for all $\gamma \in \mathrm{SL}(2, \mathbb{Z})$, the difference,

$$h_\gamma(\tau) = f(\tau) - \varepsilon(\gamma)(c\tau + d)^k f\left(\frac{a\tau + b}{c\tau + d}\right) \tag{4.7.1}$$

is an analytic function on the upper half plane. Here we have allowed for a multiplier phase $\varepsilon(\gamma)$, similar to that appearing in the transformation rules for the Dedekind η-function in Section 3.8.

It is simplest to illustrate this definition by proceeding directly to an example. Consider the following q-expansion,

$$F(q) = \sum_{n=1}^{\infty} \prod_{m=0}^{n-1} (1 - q^{m+1}) \tag{4.7.2}$$

known as *Kontsevich's strange function*. This function does not converge on any open subset of $\mathbb{C}$, but is well defined at roots of unity, that is, when $q = e^{2\pi i \tau}$ and $\tau \in \mathbb{Q}$. We now define the closely related object,

$$\phi(\tau) = q^{\frac{1}{24}} F(q) \tag{4.7.3}$$

It follows from this definition that $\phi(\tau + 1) = e^{2\pi i/24} \phi(\tau)$ and $\phi(0) = 1$.

To understand the behavior under S, we first note that the T transformation just given implies $\phi(n) = e^{2\pi i n/24} \phi(0) = e^{2\pi i n/24}$ for $n \in \mathbb{Z}$. On the other hand, Kontsevich and Zagier have shown that for $\tau = 1/n$, one has,

$$\phi\left(\frac{1}{n}\right) \sim n^{3/2} e^{2\pi i \frac{(3-n)}{24}} + \sum_{j=0}^{\infty} c_j \left(-\frac{2\pi i}{n}\right)^j \tag{4.7.4}$$

as $n \to \infty$, with the first few coefficients in the expansion being,

$$c_0 = 1 \qquad c_1 = \frac{23}{24} \qquad c_2 = \frac{1681}{1152} \tag{4.7.5}$$

---

[4] As far as we are aware, there is no direct connection between the term "quantum" used here and the usual notion of "quantum" in Physics.

Using the T transformation, the first term above can be rewritten as,

$$n^{3/2} e^{\frac{2\pi i}{8}} \phi(-n) = i^{1/2} n^{3/2} \phi(-n) \tag{4.7.6}$$

and hence we conclude that the function,

$$h_S(\tau) - \phi(\tau) + \tau^{1/2} |\tau|^{3/2} \phi(-1/\tau) \qquad \pm\tau > 0 \tag{4.7.7}$$

has a well-defined Taylor expansion at $\tau = 0$ and so is smooth there. Though we do not give details here, the function $h_S(\tau)$ can furthermore be extended to a real-analytic function on $\mathbb{C}$ (except for at $\tau = 0$), so $\phi(\tau)$ is an example of a quantum modular form.

• Bibliographical notes

Classic introductions to spectral analysis on symmetric spaces with applications to Maass forms and non-holomorphic Eisenstein series may be found in the books by Kubota [Kub73], Terras [Ter85, Ter88], and Iwaniec [Iwa02, IK21]. A discussion aimed at physicists, which includes an introduction to adeles, may be found in the book by Fleig, Gustafsson, Kleinschmidt, and Persson [FGKP18]. Group-theoretic aspects of the spectral decomposition of automorphic forms are expounded on in the book by Gel'fand, Graev, and Pyatetskii-Shapiro [GGPS69]. Faddeev's approach is based on operator perturbation theory and is reviewed in the book by Lang [Lan75]. A construction of cusp forms that are not Maass forms was given in [Bro17b]. The spectral decomposition theorem has recently been used in a physics context in [BCF+21], [CP22], and [PPR23]. Generalizations to automorphic forms under the group $\mathrm{GL}(n, \mathbb{R})$ are developed in the book by Goldfeld [Gol06].

A fundamental role was played by quasi-modular forms in Viazovska's proof of the sphere packing problem in eight dimensions, namely, that no packing is more dense than the one corresponding to the root lattice of E_8. The original paper may be found in [Via17], while the generalization to sphere packing in dimension 24 was settled in [CKM+17].

Reviews on mock modular forms include Zagier's lecture notes [Zag07] in the mathematics literature and [DMZ12, AC18] in the physics literature. The second example of a mock modular form, given in (4.6.6), appears in physics as the elliptic genus of a K3 manifold [EOTY89, EH09]. The connection with dimensions of representations of the Monster group M_{24} was observed in [EOT11], thereby initiating the program of Mathieu moonshine. Reviews of moonshine aimed at physicists may be found in [AC18], [Gan04], [Gan06], and [HHP22] and references therein. The third example of mock modular forms, given in (4.6.11), arises in physics as the partition function of $\mathcal{N} = 4$ super Yang–Mills theory with gauge group SO(3) on $\mathbb{CP}^2$ [VW94].

Quantum modular forms were introduced in mathematics in [Zag01], [Zag10], and [Ono13]. To date, their role in physics remains largely unclear, though it has been suggested that they are relevant for the study of invariants of three-manifolds [LZ99, CCPS23].

Exercises

4.1 Show that the modular transformation law for $E_2(\tau)$ of (4.1.3) and the relation (4.1.10) imply the transformation law for $E_2^*(\tau)$ of (4.1.9).

4.2 Let $\mathcal{A}_{k,N}$ be the space of almost-holomorphic modular forms of weight k and depth N, whose elements take the following form,

$$f(\tau) = \sum_{n=0}^{N} \frac{f_n(q)}{(\tau - \bar{\tau})^n} \qquad\qquad f_N \neq 0 \qquad\qquad (4.7.8)$$

Show that the operator $\nabla_{\bar{\tau}} = -2i\tau_2^2 \partial_{\bar{\tau}}$ maps $\mathcal{A}(k, N) \to \mathcal{A}(k-2, N-1)$. Verify that the function $f_N(q)$ is a holomorphic modular form of weight $k - 2N$ and that the space $\mathcal{A}(k, N)$ is in the kernel of $(\nabla_{\bar{\tau}})^{N+1}$. Obtain the modular transformation laws for $f_n(q)$ with $n = 0, \cdots, N$.

4.3 Prove Theorem 4.1.1.

4.4 Express the divisor function σ_3 in terms of σ_1 and sums of products thereof using the differential equation (4.1.20). Similarly, express σ_5 in terms of σ_1 and σ_3 using one of the equations in (4.1.17).

4.5 Define the non-holomorphic modular function $G_N(\tau)$ by the following integral representation,

$$G_N(\tau) = \sum_{m,n\in\mathbb{Z}} \int_0^{\infty} dt \, \frac{Q_N(t)}{(t+1)^{2N+1}} \, e^{-\pi t |m+n\tau|^2/\tau_2} \qquad (4.7.9)$$

where $Q_N(t)$ is a polynomial in t of degree $2N - 1$ which satisfies $t^{2N} Q_N(1/t) = Q_N(t)$. Show that $G_N(\tau)$ admits a formal expansion in terms of non-holomorphic Eisenstein series of integer weight. Calculate the inner product $g_N(s) = \langle G_N | E^*_{\frac{1}{2}+is} \rangle$ (see [PPR23] for a derivation).

5

Quantum fields on a torus

Having developed the general theory of elliptic functions and modular forms, some immediate applications to problems of physical interest are now presented, including the construction of the Green functions and functional determinants for the bc fields, the scalar field, and the spinor fields on the torus. In particular, it will be shown how the singular terms in the operator product expansion of holomorphic fields for the bc system essentially determine arbitrary correlation functions on the torus in terms of elliptic functions.

5.1 Quantum fields

A quantum field is a map from a manifold Σ to a linear operator acting on a vector space $\mathfrak{F}$ which is equipped with a Hermitian inner product. If the inner product is positive definite then $\mathfrak{F}$ is a Hilbert space, familiar from standard quantum theory, but we shall also consider quantum field theories where the inner product is not necessarily positive definite. We shall be interested in the special case where the manifold Σ is a Riemann surface which may be compact or non-compact, with or without a boundary. Quantum fields on the complex plane $\Sigma = \mathbb{C}$ describe the continuum limit and critical phenomena of two-dimensional statistical mechanics systems, such as the Ising, Potts, XY, and Heisenberg spin models. Quantum fields on a general Riemann surface Σ provide the proper setting for string perturbation theory, where probability amplitudes are obtained via an expansion in powers of the string coupling constant. The expansion is obtained by summing contributions from Riemann surfaces of all genera and, for each genus, by integrating over the corresponding moduli space of Riemann surfaces, as will be explained in detail in Chapter 12.

5.1.1 Conformal fields

A conformal field ϕ on a Riemann surface Σ is a map from Σ to the space of linear operators on a complex vector space $\mathfrak{F}$ equipped with a Hermitian inner product subject to conformal properties of ϕ to be given below.

Consider two arbitrary coordinate charts $\mathcal{U}_\alpha, \mathcal{U}_\beta \subset \Sigma$ with local complex coordinate systems $(z_\alpha, \bar{z}_\alpha)$ and $(z_\beta, \bar{z}_\beta)$. When $\mathcal{U}_\alpha$ and $\mathcal{U}_\beta$ have non-empty intersection $\mathcal{U}_\alpha \cap \mathcal{U}_\beta \neq \emptyset$, the coordinate systems in the overlap $\mathcal{U}_\alpha \cap \mathcal{U}_\beta$ are related to one another by a holomorphic transition function $z_\beta = f(z_\alpha)$ whose derivative $\partial_{z_a} f(z_\alpha)$ is non-zero throughout $\mathcal{U}_\alpha \cap \mathcal{U}_\beta$. Any such transformation of the form $z_\beta = f(z_\alpha)$ is a *locally conformal transformation*, namely it preserves all angles. For more details see Appendix B.

In the coordinate systems $(z_\alpha, \bar{z}_\alpha)$ and $(z_\beta, \bar{z}_\beta)$ the field ϕ is given by $\phi_\alpha(z_\alpha, \bar{z}_\alpha)$ and $\phi_\beta(z_\beta, \bar{z}_\beta)$, respectively. The field ϕ is a *conformal primary field* of conformal weight $(h, \tilde{h})$ provided that ϕ_α and ϕ_β are related as follows,

$$\phi_\beta(z_\beta, \bar{z}_\beta)(dz_\beta)^h (d\bar{z}_\beta)^{\tilde{h}} = \phi_\alpha(z_\alpha, \bar{z}_\alpha)(dz_\alpha)^h (d\bar{z}_\alpha)^{\tilde{h}} \qquad (5.1.1)$$

or in view of the holomorphicity of $f(z)$,

$$\left(\partial_{z_a} f(z_\alpha)\right)^h \overline{\left(\partial_{z_a} f(z_\alpha)\right)}^{\tilde{h}} \phi_\beta(z_\beta, \bar{z}_\beta) = \phi_\alpha(z_\alpha, \bar{z}_\alpha) \qquad (5.1.2)$$

Equivalently, a conformal field may be viewed as a section of an operator-valued holomorphic line bundle over Σ. See Appendix C for a discussion of holomorphic line bundles on Riemann surfaces. Clearly, h and $\tilde{h}$ must be the same for all charts $\mathcal{U} \subset \Sigma$.

Local conformal transformations form an infinite-dimensional Lie algebra acting on $\mathfrak{F}$, which is referred to as the *Virasoro algebra*. Its generators L_m for $m \in \mathbb{Z}$ obey the following structure relations,

$$[L_m, L_n] = (m - n)L_{m+n} + \frac{\hat{c}}{12} m(m^2 - 1)\delta_{m+n,0} \qquad (5.1.3)$$

where $\delta_{m,n}$ stands for the Kronecker symbol. The central charge operator $\hat{c}$ commutes with L_m for all $m \in \mathbb{Z}$ and therefore belongs to the center of the Virasoro algebra. The subalgebra of the Virasoro algebra for which the eigenvalue of $\hat{c}$ is c will be denoted Vir_c. For $c \neq 0$ the Virasoro algebra Vir_c is a central extension of the *Witt algebra* which has $c = 0$.

Actually, conformal transformations may be allowed to act independently on the holomorphic and anti-holomorphic coordinates, in which case we have two copies of the Virasoro algebra $\mathrm{Vir}_c \oplus \widetilde{\mathrm{Vir}}_c$ whose generators are denoted

L_n acting on z and $\tilde{L}_n$ acting on $\bar{z}$.[1] These two Virasoro algebras commute with one another so that $[L_m, \tilde{L}_n] = 0$ for all $m, n \in \mathbb{Z}$. The conformal weights $h, \tilde{h} \in \mathbb{R}$ characterize the representation of the Virasoro algebra under which the conformal primary field ϕ transforms,

$$[L_m, \phi(z, \bar{z})] = z^{m+1}\partial_z\psi(z, \bar{z}) + h(m+1)z^m\phi(z, \bar{z})$$
$$[\tilde{L}_m, \phi(z, \bar{z})] = \bar{z}^{m+1}\partial_{\bar{z}}\phi(z, \bar{z}) + \tilde{h}(m+1)\bar{z}^m\phi(z, \bar{z}) \qquad (5.1.4)$$

The *minimal models* correspond to the irreducible representations of the Virasoro algebra for $0 < \mathfrak{c} < 1$, for which the spectrum of allowed conformal weights h (of primary conformal fields) is finite. Minimal models have rational conformal weights $h, \tilde{h}$ and a rational central charge $\mathfrak{c}$. They are special cases of the more general class of *rational conformal field theories*, to be discussed further in Section 13.6.

For $0 < \mathfrak{c} < 1$, the *unitary minimal models* constitute a further subclass of the minimal models that are so-called *unitary* quantum field theories. In particular, the fields of a unitary quantum field theory are operators acting on a Hilbert space, while for non-unitary theories fields may act on a vector space whose inner product fails to be positive definite.

5.1.2 Free conformal fields

In the sequel, we shall limit attention to free conformal field theories. A free quantum field satisfies a linear partial differential equation on Σ, given by a differential operator $\mathcal{D}$. The order of the operator $\mathcal{D}$ is usually no larger than 2. All correlation functions of a free quantum field may be evaluated in terms of the Green function $\mathcal{D}^{-1}$ and the functional determinant of $\mathcal{D}$ which, on a genus-one Riemann surface Σ, may be expressed in terms of elliptic functions and modular forms.

It turns out that the quantization of free conformal fields already provides a sufficiently rich structure for all of string perturbation theory on an arbitrary flat space-time manifold, including Minkowski space-time with or without partial toroidal compactification, as will be further detailed in Chapters 12 and 13. Certain interacting conformal field theories, such as the non-linear sigma models, will be discussed in Chapter 12.

---

[1] More generally, we could allow the holomorphic and anti-holomorphic central charges to differ, $\mathfrak{c} \neq \tilde{\mathfrak{c}}$. This will still result in a modular invariant theory as long as $\mathfrak{c} - \tilde{\mathfrak{c}} \in 24\,\mathbb{Z}$. In fact, one can further generalize to theories whose modular non-invariance can be fixed by coupling to a (2+1)d bosonic invertible theory, which allows one to relax the constraint to $\mathfrak{c} - \tilde{\mathfrak{c}} \in 8\,\mathbb{Z}$. The $(\mathfrak{e}_8)_1$ vertex operator algebra is a well-known example.

5.2 The bc system

An important example of a free conformal field theory that has an immediate connection with elliptic functions and modular forms is provided by a system of two quantum fields b and c on a torus Σ with local complex coordinates $z, \bar{z}$, governed by a first order action $I[b, c]$,

$$I[b, c] = \frac{1}{2\pi} \int_\Sigma d^2z\, b(z, \bar{z}) \partial_{\bar{z}} c(z, \bar{z}) \qquad d^2z = \tfrac{i}{2} dz \wedge d\bar{z} \qquad (5.2.1)$$

Invariance of $I[b, c]$ under local conformal transformations requires both fields to have vanishing conformal weight $\tilde{h} = 0$ and their remaining conformal weights h to be conjugate: if b has conformal weight $(h, 0)$ then c must have conformal weight $(1 - h, 0)$. The field equations for b and c are obtained using the variational principle for $I[b, c]$, and are given by,

$$\partial_{\bar{z}} b = \partial_{\bar{z}} c = 0 \qquad (5.2.2)$$

whose solutions b and c are locally holomorphic functions of z on Σ. The quantum field theory with these assignments is referred to as the bc system of weight h. Below we shall consider the bc system on the annulus and on the torus. The bc system on a Riemann surface of arbitrary genus will be discussed in Appendix D.7.

5.2.1 The bc system on an annulus

An annulus is conformally isomorphic to a cylinder. Considering a conformal field theory on an annulus will turn out to be particularly convenient in order to relate correlators to their Hilbert space interpretation. Choosing the local complex coordinate $z' = x + iy$ on the cylinder with $x, y \in \mathbb{R}$ and the periodic identification $x \approx x + 1$, the coordinate z on the annulus is given by the conformal map $z = e^{2\pi i z'} = e^{2\pi i(x+iy)}$. Holomorphicity and periodicity of the fields b and c on the cylinder produce the following Fourier expansions in the coordinate $z' = x + iy$,[2]

$$b'(z') \sim \sum_{n \in \mathbb{Z}} b_n e^{-2\pi i n(x+iy)} \qquad c'(z') \sim \sum_{n \in \mathbb{Z}} c_n e^{-2\pi i n(x+iy)} \qquad (5.2.3)$$

If the fields b and c have conformal weights $(h, 0)$ and $(1 - h, 0)$ respectively, then the fields on the annulus are obtained by the conformal transformation (5.1.2), and the Fourier expansions become Laurent expansions in the

---

[2] Here, b' and c' denote the fields in the coordinate z' which, given their conformal weights, are related to $b(z)$ and $c(z)$ by $(\partial z'/\partial z)^h b'(z') = b(z)$ and $(\partial z'/\partial z)^{1-h} c'(z') = c(z)$. We use proportionality signs for the expansions of $b'(z')$ and $c'(z')$ on the cylinder and will choose a convenient normalization for the fields $b(z)$ and $c(z)$ on the annulus, to be given below.

complex coordinate $z = e^{2\pi i z'}$ given as follows,

$$b(z) = \sum_{n \in \mathbb{Z}} b_n \, z^{-n-h} \qquad\qquad c(z) = \sum_{n \in \mathbb{Z}} c_n z^{-n-1+h} \qquad (5.2.4)$$

Considering the fields b and c as quantum operators on the vector space $\mathfrak{F}$, the coefficients b_n and c_n are operators as well. The closest connection with holomorphic and meromorphic functions will be obtained by requiring the operators to satisfy anti-commutation relations,

$$\{b_m, b_n\} = \{c_m, c_n\} = 0 \qquad\qquad \{b_m, c_n\} = \delta_{m+n,0} \qquad (5.2.5)$$

where $\{A, B\} = AB + BA$ is the anti-commutator and $\delta_{m+n,0}$ is the Kronecker symbol. We choose a polarization in which b_n and c_n are annihilation operators for $n > 0$ while $b_n^\dagger = b_{-n}$ and $c_n^\dagger = c_{-n}$ are creation operators for $n > 0$. The anti-commutation relations for b_0, c_0 define a Clifford-Dirac algebra whose unique irreducible representation has dimension 2. The ground state $|0\rangle$ is defined to be annihilated by all b_n, c_n for $n > 0$,

$$b_0|0\rangle = \langle 0|c_0 = 0 \qquad\qquad b_n|0\rangle = c_n|0\rangle = 0 \qquad\qquad n \geq 1 \qquad (5.2.6)$$

and to have unit norm $\langle 0|0\rangle = 1$. The Fock space construction of the vector space $\mathfrak{F}$ is obtained by applying arbitrary polynomials in $b_n^\dagger$ and $c_n^\dagger$ for all possible values of $n > 0$ to the ground state $|0\rangle$. A convenient basis for $\mathfrak{F}$ is obtained as follows,

$$b_{m_p}^\dagger \cdots b_{m_1}^\dagger c_{n_q}^\dagger \cdots c_{n_1}^\dagger |0\rangle \qquad (5.2.7)$$

where $p, q = 0, 1, 2, \cdots$ and the indices of $b^\dagger$ and $c^\dagger$ may be ordered as follows, $m_p > m_{p-1} > \cdots > m_1 > 0$ and $n_p > n_{p-1} > \cdots > n_1 > 0$ in view of the anti-commutation relations of (5.2.5). For general conformal weight h, the Fock states constructed in (5.2.7) span a Fock space $\mathfrak{F}$ that is not a Hilbert space so that the corresponding conformal field theory is not a unitary theory. This may be seen, for example, from the fact that the states $b_n^\dagger|0\rangle$ are non-zero but have vanishing norm. However, when $h = \frac{1}{2}$ we may identify $c_n = b_n$ to get a single spin $\frac{1}{2}$ fermion whose Fock space is a Hilbert space and whose conformal field theory is unitary; see Section 5.7.1.

5.2.2 *The operator product expansion*

The fields $b(z)$ and $c(z)$ are holomorphic in z on the annulus. The product $c(z)b(w)$ of operators, however, is singular as $z \to w$. This may be seen by applying the operator to the ground state and using the fact that $|0\rangle$ is annihilated by all b_n with $n > 0$,

$$c(z)b(w)|0\rangle = \sum_{m \in \mathbb{Z}} c_m z^{-m-1+h} \sum_{n \leq 0} b_n w^{-n-h}|0\rangle \qquad (5.2.8)$$

The contributions from $m \leq 0$ are non-singular as $z \to w$ while for $m > 0$ we may replace the product $c_m b_n$ by the anti-commutator $\{c_m, b_n\} = \delta_{m+n,0}$ in view of $c_m|0\rangle = 0$. The resulting series is geometric in w/z, absolutely convergent for $|z/w| < 1$, and may be analytically continued in z and w throughout the annulus,

$$c(z)b(w)|0\rangle = \sum_{m > 0} z^{-m-1+h} w^{m-h}|0\rangle + \text{ regular}$$

$$= \frac{1}{z-w}|0\rangle + \text{ regular} \qquad (5.2.9)$$

where "regular" stands for all contributions that have a finite limit as $z \to w$. The pole in $z-w$ arises due to the infinite nature of the series and is universal in the following sense. If we apply the operator $c(z)b(w)$ to an arbitrary Fock space state, obtained by acting on $|0\rangle$ with an arbitrary polynomial in $b_n^\dagger$ and $c_n^\dagger$ with $n > 0$, then only a finite number of terms in the infinite series (5.2.9) will be affected, which leaves the singularity unchanged. For this reason, one promotes the singularity to a statement relating operators,

$$c(z)b(w) = \frac{I_{\mathfrak{F}}}{z-w} + \text{ regular} \qquad (5.2.10)$$

where $I_{\mathfrak{F}}$ is the identity operator in the Fock space $\mathfrak{F}$. By the same arguments one shows that the Laurent series for the operator $b(z)c(w)$ is absolutely convergent for $|w/z| < 1$, produces the pole $I_{\mathfrak{F}}/(z-w)$, and may also be analytically continued to arbitrary z, w in the annulus,

$$b(z)c(w) = \frac{I_{\mathfrak{F}}}{z-w} + \text{ regular} \qquad (5.2.11)$$

Comparing with the pole in $c(z)b(w)$, we observe that the sign reversal is consistent with the fact that the Laurent coefficients of b and c satisfy the anti-commutation relations of (5.2.5).

Actually, the pole term is only the first in an infinite Laurent series expansion of the operator $c(z)b(w)$ in powers of $z - w$. The general expansion is referred to as the *operator product expansion* of $c(z)$ and $b(w)$,

$$c(z)b(w) = \frac{I_{\mathfrak{F}}}{z-w} + \sum_{n=0}^{\infty} (z-w)^n j^{(n+1)}(w) \qquad (5.2.12)$$

The coefficients $j^{(n+1)}(w)$ for $n > 0$ are conformal primary fields of weight

$(n+1,0)$. The transformation law for $j^{(1)}$ is more complicated and will be analyzed in Exercise 5.2. The operators $b(z)b(w)$ and $c(z)c(w)$ similarly have operator product expansions. In view of the anti-commutation relations (5.2.5) we have $b(z)^2 = c(z)^2 = 0$ and,

$$b(z)b(w) = -(z-w)b(w)\,\partial_w b(w) + \mathcal{O}\big((z-w)^2\big)$$
$$c(z)c(w) = -(z-w)c(w)\,\partial_w c(w) + \mathcal{O}\big((z-w)^2\big) \qquad (5.2.13)$$

so that the operator product expansion is entirely regular.

5.3 Correlators of the bc system on the torus

We now consider the bc system on a torus $\Sigma = \mathbb{C}/\Lambda$ with $\Lambda = \mathbb{Z} \oplus \mathbb{Z}\tau$ and arbitrary modulus τ with $\mathrm{Im}\,(\tau) > 0$. The fields b and c are meromorphic and doubly periodic, and are thus elliptic functions. On the torus the bc systems for different integer values of h are all isomorphic to one another, in view of the fact that the torus has a holomorphic differential dz which is nowhere vanishing. As a result, a field $\phi(z)(dz)^h$ of conformal weight $(h,0)$ is isomorphic to a scalar field $\phi(z)$ of weight $(0,0)$. This isomorphism allows for a direct evaluation of all correlators of b,c fields on the torus. A correlator is the matrix element in the vacuum state $|0\rangle$ of a product of b and c fields evaluated at distinct points on Σ,

$$\mathcal{A} = \langle 0|c(z_1)\cdots c(z_N)b(w_1)\cdots b(w_M)|0\rangle \qquad (5.3.1)$$

The correlators $\mathcal{A}$ are meromorphic in z_n for all $n = 1,\cdots,N$ and in w_m for all $m = 1,\cdots,M$. We shall now show that the positions of their zeros and poles in each variable are known from the operator product expansion. In view of (5.2.13), the correlator has a simple zero when two points z_n and $z_{n'}$ for $n' \neq n$ coincide, as well as when two points w_m and $w_{m'}$ for $m' \neq m$ coincide. In view of (5.2.10), the correlator has a simple pole in z_n at w_m, whose residue is given by a correlator with one fewer b and one fewer c,

$$\mathcal{A} = \frac{(-)^{N-1+m-n}}{z_n - w_m}\,\langle 0|c(z_1)\cdots \widehat{c(z_n)}\cdots c(z_N) \qquad (5.3.2)$$
$$\times b(w_1)\cdots \widehat{b(w_m)}\cdots b(w_M)|0\rangle + \text{ regular}$$

The sign factor arises from permuting the positions of the fields, the hatted operators are to be omitted, and "regular" includes all terms that are non-singular as $z_n \to w_m$.

Considered as a function of z_n, the correlator $\mathcal{A}$ of (5.3.1) has precisely M simple poles at $w_1,\cdots,w_M$. In view of Theorem 2.3.2, it must also have M

zeros in z_n. These zeros must include the $N-1$ points $z_{n'}$ with $n' \neq n$, so we must have $M \geq N-1$. Similarly, as a function of w_m, the correlator $\mathcal{A}$ has precisely N simple poles at $z_1, \cdots, z_N$. In view of Theorem 2.3.2 it must also have N zeros in w_m which must include the $M-1$ points $w_{m'}$ with $m' \neq m$, so we must also have $N \geq M-1$. Combining both inequalities allows for either $M = N$ or $M - N = \pm 1$. The latter is excluded by symmetry under reversing the sign of both b and c. Thus only $M = N$ gives a non-vanishing correlator. We shall now evaluate the correlators $\mathcal{A}$ for $M = N$ in terms of ϑ-functions using Theorems 2.3.2, 2.6.1, and 2.6.2.

5.3.1 The 4-point correlator

We begin with the correlator $\mathcal{A} = \langle 0|c(z_1)c(z_2)b(w_1)b(w_2)|0\rangle$. As a function of z_1, it has a simple zero at z_2, simple poles at w_1 and w_2, and no other poles. In view of the second relation in Theorem 2.3.2, it must have a second zero, whose position is determined by the third relation in Theorem 2.3.2 to be given by $-z_2 + w_1 + w_2$ (mod Λ). Proceeding analogously for z_2 using Theorem 2.6.1 we obtain the following expression for the correlator,

$$\mathcal{A} = A_0 \frac{\vartheta_1(z_1 + z_2 - w_1 - w_2)\vartheta_1(z_1 - z_2)\vartheta_1(w_1 - w_2)}{\vartheta_1(z_1 - w_1)\vartheta_1(z_1 - w_2)\vartheta_1(z_2 - w_1)\vartheta_1(z_2 - w_2)} \qquad (5.3.3)$$

Here, A_0 is independent of z_1, z_2, w_1, w_2 and all dependence on τ has been suppressed, as usual. We may use this expression to obtain the correlator $\langle 0|c(z_1)b(w_2)|0\rangle$ by letting $z_2 \to w_1$,

$$\mathcal{A} = \frac{1}{z_2 - w_1}\frac{A_0}{\vartheta_1'(0)} + \text{regular} \qquad (5.3.4)$$

As a result, the 2-point correlator is constant and is "saturated" by the zero modes of the c and b fields as follows,

$$\langle 0|c(z_1)b(w_2)|0\rangle = \frac{A_0}{\vartheta_1'(0)} \qquad (5.3.5)$$

The constancy of this correlator is consistent with the fact that no function on a torus can have a single simple pole. Alternatively, we may use Theorem 2.6.2 to express the correlator $\mathcal{A}$ in terms of the Weierstrass ζ-function and its derivatives. Since the poles in z_1, z_2 of the correlator at w_1, w_2 are simple, and must have opposite residue, we have the following general expression,

$$\mathcal{A} = A_3 + \sum_{i=1}^{2} A_i\big\{\zeta(z_i - w_1) - \zeta(z_i - w_2)\big\} \qquad (5.3.6)$$

where A_1, A_2, A_3 are independent of z_1, z_2, w_1, w_2. Matching the residues with (5.3.4), we find $A_1 = -A_2 = -A_0/\vartheta_1'(0)$. This condition also shows that the correlator has zeros at $z_1 = z_2$ and $w_1 = w_2$ provided $A_3 = 0$.

5.3.2 Arbitrary correlators

Arbitrary correlators $\mathcal{A}$ in (5.3.1) with $M = N$ may be evaluated iteratively. As a function of z_1 the correlator $\mathcal{A}$ has N simple poles at $w_1, \cdots, w_N$, and $N - 1$ known zeros at $z_2, \cdots, z_N$. The missing zero is given by Theorems 2.3.2 and 2.6.1 by the sum $-\sum_{n=2}^{N} z_n + \sum_{n=1}^{N} w_n$. Putting all together,

$$\mathcal{A} = \hat{A}_1 \frac{\vartheta_1(D) \prod_{n=2}^{N} \vartheta_1(z_1 - z_n)}{\prod_{n=1}^{N} \vartheta_1(z_1 - w_n)} \qquad D = \sum_{n=1}^{N}(z_n - w_n) \quad (5.3.7)$$

where $\hat{A}_1$ is independent of z_1, but depends on all other z_n and w_n. Again, all dependence on τ has been suppressed. Proceeding iteratively on the points z_n we find that the factor $\vartheta_1(D)$ always accounts for the missing zero. In total, we obtain,

$$\mathcal{A} = A_0 \frac{\vartheta_1(D) \prod_{m<n} \vartheta_1(z_m - z_n)\vartheta_1(w_m - w_n)}{\prod_{m,n} \vartheta_1(z_m - w_n)} \qquad (5.3.8)$$

where A_0 depends only on τ. Identities that relate the correlators of the b and c fields in the presence of the operator $j(z) = j^{(1)}(z)$, which was defined in (5.2.12) via the operator product expansion, to those without this operator are the subject of Exercise 5.1, while the behavior of the operator $j(z)$ under local conformal transformations is derived in Exercise 5.2.

5.4 The scalar field

We next consider the example of a scalar field. A real-valued scalar field φ is governed by a second order differential operator on the torus, namely the Laplacian. In local complex coordinates $(z, \bar{z})$ on the torus $\Sigma = \mathbb{C}/\Lambda$ with $\Lambda = \mathbb{Z} \oplus \mathbb{Z}\tau$, its action may be expressed as follows,

$$I[\varphi] = \frac{1}{2\pi} \int_{\Sigma} d^2 z \, \partial_z \varphi \, \partial_{\bar{z}} \varphi \qquad\qquad d^2 z = \tfrac{i}{2} dz \wedge d\bar{z} \qquad (5.4.1)$$

The corresponding field equation $\partial_z \partial_{\bar{z}} \varphi = 0$ implies that $\partial_z \varphi$ is a locally holomorphic field while $\partial_{\bar{z}} \varphi$ is a locally anti-holomorphic field. On an annulus, these quantum fields admit Laurent expansions,

$$i\partial_z \varphi(z, \bar{z}) = \sum_{m \in \mathbb{Z}} \varphi_m \, z^{-m-1} \qquad\qquad i\partial_{\bar{z}} \varphi(z, \bar{z}) = \sum_{m \in \mathbb{Z}} \tilde{\varphi}_m \, \bar{z}^{-m-1} \quad (5.4.2)$$

where the coefficient operators φ_m and $\tilde{\varphi}_m$ satisfy commutation relations,

$$[\varphi_m, \varphi_n] = m\delta_{m+n,0} \qquad\qquad [\tilde{\varphi}_m, \tilde{\varphi}_n] = m\delta_{m+n,0} \qquad (5.4.3)$$

The commutation relations show that the operators φ_0 and $\tilde{\varphi}_0$ commute with the operators φ_n and $\tilde{\varphi}_n$ for all $n \in \mathbb{Z}$ and with one another. Furthermore, the reality of φ as a classical field requires the operators to be self-adjoint, $\varphi_0^\dagger = \varphi_0$ and $\tilde{\varphi}_0^\dagger = \tilde{\varphi}_0$. Therefore, the operators φ_0 and $\tilde{\varphi}_0$ may be diagonalized simultaneously and have real eigenvalues, in terms of which we shall label their joint eigenstates by $p, \tilde{p} \in \mathbb{R}$ as follows,

$$\varphi_0|p, \tilde{p}\rangle = p|p, \tilde{p}\rangle$$
$$\tilde{\varphi}_0|p, \tilde{p}\rangle = \tilde{p}|p, \tilde{p}\rangle \qquad (5.4.4)$$

The eigenstates are orthonormal in the following sense,

$$\langle p', \tilde{p}'|p, \tilde{p}\rangle = \delta(p - p')\delta(\tilde{p}' - \tilde{p}) \qquad (5.4.5)$$

For the operators φ_n and $\tilde{\varphi}_n$ we choose a polarization such that $\varphi_{-n} = \varphi_n^\dagger$ and $\tilde{\varphi}_{-n} = \tilde{\varphi}_n^\dagger$ for $n > 0$, and define the states $|p, \tilde{p}\rangle$ to satisfy,

$$\varphi_n|p, \tilde{p}\rangle = \tilde{\varphi}_n|p, \tilde{p}\rangle = 0 \qquad \text{for } n > 0 \qquad (5.4.6)$$

The Fock space is constructed by applying arbitrary polynomials in $\varphi_n^\dagger$ and $\tilde{\varphi}_n^\dagger$ for $n > 0$ to $|p, \tilde{p}\rangle$ and a convenient basis may be chosen as follows,

$$\varphi_{n_r}^\dagger \cdots \varphi_{n_1}^\dagger \tilde{\varphi}_{m_s}^\dagger \cdots \tilde{\varphi}_{m_1}^\dagger |p, \tilde{p}\rangle \qquad (5.4.7)$$

Each such state has positive norm so that the Fock space for a scalar field is actually a Hilbert space, which we denote by $\mathfrak{F}$.

The scalar field theory is conformal and has two commuting Virasoro algebras of central charge one, $\mathrm{Vir}_1 \oplus \widetilde{\mathrm{Vir}}_1$, generated by L_m and $\tilde{L}_m$ whose expressions in terms of φ_m and $\tilde{\varphi}_m$ are given as follows,

$$L_0 = -\frac{1}{24} + \frac{1}{2}\varphi_0^2 + \sum_{n=1}^\infty \varphi_{-n}\varphi_n$$

$$L_m = \frac{1}{2}\sum_{n\in\mathbb{Z}} \varphi_{m-n}\varphi_n \qquad \text{for } m \neq 0 \qquad (5.4.8)$$

and similarly for $\tilde{L}_m$. The offset $-\frac{1}{24}$ in L_0 arises from the conformal transformation of the expression on the cylinder (which has no such offset) to the annulus. It may also be viewed as the sum of the zero-point energies of the harmonic oscillators, using the fact that the sum of all positive integers m gives $\zeta(-1) = -\frac{1}{12}$, c.f. (2.2.16).

The operators $\partial_z\varphi$ and $\partial_{\bar{z}}\varphi$ are conformal primary fields of weights $(1,0)$ and $(0,1)$ respectively, and satisfy the corresponding transformations given in (5.1.4). The operator product expansion of holomorphic fields is given by the following Laurent expansion,

$$\partial_z\varphi(z)\partial_w\varphi(w) = \frac{1}{(z-w)^2} - 2T(w) + \mathcal{O}(z-w) \qquad (5.4.9)$$

where $T(w)$ is the *stress tensor* for the holomorphic component of the scalar field. The stress tensor $T(w)$ is given in terms of the Virasoro generators L_m,

$$T(w) = \sum_{n\in\mathbb{Z}} L_n\, w^{-n-2} \qquad (5.4.10)$$

and similarly for the anti-holomorphic field $\partial_{\bar{z}}\varphi$ and the generators $\tilde{L}_m$.

We may attempt to use the operator product expansion (5.4.9) to evaluate the correlator of holomorphic fields $\partial_z\varphi(z)$, just as we had in the case of the bc system. For simplicity, we choose matrix elements between the ground state $|0\rangle = |0,0\rangle$ on the torus Σ,

$$\langle 0|\partial_z\varphi(z_1)\cdots\partial_z\varphi(z_N)|0\rangle \qquad (5.4.11)$$

The correlator vanishes for odd N and is meromorphic in all its points z_n for even N. As a function of z_n, the correlator has a double pole at every point z_m with $m \neq n$. For example, for the 2-point correlator, the double pole shows that the correlator must be of the form,

$$\langle 0|\partial_z\varphi(z_1)\partial_z\varphi(z_2)|0\rangle = \wp(z_1-z_2) + A_4 \qquad (5.4.12)$$

where A_4 is independent of z_1, z_2. In contrast with the bc system, the zeros of the correlator are not manifest now, and the constant A_4 cannot be fixed by using known zeros. It may be determined, however, by using not just the pole term in the operator product expansion, but also the term proportional to the stress tensor T, and we find $A_4 = 0$. For higher point correlators, however, the corresponding constant terms will not be so easy to evaluate. Instead we shall turn to evaluating the full correlators of the field φ itself.

5.5 The scalar Green function on the torus

The scalar Laplacian on the torus has a zero mode and is not invertible on all scalar functions. However, an inverse may be defined on its range, namely on functions orthogonal to constant functions on Σ. To construct this inverse, we shall present two different approaches, the first based on Fourier series, and the second using ϑ-functions.

5.5.1 The Laplacian on the torus

An arbitrary Riemannian metric on the torus Σ may be rescaled to a flat metric of unit area using a Weyl transformation, as shown in Appendix B. This reduced flat metric depends on a single complex parameter $\tau \in \mathcal{H}$, and we represent the torus by the quotient $\Sigma = \mathbb{C}/\Lambda$ for the lattice $\Lambda = \mathbb{Z} + \mathbb{Z}\tau$. The points on Σ may be parametrized by a local complex coordinate $z = x + y\tau$ for $x, y \in \mathbb{R}/\mathbb{Z}$, for example in the range $0 \leq x, y \leq 1$ (see Figure 2.2). In these coordinates, the flat metric $\mathfrak{g}$ and associated volume form $d\mu_\mathfrak{g} = d\mu_\mathfrak{g}(z)$ of unit area are given by,

$$\mathfrak{g} = \frac{|dz|^2}{\tau_2} = \frac{|dx + \tau dy|^2}{\tau_2} \qquad d\mu_\mathfrak{g} = \frac{i}{2}\frac{dz \wedge d\bar{z}}{\tau_2} = dx \wedge dy \quad (5.5.1)$$

The inner product $\langle f_1 | f_2 \rangle$ on the space of complex-valued functions $f_1, f_2 : \Sigma \mapsto \mathbb{C}$ is defined as follows,

$$\langle f_1 | f_2 \rangle = \int_\Sigma d\mu_\mathfrak{g}\, \bar{f}_1\, f_2 \qquad\qquad (5.5.2)$$

making the space of bounded functions on Σ into an $L^2(\Sigma)$ Hilbert space. The Dirac δ-function $\delta_\mathfrak{g}$ with respect to the form $d\mu_\mathfrak{g}$ is defined by,

$$\int_\Sigma d\mu_\mathfrak{g}(z)\, \delta_\mathfrak{g}(z - w) f(z) = f(w) \qquad\qquad (5.5.3)$$

for an arbitrary test function f, and may be given explicitly in terms of the Dirac δ-function on real or complex coordinates by,

$$\delta_\mathfrak{g}(z - w) = \delta(x - u)\delta(y - v) = \tau_2 \delta^{(2)}(z - w) \qquad (5.5.4)$$

where $z = x + y\tau$ and $w = u + v\tau$ with $x, y, u, v \in \mathbb{R}/\mathbb{Z}$ and $\delta^{(2)}(z - w)$ is the coordinate δ-function normalized by $\int_\Sigma d^2z\, \delta^{(2)}(z - w) f(z) = f(w)$.

The Cauchy–Riemann operators in these coordinates take the form,

$$\partial_z = \frac{1}{2\tau_2}(-i\partial_y + \bar{\tau}\, i\partial_x) \qquad\qquad \partial_{\bar{z}} = \frac{1}{2\tau_2}(i\partial_y - \tau\, i\partial_x) \quad (5.5.5)$$

while the Laplace–Beltrami operator, or simply Laplacian, is given by,

$$\Delta_\mathfrak{g} = 4\tau_2 \partial_{\bar{z}} \partial_z = \frac{1}{\tau_2}(\partial_y - \tau\partial_x)(\partial_y - \bar{\tau}\partial_x) \qquad (5.5.6)$$

The Laplacian $\Delta_\mathfrak{g}$ is self-adjoint with respect to the inner product (5.5.2) on $L^2(\Sigma)$ so that its eigenvalues are real. The eigenvalues of $\Delta_\mathfrak{g}$ are negative or zero since, on the torus Σ, we have for an arbitrary test function,

$$\langle f | \Delta_\mathfrak{g} f \rangle = -4 \int_\Sigma d^2z\, |\partial_z f|^2 \leq 0 \qquad\qquad (5.5.7)$$

Furthermore, $\Delta_{\mathfrak{g}}$ is invariant under translations in x and y and thus commutes with the operators $i\partial_x$ and $i\partial_y$ that generate translations in x and y, respectively. For fixed τ, the Laplacian $\Delta_{\mathfrak{g}}$ has a discrete spectrum with a single zero mode, namely the constant function on Σ, so that $\mathrm{Ker}\Delta_{\mathfrak{g}} = \mathbb{C}$. Its range is the set of functions orthogonal to constant functions

5.5.2 The scalar Green function via Fourier series

Since the self-adjoint operators $i\partial_x$ and $i\partial_y$ mutually commute and commute with $\Delta_{\mathfrak{g}}$, the Laplacian is diagonal in a basis where $i\partial_x$ and $i\partial_y$ are simultaneously diagonal. Such a basis is given by the following orthonormal basis of doubly periodic functions on Σ, parametrized by $z = x + \tau y$ for $x, y \in \mathbb{R}/\mathbb{Z}$,

$$f_{m,n}(x,y) = e^{2\pi i(nx - my)}$$
$$\langle f_{m',n'} | f_{m,n} \rangle = \delta_{m',m}\,\delta_{n',n} \qquad\qquad m, m', n, n' \in \mathbb{Z} \quad (5.5.8)$$

They also satisfy the completeness relation,

$$\sum_{m,n\in\mathbb{Z}} \overline{f_{m,n}(u,v)} f_{m,n}(x,y) = \delta(x-u)\delta(y-v) \qquad (5.5.9)$$

where the Dirac δ-functions on the right are periodic in their argument with period 1. Actually, it is convenient and standard practice to define the Green function as the inverse of the rescaled positive Laplacian $\hat{\Delta}_{\mathfrak{g}} = -\Delta_{\mathfrak{g}}/4\pi$ rather than the original negative Laplacian $\Delta_{\mathfrak{g}}$. The extra factor of -4π will simplify many important formulas. Correspondingly, we evaluate the eigenvalues of $\hat{\Delta}_{\mathfrak{g}}$ for the eigenfunctions (5.5.8), and obtain,

$$\hat{\Delta}_{\mathfrak{g}} f_{m,n} = \lambda_{m,n}(\tau) f_{m,n} \qquad \lambda_{m,n}(\tau) = \pi \frac{|m + n\tau|^2}{\tau_2} \qquad (5.5.10)$$

The zero eigenvalue corresponds to the constant function with $m = n = 0$, while all other eigenvalues are strictly positive, as expected. The Green function is defined as the inverse of $\hat{\Delta}_{\mathfrak{g}}$ on the space of functions orthogonal to constants. In view of translation invariance, we have $G(z, w|\tau) = G(z - w|\tau)$ with $z = x + y\tau$, and $G(z|\tau)$ is given by,

$$G(z|\tau) = \sum_{(m,n)\neq(0,0)} \frac{\tau_2}{\pi|m + n\tau|^2} e^{2\pi i(nx - my)} \qquad (5.5.11)$$

Applying $\hat{\Delta}_{\mathfrak{g}}$ using the completeness relation (5.5.8), one verifies,

$$\hat{\Delta}_{\mathfrak{g}} G(z - w|\tau) = \tau_2\,\delta^{(2)}(z - w) - 1 \qquad (5.5.12)$$

where the δ-function was defined in (5.5.4). By construction, the right side of (5.5.12) is orthogonal to constant functions. Actually, the Green function obtained in (5.5.11) itself is orthogonal to constant functions,

$$\int_{\Sigma} d\mu_{\mathfrak{g}}(z)\, G(z - w|\tau) = 0 \qquad (5.5.13)$$

since the zero mode $(m, n) = (0, 0)$ was removed from the sum in (5.5.11).

5.5.3 The scalar Green function via ϑ-functions

We may take equations (5.5.12) and (5.5.13) as defining G and express both in terms of the complex variables z, w,

$$\partial_z \partial_{\bar{z}} G(z - w|\tau) = -\pi \delta^{(2)}(z - w) + \frac{\pi}{\tau_2} \qquad (5.5.14)$$

The solution for G is easy to construct in terms of ϑ-functions, as we shall now show. This method will generalize to constructing the Green function on Riemann surfaces of arbitrary genus, whereas the method of Fourier series of the preceding section does not generalize since higher genus Riemann surfaces have no continuous isometries.

We proceed as follows for fixed $\tau \in \mathcal{H}$. For $|z - w| \ll 1$ and $|z - w| \ll \tau_2$, the global structure of the torus is negligible and the Green function must approach its expression on $\mathbb{C}$ given by $G(z - w|\tau) \approx -\ln|z - w|^2$. The function $\vartheta_1(z - w|\tau)/\vartheta_1'(0|\tau)$ behaves as $z - w$ for small $|z - w|$ and is therefore a reasonable candidate to play the role of $z - w$ on the torus. However, $\vartheta_1(z - w|\tau)$ cannot be a doubly periodic function as it is holomorphic and would then have to be constant by Liouville's theorem. Still, using the transformation laws of $\vartheta_1(z - w|\tau)$ under $z \to z + 1$ and $z \to z + \tau$ given in (2.6.8), one readily finds a simple additive term to restore double periodicity, and $G(z, w|\tau) = G(z - w|\tau)$ is given by,

$$G(z|\tau) = -\ln \left| \frac{\vartheta_1(z|\tau)}{\vartheta_1'(0|\tau)} \right|^2 + 2\pi \frac{(\mathrm{Im}\, z)^2}{\tau_2} + G_0(\tau) \qquad (5.5.15)$$

Applying the Laplace operator, the first term produces the δ-function, while the second term produces the constant on the right side of (5.5.14). The z-independent term $G_0(\tau)$ may be determined by imposing the normalization condition (5.5.13) and is evaluated in Exercise 5.3 to give,

$$G(z|\tau) = -\ln \left| \frac{\vartheta_1(z|\tau)}{\eta(\tau)} \right|^2 + 2\pi \frac{(\mathrm{Im}\, z)^2}{\tau_2} \qquad (5.5.16)$$

The expression (5.5.11) for G may be related to (5.5.16) by carrying out the sum over m first in (5.5.11) using the following Lemma.

Lemma 5.5.1. *The Lipschitz formula for* $\text{Im}(z) > 0$, $\text{Re}(s) > 0$ *and* $0 < \alpha < 1$ *is given by,*

$$\sum_{n \in \mathbb{Z}} \frac{e^{2\pi i n \alpha}}{(z+n)^s} = \frac{(-2\pi i)^s}{\Gamma(s)} \sum_{m=0}^{\infty} (m+\alpha)^{s-1} e^{2\pi i z(m+\alpha)} \tag{5.5.17}$$

When $\text{Im}(z) > 0$ *and* $\text{Re}(s) > 1$, *the formula may be extended to* $\alpha \to 1$,

$$\sum_{n \in \mathbb{Z}} \frac{1}{(z+n)^s} = \frac{(-2\pi i)^s}{\Gamma(s)} \sum_{m=1}^{\infty} m^{s-1} e^{2\pi i m z} \tag{5.5.18}$$

The proof may be found in Rademacher [Rad70]. In the special case where s is a positive integer, the second formula may be derived from the equality between (2.2.1) and (2.2.3) by differentiating $s-1$ times on both sides.

The scalar Green function on the torus may be used to evaluate the correlators of the scalar field φ, as illustrated in Exercise 5.4, and to solve problems of electro-statics in two dimensions, as shown in Exercise 5.5.

5.5.4 Modular properties

A modular transformation $\gamma \in \text{SL}(2, \mathbb{Z})$ leaves the metric $\mathfrak{g}$ of (5.5.1) and the Laplacians $\Delta_{\mathfrak{g}}$ and $\hat{\Delta}_{\mathfrak{g}}$ invariant provided τ and the complex coordinate $z = x + y\tau$ transform according to the following rules,

$$\tilde{\tau} = \gamma\tau = \frac{a\tau + b}{c\tau + d} \qquad \tilde{z} = \gamma z = \frac{z}{c\tau + d} \qquad \gamma = \begin{pmatrix} a & b \\ c & d \end{pmatrix} \tag{5.5.19}$$

The transformations on the real coordinates x, y are as follows,

$$\begin{pmatrix} \tilde{x} \\ \tilde{y} \end{pmatrix} = \begin{pmatrix} a & -b \\ -c & d \end{pmatrix} \begin{pmatrix} x \\ y \end{pmatrix} \tag{5.5.20}$$

The spectrum of the Laplacian is invariant under $\text{SL}(2, \mathbb{Z})$, as the eigenvalues $\lambda_{m,n}(\tau)$ of $\hat{\Delta}_{\mathfrak{g}}$ transform as follows,

$$\lambda_{m,n}(x, y) = \lambda_{\tilde{m}, \tilde{n}}(\tilde{x}, \tilde{y}) \qquad \begin{pmatrix} \tilde{m} \\ \tilde{n} \end{pmatrix} = \begin{pmatrix} a & -b \\ -c & d \end{pmatrix} \begin{pmatrix} m \\ n \end{pmatrix} \tag{5.5.21}$$

in view of the relation $\tilde{n}\tilde{x} - \tilde{m}\tilde{y} = nx - my$. It readily follows that also the normalized Green function, given in (5.5.11), is modular invariant,

$$G(\gamma z - \gamma w | \gamma \tau) = G(z - w | \tau) \tag{5.5.22}$$

as may be verified directly on the expression for G in terms of ϑ-functions in (5.5.16) by using the transformation laws of ϑ_1 given in (3.7.5).

5.6 Scalar determinant and Kronecker's first limit formula

Just as it was natural to define the Green function G as the inverse of the operator $\hat{\Delta}_{\mathfrak{g}} = -\Delta_{\mathfrak{g}}/4\pi$, it is also natural to evaluate the functional determinant of $\hat{\Delta}_{\mathfrak{g}}$ rather than the determinant of $\Delta_{\mathfrak{g}}$. The functional determinant of $\hat{\Delta}_{\mathfrak{g}}$, considered as an operator on the subspace of functions orthogonal to constants, is formally given by the product of the eigenvalues $\lambda_{m,n}(\tau)$ of (5.5.10). However, this product does not converge. To define the determinant we introduce the ζ-function associated with $\hat{\Delta}_{\mathfrak{g}}$,

$$\zeta_{\hat{\Delta}_{\mathfrak{g}}}(\tau, s) = \sum_{(m,n)\neq(0,0)} \frac{1}{(\lambda_{m,n})^s} = E_s(\tau) \tag{5.6.1}$$

where $E_s(\tau)$ is the non-holomorphic Eisenstein series introduced in Section 4.2. The series for $E_s(\tau)$ is absolutely convergent for $\mathrm{Re}\,(s) > 1$. Its analytic continuation to $s \in \mathbb{C}$ has simple poles at $s = 0$ and $s = 1$, is modular invariant, and satisfies the functional relation (4.2.10). Taking the limit $s \to 0$ of (4.2.17) gives $\zeta_{\hat{\Delta}_{\mathfrak{g}}}(\tau, 0) = -1$ and the determinant,

$$\ln \mathrm{Det}'\hat{\Delta}_{\mathfrak{g}} = -\zeta'_{\hat{\Delta}_{\mathfrak{g}}}(\tau, 0) = -\partial_s E_s(\tau)\Big|_{s=0} \tag{5.6.2}$$

To compute the determinant, we use the expressions for $E_s(\tau)$ in terms of the functions $A_s(\tau)$ and $B_s(\tau)$ defined in (4.2.14) and (4.2.15),

$$\zeta'_{\hat{\Delta}_{\mathfrak{g}}}(\tau, 0) = A_0(\tau) + \frac{\pi\tau_2}{3} - \ln\tau_2 - \ln(4\pi) \tag{5.6.3}$$

The last three terms on the right side arise from differentiating $B_s(\tau)/\Gamma(s)$ at $s = 0$. To compute $A_0(\tau)$, we simplify the summand in (4.2.14) by using the expression for the modified Bessel function in terms of elementary functions, $K_{-\frac{1}{2}}(x) = \sqrt{\pi/2x}\,e^{-x}$. We discover that $A_0(\tau)$ simplifies considerably,

$$A_0(\tau) = \sum_{m,n\neq 0} \frac{1}{|m|} \exp\left\{2\pi i m n\tau_1 - 2\pi|mn|\tau_2\right\} \tag{5.6.4}$$

and may be reexpressed as a sum of products of holomorphic times anti-holomorphic dependence on τ,

$$A_0(\tau) = -2\ln \prod_{m=1}^{\infty} (1 - q^m)(1 - \bar{q}^m) \tag{5.6.5}$$

We may express the final result in terms of the Dedekind $\eta(\tau)$ function,

$$\zeta'_{\hat{\Delta}_g}(\tau, 0) = -\ln\left(4\pi\tau_2|\eta(\tau)|^4\right) \tag{5.6.6}$$

Combining this with $\zeta_{\hat{\Delta}_g}(\tau, 0) = -1$ and (5.6.2), we obtain,

$$\text{Det}'\hat{\Delta}_g = 4\pi\tau_2|\eta(\tau)|^4 = -4\pi\text{Det}'\Delta_g \tag{5.6.7}$$

which is easily shown to be modular invariant, using the transformation law of (3.8.3) for the Dedekind η-function and that of τ_2.

5.6.1 The first Kronecker limit formula

The first Kronecker limit formula gives the residue and the finite part of $\zeta_{\hat{\Delta}_g}(\tau, s)$ at its pole $s = 1$ by the following expression,

$$\zeta_{\hat{\Delta}_g}(\tau, s) = \frac{a_{-1}}{s-1} + a_0 + \mathcal{O}(s-1) \tag{5.6.8}$$

Using the functional relation (4.2.10), it is immediate that $a_{-1} = -1$. From the same formula we get also a_0 by,

$$a_0 = \frac{d}{ds}\left((s-1)E_s(\tau)\right)\Big|_{s=1} = -\ln\left(4\pi\tau_2|\eta(\tau)|^4\right) + 2\gamma \tag{5.6.9}$$

where γ is Euler's constant as it arises in $\Gamma'(1) = -\gamma$.

5.6.2 The scalar partition function

The scalar determinant is intimately related to the partition function for the free scalar field in statistical mechanics. A partition function $Z(\beta)$ for a Hamiltonian H acting on a Hilbert space $\mathfrak{F}$, evaluated at an inverse temperature β, is given by,

$$Z(\beta) = \text{Tr}_{\mathfrak{F}}\left(e^{-\beta H}\right) \tag{5.6.10}$$

where $\text{Tr}_{\mathfrak{F}}$ is the trace evaluated in the Hilbert space $\mathfrak{F}$. The scalar field has two independent sets of oscillators, namely, φ_m and $\tilde{\varphi}_m$, with independent Virasoro generators L_n and $\tilde{L}_n$. In particular, the generators L_0 and $\tilde{L}_0$ are given by a sum of decoupled harmonic oscillator Hamiltonians plus the term due to the zero-mode. Instead of the real-valued inverse temperature $\beta > 0$, we shall complexify β and let $\beta \to -2\pi i\tau$ whose real part is still positive in view of $\text{Im}(\tau) > 0$. The partition function for L_0 is then given as follows,

$$\text{Tr}_{\text{osc}}\left(q^{L_0}\right) = q^{-\frac{1}{24}+\frac{1}{2}\varphi_0^2}\prod_{n=1}^{\infty}\frac{1}{(1-q^n)} = q^{\frac{1}{2}\varphi_0^2}\eta(\tau)^{-1} \tag{5.6.11}$$

where the trace Tr_{osc} is taken only over the oscillator part of the Hilbert space $\mathfrak{F}$, thereby leaving the operator φ_0 untraced over. Proceeding similarly for $\tilde{L}_0$ and complexifying its inverse temperature $\beta \to 2\pi i \bar{\tau}$, we obtain the combined partition function over the oscillator part of $\mathfrak{F}$,

$$\text{Tr}_{\text{osc}}\left(q^{L_0}\bar{q}^{\tilde{L}_0}\right) = q^{\frac{1}{2}\varphi_0^2}\bar{q}^{\frac{1}{2}\tilde{\varphi}_0^2}\frac{1}{|\eta(\tau)|^2} \qquad (5.6.12)$$

The zero-mode operators φ_0 and $\tilde{\varphi}_0$ are self-adjoint and have a continuous spectrum of real eigenvalues, denoted p and $\tilde{p}$, respectively. Physically, these eigenvalues are the momenta of left and right movers. Periodicity under $\tau_1 \to \tau_1 + 1$ requires $p^2 - \tilde{p}^2 \in 2\mathbb{Z}$. When the field φ takes values in $\mathbb{R}$, we require the eigenvalues p and $\tilde{p}$ to be equal to one another and to range over the full real line $\mathbb{R}$. The full trace is given by,

$$\text{Tr}_{\mathfrak{F}}\left(q^{L_0}\bar{q}^{\tilde{L}_0}\right) = \frac{1}{|\eta(\tau)|^2}\int_{\mathbb{R}} dp\, |q|^{p^2}$$

$$= \frac{1}{\sqrt{2\tau_2}\,|\eta(\tau)|^2} = \sqrt{2\pi}\left(\text{Det}'\hat{\Delta}_{\mathfrak{g}}\right)^{-\frac{1}{2}} \qquad (5.6.13)$$

which is the modular invariant partition function evaluated using the functional integral, up to an overall constant normalization factor.

5.7 Spinor fields

The Lie algebra $\text{SO}(d)$ has spinor representations in addition to its tensorial representations. A spinor representation may be viewed either as a double-valued representation of the group $\text{SO}(d)$ or as a single-valued representation of the double cover $\text{Spin}(d)$ of $\text{SO}(d)$. A spinor field is a map from Σ to a linear operator on a Fock space $\mathfrak{F}$ that transforms under a spinor representation of $\text{Spin}(d)$. A manifold supports spinor fields provided it is orientable and its second Stiefel–Whitney class vanishes. These conditions are satisfied in $d = 2$ for Riemann surfaces.

To realize spinor fields in terms of line bundles, one decomposes the cotangent bundle $T^*\Sigma$ into a direct sum of its holomorphic and anti-holomorphic parts $T^*\Sigma = K \oplus K^*$, where K is referred to as the canonical bundle. A spin bundle S is a holomorphic line bundle whose square is isomorphic to the canonical bundle $S^{\otimes 2} \approx K$. A spinor field is then an operator-valued section of S, referred to in physics as a spin $\frac{1}{2}$-field. More general spinor fields, of spin $n + \frac{1}{2}$ with $n \in \mathbb{Z}$, may be constructed from the tensor product of S with powers of K to produce holomorphic line bundles $S \otimes K^n$ with first Chern class $c_1(S \otimes K^n) = (2n + 1)(g - 1)$. For more details on Riemann surfaces, see Appendix B, and for the general construction of line bundles on an arbitrary Riemann surface, see Appendix C.

Different realizations of spinor fields on a given Riemann surface Σ with given Chern class correspond to different spin bundles that are distinguished by their *spin structure* and are labeled by the first homology group of Σ with $\mathbb{Z}_2$ coefficients, namely, $H_1(\Sigma, \mathbb{Z}_2)$. For example, on a compact Riemann surface of genus g, this homology group is isomorphic to $\mathbb{Z}_2^{2g}$ and the different spin structures are labeled by an element of $\mathbb{Z}_2$ on each one of the $2g$ independent homology cycles.

5.7.1 Spinor fields on the annulus

The annulus $\Sigma = \mathbb{C}/\mathbb{Z}$ has a single homology generator, and thus two different spin structures, referred to in physics as the Neveu–Schwarz (NS) and Ramond (R) sectors. Denoting a holomorphic spinor field of conformal weight $(h, 0)$ by ψ, where $h = n + \frac{1}{2}$, its Laurent series on the cylinder and on the annulus, with coordinates $z' = x + iy$ and $z = e^{2\pi i z'}$, respectively, are given as follows for the NS field,

$$\psi'(z') = \sum_{r \in \mathbb{Z} + \frac{1}{2}} \psi_r \, e^{-2\pi i r(x+iy)} \qquad \psi(z) = \sum_{r \in \mathbb{Z} + \frac{1}{2}} \psi_r \, z^{-r-h} \quad (5.7.1)$$

and for the R field,

$$\psi'(z') = \sum_{n \in \mathbb{Z}} \psi_n \, e^{-2\pi i n(x+iy)} \qquad \psi(z) = \sum_{n \in \mathbb{Z}} \psi_n \, z^{-n-h} \quad (5.7.2)$$

A free spinor field on a Riemann surface Σ is governed by a first-order differential operator, namely, by a generalized bc system in which the fields b and c are allowed either NS or R spin structures. An important special case is that of a spinor field of conformal weight $(\frac{1}{2}, 0)$ for which the fields b and c have the same conformal weight $(\frac{1}{2}, 0)$ and are quantized using anticommutation relations as in the bc system we discussed earlier.[3] In this case, the NS and R spin structures make the field $\psi(z)$ single-valued and double-valued on the annulus, respectively.

5.7.2 Spinor fields on the torus

The torus $\Sigma = \mathbb{C}/(\mathbb{Z} \oplus \mathbb{Z}\tau)$ has two homology generators $\mathfrak{A}$ and $\mathfrak{B}$, given by the identifications $z \to z + 1$ and $z \to z + \tau$, respectively, and thus four

---

[3] Another important case is the superghost $\beta\gamma$ system, where $b = \beta$ has weight $(\frac{3}{2}, 0)$ and $c = \gamma$ has weight $(-\frac{1}{2}, 0)$ and the oscillators obey commutation relations.

different spin structures. The four spin structures are distinguished according to whether the field is periodic or antiperiodic around each cycle. The signs are inherited from the cylinder and may be parametrized as follows,

$$\psi(z + 1) = -e^{2\pi i \alpha} \psi(z)$$
$$\psi(z + \tau) = -e^{2\pi i \beta} \psi(z) \qquad (5.7.3)$$

for $\alpha, \beta \in \{0, \frac{1}{2}\}$ (mod 1) with the values 0 and $\frac{1}{2}$ corresponding to NS and R, respectively. Under the modular transformations S and T the four spin structures transform as follows,

$$S : \begin{bmatrix} \alpha \\ \beta \end{bmatrix} \to \begin{bmatrix} \beta \\ \alpha \end{bmatrix} \qquad\qquad T : \begin{bmatrix} \alpha \\ \beta \end{bmatrix} \to \begin{bmatrix} \alpha \\ \alpha + \beta + \frac{1}{2} \end{bmatrix} \qquad (5.7.4)$$

where the addition $\alpha + \beta + \frac{1}{2}$ is evaluated mod 1. The spin structure $\alpha = \beta = \frac{1}{2}$ is referred to as the *odd spin structure*, while the three other spin structures are referred to as *even spin structures*. Under modular transformations, the odd spin structure maps to itself, while the three even spin structures are permuted among one another. This transformation rule precisely corresponds to the transformation rule for Jacobi ϑ-functions with half-integer characteristics given in (3.7.5). This is no accident, as will be further clarified below. Furthermore, in view of the existence of the holomorphic and nowhere vanishing differential dz, spinor fields on the torus for a given spin structure but different conformal weights $(h, 0)$ are all isomorphic to one another. Therefore, we may focus attention on the case $h = \frac{1}{2}$. For two functions f_1 and f_2 with the same spin structure (or more accurately for two sections of a given spin bundle S), we define a Hermitian inner product,

$$\langle f_1 | f_2 \rangle = \int_\Sigma d\mu_g \bar{f}_1 f_2 \qquad (5.7.5)$$

which has a positive definite norm and produces a Hilbert space $L^2(\Sigma, S)$.

5.8 Spinor Green functions on the torus

In this section, we shall derive the Green function for the spinor field on a torus $\Sigma = \mathbb{C}/\Lambda$ with $\Lambda = \mathbb{Z} \oplus \mathbb{Z}\tau$ for each one of the four spin structures. Just as we did for the scalar field, we offer two perspectives: a calculation via Fourier analysis and a calculation directly in terms of ϑ-functions.

5.8.1 Odd spin structure

For the odd spin structure $\alpha = \beta = \frac{1}{2}$, the Green function may be obtained from the results of the scalar case, because the spinor field is doubly periodic and has the constant spinor as a zero mode. We define the Green function $S_1(z|\tau)$ that satisfies the following equation,

$$\partial_{\bar{z}} S_1(z - w|\tau) = \pi \delta^{(2)}(z - w) - \frac{\pi}{\tau_2} \qquad (5.8.1)$$

by analogy with the equation satisfied by the scalar Green function in (5.5.14). The solution is given in terms of the Green function $G(z - w|\tau)$ by,

$$S_1(z - w|\tau) = -\partial_z G(z - w|\tau)$$
$$= \partial_z \ln \vartheta_1(z - w|\tau) + \frac{2\pi i}{\tau_2} \operatorname{Im}(z - w) \qquad (5.8.2)$$

up to an additive constant that amounts to adding a term proportional to the Dirac zero mode. Since $S_1(z - w|\tau)$ has a single simple pole in z at w and is doubly periodic, it cannot be meromorphic. The second term on the right is present to compensate. Alternatively, a term-by-term differentiation of the Kronecker–Eisenstein series of $G(z - w|\tau)$ in (5.5.11) gives the following expression for S_1,

$$S_1(z - w|\tau) = \sum_{(m,n) \neq (0,0)} \frac{e^{2\pi i(nx - my)}}{m + n\tau} \qquad (5.8.3)$$

The sum is only conditionally convergent and requires regularization. An alternative Green function $\tilde{S}_1(z; w_1, w_2|\tau)$ allows for two poles in z, namely, at w_1 and w_2 with opposite residue. Requiring antisymmetry in w_1, w_2 produces a unique Green function,

$$\tilde{S}_1(z; w_1, w_2|\tau) = S_1(z - w_1|\tau) + S_1(w_1 - w_2|\tau) + S_1(w_2 - z|\tau)$$
$$= \zeta(z - w_1|\tau) + \zeta(w_1 - w_2|\tau) + \zeta(w_2 - z|\tau) \qquad (5.8.4)$$

where the expression is recast in terms of the Weierstrass ζ-function on the second line above.

5.8.2 Even spin structure via Fourier analysis

The chiral Dirac operator $\partial_{\bar{z}}$ is not self-adjoint. However, it is a linear combination with constant complex-valued coefficients of two self-adjoint infinitesimal translation operators $i\partial_x$ and $i\partial_y$ of the torus,

$$\partial_{\bar{z}} = \frac{1}{2\tau_2}(i\partial_y - \tau\, i\partial_x) \qquad (5.8.5)$$

Since the operators $i\partial_x$ and $i\partial_y$ mutually commute, the operator $\partial_{\bar{z}}$ may be diagonalized in the basis in which $i\partial_x$ and $i\partial_y$ are diagonal. However, since the coefficients of the linear combination in (5.8.5) are complex, the eigenvalues of $\partial_{\bar{z}}$ will be complex.

For a given even spin structure $[\kappa] = [{\alpha \atop \beta}]$, the operators $i\partial_x$ and $i\partial_y$ are diagonalized in the following basis,

$$f[\kappa]_{m,n}(x,y) = \exp\left\{2\pi i((n+\beta+\tfrac{1}{2})x - (m+\alpha+\tfrac{1}{2})y)\right\} \qquad (5.8.6)$$

The eigenvalues of the operator $\partial_{\bar{z}}$ are given by,

$$\lambda[\kappa]_{m,n} = \frac{\pi}{\tau_2}\left(m+\alpha+\tfrac{1}{2} + (n+\beta+\tfrac{1}{2})\tau\right) \qquad (5.8.7)$$

Since $\mathrm{Im}\,(\tau) > 0$ and α, β do not both equal $\tfrac{1}{2}$ mod 1, none of the eigenvalues vanishes. The Green function for the operator $\tau_2\partial_{\bar{z}}$ is then given by,

$$S[\kappa](z|\tau) = \frac{1}{\tau_2} \sum_{m,n\in\mathbb{Z}} \frac{f[\kappa]_{m,n}(x,y)}{\lambda[\kappa]_{m,n}}$$

$$= \sum_{m,n\in\mathbb{Z}} \frac{e^{2\pi i\{(n+\beta+\frac{1}{2})x - (m+\alpha+\frac{1}{2})y\}}}{m+\alpha+\tfrac{1}{2} + (n+\beta+\tfrac{1}{2})\tau} \qquad (5.8.8)$$

so that $S[\kappa]$ satisfies,

$$\partial_{\bar{z}}S[\kappa](z-w|\tau) = \pi\delta(z-w) \qquad (5.8.9)$$

where the integral of $\delta(z-w)$ over Σ is normalized to τ_2.

5.8.3 Even spin structure via ϑ-functions

We seek a meromorphic function $S[\kappa](z-w|\tau)$, which has a single simple pole in a fundamental domain for the lattice $\Lambda = \mathbb{Z} \oplus \mathbb{Z}\tau$, is doubly periodic with periods in 2Λ, and behaves as follows under translations in $\Lambda/(2\Lambda)$,

$$\begin{aligned}S[\kappa](z+1|\tau) &= -e^{2\pi i\alpha}\,S[\kappa](z|\tau) & [\kappa] = [{\alpha \atop \beta}] \\ S[\kappa](z+\tau|\tau) &= -e^{2\pi i\beta}\,S[\kappa](z|\tau)\end{aligned} \qquad (5.8.10)$$

For even spin structures κ, the simple pole in $\mathbb{C}/\Lambda$ and the above translations are realized uniquely by the ratio $\vartheta[\kappa](z|\tau)/\vartheta_1(z|\tau)$, and the normalization of the residue at $z = w$ gives,

$$S[\kappa](z-w|\tau) = \frac{\vartheta[\kappa](z-w|\tau)\,\vartheta_1'(0|\tau)}{\vartheta[\kappa](0|\tau)\,\vartheta_1(z-w|\tau)} \qquad (5.8.11)$$

The Green function $S[\kappa](z - w|\tau)$ is referred to as the *Szegö kernel* on the torus for even spin structure κ. As an elliptic function in $\mathbb{C}/(2\Lambda)$, the Szegö kernel $S[\kappa](z - w|\tau)$ has four simple poles, namely, at $z \equiv 0, \tau + 1$ with residue 1, and at $z \equiv 1, \tau$ with residue -1.

5.9 Spinor determinant and Kronecker's second limit formula

The eigenvalues of the chiral Dirac operator $\partial_{\bar{z}}$, given in (5.8.7), are complex, rendering the ζ-function approach to evaluating its determinant inapplicable. Instead we shall make use of the fact that the dependence of the operator $\tau_2 \partial_{\bar{z}}$ on the modulus τ is holomorphic, and we shall seek to construct an object $\mathrm{Det}(\partial_{\bar{z}})$ that is holomorphic in τ.

What we do know how to evaluate using ζ-function methods is the determinant of the (positive) Laplace operator $\hat{\Delta}_\kappa = -\tau_2/\pi\, \partial_z \partial_{\bar{z}}$ on a spinor with even spin structure κ. The Laplace operator is self-adjoint, has positive spectrum, and no zero modes,[4]

$$\lambda[\kappa]_{m,n} = \frac{\pi}{\tau_2}\left|m + \alpha + (n + \beta)\tau\right|^2 \qquad \kappa = \begin{bmatrix} \frac{1}{2} + \alpha \\ \frac{1}{2} + \beta \end{bmatrix} \qquad (5.9.1)$$

In fact we may generalize the problem by taking the characteristics α, β to be arbitrary with the sole restriction that $(\alpha, \beta) \not\equiv (0, 0) \pmod 1$. The corresponding ζ-function is defined by the Kronecker–Eisenstein sum,

$$\zeta_{\hat{\Delta}_\kappa}(\tau, s) = \sum_{m,n\in\mathbb{Z}} \frac{\tau_2^s}{\pi^s |m + \alpha + (n + \beta)\tau|^{2s}} \qquad (5.9.2)$$

which is absolutely convergent for $\mathrm{Re}\,(s) > 1$. We shall show below that $\zeta_{\hat{\Delta}_\kappa}(\tau, s)$ may be analytically continued in s throughout the complex plane. The determinant is obtained from this analytic continuation as usual,

$$\ln \mathrm{Det}\hat{\Delta}_\kappa = -\zeta'_{\hat{\Delta}_\kappa}(\tau, 0) \qquad (5.9.3)$$

We recast the Kronecker–Eisenstein sum in terms of the following heat-kernel representation,

$$\Gamma(s)\zeta_{\hat{\Delta}_\kappa}(\tau, s) = \int_0^\infty \frac{dt}{t} t^s \sum_{m,n\in\mathbb{Z}} \exp\left\{-\frac{\pi t}{\tau_2}|m + \alpha + (n + \beta)\tau|^2\right\} \qquad (5.9.4)$$

and Poisson resum either in the summation variable m if $\beta \neq 0$ or in n if

---

[4] To save unnecessarily lengthy formulas, it is convenient to parametrize the spin structure κ in terms of shifted characteristics α, β.

$\alpha \neq 0$ (but not in both). We shall work out the case $\beta \neq 0$, the other case being analogous, using the formula,

$$\sum_{m,n \in \mathbb{Z}} e^{-\pi t |m+\alpha+(n+\beta)\tau|^2/\tau_2}$$

$$= \sqrt{\frac{\tau_2}{t}} \sum_{m,n \in \mathbb{Z}} e^{2\pi i m(\alpha+(n+\beta)\tau_1)} e^{-\pi m^2 \tau_2/t - \pi(n+\beta)^2 t \tau_2} \qquad (5.9.5)$$

Substituting this result into (5.9.4) and splitting the sum over m, n into contributions for $m \neq 0$ and $m = 0$, we have.

$$\Gamma(s)\zeta_{\hat{\Delta}_\kappa}(\tau, s) = A_s(\tau) + B_s(\tau) \qquad (5.9.6)$$

with,

$$A_s(\tau) = \sqrt{\tau_2} \sum_{m \neq 0} \sum_n e^{2\pi i m(\alpha+(n+\beta)\tau_1)} \int_0^\infty \frac{dt}{t} t^{s-\frac{1}{2}} e^{-\pi \tau_2 (m^2/t + (n+\beta)^2 t)}$$

$$B_s(\tau) = \sqrt{\tau_2} \int_0^\infty \frac{dt}{t} t^{s-\frac{1}{2}} \sum_n e^{-\pi(n+\beta)^2 t \tau_2} \qquad (5.9.7)$$

The integral $A_s(\tau)$ is manifestly holomorphic in s throughout $\mathbb{C}$ and may be evaluated by performing the t-integral in terms of a Bessel function,

$$A_s(\tau) = 2\sqrt{\tau_2} \sum_{m \neq 0} \sum_n \left| \frac{m}{n+\beta} \right|^{s-\frac{1}{2}} e^{2\pi i m(\alpha+(n+\beta)\tau_1)}$$

$$\times K_{s-\frac{1}{2}} \left(2\pi \tau_2 |m(n+\beta)| \right) \qquad (5.9.8)$$

To evaluate the determinant, we only need $A_0(\tau)$, which may be read off from the above formula by setting $s = 0$ and using the expression for the spherical Bessel $K_{-\frac{1}{2}}(x) = \sqrt{\pi/2x}\, e^{-x}$,

$$A_0(\tau) = \sum_{m=1}^\infty \sum_n \frac{1}{m} e^{-2\pi \tau_2 m |n+\beta|} \left(e^{2\pi i m(\alpha+(n+\beta)\tau_1)} + \text{c.c.} \right) \qquad (5.9.9)$$

Without loss of generality, we take $0 < \beta < 1$ to carry out the sum,

$$A_0(\tau) = -\sum_{n=0}^\infty \ln \left| 1 - q^{n+\beta} e^{2\pi i \alpha} \right|^2 - \sum_{n=1}^\infty \ln \left| 1 - q^{n-\beta} e^{-2\pi i \alpha} \right|^2 \qquad (5.9.10)$$

Finally, we compute $B_s(\tau)$ by performing the integrals under the sum,

$$B_s(\tau) = \Gamma(s-\tfrac{1}{2})\pi^{\frac{1}{2}-s}\tau_2^{1-s} \sum_n \frac{1}{|n+\beta|^{2s-1}} \qquad (5.9.11)$$

The sum is absolutely convergent for $\text{Re}\,(s) > 1$ and may be evaluated in terms of Hurwitz ζ-functions. The analytic continuation of the Hurwitz ζ-function is well known, and the final result is given by,

$$\zeta'_{\hat{\Delta}_\kappa}(\tau,0) = -\ln\left|\frac{\vartheta\left[\begin{smallmatrix}\alpha\\\beta\end{smallmatrix}\right](0|\tau)}{\eta(\tau)}\right|^2 \tag{5.9.12}$$

This result is a version of the *second Kronecker limit formula*.

The determinant obtained above is for the bc system in which both b and c are conformal spinor fields of weight $(\frac{1}{2},0)$, along with their complex conjugate fields. The natural candidate for the determinant of the chiral operator $\partial_{\bar{z}}$ is obtained by retaining only the part with holomorphic dependence on τ,

$$\mathrm{Det}\,\partial_{\bar{z}}\Big|_{bc} = \frac{\vartheta\left[\begin{smallmatrix}\alpha\\\beta\end{smallmatrix}\right](0|\tau)}{\eta(\tau)} \tag{5.9.13}$$

The spinor fields that are required in string theory for the worldsheet matter fermions are further restricted to $\psi = b = c$, so that the b and c fields are actually identical fields. The functional chiral determinant of the operator $\partial_{\bar{z}}$ in this case is the square root of the determinant of the bc system,

$$\mathrm{Det}\,\partial_{\bar{z}}\Big|_{\psi} = \frac{\vartheta\left[\begin{smallmatrix}\alpha\\\beta\end{smallmatrix}\right](0|\tau)^{\frac{1}{2}}}{\eta(\tau)^{\frac{1}{2}}} \tag{5.9.14}$$

Neither determinant is modular invariant or even covariant. Physicists say that there is an anomaly, which manifests the incompatibility between maintaining holomorphicity and modular invariance. We may try a combination similar to the one we established in the non-chiral case,

$$\sum_{i=2}^{4}\frac{\vartheta_i(0|\tau)^{\frac{N}{2}}}{\eta(\tau)^{\frac{N}{2}}} = \frac{\vartheta_2(0|\tau)^{\frac{N}{2}}}{\eta(\tau)^{\frac{N}{2}}} + \frac{\vartheta_3(0|\tau)^{\frac{N}{2}}}{\eta(\tau)^{\frac{N}{2}}} + \frac{\vartheta_4(0|\tau)^{\frac{N}{2}}}{\eta(\tau)^{\frac{N}{2}}} \tag{5.9.15}$$

The smallest value of N for which we have good modular transformations is $N = 16$, and more generally any multiple of 16. In those cases, the sum of the numerators transforms as a modular form of weight $N/4$. The first modular invariant is encountered as $N = 24$ for 48 chiral fermions.

• Bibliographical notes

A systematic introduction to quantum field theory and conformal field theory in two dimensions is presented in the book by Di Francesco, Mathieu, and Sénéchal [DFMS97], which also has an extensive bibliography. The lecture notes in [Gin88] provide an excellent introduction, while those in [Gaw99] offer a more mathematical perspective.

The development of two-dimensional conformal field theory and the use of radial quantization date back to [FHJ73]. The modern approach, based on the representation theory of the Virasoro algebra, was developed in [BPZ84], where minimal models were also introduced. The bc system was discussed in the context of string perturbation theory in [FMS86]. Functional determinants of the Laplace and Cauchy–Riemann operators on the torus are evaluated there and in the classic paper by Polchinski [Pol86]. The Kronecker limit formulas are discussed in the lecture notes by Siegel [Sie61], as well as in the book by Iwaniec [Iwa02].

On higher genus Riemann surfaces, functional determinants of Laplace operators were considered in [RS71] in connection with Ray–Singer torsion, in [Hej83] and [Fri86] in connection with Reidemeister torsion, and were evaluated in terms of the Selberg zeta function in [DP86b, Sar87]. An overview is presented in the book by Jorgensen and Lang [JL93]. Functional determinants of Cauchy–Riemann operators were evaluated in terms of Riemann ϑ-functions via chiral bosonization in [VV87] and [AGBM$^+$87] by building on the properties of the holomorphic anomaly [BK86] and [BJ86].

Exercises

5.1 Using the definition of the operator $j(z) = j^{(1)}(z)$ in (5.2.12) for the bc system on a torus of modulus τ, determine the singularities in z of the following correlators of $j(z)$ and c and b fields,

$$A(z; x, y) = \langle 0|j(z)c(x_1) \cdots c(x_n)b(y_1) \cdots b(y_n)|0\rangle \qquad (5.9.16)$$

in terms of $B(x, y) = \langle 0|c(x_1) \cdots c(x_n)b(y_1) \cdots b(y_n)|0\rangle$. Does this result completely determine $A(z; x, y)$ in terms of $B(x, y)$? Explain.

5.2 Compute the transformation law of the operator $j(z)$, discussed in Exercise 5.1, under local conformal transformations $z \to z' = f(z)$ for which $f(z)$ is locally holomorphic.

5.3 Prove formula (5.5.16) starting from (5.5.15).

5.4 Compute the correlator $\langle \partial_{z_1}\varphi(z_1)\partial_{z_2}\varphi(z_2)\partial_{z_3}\varphi(z_3)\partial_{z_4}\varphi(z_4)\rangle$ for a real massless scalar field in terms of the Green function G.

5.5 Evaluate the electric potential $\Phi(z)$ for a distribution of electric charges q_n at points z_n with $n = 1, \cdots, N$ on a parallelogram with vertices $0, 1, \tau, 1 + \tau$ and $\mathrm{Im}\,\tau > 0$ in the following three cases: (a) with doubly periodic boundary conditions so that we have a torus with modulus τ; (b) with singly periodic boundary conditions under $z \to z + 1$ and grounded at $\mathrm{Im}\,z = 0, \tau_2$; (c) grounded at all four edges.

6

Congruence subgroups and modular curves

Congruence subgroups, which form a countable infinite class of discrete non-Abelian subgroups of $\mathrm{SL}(2,\mathbb{Z})$, play a particularly prominent role in deriving the arithmetic properties of modular forms. In this chapter, we study various aspects of congruence subgroups, including their elliptic points, cusps, and topological properties of the associated modular curve. Jacobi ϑ-functions, ϑ-constants, and the Dedekind η-function are used as examples of modular forms under congruence subgroups which are not modular forms under the full modular group $\mathrm{SL}(2,\mathbb{Z})$.

6.1 Definition of congruence subgroups of $\mathrm{SL}(2,\mathbb{Z})$

The *principal congruence subgroup* of level $N \geq 1$ is defined as follows,

$$\Gamma(N) = \left\{ \begin{pmatrix} a & b \\ c & d \end{pmatrix}, \quad a, d \equiv 1 \ (\mathrm{mod}\, N), \quad b, c \equiv 0 \ (\mathrm{mod}\, N) \right\} \quad (6.1.1)$$

Clearly, $\Gamma(1) = \mathrm{SL}(2,\mathbb{Z})$ and $\Gamma(N)$ is a subgroup of $\mathrm{SL}(2,\mathbb{Z})$ for any N. Actually, $\Gamma(N)$ is a normal subgroup of $\mathrm{SL}(2,\mathbb{Z})$ so that the quotient is itself a group. As shown in Exercise 6.1, the quotient group is isomorphic to $\mathrm{SL}(2,\mathbb{Z}_N)$ (to physicists, isomorphism is synonymous with equality),

$$\mathrm{SL}(2,\mathbb{Z})/\Gamma(N) = \mathrm{SL}(2,\mathbb{Z}_N) \quad (6.1.2)$$

namely the group of 2×2 matrices whose entries are integers mod N and whose determinant is 1 mod N. Equivalently, the group $\Gamma(N)$ may be identified with the kernel of the homomorphism $\mathbb{Z} \to \mathbb{Z}_N$ of reduction mod N. The index $[\mathrm{SL}(2,\mathbb{Z}) : \Gamma(N)]$ of $\Gamma(N)$ equals the order of $\mathrm{SL}(2,\mathbb{Z}_N)$, which is finite. It will be determined below shortly.

By definition, a group Γ is a *congruence subgroup of $SL(2,\mathbb{Z})$ of level N* provided there exists an integer $N \geq 1$ such that $\Gamma(N) \subset \Gamma \subset \mathrm{SL}(2,\mathbb{Z})$.

Since the index of $\Gamma(N)$ is finite, it follows that the index $[\mathrm{SL}(2,\mathbb{Z}):\Gamma]$ is finite for any congruence subgroup Γ. Note that Γ need not be a normal subgroup of $\mathrm{SL}(2,\mathbb{Z})$ so that $\mathrm{SL}(2,\mathbb{Z})/\Gamma$ need not be a group.

6.2 The classic congruence subgroups

Besides the principal congruence subgroups $\Gamma(N)$, two other fundamental congruence subgroups of level N are defined as follows,

$$\Gamma_1(N) = \left\{ \begin{pmatrix} a & b \\ c & d \end{pmatrix}, \quad a,d \equiv 1 \ (\mathrm{mod}\,N), \quad c \equiv 0 \ (\mathrm{mod}\,N) \right\}$$

$$\Gamma_0(N) = \left\{ \begin{pmatrix} a & b \\ c & d \end{pmatrix}, \quad c \equiv 0 \ (\mathrm{mod}\,N) \right\} \tag{6.2.1}$$

with the following manifest subgroup inclusions,

$$\Gamma(N) \subset \Gamma_1(N) \subset \Gamma_0(N) \subset \mathrm{SL}(2,\mathbb{Z}) \tag{6.2.2}$$

For $N \geq 2$, none of the groups $\Gamma(N), \Gamma_1(N), \Gamma_0(N)$ contains the generator S of $\mathrm{SL}(2,\mathbb{Z})$ and, while $\Gamma_1(N)$ and $\Gamma_0(N)$ contain the translation generator T, the principal congruence subgroup $\Gamma(N)$ contains T^N but not T. The groups of transposes are,

$$\Gamma^1(N) = \{\gamma | \gamma^t \in \Gamma_1(N)\} \qquad \Gamma^0(N) = \{\gamma | \gamma^t \in \Gamma_0(N)\} \tag{6.2.3}$$

The following congruence subgroup,

$$\Gamma_0(M,N) = \left\{ \begin{pmatrix} a & b \\ c & d \end{pmatrix}, \quad b \equiv 0 \ (\mathrm{mod}\,M), \quad c \equiv 0 \ (\mathrm{mod}\,N) \right\} \tag{6.2.4}$$

interpolates between $\Gamma_0(N) = \Gamma_0(1,N)$ and $\Gamma^0(M) = \Gamma_0(M,1)$.

Recall that $\Gamma(N)$ is a normal subgroup of $\mathrm{SL}(2,\mathbb{Z})$ induced by the homomorphism $\mathbb{Z} \to \mathbb{Z}_N$ on all its entries. Similarly, $\Gamma(N)$ is a normal subgroup of $\Gamma_1(N)$ induced by $\mathbb{Z} \to \mathbb{Z}_N$ on the entry b. The quotient $\Gamma_1(N)/\Gamma(N)$ is isomorphic to $\mathbb{Z}_N$. Finally, $\Gamma_1(N)$ is a normal subgroup of $\Gamma_0(N)$. The quotient $\Gamma_0(N)/\Gamma_1(N)$ is isomorphic to the group $\mathbb{Z}_N^*$ of multiplicatively invertible elements in $\mathbb{Z}_N$. In summary then, we have the following quotients,

$$\mathrm{SL}(2,\mathbb{Z})/\Gamma(N) = \mathrm{SL}(2,\mathbb{Z}_N)$$
$$\Gamma_1(N)/\Gamma(N) = \mathbb{Z}_N$$
$$\Gamma_0(N)/\Gamma_1(N) = \mathbb{Z}_N^* \tag{6.2.5}$$

The order of $\mathrm{SL}(2,\mathbb{Z}_N)$ is denoted $2d_N$ and will be evaluated in Section 6.3. The order of $\mathbb{Z}_N$ is N. The order of $\mathbb{Z}_N^*$ equals the Euler totient function

$\phi(N)$, which counts the number of integers in $\mathbb{Z}_N$ that are relatively prime to N, and is given by the following formula (see also Appendix A),

$$|\mathbb{Z}_N^*| = \phi(N) = N \prod_i \left(1 - \frac{1}{p_i}\right) \qquad\qquad N = \prod_i p_i^{e_i} \qquad (6.2.6)$$

where the product is over distinct primes p_i, and $e_i \geq 1$ for all i. For the special case where $N = p$ is prime, we have $\phi(p) = p - 1$. The indices of these subgroups are related to the orders of the quotient groups as follows,

$$[\mathrm{SL}(2, \mathbb{Z}) : \Gamma(N)] = |\mathrm{SL}(2, \mathbb{Z}_N)| = 2d_N$$
$$[\Gamma_1(N) : \Gamma(N)] = |\mathbb{Z}_N| = N$$
$$[\Gamma_0(N) : \Gamma_1(N)] = |\mathbb{Z}_N^*| = \phi(N) \qquad (6.2.7)$$

We obtain the remaining indices,

$$[\mathrm{SL}(2, \mathbb{Z}) : \Gamma_0(N)] = 2d_N/(N\phi(N))$$
$$[\mathrm{SL}(2, \mathbb{Z}) : \Gamma_1(N)] = 2d_N/N \qquad (6.2.8)$$

using the following multiplicative property for groups $K \subset H \subset G$,

$$[G : H] \times [H : K] = [G : K] \qquad (6.2.9)$$

6.3 Computing the order of SL(2, $\mathbb{Z}_N$)

We shall now evaluate the order of SL(2, $\mathbb{Z}_N$), closely following [DS05]. The order will enter crucially in the valence formulas for congruence subgroups. To compute it, we shall make use of the Chinese remainder theorem (reviewed and proven in Appendix A), which provides a fundamental isomorphism for $\mathbb{Z}_N$ in terms of its prime decomposition factors.

Let N have the following prime number decomposition $N = \prod_i p_i^{e_i}$, where the p_i are distinct primes and $e_i \geq 1$ are positive integer exponents. Then by the Chinese remainder theorem, we have the following isomorphism,

$$\mathrm{SL}(2, \mathbb{Z}_N) = \prod_i \mathrm{SL}(2, \mathbb{Z}_{p_i^{e_i}}) \qquad (6.3.1)$$

Thus, the order $|\mathrm{SL}(2, \mathbb{Z}_N)|$ of the group SL(2, $\mathbb{Z}_N$) is given by,

$$|\mathrm{SL}(2, \mathbb{Z}_N)| = \prod_i |\mathrm{SL}(2, \mathbb{Z}_{p_i^{e_i}})| \qquad (6.3.2)$$

To compute the order of each component SL(2, $\mathbb{Z}_{p^e}$), we proceed by induction on e. We begin by computing the order of $\mathrm{GL}(2, \mathbb{Z}_p)$ for p prime. To do so, consider the set of all matrices γ,

$$\gamma = \begin{pmatrix} a & b \\ c & d \end{pmatrix} \qquad (6.3.3)$$

whose entries $a, b, c, d \in \mathbb{Z}_p$ are parametrized by $a, b, c, d \in \{0, 1, \cdots, p-1\}$. To belong to $\mathrm{GL}(2, \mathbb{Z}_p)$, the entries must satisfy $ad - bc \neq 0 \pmod{p}$. To enumerate all its elements, we proceed by listing the number of elements in the four subsets of possible values of the doublet (a, b). No matrices with $a = b = 0$ belong to $\mathrm{GL}(2, \mathbb{Z}_p)$; there are $p-1$ doublets with $a \neq 0, b = 0$ and for each nonzero value of a the $p-1$ nonzero values of d and the p arbitrary values c give $(p-1)^2 p$ elements $\gamma \in \mathrm{GL}(2, \mathbb{Z}_p)$. For the case $a = 0, b \neq 0$, the roles of c and d are reversed, and we obtain another $(p-1)^2 p$ elements. Finally, for each one of the $(p-1)^2$ doublets with $a, b \neq 0$, the entry c may take all p values, while d may take all p values except the one that gives zero determinant, producing $(p-1)^3 p$ contributions. Putting all together,

$$|\mathrm{GL}(2, \mathbb{Z}_p)| = (p-1)^3 p + 2(p-1)^2 p = (p-1)p^3 \left(1 - \frac{1}{p^2}\right) \quad (6.3.4)$$

Any $\gamma \in \mathrm{GL}(2, \mathbb{Z}_p)$ satisfies $\det \gamma \neq 0$, and taking the quotient by the determinant normal subgroup isomorphic to the multiplicative group $(\mathbb{Z}_p)^*$ with $p-1$ elements, we obtain,

$$|\mathrm{SL}(2, \mathbb{Z}_p)| = p^3 \left(1 - \frac{1}{p^2}\right) \qquad (6.3.5)$$

To proceed by induction on the exponent e, we consider an arbitrary matrix $\gamma' \pmod{p^{e+1}}$ and parametrize it uniquely $\pmod{p^e}$,

$$\gamma' = \gamma + p^e \begin{pmatrix} a' & b' \\ c' & d' \end{pmatrix} \qquad\qquad \gamma = \begin{pmatrix} a & b \\ c & d \end{pmatrix} \qquad (6.3.6)$$

where $a, b, c, d \in \mathbb{Z}_{p^e}$ and $a', b', c', d' \in \mathbb{Z}_p$. Requiring $\det \gamma \equiv 1 \pmod{p^{e+1}}$ requires $ad - bc \equiv 1 \pmod{p^e}$ so that $\gamma \in \mathrm{SL}(2, \mathbb{Z}_{p^e})$. Now for an arbitrary matrix $\gamma \in \mathrm{SL}(2, \mathbb{Z}_{p^e})$, we obtain $\gamma' \in \mathrm{SL}(2, \mathbb{Z}_{p^{e+1}})$ if and only if,

$$ad' + da' - bc' - cb' \equiv 0 \pmod{p} \qquad (6.3.7)$$

At least two entries in γ are nonzero $\pmod{p}$ since $ad - bc \equiv 1 \pmod{p}$. If $a \neq 0$, we let a', b', and c' take independent arbitrary values in $\mathbb{Z}_p$ and determine the number d' uniquely by solving equation (6.3.7). If $a = 0$, then we must have $b \not\equiv 0 \pmod{p}$, so we take the values of a', b', and d' arbitrary and solve uniquely for c' using (6.3.7). In each case, to every $\gamma \in \mathrm{SL}(2, \mathbb{Z}_{p^e})$ there correspond p^3 elements $\gamma' \in \mathrm{SL}(2, \mathbb{Z}_{p^{e+1}})$. Hence we have,

$$|\mathrm{SL}(2, \mathbb{Z}_{p^e})| = p^{3e}\left(1 - \frac{1}{p^2}\right) \qquad (6.3.8)$$

Thus, for any $N \geq 2$ with prime number decomposition $N = \prod_i p_i^{e_i}$, we have,

$$2d_N = |\mathrm{SL}(2, \mathbb{Z}_N)| = N^3 \prod_i \left(1 - \frac{1}{p_i^2}\right) \qquad (6.3.9)$$

The order of the simplest nontrivial group $\mathrm{SL}(2, \mathbb{Z}_2)$ is 6 by the above formula. In fact, we can write its generators in terms of the generators S and T of $\mathrm{SL}(2, \mathbb{Z})$, and we find,

$$\mathrm{SL}(2, \mathbb{Z}_2) = \{I,\, S,\, T,\, ST,\, TS,\, TST\} \pmod{2} \qquad (6.3.10)$$

Opposite elements are not included in the set since $I \equiv -I \pmod{2}$.

6.4 Modular curves

Recall that the quotient $\mathrm{SL}(2, \mathbb{Z})\backslash\mathcal{H}$ of the upper half plane $\mathcal{H}$ by $\mathrm{SL}(2, \mathbb{Z})$ is topologically a sphere with one puncture corresponding to the unique cusp. The quotient $\mathrm{SL}(2, \mathbb{Z})\backslash\mathcal{H}$ may be represented by a choice of fundamental domain in $\mathcal{H}$ with boundary sides identified pairwise under generators of $\mathrm{SL}(2, \mathbb{Z})$. The standard choice F of fundamental domain for $\mathrm{SL}(2, \mathbb{Z})$ was defined in (3.2.6) and depicted in Figure 3.2.

For an arbitrary congruence subgroup Γ the quotient is the *modular curve* $Y(\Gamma) = \Gamma\backslash\mathcal{H}$, which is again a Riemann surface. However, $Y(\Gamma)$ may be a sphere or a surface of genus greater than zero and may have several punctures corresponding to the several cusps of Γ.

6.4.1 Fundamental domain for a congruence subgroup Γ

The modular curve $Y(\Gamma)$ may again be represented by a choice of fundamental domain F_Γ in $\mathcal{H}$ with opposite sides identified pairwise under generators of Γ. Given the fundamental domain F for $\mathrm{SL}(2, \mathbb{Z})$, the fundamental domain F_Γ may be obtained by applying the cosets of $\mathrm{SL}(2, \mathbb{Z})/\Gamma$ to F,

$$\mathrm{SL}(2, \mathbb{Z})/\Gamma : F \to F_\Gamma \qquad (6.4.1)$$

Therefore F_Γ is the union of a number of copies of F equal to the index $[\mathrm{SL}(2, \mathbb{Z}) : \Gamma]$. The elliptic points of Γ are the images of the elliptic points i and ρ of $\mathrm{SL}(2, \mathbb{Z})$ under the cosets $\mathrm{SL}(2, \mathbb{Z})/\Gamma$ and, similarly, the cusps of Γ

are the images of the cusp $i\infty$ of F under the cosets $\mathrm{SL}(2,\mathbb{Z})/\Gamma$. Thus, the number of cusps in $Y(\Gamma)$ equals the index $[\mathrm{SL}(2,\mathbb{Z}):\Gamma]$. For a nontrivial congruence subgroup $\Gamma \neq \mathrm{SL}(2,\mathbb{Z})$, the index is strictly greater than 1, as may be verified for the index of $\Gamma = \Gamma(N), \Gamma_1(N), \Gamma_0(N)$ evaluated in (6.2.7) and (6.2.8). Since $Y(\Gamma)$ has at least one puncture, it is always noncompact.

There is a useful alternative way of looking at the cusps of a congruence subgroup Γ of $\mathrm{SL}(2,\mathbb{Z})$. Recall that the Borel subgroup Γ_∞ of $\mathrm{SL}(2,\mathbb{Z})$ is defined by $\Gamma_\infty = \{\pm T^n, n \in \mathbb{Z}\}$ and that $\mathrm{SL}(2,\mathbb{Z})/\Gamma_\infty$ acts transitively on $\mathbb{Q} \cup \{\infty\}$. This property results from the fact that every rational number $r \in \mathbb{Q}$ is the image of ∞ under a unique coset in $\mathrm{SL}(2,\mathbb{Z})/\Gamma_\infty$.

To show this, we distinguish two cases: $r = 0$ is the image of ∞ under the transformation $S \in \mathrm{SL}(2,\mathbb{Z})/\Gamma_\infty$, while for $r \neq 0$, we write $r = a/c$ uniquely with $a, c \in \mathbb{Z}^*$, $\gcd(a,c) = 1$, and $c > 0$. By Bézout's lemma there exists a particular solution (b_0, d_0) for b and d to the equation $ad - bc = 1$. The general solution, given by $b = b_0 + ka$ and $d = d_0 + kc$ for $k \in \mathbb{Z}$, is equivalent to the particular solution by right multiplication by T^k, which proves the assertion. The cusps of $Y(\Gamma)$ may now be viewed as those points in $\mathbb{Q} \cup \{\infty\}$, which are inequivalent under the action of the congruence group Γ, or equivalently,[1]

$$\{\text{cusps of } \Gamma\} = \big(\Gamma \backslash \mathrm{SL}(2,\mathbb{Z})/\Gamma_\infty\big)\{\infty\} \qquad (6.4.2)$$

The number of cusps, counted with multiplicity, is given by $[\mathrm{SL}(2,\mathbb{Z}):\Gamma]$.

6.4.2 *The example of* $\Gamma(2)$

The fundamental domain $F_{\Gamma(2)}$ of $\Gamma = \Gamma(2)$ is the union of the six images of F under the elements of $\mathrm{SL}(2,\mathbb{Z}_2) = \{I, S, T, ST, TS, TST\}$ and is depicted in Figure 6.1. The area formula (3.2.4) gives $\mathrm{area}(F_{\Gamma(2)}) = 6\,\mathrm{area}(F) = 2\pi$, as expected given the order of $\mathrm{SL}(2,\mathbb{Z}_2)$. The elliptic points i and ρ in $\mathrm{SL}(2,\mathbb{Z})$ are no longer clliptic points of $\Gamma(2)$, as the corresponding generators S and ST do not belong to $\Gamma(2)$. Their images i, i', i'', i''', and ρ, ρ' under $\mathrm{SL}(2,\mathbb{Z}_2)$, which are depicted in Figure 6.1 and in Table 6.1, and are also no longer elliptic points of $\Gamma(2)$. Thus, $\Gamma(2)$ has no elliptic points, as will be confirmed shortly in Theorem 6.7.1. The three cusps $0, 1, \infty$ of $\Gamma(2)$ each have multiplicity 2.

---

[1] Throughout, it will be convenient to use the notation $\Gamma z = \{\gamma z \text{ for all } \gamma \in \Gamma\}$ for the set of all images of the point z under the action of the group Γ.

	I	S	T	ST	TS	TST
i	i	i	i'	i'''	i'	i''
ρ	ρ	ρ'	ρ'	ρ	ρ	ρ'
∞	∞	0	∞	0	1	1

Table 6.1 The images of the points i, ρ, and ∞ under $\mathrm{SL}(2, \mathbb{Z}_2)$.

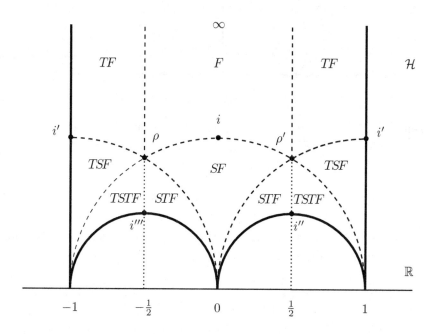

Figure 6.1 The fundamental domain $F_{\Gamma(2)}$ for $\Gamma(2)$ is the union of the images of F under the elements of $\mathrm{SL}(2, \mathbb{Z}_2) = \{I, S, T, ST, TS, TST\}$. The elliptic points i, i', i'', i''', ρ, and ρ' of $\mathrm{SL}(2, \mathbb{Z})$ are not elliptic points of $\Gamma(2)$. The cusps of $\Gamma(2)$ are $0, 1, \infty$ each with multiplicity 2.

6.5 Compactification of the modular curves

To compactify $Y(\mathrm{SL}(2, \mathbb{Z}))$, one adjoins the point $\{\infty\}$ to the fundamental domain F of $\mathrm{SL}(2, \mathbb{Z})$. To compactify $Y(\Gamma)$ for an arbitrary congruence subgroup Γ, one similarly adjoins $\{\text{cusp of } \Gamma\} = \Gamma \backslash \mathrm{SL}(2, \mathbb{Z})/\Gamma_\infty \{\infty\}$ to $Y(\Gamma)$. For an arbitrary congruence subgroup Γ, the resulting *modular curve* $X(\Gamma)$ is a compact Riemann surface.

An alternative but equivalent way to obtain the compact modular curve $X(\Gamma)$ is to start from the union $\mathcal{H}^* = \mathcal{H} \cup \{\infty\} \cup \mathbb{Q}$ of the upper half plane with the rational numbers and infinity. One chooses a topology for $\mathcal{H}^*$ whose open sets include all the standard open sets of $\mathcal{H}$ (which are generated by

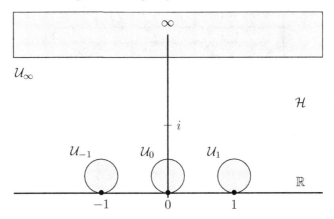

Figure 6.2 The gray rectangle $\mathcal{U}_\infty$ represents a neighborhood of the cusp ∞ of SL$(2, \mathbb{Z})$ while the gray discs $\mathcal{U}_{-1}, \mathcal{U}_0, \mathcal{U}_1$ represent the neighborhoods of the additional cusps of $\Gamma(2)$.

the open coordinate discs whose closure lies entirely in $\mathcal{H}$), as well as the open sets generated by discs that are open in $\mathcal{H}$ and intersect $\mathbb{R}$ at a single rational number. The latter provide an open neighborhood for every point in $\{\infty\} \cup \mathbb{Q}$ and are depicted in gray in Figure 6.2 for the discs at infinity $\mathcal{U}_\infty$ and the discs $\mathcal{U}_{\pm 1}$ and $\mathcal{U}_0$ intersecting the real axis at the rational points ± 1 and 0. The topological space $\mathcal{H}^*$ obtained this way is not a Haussdorff space. Recall that a topological space $\mathcal{T}$ is a Haussdorff space if for any two distinct points of $\mathcal{T}$ there exist neighborhoods of each point that are disjoint from each other.

The modular curve $X(\Gamma)$ may be obtained from $\mathcal{H}^*$ by taking the *extended quotient*, which includes the action of Γ on $\mathbb{Q} \cup \{\infty\}$,

$$X(\Gamma) = \Gamma\backslash\mathcal{H}^* \tag{6.5.1}$$

The modular curve $X(\Gamma)$ is Haussdorff even though $\mathcal{H}^*$ is not. The compact modular curves for the classic congruence subgroups are denoted as follows,

$$X(N) = X\left(\Gamma(N)\right)$$
$$X_1(N) = X\left(\Gamma_1(N)\right)$$
$$X_0(N) = X\left(\Gamma_0(N)\right) \tag{6.5.2}$$

A modular form $f(\tau)$ of weight k for a congruence subgroup $\Gamma \subset \text{SL}(2, \mathbb{Z})$ corresponds to a Γ-invariant differential form $\mathfrak{f} = f(\tau)d\tau^{\frac{k}{2}}$ on $\mathcal{H}^*$ and thus pulls back to a well-defined differential on $X(\Gamma) = \Gamma\backslash\mathcal{H}^*$. To obtain the dimensions of the spaces $\mathcal{M}_k(\Gamma)$ and $\mathcal{S}_k(\Gamma)$ of modular forms and cusp forms of weight k, we can now use the machinery of line bundles on the compact Riemann surfaces $X(\Gamma)$, reviewed in Appendix C.

6.6 The genus of $X(\Gamma)$

To derive the genus of $X(\Gamma)$, we shall view $X(\Gamma)$ as a branched covering of the sphere $X(\mathrm{SL}(2,\mathbb{Z}))$ by a nonconstant holomorphic function $f : X(\Gamma) \to X(\mathrm{SL}(2,\mathbb{Z}))$. We provide a summary of various relevant topological aspects of Riemann surfaces in Appendix B.

6.6.1 The Hurwitz formula for branched coverings

Consider a nonconstant holomorphic map $f : X \to Y$ between two compact Riemann surfaces X, Y. The degree of f is the cardinality of f^{-1} denoted $d = |f^{-1}(y)|$ for all but finitely many points $y \in Y$. A formula valid for all points in Y is obtained in terms of the *degree of ramification* e_x of f at a point $x \in X$ which, in local coordinates, is given by $f(z) - f(x) \sim (z - x)^{e_x}$. The degree of f may then be defined for all $y \in Y$ by the formula,

$$d = \sum_{x \in f^{-1}(y)} e_x \tag{6.6.1}$$

Theorem 6.6.1. *The Riemann–Hurwitz formula relates the genera $g(X)$ and $g(Y)$ of X and Y, respectively, to the degree of f and the deviations from 1 of the degree of ramification,*

$$2g(X) - 2 = d\bigl(2g(Y) - 2\bigr) + \sum_{x \in X} (e_x - 1) \tag{6.6.2}$$

A proof of the Riemann–Hurwitz formula is given in Appendix B.

6.6.2 Calculating the genus of $X(\Gamma)$

To calculate the genus of $X(\Gamma)$, we return to the map $f : X(\Gamma) \to X(\mathrm{SL}(2,\mathbb{Z}))$ and obtain an alternative formula for its degree in terms of the index of the congruence subgroup $\Gamma \subset \mathrm{SL}(2,\mathbb{Z})$,

$$d = \begin{cases} [\mathrm{SL}(2,\mathbb{Z}) : \Gamma] & \text{if } -I \in \Gamma \\ \frac{1}{2}[\mathrm{SL}(2,\mathbb{Z}) : \Gamma] & \text{if } -I \notin \Gamma \end{cases} \tag{6.6.3}$$

We shall collectively denote the elliptic points of $\mathrm{SL}(2,\mathbb{Z})$ of order $2h$ by $y_h \in \mathrm{SL}(2,\mathbb{Z})\{e^{\pi i/h}\}$ for $h = 2, 3$ and the cusps by $y_\infty \in \mathrm{SL}(2,\mathbb{Z})\{\infty\}$. The number of elliptic points of order $2h$ in $X(\Gamma)$ will be denoted by ε_h, and the number of cusps in $X(\Gamma)$ by ε_∞.

Theorem 6.6.2. *The genus* $g = g(X(\Gamma))$ *of* $X(\Gamma)$ *is given by,*

$$g = 1 + \frac{d}{12} - \frac{\varepsilon_2}{4} - \frac{\varepsilon_3}{3} - \frac{\varepsilon_\infty}{2} \qquad (6.6.4)$$

where d is given in (6.6.3), ε_h *denotes the number of elliptic points of order* $2h$, *and* ε_∞ *denotes the number of cusps in* $X(\Gamma)$.

To prove this result using the Riemann–Hurwitz formula (6.6.2), we evaluate the contributions from the points $x \in X(\Gamma)$ for which $e_x > 1$, namely, from the points $f^{-1}(y_h)$ and the cusps of $X(\Gamma)$. If $f^{-1}(y_h)$ is a ramification point of $X(\Gamma)$ then $e_{y_h} = 1$, while if it is not a ramification point of $X(\Gamma)$ then $e_{y_h} = h$. Using the degree formula (B.1.3) at the points y_h, we have,

$$\sum_{x \in f^{-1}(y_h)} e_x = \varepsilon_h + h(|f^{-1}(y_h)| - \varepsilon_h) \qquad (6.6.5)$$

where the first term arises from those points y_h that are ramification points of $X(\Gamma)$, while the second term arises from those points in $f^{-1}(y_h)$ that are not ramification points of $X(\Gamma)$, in each case weighted by their respective ramification degrees 1 and h. Using the degree formula again to eliminate the number $|f^{-1}(y_h)|$, one obtains,

$$\sum_{x \in f^{-1}(y_h)} (e_x - 1) = \frac{h-1}{h}(d - \varepsilon_h) \qquad (6.6.6)$$

The formula holds for the cusps by setting $h = \infty$. Using the Riemann–Hurwitz formula (6.6.2), we set $g(Y) = 0$ for the sphere $Y = X(\mathrm{SL}(2, \mathbb{Z}))$, $g = g(X(\Gamma))$, and substitute the contributions to the sum of $(e_x - 1)$ obtained in (6.6.6) to obtain (6.6.4), thereby completing the proof of Theorem 6.6.2.

6.7 Formulas for d, $\varepsilon_2, \varepsilon_3$, and ε_∞ for $\Gamma(N), \Gamma_1(N)$, and $\Gamma_0(N)$

In order to obtain formulas for the genus of $X(\Gamma)$, and later on the dimensions of the spaces of modular and cusp forms on $X(\Gamma)$, we need to obtain the degree d, and formulas for $\varepsilon_2, \varepsilon_3$, and ε_∞ for an arbitrary congruence subgroup Γ. But first, it will be useful to obtain these formulas for the special cases of the congruence subgroups $\Gamma(N), \Gamma_1(N)$, and $\Gamma_0(N)$. To do so, we begin by assembling all the partial results that are required.

6.7.1 The degree

The general formula for the degree in (6.6.3) may be worked out explicitly for $\Gamma(N), \Gamma_1(N)$, and $\Gamma_0(N)$ by introducing the function d_N for $N \geq 2$ by,

$$d_2 = 6 \qquad\qquad d_{N \geq 3} = \frac{1}{2}N^3 \prod_i \left(1 - \frac{1}{p_i^2}\right) \qquad (6.7.1)$$

for $N = \prod_i p_i^{e_i}$ with p_i distinct primes and $e_i \geq 1$. The other degrees may then be read off from (6.2.8),

$$d\big(\Gamma(N)\big) = d_N$$
$$d\big(\Gamma_1(N)\big) = d_N/N$$
$$d\big(\Gamma_0(N)\big) = 2d_N/(N\phi(N)) \qquad (6.7.2)$$

taking into account the fact that $-I$ is an element of the groups $\Gamma_0(N)$ for all $N \geq 2$; of $\Gamma_1(2)$ and $\Gamma(2)$; but not of $\Gamma(N)$ and $\Gamma_1(N)$ for $N \geq 3$.

6.7.2 The absence of elliptic points

Recall that the elliptic points of $\mathrm{SL}(2,\mathbb{Z})$ are the points of order four, namely, $\mathrm{SL}(2,\mathbb{Z})\{i\}$, and the points of order 6, namely, $\mathrm{SL}(2,\mathbb{Z})\{\rho = e^{2\pi i/6}\}$. In this section and Section 6.7.3, we shall count the number of elliptic points of orders 4 and 6 of the congruence subgroups $\Gamma(N), \Gamma_1(N)$, and $\Gamma_0(N)$. We begin with the following theorem.

Theorem 6.7.1. *The following congruence groups have no elliptic points,*

1. $\Gamma(N)$ for $N \geq 2$;
2. $\Gamma_1(N)$ for $N \geq 4$;
3. $\Gamma_0(N)$ for any N divisible by a prime p which obeys $p \equiv -1 \pmod{12}$.

To prove 1, we parametrize $\gamma \in \Gamma(N)$ as follows,

$$\gamma = \begin{pmatrix} a & b \\ c & d \end{pmatrix} = \begin{pmatrix} 1 + Na' & Nb' \\ Nc' & 1 + Nd' \end{pmatrix} \qquad (6.7.3)$$

For an elliptic γ, we must have $|a + d| < 2$, following the analysis of Section 3.2.3. The condition $|a+d| = |2+N(a'+d')| < 2$ allows for three possibilities, namely, $N(a'+d') = -1, -2, -3$, which are all ruled out for $N \geq 4$. For both $N = 2, 3$, the only solution is given by $a'+d' = -1$. Combining this condition with $\det \gamma = 1$, we find that we must have $N(a'd' - b'c') = 1$ which has no solutions for $N \geq 2$. To prove 2, we parametrize $\gamma \in \Gamma_1(N)$ by,

$$\gamma = \begin{pmatrix} 1 + Na' & b \\ Nc' & 1 + Nd' \end{pmatrix} \qquad (6.7.4)$$

The condition $|a + d| = |2 + N(a' + d')| < 2$ for γ to be elliptic again allows for three possibilities, namely, $N(a' + d') = -1, -2, -3$, which are all ruled out for $N \geq 4$. To prove 3, we parametrize $\gamma \in \Gamma_0(N)$ as follows,

$$\gamma = \begin{pmatrix} a & b \\ Nc' & d \end{pmatrix} \tag{6.7.5}$$

The condition $|a + d| < 2$ again leaves three possibilities: $a + d = 0, \pm 1$.

- For $a + d = 0$, we have $-a^2 - Nbc' = 1$, so that $a^2 \equiv -1 \pmod{N}$.
- For $a + d = \pm 1$, we have $-a^2 \pm a - Nbc' = 1$ so that $a^2 \pm a \equiv -1 \pmod{N}$.

Now if $p \equiv -1 \pmod{12}$ then we have $p \equiv -1 \pmod{3}$ and $p \equiv -1 \pmod{4}$. But $a^2 \equiv -1 \pmod{p}$ has no solutions if $p \equiv -1 \pmod{4}$ in view of (A.4.9), so that the case $a + d = 0$ is ruled out. Similarly, $a^2 + a \equiv -1 \pmod{p}$ has no solutions in view (A.4.12), so that also the cases $a + d = \pm 1$ are ruled out, which completes the proof of point 3 and of Theorem 6.7.1.

6.7.3 Counting elliptic points for $\Gamma_0(N)$

We now count elliptic points for $\Gamma_0(N)$, at the values of N for which they exist. There are two cases, namely elliptic points of orders 4 and 6, for which the transformation in $\mathrm{SL}(2, \mathbb{Z})$ is conjugate to either,

$$S = \begin{pmatrix} 0 & -1 \\ 1 & 0 \end{pmatrix} \qquad ST = \begin{pmatrix} 0 & -1 \\ 1 & 1 \end{pmatrix} \tag{6.7.6}$$

with $S^4 = (ST)^6 = I$, $\mathrm{tr}(S) = 0$, and $\mathrm{tr}(ST) = 1$. Given the conditions on the traces, the representatives of S and ST in $\Gamma_0(N)$ are of the form,

$$\gamma_4 = \begin{pmatrix} a & b \\ Nc' & -a \end{pmatrix} \qquad \gamma_6 = \begin{pmatrix} a & b \\ Nc' & 1 - a \end{pmatrix} \tag{6.7.7}$$

with $-a^2 - bc'N = 1$ and $a - a^2 - bc'N = 1$, respectively. We shall now assume the prime decomposition $N = \prod_i p_i^{e_i}$ for distinct primes p_i and $e_i \geq 1$. For the elliptic points of order 4, the congruence equation $a^2 \equiv -1 \pmod{N}$ is equivalent to the congruence equations $a^2 \equiv -1 \pmod{p_i^{e_i}}$. Using the results of Section A.4.1, the number of solutions is given by (A.4.9),

$$\varepsilon_2(\Gamma_0(N)) = \begin{cases} 0 & 4 | N \\ \prod_i \left(1 + (-1|p_i)\right) & 4 \nmid N \end{cases} \tag{6.7.8}$$

where $(-1|p) = \pm 1$ for $p \equiv \pm 1 \pmod{4}$ and vanishes for $p = 2$. For the elliptic points of order 6, the congruence equation, $a^2 + a \equiv -1 \pmod{N}$

is equivalent to $(2a+1)^2 + 3 \equiv 0 \pmod{4N}$. The number of solutions was obtained in Section A.4.2, and is given by,

$$\varepsilon_3(\Gamma_0(N)) = \begin{cases} 0 & 9 \mid N \\ \prod_i \left(1 + (-3|p_i)\right) & 9 \nmid N \end{cases} \tag{6.7.9}$$

where $(-3|p) = \pm 1$ for $p \equiv \pm 1 \pmod 3$ and vanishes for $p = 3$.

6.7.4 Counting the number of cusps

The cusps of a modular curve $X(\Gamma)$ for a congruence subgroup $\Gamma \subset \mathrm{SL}(2,\mathbb{Z})$ are given by the cusps $\{\infty\} \cup \mathbb{Q} = \mathrm{SL}(2,\mathbb{Z})\infty$ modulo equivalence under Γ. In this section, we shall count the number ε_∞ of cusps of $X(\Gamma)$ for the classic congruence subgroups $\Gamma = \Gamma(N), \Gamma_1(N)$, and $\Gamma_0(N)$.

We begin by considering the case $\Gamma = \Gamma(N)$. Two cusps $r = m/n$ and $r' = m'/n'$ with $\gcd(m,n) = \gcd(m',n') = 1$ are equivalent under $\Gamma(N)$ provided there exists a $\gamma \in \Gamma(N)$ such that $r' = \gamma r$, or more explicitly,

$$\frac{m'}{n'} = \frac{(1+Na')m + Nb'n}{Nc'm + (1+Nd')n} \qquad \gamma = \begin{pmatrix} 1+Na' & Nb' \\ Nc' & 1+Nd' \end{pmatrix} \tag{6.7.10}$$

The equation on the left side is equivalent to $mn' \equiv m'n \pmod N$, whose solution is given by $(m',n') = \pm(m,n) \pmod N$ in view of the fact that both pairs (m,n) and (m',n') are relatively prime. Therefore, the cusps for $\Gamma(N)$ are labeled by all pairs $(m,n) \pmod N$ with $\gcd(m,n) = 1$.

An alternative perspective on this problem uses *double cosets* and will lend itself more directly to the cases $\Gamma_1(N)$ and $\Gamma_0(N)$. Double cosets will be used in the construction of Hecke operators in Chapter 10, where a more extensive discussion will be provided. Suffice it here to say that if G is a group and H_1, H_2 are subgroups of G, then a double coset is an orbit under the left action of H_1 and the right action of H_2 containing a given element $g \in G$, and is of the form $H_1 g H_2$. The set of all double cosets is denoted,

$$H_1 \backslash G / H_2 = \{H_1\, g\, H_2,\ g \in G\} \tag{6.7.11}$$

Applying the construction to computing the cusps of $\Gamma(N)$, we set $G = \mathrm{SL}(2,\mathbb{Z})$, $H_1 = \Gamma(N)$, and $H_2 = \Gamma_\infty = \{\pm \left(\begin{smallmatrix} 1 & b \\ 0 & 1 \end{smallmatrix}\right), b \in \mathbb{Z}\}$, as was already argued in (6.4.2). The simple coset $\mathrm{SL}(2,\mathbb{Z})/\Gamma_\infty$ provides a one-to-one map from ∞ to all the cusps of $\mathrm{SL}(2,\mathbb{Z})$ while the further left cosets identified cusps that are equivalent under $\Gamma(N)$. Hence we have,

$$\varepsilon_\infty(\Gamma(N)) = |\Gamma(N)\backslash\mathrm{SL}(2,\mathbb{Z})/\Gamma_\infty| \tag{6.7.12}$$

Since $\Gamma(N)$ is a normal subgroup of $\mathrm{SL}(2,\mathbb{Z})$ and we have the relation $\Gamma(N)\backslash\mathrm{SL}(2,\mathbb{Z}) = \mathrm{SL}(2,\mathbb{Z}_N)$, it follows that,

$$\Gamma(N)\backslash\mathrm{SL}(2,\mathbb{Z})/\bar{\Gamma}_\infty = \mathrm{SL}(2,\mathbb{Z}_N)/\bar{\Gamma}_\infty \qquad (6.7.13)$$

where $\bar{\Gamma}_\infty = \{\pm\left(\begin{smallmatrix}1 & b\\ 0 & 1\end{smallmatrix}\right), b \in \mathbb{Z}_N\}$. Therefore the number of cusps is given by,

$$\varepsilon_\infty(\Gamma(N)) = |\mathrm{SL}(2,\mathbb{Z}_N)|/|\bar{\Gamma}_\infty| = d_N/N \qquad (6.7.14)$$

in view of (6.3.9) and the fact that $|\bar{\Gamma}_\infty| = 2N$.

The case of an arbitrary congruence subgroup Γ may similarly be treated with the double coset construction, and we have,

$$\varepsilon_\infty(\Gamma) = |\Gamma\backslash\mathrm{SL}(2,\mathbb{Z})/\Gamma_\infty| \qquad (6.7.15)$$

The difference with the case of $\Gamma(N)$ is that an arbitrary congruence subgroup Γ is not a normal subgroup of $\mathrm{SL}(2,\mathbb{Z})$, and this is indeed the case for $\Gamma_1(N)$ and $\Gamma_0(N)$. The calculation of the number of cusps in these cases is more involved, and we refer to [Shi71, DS05] for a detailed exposition and here only quote the result,

$$N \geq 5 \qquad \varepsilon_\infty(\Gamma_1(N)) = \frac{1}{2}\sum_{ad=N}\phi(a)\phi(d)$$

$$N \geq 3 \qquad \varepsilon_\infty(\Gamma_0(N)) = \sum_{ad=N}\phi\big(\gcd(a,d)\big) \qquad (6.7.16)$$

where the sums are over $a, d > 0$ and ϕ is Euler's totient function given in (6.2.6). The remaining cases may be worked out as well and we have,

$$\varepsilon_\infty(\Gamma_1(2)) = \varepsilon_\infty(\Gamma_1(3)) = \varepsilon_\infty(\Gamma_0(2)) = 2$$

$$\varepsilon_\infty(\Gamma_1(4)) = 3 \qquad (6.7.17)$$

For example, when $N = p \geq 5$ is a prime the above formulas simplify considerably, and we have $\varepsilon_\infty(\Gamma_1(p)) = p - 1$ and $\varepsilon_\infty(\Gamma_0(p)) = 2$. Using the results for $d, \varepsilon_2, \varepsilon_3$ and ε_∞ along with (6.6.4), the genus of $X(\Gamma)$ may be computed and the results are displayed in Tables 6.2 and 6.3 for $N \leq 25$.

6.8 Explicit expressions for the genus

The formulas for the genus of $X(\Gamma)$ are particularly simple for the cases when no elliptic points occur, as specified in Theorem 6.7.1. In these cases, the genus is obtained by combining (6.6.4) with the formula for the degree in (6.7.2) with the expressions of the indices given in (6.2.7), (6.2.8), and the formulas for the number of cusps in (6.7.14), (6.7.16), (6.7.17),

$$N \geq 3: \quad g(X(N)) = 1 + \frac{d_N(N-6)}{12N} \tag{6.8.1}$$

$$N \geq 5: \quad g(X_1(N)) = 1 + \frac{d_N}{12N} - \frac{1}{4}\sum_{ad=N}\phi(a)\phi(d)$$

$$N \geq 2: \quad g(X_0(N)) = 1 + \frac{d_N}{6N\phi(N)} - \frac{\varepsilon_2}{4} - \frac{\varepsilon_3}{3} - \frac{1}{2}\sum_{ad=N}\phi\big(\gcd(a,d)\big)$$

where $\varepsilon_2 = \varepsilon_2(\Gamma_0(N))$ and $\varepsilon_3 = \varepsilon_3(\Gamma_0(N))$ were given in (6.7.8) and (6.7.9). For low values of N, d_N/N, and the genera are given in Tables 6.2 and 6.3.

N	2	3	4	5	6	7	8	9	10	11	12	13
d_N/N	$\frac{3}{2}$	4	6	12	12	24	24	36	36	60	48	84
$g(X(N))$	0	0	0	0	1	3	5	10	13	26	25	50
$g(X_1(N))$	0	0	0	0	0	0	0	0	0	1	0	2
$g(X_0(N))$	0	0	0	0	0	0	0	0	0	1	0	0
$\varepsilon_2, \varepsilon_3$	1,0	0,1	0,0	2,0	0,0	0,2	0,0	0,0	2,0	0,0	0,0	2,2

Table 6.2   Genera for $X(N)$, $X_1(N)$, $X_0(N)$ for $2 \leq N \leq 13$ and the number of elliptic points ε_2 and ε_3 in $X_0(N)$.

N	14	15	16	17	18	19	20	21	22	23	24	25
d_N/N	72	96	96	144	108	180	144	192	180	264	192	300
$g(X(N))$	49	73	81	133	109	196	169	241	241	375	289	476
$g(X_1(N))$	1	1	2	5	2	7	3	5	6	12	5	12
$g(X_0(N))$	1	1	0	1	0	1	1	1	2	2	1	0
$\varepsilon_2, \varepsilon_3$	0,0	0,0	0,0	2,0	0,0	0,2	0,0	0,2	0,0	0,0	0,0	2,0

Table 6.3   Genera for $X(N)$, $X_1(N)$, $X_0(N)$ for $14 \leq N \leq 25$ and the number of elliptic points ε_2 and ε_3 in $X_0(N)$.

• Bibliographical notes

The books by Shimura [Shi71] and by Diamond and Shurman [DS05] stand out for their clear and useful introductions to congruence subgroups and modular curves. Our summary here closely follows their presentation.

Exercises

6.1   Show that $\Gamma(N)$ is a normal subgroup of $\mathrm{SL}(2,\mathbb{Z})$, and that its quotient $\mathrm{SL}(2,\mathbb{Z})/\Gamma(N)$ is isomorphic to $\mathrm{SL}(2,\mathbb{Z}_N)$.

6.2   The compactified modular curve $X(\Gamma)$ for a congruence subgroup Γ is a compact Riemann surface, whose genus was given in Section 6.8. The function theory on $X(\Gamma)$ may be developed using Poincaré series for Γ applied to functions on the Poincaré upper half plane $\mathcal{H}$. To this end, evaluate the scalar Green function $G_{\mathcal{H}}(x,y)$ for $x,y \in \mathcal{H}$, defined to be the inverse of the scalar Laplace–Beltrami operator $\Delta = -(\tau - \bar{\tau})^2 \partial_\tau \partial_{\bar{\tau}}$ of (4.2.6) for the Poincaré metric on functions that vanish on the real line. Using the results of Exercise 3.1 of Chapter 3, express $G_{\mathcal{H}}(x,y)$ in terms of the geodesic distance between $x,y \in \mathcal{H}$.

6.3   In this exercise as well as in the subsequent two exercises, we consider the Laplacian $\Delta = -(z - \bar{z})^2 \partial_z \partial_{\bar{z}}$ acting on scalar functions on the Poincaré upper half plane $\mathcal{H}$ that vanish on $\mathbb{R} = \partial\mathcal{H}$. Show that
(a) Δ is self-adjoint;
(b) its spectrum is bounded from above by $-\frac{1}{4}$;
(c) the function f_s defined for any function $g(\xi)$ by the convolution,

$$f_s(z) = \int_{\mathbb{R}} d\xi \, C_s(z,\xi) \, g(\xi) \tag{6.8.2}$$

where C_s is given by,

$$C_s(x + iy, \xi) = \left(\frac{y}{y^2 + (x - \xi)^2} \right)^s$$

and the convolution is assumed to converge absolutely, is an eigenfunction of Δ with eigenvalue $s(s - 1)$. Properly normalized, the kernel C_s plays a fundamental role in the AdS/CFT correspondence as the bulk-to-boundary propagator.

6.4   Construct an orthonormal basis of eigenfunctions for the scalar Laplacian Δ on functions of $z = x + iy \in \mathcal{H}$ that vanish on $\mathbb{R} = \partial\mathcal{H}$, in a basis where the translation operator $-i\partial_x$ is diagonal.

6.5   The heat-kernel $\mathcal{K}_{\mathcal{H}}^t(z,z')$ is the integral operator $e^{t\Delta}$ for $t > 0$ for the Laplacian $\Delta = -(z - \bar{z})^2 \partial_{\bar{z}} \partial_z$ on functions that vanish on $\mathbb{R} = \partial\mathcal{H}$. Compute $\mathcal{K}_{\mathcal{H}}^t(z,z')$ as a function of the geodesic distance $\ell = \ell(z,z')$ between the points $z, z' \in \mathcal{H}$ using the results of Exercise 6.4. Generalize the calculation to the integral operator $h(-\Delta)$ for an arbitrary function $h(x)$ that is sufficiently tempered for large argument x.

7
Modular forms for congruence subgroups

In this chapter, we shall discuss modular forms for the congruence subgroups Γ introduced in Chapter 6. We shall obtain the dimension formulas for the corresponding rings of modular forms and cusp forms, describe the fields of modular functions on the modular curves $X(\Gamma)$ introduced in Chapter 6, and construct the Eisenstein series for Γ. Throughout, we shall make use of the correspondence between modular forms and differential forms, viewed as sections of holomorphic line bundles on the compact Riemann surface $X(\Gamma)$. We shall provide concrete examples of modular forms for the standard congruence subgroups and apply the results to representations of integers as sums of squares.

7.1 Modular forms and cusp forms with respect to Γ

One defines modular forms and cusp forms for an arbitrary congruence subgroup Γ by analogy with $\mathrm{SL}(2, \mathbb{Z})$. Let $f(\tau)$ be a meromorphic function on the upper half plane $\mathcal{H}$, which obeys the following transformation law,

$$ f(\gamma\tau) = (c\tau + d)^k f(\tau) \qquad\qquad \gamma = \begin{pmatrix} a & b \\ c & d \end{pmatrix} \qquad (7.1.1) $$

for some integer $k \in \mathbb{Z}$, and for all $\gamma \in \Gamma$. Since Γ is a congruence subgroup, there exists an integer N such that $\Gamma(N) \subset \Gamma$, and hence a smallest integer h that divides N such that $T^h \in \Gamma$. As a result, f is invariant under $\tau \to \tau + h$ and admits a Fourier expansion,

$$ f(\tau) = \sum_{n\in\mathbb{Z}} a_n q_h^n \qquad\qquad q_h = e^{2\pi i \tau / h} \qquad (7.1.2) $$

1. f *is a meromorphic modular form of weight* k *if* f *is meromorphic at every cusp of* $\Gamma\backslash\mathcal{H}$, *namely, if the Fourier expansion of* $f(\gamma\tau)$ *for every*

$\gamma \in \mathrm{SL}(2, \mathbb{Z})$ has only a finite number of nonvanishing Fourier coefficients a_n with negative n;

2. f *is a holomorphic modular form of weight* k if it is holomorphic in $\mathcal{H}$ and at every cusp of $\Gamma \backslash \mathcal{H}$, so that its Fourier coefficients a_n of $f(\gamma\tau)$ for every $\gamma \in \mathrm{SL}(2, \mathbb{Z})$ vanish for all $n < 0$, and thus f is finite at all cups;

3. f *is a cusp form of weight* k if it is a holomorphic modular form and vanishes at all cusps.

The spaces of holomorphic modular forms (or simply *modular forms*) and cusp forms of weight k are denoted $\mathcal{M}_k(\Gamma)$ and $\mathcal{S}_k(\Gamma)$, respectively. Clearly, the smaller the subgroup, the larger the space of modular forms,

$$\Gamma' \subset \Gamma \qquad \begin{cases} \mathcal{M}_k(\Gamma) \subset \mathcal{M}_k(\Gamma') \\ \mathcal{S}_k(\Gamma) \subset \mathcal{S}_k(\Gamma') \end{cases} \qquad (7.1.3)$$

In particular, a congruence subgroup Γ of level N contains $\Gamma(N)$ so that $\mathcal{M}_k(\Gamma) \subset \mathcal{M}_k(\Gamma(N))$ and $\mathcal{S}_k(\Gamma) \subset \mathcal{S}_k(\Gamma(N))$.

7.2 Dimension formulas for modular forms

In this section, we shall generalize the dimension formulas for the spaces of holomorphic modular forms for the full modular group $\mathrm{SL}(2, \mathbb{Z})$ given by Theorem 3.5.2 to the case of congruence subgroups. While the results for $\mathrm{SL}(2, \mathbb{Z})$ were derived with the help of the valence formula of Theorem 3.5.1, the results for a congruence subgroup Γ will require the more powerful Riemann–Roch Theorem C.2.1 of Appendix C for line bundles over the compact Riemann surface $X(\Gamma) = \Gamma \backslash \mathcal{H}^*$. Combining this theorem with the vanishing Theorem C.3.1 gives the general formulas (C.3.3) and (C.3.4).

The Riemann–Roch theorem applies to differential forms on the compact Riemann surface $X(\Gamma)$. The relation between modular forms and differential form, expounded in Section 3.3.1, associates to a modular form $f(\tau)$ of weight k in $\mathcal{M}_k$ the following Γ-invariant differential form $\mathfrak{f}$ of degree $\frac{k}{2}$,

$$\mathfrak{f} = f(\tau)(d\tau)^{\frac{k}{2}} \qquad (7.2.1)$$

Modular forms of even weight k correspond to differential forms of integral degree, while modular forms of odd weight k correspond to spinor forms whose treatment involves additional technical machinery for which we refer to [DS05]. Immediate results are as follows,

$$\begin{aligned} \mathcal{S}_0(\Gamma) &= \{0\} & \mathcal{S}_{k<0}(\Gamma) &= \{0\} \\ \mathcal{M}_0(\Gamma) &= \mathbb{C} & \mathcal{M}_{k<0}(\Gamma) &= \{0\} \end{aligned} \qquad (7.2.2)$$

The results on the left follow from the fact that a holomorphic function on a compact Riemann surface must be constant. The first result on the right is obtained by noting that if $f \in \mathcal{M}_{k<0}$ then $f^{12}\Delta^{-k}$ is a holomorphic function in $\mathcal{S}_0(\Gamma)$ which must vanish by the left side.

7.2.1 Modular forms of even weight, $k \geq 2$

Before discussing the case of arbitrary even weight k and arbitrary congruence subgroup Γ, we consider the special case of $k = 2$ for which $\mathfrak{f} = f(\tau)d\tau$ is a holomorphic 1-form. The Riemann–Roch theorem tells us that the number of holomorphic 1-forms $\mathfrak{f}$ on $X(\Gamma)$ equals the genus $g = g(X(\Gamma))$. Obtaining holomorphic modular forms f in $\mathcal{M}_2(\Gamma)$ and $\mathcal{S}_2(\Gamma)$ from holomorphic differentials $\mathfrak{f}$ requires dividing $\mathfrak{f}$ by the 1-form $d\tau$. In the absence of elliptic points, namely, when $\varepsilon_2(\Gamma) = \varepsilon_3(\Gamma) = 0$, the form $d\tau$ has a simple pole at each cusp and is holomorphic elsewhere. To see this, we use a regular coordinate at each cusp, such as $q = e^{2\pi i \tau}$ near the cusp $i\infty$ where $q \to 0$. The form $d\tau = dq/(2\pi i q)$ has a simple pole at the cusp $i\infty$, and similarly $d\tau$ has a simple pole at every cusp of Γ. But then $f(\tau) = \mathfrak{f}/d\tau$ vanishes at every cusp and therefore is a cusp form. Thus the dimensions are given by,

$$\dim \mathcal{S}_2(\Gamma) = g$$
$$\dim \mathcal{M}_2(\Gamma) = g + \varepsilon_\infty(\Gamma) - 1 \qquad (7.2.3)$$

The result for $\mathcal{M}_2(\Gamma)$ is obtained by allowing for simple poles in $\mathfrak{f}$ at the cusps for which the sum of the residues vanishes. The above construction readily generalizes, in the absence of elliptic points, to the dimensions of $\mathcal{M}_k(\Gamma)$ and $\mathcal{S}_k(\Gamma)$ from the Riemann–Roch theorem for even $k \geq 4$,

$$\dim \mathcal{S}_k(\Gamma) = (k-1)(g-1) + (\tfrac{k}{2} - 1)\varepsilon_\infty(\Gamma)$$
$$\dim \mathcal{M}_k(\Gamma) = (k-1)(g-1) + \tfrac{k}{2}\varepsilon_\infty(\Gamma) \qquad (7.2.4)$$

For example, for the congruence subgroup $\Gamma(N)$ with $N \geq 2$, we have $\varepsilon_2 = \varepsilon_3 = 0$ and obtain the following dimension formulas for even $k \geq 4$,

$$\dim \mathcal{S}_k(\Gamma(N)) = \frac{d_N}{12N}(N(k-1) - 6)$$
$$\dim \mathcal{M}_k(\Gamma(N)) = \frac{d_N}{12N}(N(k-1) + 6) \qquad (7.2.5)$$

where the values of d_N/N for $N \leq 25$ may be found in Tables 6.2 and 6.3.

Still for even $k \geq 4$ but now with $\varepsilon_2(\Gamma)$ and/or $\varepsilon_3(\Gamma)$ nonzero, the differential $d\tau^{\frac{k}{2}}$ develops singularities at the corresponding elliptic points. To analyze the behavior near the elliptic points, one introduces regular local coordinates. For example, near $\tau = i, \rho$, we introduce,

$$
\begin{aligned}
w_i &= (\tau - i)^2 & d\tau &= \tfrac{1}{2}\, dw_i/\sqrt{w_i} \\
w_\rho &= (\tau - \rho)^3 & d\tau &= \tfrac{1}{3}\, dw_\rho/\sqrt{w_\rho^{3}} & \quad (7.2.6)
\end{aligned}
$$

A holomorphic modular form $f \in \mathcal{M}_k$ is again constructed from the associated differential form $\mathfrak{f}$ by $f = \mathfrak{f}/(d\tau)^{\frac{k}{2}}$, which now allows $\mathfrak{f}$ to have poles at the elliptic points of order the integral parts of the powers of w_i and w_ρ, respectively. The result is the following formulas for even $k \geq 4$,

$$
\begin{aligned}
\dim \mathcal{S}_k(\Gamma) &= (k-1)(g-1) + \lfloor \tfrac{k}{4} \rfloor \varepsilon_2(\Gamma) + \lfloor \tfrac{k}{3} \rfloor \varepsilon_3(\Gamma) + (\tfrac{k}{2}-1)\varepsilon_\infty(\Gamma) \\
\dim \mathcal{M}_k(\Gamma) &= (k-1)(g-1) + \lfloor \tfrac{k}{4} \rfloor \varepsilon_2(\Gamma) + \lfloor \tfrac{k}{3} \rfloor \varepsilon_3(\Gamma) + \tfrac{k}{2}\varepsilon_\infty(\Gamma) \quad (7.2.7)
\end{aligned}
$$

where $\lfloor x \rfloor$ for $x \in \mathbb{R}$ is the integer n such that $n \leq x < n+1$.

The holomorphic forms f in the space of Eisenstein series, defined by,

$$
\begin{aligned}
\mathcal{E}_k(\Gamma) &= \mathcal{M}_k(\Gamma)/\mathcal{S}_k(\Gamma) \\
\dim \mathcal{E}_k(\Gamma) &= \varepsilon_\infty(\Gamma) \quad (7.2.8)
\end{aligned}
$$

correspond to differential forms $\mathfrak{f}$ that have simple poles at the cusps whose residues sum to zero. Actually, the stronger assertion $\mathcal{M}_k(\Gamma) = \mathcal{S}_k(\Gamma) \oplus \mathcal{E}_k(\Gamma)$ holds, where the subspaces $\mathcal{S}_k(\Gamma)$ and $\mathcal{E}_k(\Gamma)$ are mutually orthogonal with respect to the Petersson inner product, introduced in (4.4.1).

7.2.2 Modular forms of odd weight, $k \geq 1$

The full modular group $\mathrm{SL}(2, \mathbb{Z})$ has no nonzero modular forms of odd weight k because the transformation law (7.1.1) for $-I \in \mathrm{SL}(2, \mathbb{Z})$ shows that $f(\tau) = 0$. The same conclusion holds for any congruence subgroup Γ that contains $-I$. The congruence subgroups $\Gamma(2), \Gamma_1(2) = \Gamma_0(2)$, and $\Gamma_0(N)$ for every N contain $-I$ and thus have no modular forms of odd weight k.

We shall now concentrate on congruence subgroups Γ that do not contain the element $-I$, which include the classic subgroups $\Gamma(N)$ for $N \geq 2$ and $\Gamma_1(N)$ for $N > 2$. When $k < 0$, the corresponding spaces $\mathcal{M}_k(\Gamma)$ and $\mathcal{S}_k(\Gamma)$ continue to be empty, just as when k was even. But for odd $k \geq 1$, the corresponding spaces of holomorphic modular forms $\mathcal{M}_k(\Gamma)$ and of cusp forms $\mathcal{S}_k(\Gamma)$ are not necessarily trivial. The dimension formulas for these spaces when $k \geq 3$ may be derived from the Riemann–Roch theorem (Theorem C.2.1 of Appendix C) along the same lines as for k even, except for the fact that a distinction needs to be made between *regular cusps* and *irregular cusps*, a distinction that we shall explain later in this section. The result for an arbitrary congruence subgroup Γ is as follows for odd $k \geq 3$,

$$\dim \mathcal{S}_k(\Gamma) = (k-1)(g-1) + \lfloor \tfrac{k}{4} \rfloor \varepsilon_2(\Gamma) + \lfloor \tfrac{k}{3} \rfloor \varepsilon_3(\Gamma)$$
$$+ \tfrac{k-2}{2} \varepsilon_\infty^{\mathrm{reg}}(\Gamma) + \tfrac{k-1}{2} \varepsilon_\infty^{\mathrm{irreg}}(\Gamma)$$
$$\dim \mathcal{M}_k(\Gamma) = (k-1)(g-1) + \lfloor \tfrac{k}{4} \rfloor \varepsilon_2(\Gamma) + \lfloor \tfrac{k}{3} \rfloor \varepsilon_3(\Gamma)$$
$$+ \tfrac{k}{2} \varepsilon_\infty^{\mathrm{reg}}(\Gamma) + \tfrac{k-1}{2} \varepsilon_\infty^{\mathrm{irreg}}(\Gamma) \qquad (7.2.9)$$

For the classic series of congruence subgroups $\Gamma(N)$ for $N \geq 2$ and $\Gamma_1(N)$ for $N > 2$ which do not contain $-I$, only $\Gamma_1(4)$ possess an irregular cusp, namely, at $\tau = \tfrac{1}{2}$ (see exercise 3.2.5 of [DS05]).

The distinction between regular and irregular cusps may be understood as follows. Let s be a cusp of the congruence subgroup Γ, and let f be a modular form of Γ of odd weight k. Define the stability group of s by $\Gamma_s = \{\gamma \in \Gamma | \gamma(s) = s\}$. Now take an element $\sigma \in \mathrm{SL}(2,\mathbb{Z})$ to map s to ∞, namely, $\sigma(s) = \infty$ so that $\sigma \Gamma_s \sigma^{-1}$ is the stability group of the corresponding cusp at ∞. By the argument given to derive (7.1.2), there exists a positive integer h such that,

$$\sigma \Gamma_s \sigma^{-1} = \left\{ \pm (T^h)^m, \, m \in \mathbb{Z} \right\} \qquad T = \begin{pmatrix} 1 & 1 \\ 0 & 1 \end{pmatrix} \qquad \sigma = \begin{pmatrix} a & b \\ c & d \end{pmatrix} \quad (7.2.10)$$

When k is even, the $\pm$ factor leaves f inert and the cusp is *regular*. But when $-I \notin \Gamma$, the orbits $+(T^h)^m$ and $-(T^h)^m$ must be distinguished. The cusps corresponding to the orbit $+(T^h)^m$ are *regular cusps*, and the modular form $\varphi(z) = (c\tau + d)^{-k} f(\rho^{-1}z)$ is invariant under $z \to z + h$. The cusps corresponding to the orbit $-(T^h)^m$ are *irregular cusps* and the modular form $\varphi(z) = (c\tau + d)^{-1} f(\rho^{-1}z)$ changes sign as follows, $\varphi(z + h) = -\varphi(z)$ (see [Shi71] and [DS05] for more details). In the language of string theory, regular cusps correspond to *Neveu–Schwarz punctures* while irregular cusps correspond to *Ramond punctures*; see Subsection 5.7.1 for the definitions of the Neveu–Schwarz and Ramond sectors, and Section 4 of [Wit19] for accessible definitions of the corresponding punctures.

Finally, modular forms of weight $k = 1$ correspond to spinor differentials of weight $\tfrac{1}{2}$. It is well known that the dimensions of holomorphic differentials of weight $\tfrac{1}{2}$ may jump at certain divisors in Riemann surfaces of genus $g \geq 3$. However, the difference,

$$\dim \mathcal{M}_1(\Gamma) - \dim \mathcal{S}_1(\Gamma) = \tfrac{1}{2} \varepsilon_\infty^{\mathrm{reg}}(\Gamma) \qquad (7.2.11)$$

remains valid throughout. For $\varepsilon_\infty^{\mathrm{reg}}(\Gamma) > 2g - 2$, a vanishing theorem gives $\dim \mathcal{S}_1(\Gamma) = 0$ so that $\dim \mathcal{M}_1(\Gamma) = \tfrac{1}{2} \varepsilon_\infty^{\mathrm{reg}}(\Gamma)$.

7.3 The fields of modular functions on $X(\Gamma)$

The space of modular functions of $\mathrm{SL}(2,\mathbb{Z})$ is the field $\mathbb{C}(j)$ of rational functions of the j-function by Theorem 3.6.1, and the corresponding modular curve $\mathrm{SL}(2,\mathbb{Z})\backslash\mathcal{H}^*$ is nothing other than the Riemann sphere. The space of modular functions on the modular curve $X(\Gamma) = \Gamma\backslash\mathcal{H}^*$ for a general congruence subgroup Γ is an extension of $\mathbb{C}(j)$ that we shall now construct explicitly for the congruence subgroups $\Gamma(N), \Gamma_1(N)$, and $\Gamma_0(N)$.

For the case $\Gamma(N)$, we consider the following combinations of the Weierstrass $\wp$ function and the modular forms $g_2(\tau)$ and $g_3(\tau)$,[1]

$$h_N(v|\tau) = \frac{g_2(\tau)}{g_3(\tau)}\, \wp\left(\frac{\mu\tau + \nu}{N}\Big|\tau\right) \tag{7.3.1}$$

where the row vector $v \equiv (\mu\ \nu) \pmod{N}$ with $\mu, \nu \in \mathbb{Z}$ and $v \not\equiv 0 \pmod{N}$. Under a transformation $\gamma \in \mathrm{SL}(2,\mathbb{Z})$, the $\wp$-function transforms as follows,

$$\wp\left(\frac{z}{c\tau + d}\Big|\gamma\tau\right) = (c\tau + d)^2 \wp(z|\tau) \qquad\qquad \gamma = \begin{pmatrix} a & b \\ c & d \end{pmatrix} \tag{7.3.2}$$

so that the transformation law of $h_N(v|\tau)$ under $\gamma \in \mathrm{SL}(2,\mathbb{Z})$ is given by,

$$h_N(v|\gamma\tau) = h_N(v\gamma|\tau) \qquad\qquad v\gamma = (a\mu + c\nu\ \ b\mu + d\nu) \tag{7.3.3}$$

Under a transformation $\gamma \in \Gamma(N)$, we have $v\gamma \equiv v \pmod{N}$, so that $h_N(v|\tau)$ is invariant for each v. Exercises 16.4 and 16.5 will further examine the algebraic properties of the functions $h_N(v|\tau)$. Under $\gamma \in \Gamma_1(N) : v \to v\gamma = v + (0\ \ b\mu)$, we have invariance of each $h_N(v|\tau)$ for $\mu = 0$. Finally, under $\Gamma_0(N)$ with $\mu = 0$, we have $v\gamma = (0\ \ d\nu)$ so that the combination,

$$\sum_{d\in\mathbb{Z}_N^*} h_N(0\ \ d\nu|\tau) \tag{7.3.4}$$

is invariant under $\Gamma_0(N)$. The latter is related to the modular function $j(N\tau)$, which is easily seen to be invariant under $\Gamma_0(N)$

7.4 Holomorphic Eisenstein series for $\Gamma(N)$

The holomorphic Eisenstein series G_k for $\mathrm{SL}(2,\mathbb{Z})$ of integer weight $k \geq 3$ was defined in (2.4.5) and is given for the lattice $\mathbb{Z} \oplus \mathbb{Z}\tau$ as follows,

$$G_k(\tau) = \sideset{}{'}\sum_{m,n\in\mathbb{Z}} \frac{1}{(m\tau + n)^k} \tag{7.4.1}$$

---

[1] This normalization applies when $g_2(\tau), g_3(\tau) \neq 0$; near $\tau = i$ or $\tau = \rho$ the prefactor should be changed instead to $g_2(\tau)^{-\frac{1}{2}}$ or $g_3(\tau)^{-\frac{1}{3}}$, respectively.

The prime on the sum instructs us to omit any contribution that makes the summand singular, in this case $(m,n) = (0,0)$. Clearly $G_k(\tau)$ vanishes for odd k, and is a modular form of weight k for even k, transforming as follows,

$$G_k(\gamma\tau) = (c\tau + d)^k G_k(\tau) \qquad \gamma - \begin{pmatrix} a & b \\ c & d \end{pmatrix} \subset \mathrm{SL}(2,\mathbb{Z}) \qquad (7.4.2)$$

The ring of modular forms of $\mathrm{SL}(2,\mathbb{Z})$ is polynomial and is generated by the modular forms G_4 and G_6 of weight 4 and 6, respectively.

To construct modular forms for a congruence subgroup $\Gamma(N)$ with $N \geq 2$, we parametrize the elements of $\mathbb{Z}_N \times \mathbb{Z}_N$ by a row vector $v = (\mu \ \nu)$ with $\mu, \nu \in \mathbb{Z}_N$ and introduce the following Kronecker–Eisenstein sum,[2]

$$G_k(v|\tau) = \sum_{m,n\in\mathbb{Z}}{}' \frac{1}{\left(m\tau + \frac{\mu}{N}\tau + n + \frac{\nu}{N}\right)^k} = \sum_{(m,n)\equiv v\,(N)}{}' \frac{N^k}{(m\tau + n)^k} \qquad (7.4.3)$$

For $v \equiv 0 \pmod{N}$, we recover $G_k(0|\tau) = G_k(\tau)$ of (7.4.1), while for $v \not\equiv 0 \pmod{N}$, the denominators are nonvanishing and the prime on the sum may be omitted. The double sum is absolutely convergent for integer $k \geq 3$, while for $k = 1, 2$, it is defined by the Eisenstein summation convention,

$$G_k(v|\tau) = \sum_{n\equiv\nu\,(N)}{}' \frac{N^k}{(\mu\tau + n)^k} + \sum_{\substack{m\neq 0 \\ m\equiv\mu\,(N)}} \sum_{n\equiv\nu\,(N)} \frac{N^k}{(m\tau + n)^k} \qquad (7.4.4)$$

Under a transformation $\gamma \in \mathrm{SL}(2,\mathbb{Z})$, for $k \geq 3$, we have,

$$G_k(v|\gamma\tau) = (c\tau + d)^k G_k(v\gamma|\tau) \qquad \gamma = \begin{pmatrix} a & b \\ c & d \end{pmatrix} \qquad (7.4.5)$$

where the ingredients are defined as follows,

$$\gamma\tau = \frac{a\tau + b}{c\tau + d} \qquad v\gamma = (a\mu + c\nu \ \ b\mu + d\nu) \qquad (7.4.6)$$

The transformation law for $k = 1, 2$ is more complicated; see Theorem 7.7.1 for the case of $k = 2$. The action of $\gamma \in \mathrm{SL}(2,\mathbb{Z})$ may map a vector v to itself or to a different vector. For example, $(0 \ 0)$ maps to itself under any $\gamma \in \mathrm{SL}(2,\mathbb{Z})$, while $(0 \ \nu_0)$ for $\nu_0 \not\equiv 0 \pmod{N}$ maps to $(\mu \ \nu) \equiv (c\nu_0 \ d\nu_0) \pmod{N}$. Conversely, given $v = (\mu \ \nu)$ with $\gcd(\gcd(\mu,\nu),N) = 1$, there exists a $\gamma \in \mathrm{SL}(2,\mathbb{Z})$ which maps the vector v to $(0 \ \nu_0)$ with $\nu_0 \equiv \gcd(\mu,\nu) \pmod{N}$.

---

[2] To shorten notation in large formulas, we shall often abbreviate $\pmod{N}$ to simply (N).

7.4.1 Invariance under $\Gamma(N)$

Using the transformation law (7.4.5) of $G_k(v|\tau)$ under an arbitrary $\gamma \in$ SL$(2, \mathbb{Z})$, we may deduce the subgroup of SL$(2, \mathbb{Z})$ under which $G_k(v|\tau)$ is a modular form of weight k. The condition under which $\gamma \in$ SL$(2, \mathbb{Z})$ maps $G_k(v|\tau)$ to a multiple of itself is $v\gamma - v \equiv 0 \pmod{N}$, or more explicitly,

$$(a - 1)\mu + c\nu \equiv 0 \pmod{N}$$
$$b\mu + (d - 1)\nu \equiv 0 \pmod{N} \qquad (7.4.7)$$

Invariance of all pairs $(\mu \ \nu) \in (\mathbb{Z}_N)^2$ requires $\gamma \in \Gamma(N)$. To see this, it suffices to choose $(\mu \ \nu) = (1 \ 0) \pmod{N}$ and $(\mu \ \nu) = (0 \ 1) \pmod{N}$ and then derive the conditions $a, d \equiv 1 \pmod{N}$ and $b, c \equiv 0 \pmod{N}$ by solving (7.4.7) so that $\gamma \in \Gamma(N)$. A given pair $v = (\mu \ \nu)$ may be invariant, however, under a subgroup of SL$(2, \mathbb{Z})$ that is larger than $\Gamma(N)$. For example, $(0 \ 0)$ is invariant under all of SL$(2, \mathbb{Z})$, while $(0 \ 1)$ is invariant under $\Gamma_1(N)$.

7.4.2 Asymptotics near the cusps

The cusps of $\Gamma(N)$ are given by the double coset,

$$\Gamma(N)\backslash\text{SL}(2, \mathbb{Z})/\Gamma_\infty \qquad (7.4.8)$$

acting on ∞, namely, by all the images under SL$(2, \mathbb{Z})$ of the cusp at infinity modulo equivalence under $\Gamma(N)$. The number of cusps is denoted $\varepsilon_\infty(\Gamma(N))$ and was given in (6.7.14). Since every cusp may be mapped to ∞ under a suitable $\gamma \in$ SL$(2, \mathbb{Z})$, under which $G_k(v|\tau)$ maps to $G_k(v\gamma|\gamma\tau)$, it suffices to examine the asymptotics of $G_k(v|\tau)$ at the cusp ∞, the asymptotics at the other cusps being deduced by mapping back with γ^{-1}.

The asymptotics at the cusp ∞ of $G_k(v|\tau)$ for $v \not\equiv 0 \pmod{N}$ is qualitatively different for the cases $\mu \equiv 0 \pmod{N}$ where the limit is nonzero, and $\mu \not\equiv 0 \pmod{N}$ where the limit vanishes by inspection of (7.4.1),

$$\lim_{\tau \to i\infty} G_k(v|\tau) = \begin{cases} G_k^\infty(\nu) & \mu \equiv 0, \ \nu \not\equiv 0 \pmod{N} \\ 0 & \mu \not\equiv 0 \pmod{N} \end{cases} \qquad (7.4.9)$$

Here $G_k^\infty(\nu)$ is given in terms of the Hurwitz zeta function $\zeta(s, x)$ by,

$$G_k^\infty(\nu) = \sum_{n \equiv \nu \ (N)} \frac{N^k}{n^k} = \zeta(k, \tfrac{\nu}{N}) + (-)^k \zeta(k, 1 - \tfrac{\nu}{N}) \qquad (7.4.10)$$

More generally, however, $G_k(v|\tau)$ will be nonvanishing at cusps other than ∞ as well. It is shown in [DS05] that the Eisenstein series form a basis for the Eisenstein space, which is defined to be the quotient,

$$\mathcal{E}_k\big(\Gamma(N)\big) = \mathcal{M}_k\big(\Gamma(N)\big)/\mathcal{S}_k\big(\Gamma(N)\big) \qquad (7.4.11)$$

where $\mathcal{M}_k(\Gamma(N))$ is the vector space of all holomorphic modular forms of weight k and $\mathcal{S}_k(\Gamma(N))$ is the sub-space of cusp forms. For even k, the dimension of the Eisenstein space is given by the number of cusps $\varepsilon_\infty(\Gamma(N))$,

$$\dim \mathcal{E}_k\big(\Gamma(N)\big) = \varepsilon_\infty\big(\Gamma(N)\big) \qquad (7.4.12)$$

where $\varepsilon_\infty(\Gamma(N))$ was evaluated in Section 6.7.4.

7.5 Fourier decomposition of Eisenstein series for $\Gamma(N)$

Given the representation of the Eisenstein series $G_k(v|\tau)$ for $\Gamma(N)$ by a Kronecker–Eisenstein series in (7.4.3), we may obtain its Fourier decomposition along the same lines as for the Eisenstein series for $\mathrm{SL}(2,\mathbb{Z})$ in Section 3.4.1. The result may be summarized by the following theorem.

Theorem 7.5.1. *The Fourier decomposition of the Eisenstein series* $G_k(v|\tau)$ *for the principal congruence subgroup* $\Gamma(N)$ *with* $k \geq 2$, $v = (\mu\ \nu)$ *for* $\mu, \nu \in \mathbb{Z}_N = \{0, 1, \cdots, N-1\}$ *and* $v \not\equiv 0 \pmod{N}$ *is given as follows,*

$$G_k(v|\tau) = \delta_{\mu,0}\, G_k^\infty(v) + \frac{(-2\pi i)^k}{\Gamma(k)} \sum_{n=1}^{\infty} \sigma_{k-1}^v(n)\, q^{n/N} \qquad (7.5.1)$$

The function $G_k^\infty(v|\tau)$ *was given in (7.4.10) and the* v-*dependent sum of divisor function* $\sigma_{k-1}^v(n)$ *is defined as follows,*

$$\sigma_{k-1}^v(n) = \sum_{\substack{m|n \\ n/m \equiv \mu\,(N)}} \mathrm{sign}(m)\, m^{k-1} e^{2\pi i\, m\, \nu/N} \qquad (7.5.2)$$

where the sum over m *runs over both positive and negative divisors of* n.

To prove the theorem, we start from the definition (7.4.3) and assume that representatives for μ and ν are chosen such that $\mu, \nu \in \{0, 1, \cdots, N-1\}$. For $k \geq 3$ the series is absolutely convergent, while for $k = 2$ it may be defined by the Eisenstein summation prescription. For $\mu \not\equiv 0 \pmod{N}$, we split the sum over m into the contributions from $m \geq 0$ and $m < 0$, factor out N^k, and carry out the sum over n using (3.4.9),

$$G_k(v|\tau) = \frac{(-2\pi i)^k}{\Gamma(k)} \sum_{m=0}^{\infty} \sum_{p=1}^{\infty} p^{k-1}\, e^{2\pi i p(mN\tau + \mu\tau + \nu)/N}$$

$$- \frac{(+2\pi i)^k}{\Gamma(k)} \sum_{m=-1}^{-\infty} \sum_{p=1}^{\infty} p^{k-1}\, e^{-2\pi i p(mN\tau + \mu\tau + \nu)/N} \qquad (7.5.3)$$

Letting $p \to -p$ in the second line, recasting the double sum in terms of the combination $n = p(mN + \mu)$ which is a strictly positive integer in both sums, and relabeling $p \to m$ proves the theorem for $\mu \neq 0$. For the case where $\mu = 0$, the sum over m is split up instead into contributions from $m > 0$, $m = 0$, and $m < 0$. The contribution from $m = 0$ is given by the extra term $G_k^\infty(\nu)$, while the sums over $m \neq 0$ proceed as in the case $\mu \neq 0$.

7.6 Holomorphic Eisenstein series for $\Gamma_0(N)$: a first look

In view of the inclusion $\Gamma(N) \subset \Gamma_0(N)$, we have the following inclusion of holomorphic modular forms $\mathcal{M}_k(\Gamma_0(N)) \subset \mathcal{M}_k(\Gamma(N))$ by (7.1.3), and similarly for cusp forms and Eisenstein series. Thus, we may construct the Eisenstein series for $\Gamma_0(N)$ by linear combinations of those in $\Gamma(N)$. Since $\Gamma_0(N)$ contains the generator T, we deduce from the transformation laws of (7.4.5) and (7.4.6) that $v = (\mu\ \nu) \to vT = (\mu\ \nu+\mu)$ and that an Eisenstein series $G_k(v|\tau)$ for given v is invariant under T if and only if $\mu \equiv 0 \ (\mathrm{mod}\ N)$, so that its Fourier expansion will be in integer powers of q. The transformation by an arbitrary element $\gamma \in \Gamma_0(N)$ acting on $v \equiv (0\ \nu)\,(\mathrm{mod}\ N)$ is then given by (7.4.6) by $v = (0\ \nu) \to v\gamma \equiv (0\ d\nu)\,(\mathrm{mod}\ N)$. The solutions to the relation $ad \equiv 1 \ (\mathrm{mod}\ N)$ consist of the invertible elements $\mathbb{Z}_N^*$ which permute the various Eisenstein series $G_k(0\ \nu|\tau)$.

To construct $\Gamma_0(N)$-invariant Eisenstein series, we sum over the different orbits of $\Gamma_0(N)$ induced by the action of $\mathbb{Z}_N^*$. For example, $G_k(0\ 0|\tau) = G_k(\tau)$ is by itself a modular form, but $G_k(0\ 1|\tau)$ is not for $N \geq 3$. Thus, modular forms of $\Gamma_0(N)$ may be obtained by the following sums,

$$\sum_{d \in \mathbb{Z}_N^*} G_k(0\ d\nu|\tau) \tag{7.6.1}$$

When N is prime, there are only two orbits, namely, for $\nu \equiv 0$ or $\nu \equiv 1$, but for N not a prime, several different orbits emerge. For example, for $N = 4$, we have, $\mathbb{Z}_4^* = \{1,3\}$ and therefore three Eisenstein series given by,

$$G_k(0\ 0|\tau) \qquad G_k(0\ 2|\tau) \qquad G_k(0\ 1|\tau) + G_k(0\ 3|\tau) \tag{7.6.2}$$

This counting agrees with the dimension formula for $\dim(\mathcal{M}_4(\Gamma_0(4))) = 3$ obtained from (7.2.7) by using the values $d_4/4 = 4$ and $\varepsilon_2 = \varepsilon_3 = 0$ from Table 6.2 for $\Gamma_0(4)$. These modular forms will be applied to the sums of squares problem in Exercise 7.5. For arbitrary congruence subgroups, however, the above construction will not produce a full basis for $\mathcal{E}_k(\Gamma)$, and one must appeal to a more general construction, to be discussed in Section 7.7.

7.7 Holomorphic Eisenstein series for $\Gamma_1(N)$ and $\Gamma_0(N)$

The construction of holomorphic Eisenstein series for arbitrary congruence subgroups uses Dirichlet characters to symmetrize the Eisenstein series for $\Gamma(N)$ constructed in Section 7.4. As discussed in more detail in Appendix A.9, a Dirichlet character χ_n for $n \in \mathbb{N}$ is a map $\chi_n : \mathbb{Z} \to \mathbb{C}$ which is periodic with period n, namely, $\chi_n(a) = \chi_n(a+n)$, completely multiplicative, namely, $\chi_n(am) = \chi_n(a)\chi_n(m)$, and which is nonvanishing only when n and the argument a are relatively prime, namely, when $\gcd(a, n) = 1$.

One defines an Eisenstein series $G_k[\chi_n, \tilde{\chi}_{\tilde{n}}|\tau]$ for level N and characters χ_n and $\tilde{\chi}_{\tilde{n}}$ with $n\tilde{n} = N$ by the following sums over the Eisenstein series $G_k(\mu \ \nu|\tau)$ of $\Gamma(N)$, defined in (7.4.3),

$$G_k[\chi_n, \tilde{\chi}_{\tilde{n}}|\tau] = \sum_{m=0}^{n-1}\sum_{\tilde{m}=0}^{\tilde{n}-1}\sum_{\ell=0}^{n-1} \chi_n(m)\tilde{\chi}_{\tilde{n}}(\tilde{m})G_k(m\tilde{n} \ \ \tilde{m}+\ell\tilde{n}|\tau) \quad (7.7.1)$$

The transformation of $G_k[\chi_n, \tilde{\chi}_{\tilde{n}}|\tau]$ under $\gamma \in \Gamma_0(N)$ may be deduced from the transformation of $G_k(\mu \ \nu|\tau)$ given in (7.4.5) for an arbitrary $\gamma \in \mathrm{SL}(2, \mathbb{Z})$. In particular,

$$(m\tilde{n} \ \ \tilde{m}+\ell\tilde{n}) \to (m'\tilde{n} \ \ \tilde{m}'+\ell'\tilde{n}) = (m\tilde{n} \ \ \tilde{m}+\ell\tilde{n})\begin{pmatrix} a & b \\ c & d \end{pmatrix} \quad (7.7.2)$$

or equivalently using the fact that $c \equiv 0 \pmod{N}$,

$$m' = am \pmod{n} \qquad\qquad \ell' = d\ell + bm \pmod{n}$$
$$\tilde{m}' = d\tilde{m} \pmod{\tilde{n}} \quad (7.7.3)$$

In view of the effective multiplicative transformation rule for both m and $\tilde{m}$, the characters also transform multiplicatively, $\chi_n(m') = \chi_n(a)\chi_n(m)$ and $\tilde{\chi}_{\tilde{n}}(\tilde{m}') = \tilde{\chi}_{\tilde{n}}(d)\tilde{\chi}_{\tilde{n}}(\tilde{m})$, and we find,

$$G_k[\chi_n, \tilde{\chi}_{\tilde{n}}|\gamma\tau] = \chi_n(a)\tilde{\chi}_{\tilde{n}}(d)(c\tau + d)^k G_k[\chi_n, \tilde{\chi}_{\tilde{n}}|\tau] \quad (7.7.4)$$

Consistency of this relation under the action of $\gamma = -I \in \Gamma_0(N)$, namely, $a = d = -1$ which leaves τ invariant, requires that the characters satisfy,

$$\chi_n(-1)\tilde{\chi}_{\tilde{n}}(-1) = (-1)^k \quad (7.7.5)$$

The prefactor in (7.7.4) is a function of d only in view of $\chi_n(a) = 1/\chi_n(d)$.

To construct a basis of Eisenstein series for $\Gamma_1(N)$, it is convenient to work in terms of *primitive Dirichlet characters*, the definition of which is given in Section A.8 of Appendix A. To this end, we introduce the following space of pairs of primitive characters. For any positive integers N and $k \geq 3$, we define the set $\mathcal{C}_{N,k}$ of triples $(\chi_n, \tilde{\chi}_{\tilde{n}}, t)$ with χ_n and $\tilde{\chi}_{\tilde{n}}$ primitive Dirichlet characters and $t \in \mathbb{N}$,

$$\mathcal{C}_{N,k} = \left\{ (\chi_n, \tilde{\chi}_{\tilde{n}}, t) \,\middle|\, \chi_n(-1)\tilde{\chi}_{\tilde{n}}(-1) = (-1)^k \ \text{ and } \ n\tilde{n}t|N \right\} \quad (7.7.6)$$

The dependence of $\mathcal{C}_{N,k}$ on k is only through the parity of k. We shall now quote the following theorem and refer to [DS05, Miy06] for the proof.

Theorem 7.7.1. *The space $\mathcal{E}_k(\Gamma_1(N))$ of weight-k holomorphic Eisenstein series for $\Gamma_1(N)$ is given by,*

$$\mathcal{E}_k(\Gamma_1(N)) = \left\{ G_k[\chi_n, \tilde{\chi}_{\tilde{n}}|t\tau] \ \text{ for all } (\chi_n, \tilde{\chi}_{\tilde{n}}, t) \in \mathcal{C}_{N,k} \right\} \quad (7.7.7)$$

for $k \geq 3$. For $k = 2$, we have,

$$\mathcal{E}_2(\Gamma_1(N)) = \begin{cases} G_2(\tau) - tG_2(t\tau) & \text{for } (\chi_n, \tilde{\chi}_{\tilde{n}}) = (\chi_n^\pi, \chi_{\tilde{n}}^\pi), \ t|N \\ G_2[\chi_n, \tilde{\chi}_{\tilde{n}}|t\tau] & \text{otherwise, } (\chi_n, \tilde{\chi}_{\tilde{n}}, t) \in \mathcal{C}_{N,2} \end{cases} \quad (7.7.8)$$

where χ_n^π is the principal character for modulus n. The term "otherwise" on the second line means that when $(\chi_n, \tilde{\chi}_{\tilde{n}}) \neq (\chi_n^\pi, \chi_{\tilde{n}}^\pi)$, the set is formed by including the elements corresponding to all such primitive characters.

The Fourier expansion of $G_k[\chi_n, \tilde{\chi}|\tau]$ may be obtained from the Fourier expansion of $G_k(v|\tau)$ given in Theorem 7.5.1. The result may be expressed in terms of a generalized sum of divisors with characters,

$$\sigma_m^{\chi_n, \tilde{\chi}_{\tilde{n}}}(M) = \sum_{d|M} \chi_n(M/d)\tilde{\chi}_{\tilde{n}}(d)d^m \quad (7.7.9)$$

as well as the delta function on characters and the Dirichlet L-series,

$$\delta(\chi_n) = \begin{cases} 1 & \chi_n = 1, \\ 0 & \chi_n \neq 1, \end{cases} \qquad L(s, \tilde{\chi}_{\tilde{n}}) = \sum_{M=1}^{\infty} \frac{\tilde{\chi}_{\tilde{n}}(M)}{M^s} \quad (7.7.10)$$

In terms of these ingredients, we have the following Fourier expansions,

$$G_k[\chi_n, \tilde{\chi}_{\tilde{n}}|t\tau] = 2\delta(\chi_1)L(k, \tilde{\chi}_{\tilde{n}}) + \frac{(-2\pi i)^k}{\Gamma(k)} \sum_{M=1}^{\infty} \sigma_{k-1}^{\chi_n, \tilde{\chi}_{\tilde{n}}}(M)q^{tM} \quad (7.7.11)$$

When $\chi_n = \tilde{\chi}_{\tilde{n}} = 1$ and $k \geq 4$, the generalized divisor sums reduce to the usual divisor sums, and the L-series reduces to the Riemann-zeta function. In this case, the functions $G_k[\chi_n, \tilde{\chi}_{\tilde{n}}|t\tau]$ reduce to the usual Eisenstein series for $\mathrm{SL}(2, \mathbb{Z})$ at multiplied values of τ, that is, $G_k(t\tau)$. It is clear that for any $t|N$, the functions $G_k(t\tau)$ should be Eisenstein series for $\Gamma_1(N)$.

Note the qualitative distinction between the cases of $k = 2$ and $k > 2$, which results from the fact that there is no weight-2 modular function for

$\mathrm{SL}(2, \mathbb{Z})$. The first line of (7.7.8) is indeed consistent with the triviality of the space of weight-2 modular forms for $N = 1$.

Finally, we briefly discuss the case of $\Gamma_0(N)$. First note that if χ_n and $\tilde{\chi}_{\tilde{n}}$ are two Dirichlet characters and N is a common multiple of n and $\tilde{n}$, then it is possible to define a modulo N Dirichlet character $\hat{\chi}_N = \chi_n^{(N)} \chi_{\tilde{n}}^{(N)}$, where,

$$\chi_n^{(N)}(a) = \begin{cases} \chi_n(a) & \gcd(a, N) = 1 \\ 0 & \gcd(a, N) > 1 \end{cases} \tag{7.7.12}$$

We may now further define the following subspaces of $\mathcal{E}_k(\Gamma_1(N))$,

$$\mathcal{E}_k(N, \hat{\chi}_N) = \Big\{ G_k[\chi_n, \tilde{\chi}_{\tilde{n}} | t\tau] \text{ for all } (\chi_n, \tilde{\chi}_{\tilde{n}}, t) \in \mathcal{C}_{N,k}$$
$$\text{and } \chi_n^{(N)} \tilde{\chi}_{\tilde{n}}^{(N)} = \hat{\chi}_N \Big\} \tag{7.7.13}$$

such that $\mathcal{E}_k(\Gamma_1(N)) = \bigoplus_{\hat{\chi}_N} \mathcal{E}_k(N, \hat{\chi}_N)$. The quantity $\hat{\chi}_N$ is referred to as the *Nebentypus*. The space of Eisenstein series of $\Gamma_0(N)$ is obtained by restricting to trivial Nebentypus χ^p, namely, $\mathcal{E}_k(\Gamma_0(N)) = \mathcal{E}_k(N, \chi^p)$.

7.7.1 The rings of modular forms for congruence subgroups

Holomorphic modular forms and cusp forms for a congruence subgroup Γ form the rings $\mathcal{M}(\Gamma)$ and $\mathcal{S}(\Gamma)$ graded by the weight k,

$$\mathcal{M}(\Gamma) = \bigoplus_k \mathcal{M}_k(\Gamma) \qquad\qquad \mathcal{S}(\Gamma) = \bigoplus_k \mathcal{S}_k(\Gamma) \tag{7.7.14}$$

When $-I \in \Gamma$, the direct sums extend over even k only. When $-I \notin \Gamma$, both even and odd k are included while the sum over even k only produces a subring. As an aside, we note that $\mathcal{M}(\Gamma)$ also supports a Lie algebra structure through the Rankin–Cohen bracket between modular forms $f \in \mathcal{M}_k(\Gamma)$ and $g \in \mathcal{M}_\ell(\Gamma)$ and is given by $[f, g] = kfg' - \ell f'g \in \mathcal{M}_{k+\ell+2}(\Gamma)$ (see [Zag94]).

For the full modular group $\mathrm{SL}(2, \mathbb{Z})$, the rings $\mathcal{M}(\mathrm{SL}(2, \mathbb{Z}))$ and $\mathcal{S}(\mathrm{SL}(2, \mathbb{Z}))$ are generated by the modular forms G_4 and G_6 of weight 4 and 6, respectively. A theorem of Deligne and Rapoport [DR73] guarantees that the rings $\mathcal{M}(\Gamma)$ and $\mathcal{S}(\Gamma)$ are finitely generated for an arbitrary congruence subgroup Γ. Producing complete finite sets of generators for these rings is, however, still incompletely understood. For recent progress, including the use of toric modular forms, we refer to [BG03] and [KM11] and references therein.

7.7.2 SageMath

Let us close this subsection with a note for the practical researcher. Holomorphic Eisenstein series for various congruence subgroups, along with many of their properties, can be straightforwardly implemented in the computer program SageMath [Devrg]. For example, the simple code,

```
> E = EisensteinForms(Gamma0(9), 4, prec = 10)
> E.eisenstein_series()
```

can be used to output the Fourier expansions to order $\mathcal{O}(q^{10})$ for all generators of weight-4 Eisenstein series of $\Gamma_0(9)$. The triplets $(\chi_{a_1}, \tilde{\chi}_{a_2}, t)$ labeling each generator can be further obtained using,

```
> e = E.eisenstein_series()
> for e in E.eisenstein_series() :
      print e.parameters()
```

In practice, using such computer algebra programs to generate the spaces of modular forms at a given level can be much simpler than working through the definitions above.

7.8 Jacobi's theorem on sums of four and eight squares

We now provide a classic application of modular forms for the congruence subgroup $\Gamma_0(4)$ to the problem of representing integers by a sum of k squares. For positive integer n, we denote by $r_k(n)$ the number of ways in which n can be written as the sum of k positive, zero, or negative integers squared,

$$r_k(n) = \left| \left\{ m_1, \cdots, m_k \in \mathbb{Z} \text{ such that } n = m_1^2 + \cdots + m_k^2 \right\} \right| \quad (7.8.1)$$

where $|S|$ denotes the cardinality of the set S. The generating function for $r_k(n)$ is given by the function $\vartheta(0|2\tau)^k$ as may be seen by expanding this function in powers of $q = e^{2\pi i \tau}$,

$$\vartheta(0|2\tau)^k = \sum_{m_1, \cdots, m_k \in \mathbb{Z}} q^{m_1^2 + \cdots + m_k^2} = \sum_{n=0}^{\infty} r_k(n) q^n \quad (7.8.2)$$

Note that the argument of ϑ has been chosen to be 2τ for convenience, so that the expansion is in terms of q raised to integer powers. The function $\vartheta(0|2\tau)^k$ is invariant under $\tau \to \tau+1$, but it is not invariant under $\tau \to -1/\tau$ for $k \neq 0$ and, therefore, is not a modular form under $\mathrm{SL}(2, \mathbb{Z})$. However, it is a modular form under a certain congruence subgroup of $\mathrm{SL}(2, \mathbb{Z})$. In particular, we shall prove the following theorem of Jacobi with the help of modular forms and Eisenstein series for the congruence subgroup $\Gamma_0(4)$.

Theorem 7.8.1. *The functions $r_4(n)$ and $r_8(n)$ are given by,*

$$r_4(n) = 8 \sum_{4 \nmid d \mid n} d \qquad\qquad r_8(n) = 16 \sum_{d \mid n} (-1)^{d+n} d^3 \qquad (7.8.3)$$

where the sums are over positive divisors d of n, which for $r_4(n)$ are not divisible by 4. Jacobi's theorem implies the theorem of Lagrange that every positive integer can be written as the sum of four squares.

The remainder of this subsection will be devoted to providing a detailed proof of the first part of Theorem 7.8.1 regarding $r_4(n)$, while the proof of the second part, regarding $r_8(n)$, is relegated to Exercise 7.5.

We begin by identifying the largest modular subgroup $\Gamma \subset \mathrm{SL}(2,\mathbb{Z})$ under which $\vartheta(0|2\tau)^4$, and thus $\vartheta(0|2\tau)^{4\kappa}$ for $\kappa \in \mathbb{N}$, are modular forms. One generator of Γ may be taken to be $T : \tau \to \tau + 1$, under which $\vartheta(0|2\tau)^4$ is manifestly invariant. Under $S : \tau \to -1/\tau$ we have,

$$\vartheta(0|-2/\tau)^4 = (-i\tau/2)^2 \vartheta(0|\tau/2)^4 \qquad (7.8.4)$$

so that $\vartheta(0|2\tau)^4$ clearly is not a modular form under $\mathrm{SL}(2,\mathbb{Z})$. Instead, we look for more general modular transformations $\gamma = \left(\begin{smallmatrix} a & b \\ c & d \end{smallmatrix}\right) \in \mathrm{SL}(2,\mathbb{Z})$ under which we obtain good transformation laws. A general $\gamma \in \mathrm{SL}(2,\mathbb{Z})$ may be simplified by using the fact that we have invariance under T, and this allows us to set $b = 0$. But then we have $ad = 1$ which is solved by $a = d = 1$ and $a = d = -1$. These cases are equivalent to one another upon reversal of the sign of c, so that we consider the modular transformation with $a = d = 1$, namely, $\tilde{\tau} = \gamma\tau = \tau/(c\tau + 1)$, under which we have,

$$\vartheta(0|2\tilde{\tau})^4 = \left(i\frac{c\tau + 1}{2\tau} \right)^2 \vartheta\left(0 \middle| -\frac{c}{2} - \frac{1}{2\tau} \right)^4 \qquad (7.8.5)$$

The shift by $-c/2$ cancels in view of T-invariance if and only if $c \equiv 0 \pmod 4$ in view of the relation $\vartheta(z|\tau + 2) = \vartheta(z|\tau)$ from (3.7.4). Assuming that $c \equiv 0 \pmod 4$ the remaining argument $-1/(2\tau)$ of ϑ is independent of c. Using $\vartheta(0|-1/(2\tau))^4 = -4\tau^2 \vartheta(0|2\tau)^4$, we obtain,

$$\vartheta(0|2\tilde{\tau})^4 = (c\tau + 1)^2 \vartheta(0|2\tau)^4 \qquad (7.8.6)$$

Thus, we have identified two elements of $\mathrm{SL}(2,\mathbb{Z})$ under which $\vartheta(0|2\tau)^4$ transforms as a modular form of weight 2, namely,

$$T : \tau \to \tau + 1 \qquad\qquad \tilde{S} : \tau \to \tau/(4\tau + 1) \qquad (7.8.7)$$

The transformations T and $\tilde{S}$ generate the congruence subgroup $\Gamma_1(4)$. However, the element $-I$ acts trivially on τ and thus leaves $\vartheta(0|2\tau)^4$ invariant. Together, $-I, T, \tilde{S}$ generate the congruence subgroup $\Gamma_0(4) = \Gamma_1(4) \times \mathbb{Z}_2$, so that $\vartheta(0|2\tau)^4$ is a weight-2 modular form under $\Gamma_0(4)$. Of course, there are no modular forms of weight 2 under the full modular group $SL(2, \mathbb{Z})$. So this instance provides a simple illustration of the sequence of inclusions given in (7.1.3), namely, the "smaller" the modular subgroup of $SL(2, \mathbb{Z})$ is, the "larger" the number of modular forms it will admit for a given weight.

Now we would like to identify $\vartheta(0|2\tau)^4$ with an Eisenstein series whose Fourier expansion we can compute directly in terms of divisor functions $\sigma_\alpha(n)$. What are the modular forms of weight 2? Let us start from the form $\mathsf{E}_2(\tau)$ introduced in Subsection 4.1.1, which is not a modular form under $SL(2, \mathbb{Z})$, but rather a modular connection. Since the difference of two connections is a tensor, we investigate how $\mathsf{E}_2(n\tau)$ transforms for various values of n. Under $T : \tau \to \tau + 1$, the form $\mathsf{E}_2(n\tau)$ is invariant for any $n \in \mathbb{N}$. Next, examine its transformation properties under the transformation,

$$\tau \to \tilde{\tau} = \frac{\tau}{N\tau + 1} \tag{7.8.8}$$

Now E_2 is given in terms of Δ by (4.1.6) whose transformation law is,

$$\Delta(n\tilde{\tau}) = \left(-\frac{N\tau + 1}{n\tau}\right)^{12} \Delta\left(-\frac{N}{n} - \frac{1}{n\tau}\right) \tag{7.8.9}$$

When $n|N$, the combination $-N/n$ in the argument of Δ cancels by T-invariance of $\Delta(\tau)$. Taking the transform of the simplified result produced by the assumption $n|N$, we obtain after some simplifications,

$$\Delta(n\tilde{\tau}) = (N\tau + 1)^{12}\Delta(n\tau) \tag{7.8.10}$$

For $n|N$, the function $\Delta(n\tau)$ is a weight-12 modular form under $\Gamma_0(N)$, as proven in Exercise 7.1. Applied to the sum of squares problem we set $N = 4$, so that $\Delta(\tau), \Delta(2\tau), \Delta(4\tau)$ are modular forms of $\Gamma_0(4)$. Taking the derivatives of the logarithms on both sides, and expressing the result in terms of E_2, we find,

$$\mathsf{E}_2(n\tilde{\tau}) = (4\tau + 1)^2 \mathsf{E}_2(n\tau) + \frac{4}{n} \times \frac{12}{2\pi i}(4\tau + 1) \tag{7.8.11}$$

It follows that the two linear combinations,

$$\mathsf{E}_2(\tau) - 2\,\mathsf{E}_2(2\tau) \qquad\qquad \mathsf{E}_2(2\tau) - 2\,\mathsf{E}_2(4\tau) \tag{7.8.12}$$

are linearly independent modular forms under $\Gamma_0(4)$ of weight 2, matching the prediction of dimension 2 for the space of modular forms for $\Gamma_0(4)$

from (7.2.3) and Theorem 7.7.1. The above combinations generate this space completely. Matching the q^0, q^1 terms of the q-series expansion for E_2,

$$E_2(n\tau) = 1 - 24 \sum_{\ell=1}^{\infty} \sigma_1(\ell)q^{n\ell} \tag{7.8.13}$$

we establish the following identity,

$$\vartheta(0|2\tau)^4 = -\frac{1}{3} E_2(\tau) + \frac{4}{3} E_2(4\tau) = 1 + 8 \sum_{\ell=1}^{\infty} \sigma_1(\ell)\left(q^\ell - 4q^{4\ell}\right) \tag{7.8.14}$$

thereby completing the proof of Jacobi's theorem (7.8.3) for $r_4(n)$. This relation may be used to obtain the derivatives of ϑ_3 with respect to τ, as illustrated in Exercise 7.4.

For $r_2(n)$, the above derivation needs to be extended to modular form of $\Gamma_0(4)$ with characters, and the result is given by the following formula,

$$r_2(n) = 4 \sum_{d \text{ odd}, \, d|n} (-1)^{(d-1)/2} \tag{7.8.15}$$

In particular, if n is an odd prime such that $n \equiv 1 \pmod 4$, we obtain $r_2(n) = 8$ which shows that n is the sum of the squares of two nonzero distinct integers $(\pm r, \pm s)$ and $(\pm s, \pm r)$, while if n is an odd prime such that $n \equiv 3 \pmod 4$, we obtain $r_2(n) = 0$. The combination of these two results implies Fermat's "Little Theorem" that an odd prime p is the sum of two nonzero squares if and only if $p \equiv 1 \pmod 4$.

• Bibliographical notes

The books by Shimura [Shi71] and by Diamond and Shurman [DS05] provide beautiful and very useful introductions to modular forms for congruence subgroups. We also refer to the book by Miyake [Miy06] and the software database [Devrg]. A clear account of Jacobi's sum of four-square theorem may be found in Apostol's book [Apo76b]. Our summary closely follows their presentation. Essays geared toward applications to the proof of Fermat's Last Theorem are collected in [CY97].

Exercises

7.1 Show that the ratio $\Delta(N\tau)/\Delta(\tau)$ is invariant under $\Gamma_0(N)$ for all N and that $\varphi_N(\tau) = \eta(N\tau)^r/\eta(\tau)^r$ is invariant under $\Gamma_0(N)$ if and only if r is an even integer and $r(N-1) \equiv 0 \pmod{24}$ (see [Apo76b]).

7.2   By equating the representations for $\vartheta(z|\tau)$ in terms of a sum (2.6.1) and a product (2.6.24) prove the following formula [Köh11, Corr 1.3],

$$\prod_{n=1}^{\infty}(1-q^n)(1+q^{n-\frac{1}{2}}w)(1+q^{n-\frac{1}{2}}w^{-1}) = \sum_{n=-\infty}^{\infty} q^{\frac{1}{2}n^2} w^n \quad (7.8.16)$$

By suitably choosing of q and w prove the following formulas,

$$\prod_{n=1}^{\infty}(1-q^n) = \sum_{n=-\infty}^{\infty} (-)^n q^{\frac{1}{2}n(3n-1)}$$

$$\prod_{n=1}^{\infty}(1-q^n)^2(1-q^{2n})^{-1} = \sum_{n=-\infty}^{\infty} (-)^n q^{n^2} \quad (7.8.17)$$

due to Euler and Gauss, and the following formula due to Jacobi,

$$\prod_{n=1}^{\infty}(1-q^n)^3 = \sum_{n=0}^{\infty}(-)^n(2n+1)q^{\frac{1}{2}n(n+1)} \quad (7.8.18)$$

7.3   Use the results of Exercise 7.2 to show the following expansions for the Dedekind η-function and various powers and combinations thereof,

$$\eta(\tau) = \sum_{n=-\infty}^{\infty} (-)^n e^{2\pi i(6n-1)^2\tau/24} = \sum_{m=1}^{\infty}(12|m)\, e^{2\pi i m^2 \tau/24}$$

$$\frac{\eta(\tau)^2}{\eta(2\tau)} = \sum_{n=-\infty}^{\infty} (-)^n e^{2\pi i n^2 \tau} = \vartheta(\tfrac{1}{2}|2\tau) = \vartheta_4(0|2\tau) \quad (7.8.19)$$

$$\eta(\tau)^3 = \sum_{n=0}^{\infty}(-)^n(2n+1)e^{i\pi\tau(n+\frac{1}{2})^2} = \sum_{m=1}^{\infty}(-1|m)\, m\, e^{2\pi i m^2\tau/8}$$

Here, $(-1|m)$ and $(12|m)$ are the Jacobi symbols (see Appendix A.4) that satisfy $(-1|m) = 0$ when m is even and.

7.4   Use the relation (7.8.14) to prove the following derivative of ϑ-constants,

$$\vartheta_3(0|\tau)^4 = \frac{1}{\pi i}\frac{\partial}{\partial \tau}\ln\frac{\vartheta_2(0|\tau)^4}{\vartheta_4(0|\tau)^4} \quad (7.8.20)$$

7.5   Using the construction of Eisenstein series for the congruence subgroup $\Gamma_0(4)$ given in Section 7.6, prove the following relation,

$$\vartheta(0|2\tau)^8 = \tfrac{1}{15}\Big(E_4(\tau) - 2E_4(2\tau) + 16E_4(4\tau)\Big) \quad (7.8.21)$$

and deduce the explicit formula for $r_8(n)$ given in Theorem 7.8.1.

8

Modular derivatives and vector-valued modular forms

In this chapter, we construct differential equations in the modular parameter τ and find solutions to these equations in simple cases. The solutions can be assembled into vector-valued modular forms, which have proven fruitful in recent works in mathematics and physics. We will establish that, in general, each component of a vector-valued modular form is a modular form for a congruence subgroup.

8.1 Modular covariant derivatives

In Section 4.1, we encountered the holomorphic function $E_2(\tau)$, whose transformation under $SL(2, \mathbb{Z})$ is that of a modular connection, as given in (4.1.3). We may use this object to define a modular covariant differential operator acting on modular forms of weight k, referred to as the Serre derivative,

$$D_k = \frac{1}{2\pi i}\frac{d}{d\tau} - \frac{k}{12}E_2(\tau) \tag{8.1.1}$$

As was already shown in Subsection 4.1.3, the operator D_k maps a modular form of weight k to a modular form of weight $k + 2$,

$$D_k : \mathcal{M}_k \to \mathcal{M}_{k+2} \tag{8.1.2}$$

The first-order differential operators D_k may be composed to define modular covariant derivatives $\mathcal{D}_k^{(d)}$ of arbitrary order $d \geq 1$,

$$\mathcal{D}_k^{(d)} = D_{k+2d-2} \cdots D_{k+2}D_k \qquad \mathcal{D}_k^{(d)} : \mathcal{M}_k \to \mathcal{M}_{k+2d} \tag{8.1.3}$$

so that $\mathcal{D}_k^{(1)} = D_k$. For example, as already established in (4.1.17), the first-order differential operators on Eisenstein series give,

$$D_4 E_4 = -\frac{1}{3}E_6 \qquad D_6 E_6 = -\frac{1}{2}E_4^2 \tag{8.1.4}$$

while for the second-order differential operators we have,

$$\mathcal{D}_4^{(2)}\mathsf{E}_4 = \frac{1}{6}\mathsf{E}_4^2 \qquad\qquad \mathcal{D}_6^{(2)}\mathsf{E}_6 = \frac{1}{3}\mathsf{E}_4\mathsf{E}_6 \qquad\qquad (8.1.5)$$

The second-order derivatives follow directly from the first-order ones. Recall that proving (8.1.4) used the fact that the spaces of weight 6 and 8 modular forms are generated by E_6 and E_4^2, respectively, so the results are fixed up to normalization. To determine the normalization, one can compare the first terms in the q-expansion using (3.4.13).

8.2 Modular differential equations

Consider now an arbitrary linear modular covariant differential equation of order d in τ for a single modular form $f(\tau)$ of weight k,

$$\left[\mathcal{D}_k^{(d)} + \sum_{r=1}^{d} h_r(\tau)\mathcal{D}_k^{(d-r)}\right] f(\tau) = 0 \qquad\qquad (8.2.1)$$

which we shall refer to as a *modular differential equation* (MDE). In (8.2.1), $h_r(\tau)$ is an arbitrary meromorphic modular form of weight $2r$. Some of the simplest MDEs are obtained by taking the $h_r(\tau)$ to be holomorphic, given by polynomials in $\mathsf{E}_4(\tau)$ and $\mathsf{E}_6(\tau)$.

8.2.1 First-order MDE

The generic first-order MDE for modular weight k takes the form,

$$D_k f + h_1 f = 0 \qquad\qquad (8.2.2)$$

for a meromorphic modular form h_1 of weight 2 and a modular form f of weight k. Using the relation between E_2 and the discriminant Δ of (4.1.6),

$$\mathsf{E}_2(\tau) = \frac{1}{2\pi i}\partial_\tau \ln \Delta(\tau) \qquad\qquad (8.2.3)$$

and rescaling f as follows,

$$f(\tau) = \Delta(\tau)^{\frac{k}{12}} f_0(\tau) \qquad\qquad (8.2.4)$$

we obtain a simplified equation for the function f_0,

$$\frac{1}{2\pi i}\frac{df_0}{d\tau} + h_1(\tau)f_0(\tau) = 0 \qquad\qquad (8.2.5)$$

Note that $f_0(\tau)$ transforms as a modular function of weight 0, up to multiplication by 12th roots of unity. Furthermore, since h_1 is a meromorphic modular form of weight 2, it may be expressed as follows,

$$h_1(\tau) = \frac{E_6(\tau)}{E_4(\tau)} h_0(j) \qquad (8.2.6)$$

Here $h_0(j)$ is a modular function of weight zero in τ, which may therefore be expressed as a function of $j = j(\tau)$. Using the definition of $j(\tau)$ in terms of E_4 and E_6 given in (3.6.1), namely, $j = 12^3 E_4^3/(E_4^3 - E_6^2)$, together with (8.1.4), we obtain the derivative of $j(\tau)$ respect to τ,

$$\frac{1}{2\pi i}\frac{dj}{d\tau} = -\frac{E_6}{E_4} j \qquad (8.2.7)$$

which by the chain rule also gives,

$$\frac{1}{2\pi i}\frac{df_0}{d\tau} = -\frac{E_6}{E_4} j \frac{df_0(j)}{dj} \qquad (8.2.8)$$

Using the expression (8.2.6) for h_1 in terms of h_0, and cancelling a common factor of E_6/E_4, the equation for f_0 now becomes,

$$j\frac{df_0(j)}{dj} = h_0(j) f_0(j) \qquad (8.2.9)$$

which may be solved by quadrature. It remains to ensure that the local solution $f(\tau)$ obtained from (8.2.4) via this procedure is a proper modular form of weight k. This requirement may place conditions on the allowed values of k. Modular differential equations may be used to establish certain properties of the Fourier expansion of their solutions, as is illustrated by a first-order MDE for $\Delta(\tau)$ in Exercise 8.1.

8.2.2 Second-order MDE

Consider now the example of a second-order MDE with holomorphic coefficient functions h_r in (8.2.1), acting on a modular function f of weight zero (the case of arbitrary weight k may be treated by factoring out $\Delta(\tau)^{k/12}$, as was done in (8.2.4) for the case $d = 1$, and then solving for the corresponding f_0 of weight zero). The most general such MDE takes the form,

$$[D_2 D_0 + \gamma E_4(\tau)]f(\tau) = 0 \qquad (8.2.10)$$

for some free constant parameter γ. More explicitly we have,

$$\left[\left(\frac{1}{2\pi i}\frac{d}{d\tau}\right)^2 - \frac{1}{12\pi i}E_2(\tau)\frac{d}{d\tau} + \gamma E_4(\tau)\right] f(\tau) = 0 \qquad (8.2.11)$$

The general solution to this equation may be obtained as follows. We begin by trading the derivatives with respect to τ for derivatives with respect to j, as was done for the first-order MDE, using (8.2.7). It follows that,

$$\frac{df}{dj} = -\frac{\mathsf{E}_4}{j\,\mathsf{E}_6}\frac{1}{2\pi i}\frac{df}{d\tau} \tag{8.2.12}$$

and by similar means,

$$\frac{d^2 f}{dj^2} = \frac{\mathsf{E}_4^2}{j^2\,\mathsf{E}_6^2}\left(\left(\frac{1}{2\pi i}\frac{d}{d\tau}\right)^2 - \frac{\mathsf{E}_2}{6}\frac{1}{2\pi i}\frac{d}{d\tau}\right)f - \frac{7j - 6912}{6j(j - 1728)}\frac{df}{dj} \tag{8.2.13}$$

With these results, the equation (8.2.11) can be rewritten as,

$$\left[j(j - 1728)\frac{d^2}{dj^2} + \frac{1}{6}(7j - 6912)\frac{d}{dj} + \gamma\right]f(\tau(j)) = 0 \tag{8.2.14}$$

In terms of the variable $j(\tau)/1728$, this equation is of standard hypergeometric type. For generic choice of γ, there exist two distinct solutions given in closed form by,[1]

$$f_1(\tau) = j(\tau)^{-\frac{(1+\mu)}{12}}\,{}_2F_1\left(\tfrac{1+\mu}{12}, \tfrac{5+\mu}{12}\,;\, 1 + \tfrac{\mu}{6}\,;\, 1728\,j(\tau)^{-1}\right)$$

$$f_2(\tau) = j(\tau)^{-\frac{(1-\mu)}{12}}\,{}_2F_1\left(\tfrac{1-\mu}{12}, \tfrac{5-\mu}{12}\,;\, 1 - \tfrac{\mu}{6}\,;\, 1728\,j(\tau)^{-1}\right) \tag{8.2.16}$$

where we have defined $\mu = \sqrt{1 - 144\gamma}$. Some arithmetic properties of these solutions for rational values of μ are discussed in Exercise 8.2.

8.2.3 Third-order MDE

We may similarly analyze the case of a third-order MDE with holomorphic coefficient functions and with f of weight 0. In this case, the most general MDE depends on two free parameters α and β,

$$[D_4 D_2 D_0 + \alpha \mathsf{E}_4(\tau)D_0 + \beta \mathsf{E}_6(\tau)]\,f(\tau) = 0 \tag{8.2.17}$$

By switching to the local coordinate $j = j(\tau)$, as before, we can again reexpress this as a hypergeometric equation. For generic choices of the pa-

---

[1] When $\mu \in 6\,\mathbb{N}$, the second solution is divergent, since the third argument of the hypergeometric function is a negative integer. In this case, the second solution should be replaced by the Meijer G-function,

$$f_2(\tau) = G^{2,0}_{2,2}\left(\begin{matrix}\tfrac{2}{3} & \\ \tfrac{1-\mu}{12} & \end{matrix}\,\begin{matrix}1 \\ \tfrac{1+\mu}{12}\end{matrix}\,\middle|\, 1728\,j(\tau)^{-1}\right) \tag{8.2.15}$$

rameters α and β, there are three distinct solutions given by,

$$f_1(\tau) = j^{-\frac{\mu_1+1}{6}}\,{}_3F_2\left(\tfrac{\mu_1+1}{6}, \tfrac{\mu_1+3}{6}, \tfrac{\mu_1+5}{6}; \tfrac{\mu_1-\mu_2}{6}+1, \tfrac{\mu_1-\mu_3}{6}+1; 1728\,j^{-1}\right)$$

$$f_2(\tau) = j^{-\frac{\mu_2+1}{6}}\,{}_3F_2\left(\tfrac{\mu_2+1}{6}, \tfrac{\mu_2+3}{6}, \tfrac{\mu_2+5}{6}; \tfrac{\mu_2-\mu_1}{6}+1, \tfrac{\mu_2-\mu_3}{6}+1; 1728\,j^{-1}\right)$$

$$f_3(\tau) = j^{-\frac{\mu_3+1}{6}}\,{}_3F_2\left(\tfrac{\mu_3+1}{6}, \tfrac{\mu_3+3}{6}, \tfrac{\mu_3+5}{6}; \tfrac{\mu_3-\mu_2}{6}+1, \tfrac{\mu_3-\mu_1}{6}+1; 1728\,j^{-1}\right)$$

$$(8.2.18)$$

Here we have defined μ_i as,

$$\mu_1 = 4x_1 - 2x_2 - 2x_3$$
$$\mu_2 = 4x_2 - 2x_3 - 2x_1$$
$$\mu_3 = 4x_3 - 2x_1 - 2x_2 \qquad (8.2.19)$$

and x_i, $i = 1, 2, 3$ as the three roots of the cubic equation,

$$x^3 - \tfrac{1}{2}x^2 + \left(\alpha + \tfrac{1}{18}\right)x + \beta = 0 \qquad (8.2.20)$$

8.2.4 Example

One interesting example is when the parameters in (8.2.17) are taken to be,

$$\alpha = -\frac{107}{48^2} \qquad\qquad \beta = \frac{46}{48^3} \qquad (8.2.21)$$

In this case, the exact solutions given above have the following q-expansions,

$$f_1(\tau) = q^{\frac{23}{48}}\left(1 + q + q^2 + q^3 + 2q^4 + 2q^5 + \dots\right)$$
$$f_2(\tau) = q^{-\frac{1}{48}}\left(1 + q^2 + q^3 + 2q^4 + 2q^5 + \dots\right)$$
$$f_3(\tau) = q^{\frac{1}{24}}\left(1 + q + q^2 + 2q^3 + 2q^4 + 3q^5 + \dots\right) \qquad (8.2.22)$$

These q-expansions match those of the following expressions in terms of Jacobi ϑ-constants, whose product expansions are given in (2.6.24),

$$f_1(\tau) = \frac{1}{2}\left(\sqrt{\frac{\vartheta_3(0|\tau)}{\eta(\tau)}} - \sqrt{\frac{\vartheta_4(0|\tau)}{\eta(\tau)}}\right)$$

$$f_2(\tau) = \frac{1}{2}\left(\sqrt{\frac{\vartheta_3(0|\tau)}{\eta(\tau)}} + \sqrt{\frac{\vartheta_4(0|\tau)}{\eta(\tau)}}\right)$$

$$f_3(\tau) = \sqrt{\frac{\vartheta_2(0|\tau)}{\eta(\tau)}} \qquad (8.2.23)$$

The f_i satisfy (8.2.17), as verified directly in Exercises 8.3 and 8.4.

8.2.5 Modular invariance of solution spaces

Because MDEs are, by definition, modular invariant, so too are their solution spaces. This means that under $\mathrm{SL}(2,\mathbb{Z})$, the solutions must transform into linear combinations of themselves. It is useful to organize a basis for the solutions to an order-d MDE into a d-dimensional column vector and to represent the generators $T, S \in \mathrm{SL}(2,\mathbb{Z})$ by $d \times d$ matrices. For example, in the concrete case of (8.2.23), we can organize the solutions into a three-dimensional column vector $\mathbf{f}(\tau) = (f_1(\tau),\, f_2(\tau),\, f_3(\tau))^t$. From the known transformation properties of $\vartheta_i(0|\tau)$ given in (3.7.4), and $\eta(\tau)$ given in (3.8.3), we obtain the matrix representations of T and S on $\mathbf{f}(\tau)$ as follows,

$$
T = \begin{pmatrix} e^{\frac{23\pi i}{24}} & 0 & 0 \\ 0 & e^{-\frac{\pi i}{24}} & 0 \\ 0 & 0 & e^{\frac{\pi i}{12}} \end{pmatrix} \qquad S = \frac{1}{2}\begin{pmatrix} 1 & 1 & -1 \\ 1 & 1 & 1 \\ -2 & 2 & 0 \end{pmatrix} \qquad (8.2.24)
$$

This leads us to a fruitful generalization of the notion of modular forms, known as *vector-valued modular forms*.

8.3 Vector-valued modular forms

A d-dimensional vector-valued modular form $\mathbf{f}(\tau) = (f_1(\tau), \ldots, f_d(\tau))^t$ of weight k is defined by its modular transformation properties,

$$
\mathbf{f}(\gamma\tau) = (c\tau + d)^k \rho(\gamma)\,\mathbf{f}(\tau) \qquad \gamma = \begin{pmatrix} a & b \\ c & d \end{pmatrix} \in \mathrm{SL}(2,\mathbb{Z}) \qquad (8.3.1)
$$

where $\rho : \mathrm{SL}(2,\mathbb{Z}) \to \mathrm{GL}(d,\mathbb{C})$ is a $d \times d$ matrix representation of $\mathrm{SL}(2,\mathbb{Z})$. The function $\mathbf{f}(\tau)$ is further required to be holomorphic in the upper half plane and to have at most finite-order poles at the cusps.

8.3.1 Examples

We begin with some simple examples of vector-valued modular functions. We have already seen that solutions to an MDE of order d combine into a d-dimensional vector-valued modular form. In Subsection 8.3.3, we will show that the converse is also true, that is, the components of a d-dimensional vector-valued modular form are always solutions to a MDE of order d. In this section, however we will momentarily ignore the connection to MDEs.

Example 1. The Jacobi ϑ-constants produce a three-dimensional vector-valued modular form $(\vartheta_2^8(0|\tau), \vartheta_3^8(0|\tau), \vartheta_4^8(0|\tau))$. For simplicity, we have taken the eighth power to eliminate potential phases from the modular transformations given in (3.7.7). Under T and S transformations one has,

$$\vartheta_2^8(0|\tau+1) = \vartheta_2^8(0|\tau) \qquad\qquad \vartheta_2^8(0|-1/\tau) = \tau^4 \vartheta_4^8(0|\tau)$$
$$\vartheta_3^8(0|\tau+1) = \vartheta_4^8(0|\tau) \qquad\qquad \vartheta_3^8(0|-1/\tau) = \tau^4 \vartheta_3^8(0|\tau)$$
$$\vartheta_4^8(0|\tau+1) = \vartheta_3^8(0|\tau) \qquad\qquad \vartheta_4^8(0|-1/\tau) = \tau^4 \vartheta_2^8(0|\tau) \quad (8.3.2)$$

Clearly T and S act by permutations of the three ϑ-constants. In the usual permutation notation, the T transformation corresponds to the transformation (132), while the S transformation corresponds to (321). Together, these generate the group S_3 of permutations of three elements.

A more formal way of presenting the above result is through the existence of a surjective homomorphism $\rho : \mathrm{SL}(2, \mathbb{Z}) \to S_3$. The kernel of the homomorphism ρ is the subgroup of $\mathrm{SL}(2, \mathbb{Z})$ under which each individual ϑ-constant is a modular form,

$$\mathrm{Ker}(\rho) = \left\{ \gamma \in \mathrm{SL}(2, \mathbb{Z}) \,\middle|\, \vartheta_i^8(0|\gamma\tau) = (c\tau + d)^4 \vartheta_i^8(0|\tau), \; i = 2, 3, 4 \right\} \quad (8.3.3)$$

where as usual we set $\gamma = \left(\begin{smallmatrix} a & b \\ c & d \end{smallmatrix}\right) \in \mathrm{SL}(2, \mathbb{Z})$. Recalling the notation in terms of half-integer characteristics,

$$\vartheta_2 = \vartheta \begin{bmatrix} 1/2 \\ 0 \end{bmatrix} \qquad \vartheta_3 = \vartheta \begin{bmatrix} 0 \\ 0 \end{bmatrix} \qquad \vartheta_4 = \vartheta \begin{bmatrix} 0 \\ 1/2 \end{bmatrix} \quad (8.3.4)$$

from (3.7.7) we see that elements of $\mathrm{Ker}(\rho)$ must have,

$$a\alpha + c\beta \equiv \alpha \pmod 1 \qquad b\alpha + d\beta \equiv \beta \pmod 1 \quad (8.3.5)$$

for any choice $(\alpha, \beta) \in \left\{ (0,0), \left(\tfrac{1}{2}, 0\right), \left(0, \tfrac{1}{2}\right) \right\}$. For $\alpha = \beta = 0$, the equations are trivially satisfied while for the other cases they require,

$$(\alpha, \beta) = (\tfrac{1}{2}, 0) \qquad a \equiv 1 \;(\text{mod } 2) \qquad b \equiv 0 \;(\text{mod } 2)$$
$$(\alpha, \beta) = (0, \tfrac{1}{2}) \qquad c \equiv 0 \;(\text{mod } 2) \qquad d \equiv 1 \;(\text{mod } 2) \quad (8.3.6)$$

We thus conclude that $\mathrm{Ker}(\rho)$ is given by

$$\mathrm{Ker}(\rho) = \left\{ \begin{pmatrix} a & b \\ c & d \end{pmatrix} \in \mathrm{SL}(2, \mathbb{Z}) \,\middle|\, \begin{pmatrix} a & b \\ c & d \end{pmatrix} \equiv \begin{pmatrix} 1 & 0 \\ 0 & 1 \end{pmatrix} \;(\text{mod } 2) \right\} \quad (8.3.7)$$

In other words, we find that $\mathrm{Ker}(\rho)$ is equal to the principal congruence subgroup $\Gamma(2)$. Thus, while the vector $(\vartheta_2^8(0|\tau), \vartheta_3^8(0|\tau), \vartheta_4^8(0|\tau))$ transforms as a vector-valued modular form in a nontrivial permutation representation of $\mathrm{SL}(2, \mathbb{Z})$, we see that each of the individual components of this vector is itself a genuine modular form under $\Gamma(2)$.

Example 2. Proceeding to a four-dimensional example, we consider the vector $\mathbf{f}(\tau)$ with components,

$$f_1(\tau) = 3^{12}\,\eta(3\tau)^{24} \qquad\qquad f_3(\tau) = \eta\left(\tfrac{\tau+1}{3}\right)^{24}$$
$$f_2(\tau) = \eta\left(\tfrac{\tau}{3}\right)^{24} \qquad\qquad f_4(\tau) = \eta\left(\tfrac{\tau+2}{3}\right)^{24} \qquad (8.3.8)$$

Under T and S modular transformations, we have the following

$$\begin{aligned}
f_1(\tau+1) &= f_1(\tau) & f_1(-1/\tau) &= \tau^{12} f_2(\tau) \\
f_2(\tau+1) &= f_3(\tau) & f_2(-1/\tau) &= \tau^{12} f_1(\tau) \\
f_3(\tau+1) &= f_4(\tau) & f_3(-1/\tau) &= \tau^{12} f_4(\tau) \\
f_4(\tau+1) &= f_2(\tau) & f_4(-1/\tau) &= \tau^{12} f_3(\tau) & (8.3.9)
\end{aligned}$$

The generators T and S act on the four entries of the vector $\mathbf{f}(\tau)$ by the permutations (1342) and (2143) of the permutation group S_4 of four elements. Since (1342) and (2143) are both even permutations, they do not generate all of S_4, but only the alternating subgroup A_4 of even permutations. So in this case, we obtain a surjective homomorphism $\rho : \mathrm{SL}(2,\mathbb{Z}) \to A_4$. We may again ask about the kernel of ρ. Demanding invariance of $f_1(\tau)$ under $\left(\begin{smallmatrix} a & b \\ c & d \end{smallmatrix}\right) \in \mathrm{SL}(2,\mathbb{Z})$ is equivalent to requiring,

$$\begin{pmatrix} 3 & 0 \\ 0 & 1 \end{pmatrix}\begin{pmatrix} a & b \\ c & d \end{pmatrix}\begin{pmatrix} 3 & 0 \\ 0 & 1 \end{pmatrix}^{-1} = \begin{pmatrix} a & 3b \\ c/3 & d \end{pmatrix} \in \mathrm{SL}(2,\mathbb{Z}) \qquad (8.3.10)$$

which requires that $c \in 3\mathbb{Z}$. Similarly, invariance of $f_2(\tau)$ requires,

$$\begin{pmatrix} 1 & 0 \\ 0 & 3 \end{pmatrix}\begin{pmatrix} a & b \\ c & d \end{pmatrix}\begin{pmatrix} 1 & 0 \\ 0 & 3 \end{pmatrix}^{-1} = \begin{pmatrix} a & b/3 \\ 3c & d \end{pmatrix} \in \mathrm{SL}(2,\mathbb{Z}) \qquad (8.3.11)$$

which requires $b \in 3\mathbb{Z}$. Any permutation in A_4 that fixes $f_1(\tau)$ and $f_2(\tau)$ also fixes $f_3(\tau)$ and $f_4(\tau)$. We conclude that $\mathrm{Ker}(\rho)$ consists of all the matrices for which $b \equiv c \equiv 0 \pmod 3$, including the element $-I \in \mathrm{SL}(2,\mathbb{Z})$. Hence the kernel is not quite $\Gamma(3)$, but rather $\Gamma(3) \times \mathbb{Z}_2$, where $\mathbb{Z}_2 = \{I, -I\}$. So we have found that, while the vector $\mathbf{f}(\tau)$ transforms as a vector-valued modular form in an alternating representation of $\mathrm{SL}(2,\mathbb{Z})$, each of the individual components of this vector is itself modular under the group $\Gamma(3) \times \mathbb{Z}_2$.

For completeness, let us record here some of the isomorphisms that we have encountered, as well as some we have not yet seen,

$$\begin{aligned}
\mathrm{SL}(2,\mathbb{Z})/\Gamma(2) &\cong S_3 & \mathrm{SL}(2,\mathbb{Z})/\big(\Gamma(3)\times\mathbb{Z}_2\big) &\cong A_4 \\
& & \mathrm{SL}(2,\mathbb{Z})/\big(\Gamma(4)\times\mathbb{Z}_2\big) &\cong S_4 \\
& & \mathrm{SL}(2,\mathbb{Z})/\big(\Gamma(5)\times\mathbb{Z}_2\big) &\cong A_5 & (8.3.12)
\end{aligned}$$

The identities are consistent with the values of the index $[\mathrm{SL}(2,\mathbb{Z}) : \Gamma(N)] = |\mathrm{SL}(2,\mathbb{Z}_N)|$ given in (6.2.7) and (6.3.9), which are,

$$[\mathrm{SL}(2,\mathbb{Z}) : \Gamma(2)] = 6 \qquad [SL(2,\mathbb{Z}) : \Gamma(3)] = 24$$
$$[\mathrm{SL}(2,\mathbb{Z}) : \Gamma(4)] = 48 \qquad [\mathrm{SL}(2,\mathbb{Z}) : \Gamma(5)] = 120 \qquad (8.3.13)$$

and may be compared with the orders of the relevant permutation groups,

$$|S_3| = 6 \qquad |A_4| = 12 \qquad |S_4| = 24 \qquad |A_5| = 60 \qquad (8.3.14)$$

For $N > 2$, the match requires the presence of the factor $\mathbb{Z}_2$ since $-I \notin \Gamma(N)$, while no such factor is required for $N = 2$ since $-I \in \Gamma(2)$.

8.3.2 Integrality and $\Gamma(N)$

In both of the above examples, we found that the components of the vector-valued modular forms were themselves modular under $\Gamma(N)$ for some appropriate N. This is not true in general, and it is interesting to ask when it holds. The following theorem provides the answer to this question.

Conjecture 8.3.1 (Integrality theorem). *Consider a component $f_i(\tau)$ of a vector-valued modular form with Fourier expansion,*

$$f_i(\tau) = q^{n_0^{(i)}} \sum_{n \geq 0} a_n^{(i)} q^n \qquad (8.3.15)$$

with $n_0^{(i)} = p_i/N_i$ and $(p_i, N_i) = 1$. If $n_0^{(i)} \in \mathbb{Q}$ and all coefficients $a_n^{(i)}$ are integers, then $f_i(\tau)$ is a modular form for $\Gamma(N_i)$.

The integrality theorem states that if a vector-valued modular form transforming in a representation ρ is integral (i.e., each component of the vector has integer Fourier coefficients), then $\mathrm{Ker}(\rho)$ contains a principal congruence subgroup $\Gamma(N)$. Indeed, the two examples given above had integer Fourier coefficients and are directly in line with this result. As another example, the three-dimensional vector-valued modular function given in (8.2.23) has all integer Fourier coefficients, and indeed the components can be seen to be modular functions for $\Gamma(48)$. A simple application of the integrality theorem is given in Exercise 8.5.

One common place in physics where vector-valued modular forms appear is as characters of conformal field theories. Since the Fourier coefficients in that case are interpretable as physical degeneracies, they must be integers. The conjecture then tells us that conformal field theory characters are always modular functions for some $\Gamma(N)$. Going back to the example of (8.2.23), the functions in this case are actually characters for the Ising conformal field theory,

$$f_1(\tau) = \chi_\epsilon(\tau) \qquad f_2(\tau) = \chi_1(\tau) \qquad f_3(\tau) = \chi_\sigma(\tau) \qquad (8.3.16)$$

where χ_1, χ_ε, and χ_σ are the characters of the identity 1, the energy density ε, and the spin σ operators, respectively.

8.3.3 Relation to modular differential equations

We now show that the components of a d-dimensional vector-valued modular form $\mathbf{f}(\tau) = (f_1(\tau), \cdots, f_d(\tau))^t$ are always solutions to a MDE of order d, assuming they are linearly independent. To see this, we begin by constructing the Wronskians W_r associated to $\mathbf{f}(\tau)$,

$$W_r = \det \begin{pmatrix} f_1 & \cdots & f_d \\ \mathcal{D}_k^{(1)} f_1 & \cdots & \mathcal{D}_k^{(1)} f_d \\ \vdots & & \vdots \\ \mathcal{D}_k^{(r-1)} f_1 & \cdots & \mathcal{D}_k^{(r-1)} f_d \\ \mathcal{D}_k^{(r+1)} f_1 & \cdots & \mathcal{D}_k^{(r+1)} f_d \\ \vdots & & \vdots \\ \mathcal{D}_k^{(d-1)} f_1 & \cdots & \mathcal{D}_k^{(d-1)} f_d \\ \mathcal{D}_k^{(d)} f_1 & \cdots & \mathcal{D}_k^{(d)} f_d \end{pmatrix} \qquad (8.3.17)$$

for $r = 1, \ldots, d$. Note that the determinant in W_r is being taken of a $d \times d$ matrix in view of the fact that the row $\mathcal{D}_k^{(r)}$ is missing. From the trivial identity $W_{d-1} - (W_{d-1}/W_d) W_d = 0$, we obtain the following equation,

$$\det \begin{pmatrix} f_1 & \cdots & f_d \\ \mathcal{D}_k^{(1)} f_1 & \cdots & \mathcal{D}_k^{(1)} f_d \\ \vdots & & \vdots \\ \mathcal{D}_k^{(d-2)} f_1 & \cdots & \mathcal{D}_k^{(d-2)} f_d \\ \mathcal{D}_k^{(d)} f_1 - \frac{W_{d-1}}{W_d} \mathcal{D}_k^{(d-1)} f_1 & \cdots & \mathcal{D}_k^{(d)} f_d - \frac{W_{d-1}}{W_d} \mathcal{D}_k^{(d-1)} f_d \end{pmatrix} = 0 \qquad (8.3.18)$$

For this determinant to vanish, the bottom row must be a linear combination of the other rows, which tells us that,

$$\mathcal{D}_k^{(d)} f_i - \frac{W_{d-1}}{W_d} \mathcal{D}_k^{(d-1)} f_i + \sum_{r=2}^{d} h_r(\tau) \mathcal{D}_k^{(d-r)} f_i = 0 \qquad (8.3.19)$$

This relation is of the usual MDE type given in (8.2.1), where in particular we have the coefficient function,

$$h_1(\tau) = -\frac{W_{d-1}}{W_d} \tag{8.3.20}$$

In fact, by eliminating $\mathcal{D}_k^{(d)} f_i$ from (8.3.17) using (8.3.19), we may express all of the coefficient functions in terms of Wronskians, giving,

$$h_r(\tau) = (-1)^r \frac{W_{d-r}}{W_d} \tag{8.3.21}$$

Say that the function $f_i(\tau)$ is bounded in the interior of the fundamental domain, and has zeros at locations p of order $\mathrm{ord}_{f_i}(p)$. We define the so-called *Wronskian index* by,

$$\ell = 6 \left(\frac{1}{2} \mathrm{ord}_{W_d}(i) + \frac{1}{3} \mathrm{ord}_{W_d}(\rho) + \sum_{p \in F} \mathrm{ord}_{W_d}(p) \right) \tag{8.3.22}$$

so that $\ell/6$ is the number of zeros of W_d. The Wronskian index can take any non-negative integer value except 1. For future use, we note that the valence formula (3.5.1) implies,

$$\mathrm{ord}_{W_d}(i\infty) + \frac{\ell}{6} = \frac{d(k+d-1)}{12} \tag{8.3.23}$$

using the fact that the Wronskian W_d has modular weight $d(k+d-1)$.

For $\ell = 0$, the coefficient functions $h_r(\tau)$ are holomorphic, which is the case analyzed in detail for $d = 2, 3$ in Section 8.2. For $\ell > 0$ on the other hand, the coefficient functions will generically be singular. For example, consider the case for which $d = \ell = 2$. By definition (8.3.22), we conclude that $\mathrm{ord}_{W_2}(\rho) = 1$, and hence $W_2 \sim \mathsf{E}_4(\tau)$. The coefficient functions then take the form $h_r(\tau) \sim W_{2-r}/\mathsf{E}_4(\tau)$. The remaining τ dependence can be fixed by simply requiring that one has the correct modular weight, giving $h_1(\tau) \sim \mathsf{E}_4(\tau)$ and $h_2(\tau) \sim \mathsf{E}_6(\tau)/\mathsf{E}_4(\tau)$.

Finding a solution to an MDE with meromorphic coefficients is often significantly more difficult than for its holomorphic counterpart (though for $d = 2$ and $\ell < 6$ it turns out that the solutions can again be written as hypergeometric functions). As we discuss in Section 10.9, Hecke operators can be used to generate such solutions.

• **Bibliographical notes**

Solving modular differential equations by recasting them as hypergeometric equations has been discussed in [KK03], [BG07], and [FM16a]. Vector-valued modular forms grew out of work of Gunning [Gun61], as well as Selberg [Sel65], who used them as a way of studying Fourier coefficients of modular forms for noncongruence subgroups. For a modern review of various properties of vector-valued modular forms, see [Gan13], and for a discussion of modular forms and differential operators, see [Zag94]. The integrality theorem given in Section 8.3.2 was first put forward as a conjecture by Atkin and Swinnerton-Dyer in [ASD71] and initially proven for two-dimensional and three-dimensional vector-valued modular forms in [Mas12, FM14] and [Mar12, FM16b], respectively. The proof for generic vector-valued modular forms with entries interpretable as characters for rational conformal field theories was given in [DLN15], using modular tensor category techniques. The full proof, independent of rational conformal field theory, was given recently in [CDT21].

Finally, we note that modular differential equations have been applied to the classification of rational conformal field theories, starting with the seminal work in [MMS88] and [MMS89], and follow-up papers [Kir89, Nac89] shortly thereafter. In subsequent works, modular differential techniques were applied to the classification of bosonic rational conformal fields theories with two characters at Wronskian index $\ell = 2, 4$ [HM16a, HM16b, CM19], and to the classification of theories with three characters at $\ell = 0$ [MPS20]; for a recent review of some of this progress, see [Muk19]. As of the writing of this book, the state-of-the-art for the classification of bosonic rational conformal field theories is for theories with up to five characters (subject to restrictions on the Wronskian index), obtained in [KLPM21]. For fermionic theories on the other hand, there is a complicating factor that the relevant modular differential equations need not be invariant under $\mathrm{SL}(2, \mathbb{Z})$, but only an appropriate index 2 congruence subgroup. Works toward the classification of fermionic theories with small numbers of characters have appeared in [BDL$^+$21] and [BDL$^+$22], with others sure to follow in the coming years.

Exercises

8.1   Denote the Fourier expansion of the $k \geq 1$ power of the modular discriminant as follows,

$$\Delta(\tau)^k = (2\pi)^{12k} q^k \sum_{n=0}^{\infty} a_n^{(k)} q^n \qquad\qquad q = e^{2\pi i \tau}$$

so that $a_n^{(k)} \in \mathbb{Z}$. Use modular differential equations to prove that $a_n^{(k)}$ is divisible by $24/\gcd(24, n)$.

8.2 In (8.2.16) we gave the general solution to a modular differential equation of degree 2 with holomorphic coefficient functions. Denote the Fourier expansion of the solution $f_2(\tau)$ by,

$$f_2(\tau) = q^{\frac{1-\mu}{12}} \sum_{n=0}^{\infty} m_n q^n \qquad (8.3.24)$$

In the context of *rational conformal field theory*, the parameters in this expansion are required to obey $\mu \in \mathbb{Q}$ and $m_n \in \mathbb{Z}$. Show that there are a finite number of such solutions, and give the full list of possible values $2(\mu - 1)$ (checking $m_n \in \mathbb{Z}$ up to $n = 15$).

8.3 Obtain the derivatives with respect to τ of the functions $\vartheta_i(0|\tau)/\eta(\tau)$ for $i = 2, 3, 4$ using the results of Exercise 7.4 and express the results in terms of powers of ϑ-constants.

8.4 Use the results of Exercise 8.3 to prove that the functions $f_1, f_2,$ and f_3 of (8.2.23) solve the MDE (8.2.17).

8.5 Consider a meromorphic vector-valued modular function $\mathbf{f}(\tau)$ with components $f_i(\tau)$ for $i = 1, \ldots, d$, given by,

$$f_i(\tau) = q^{n_0^{(i)}} \sum_{n \geq 0} a_n^{(i)} q^n \qquad a_0^{(i)} \neq 0 \qquad (8.3.25)$$

We define $n_0 = \min(n_0^{(1)}, \ldots, n_0^{(d)})$. Show that if at least one component $f_i(\tau)$ is a modular function for a congruence subgroup, then $n_0 < 0$.

9
Modular graph functions and forms

Modular graph functions and modular graph forms map (decorated) graphs to complex-valued functions on the Poincaré upper half plane $\mathcal{H}$ with definite transformation properties under $SL(2,\mathbb{Z})$. Specifically, modular graph functions are $SL(2,\mathbb{Z})$-invariant functions on $\mathcal{H}$, while modular graph forms may be identified with $SL(2,\mathbb{Z})$-invariant differential forms. Modular graph functions and forms generalize, and at the same time unify, holomorphic and non-holomorphic Eisenstein series, multiple zeta-functions, and iterated modular integrals. In particular, the non-holomorphic Eisenstein series introduced in Section 4.2 may be associated with simple one-loop graphs and thus represent a special class of modular graph functions. The expansion of modular graph forms at the cusp includes Laurent polynomials whose coefficients are combination of Riemann zeta-values and multiple zeta-values, while each modular graph form may be expanded in a basis of iterated modular integrals. Eisenstein series and modular graph functions and forms beyond Eisenstein series occur naturally and pervasively in the study of the low energy expansion of superstring amplitudes, as will be made clear in Chapter 12. Here we shall present a purely mathematical approach with only minimal reference to physics.

9.1 Non-holomorphic modular forms of arbitrary weight

The terminology of modular graph *functions* versus modular graph *forms* follows the terminology adopted for automorphic functions and forms in the holomorphic and non-holomorphic categories discussed in Subsection 3.3.1. It will be convenient to introduce the spaces $\mathcal{M}_{k,\ell}$ of complex-valued smooth functions $f : \mathcal{H} \to \mathbb{C}$ with the following $SL(2,\mathbb{Z})$ transformation properties,

$$\mathcal{M}_{k,\ell} = \left\{ f(\gamma\tau) = (c\tau + d)^k (c\bar{\tau} + d)^\ell f(\tau) \text{ for all } \gamma \in SL(2,\mathbb{Z}) \right\} \quad (9.1.1)$$

using the customary parametrization of $\gamma \in \mathrm{SL}(2,\mathbb{Z})$,

$$\gamma = \begin{pmatrix} a & b \\ c & d \end{pmatrix} \tag{9.1.2}$$

for $k, \ell \in \mathbb{C}$, subject to the condition $k - \ell \in \mathbb{Z}$. The latter condition is imposed to ensure that the functions are single-valued on $\mathcal{H}$. By definition, the *modular weight* of a function $f \in \mathcal{M}_{k,\ell}$ is the pair (k, ℓ), where k and ℓ are referred to as the holomorphic and anti-holomorphic modular weight of f, respectively. The functions of interest to us here will have at most polynomial growth at the cusp $\tau_2 \to \infty$, and we shall impose this property also on the functions in $\mathcal{M}_{k,\ell}$. Multiplication by a factor of τ_2^s for $s \in \mathbb{C}$ induces an isomorphism between the spaces $\mathcal{M}_{k+s,\ell+s}$ and $\mathcal{M}_{k,\ell}$, and we may identify the functions related in this manner. In particular, when $k = \ell$, the spaces $\mathcal{M}_{k,k}$ are isomorphic to $\mathcal{M}_{0,0}$.

For integer k, ℓ, we may equivalently view the functions in $\mathcal{M}_{k,\ell}$ as differential $(\frac{k}{2}, \frac{\ell}{2})$-forms, following the identification of (3.3.7). A modular graph form of modular weight (k, ℓ) is then a map from a certain (decorated) graph to $\mathcal{M}_{k,\ell}$, as we shall see below.

9.2 One-loop modular graph functions are Eisenstein series

We begin by considering the scalar Green function $G(z - w | \tau)$ on the torus $\Sigma = \mathbb{C}/\Lambda$ for a lattice $\Lambda = \mathbb{Z} + \mathbb{Z}\tau$ with modulus $\tau \in \mathcal{H}$ and $z, w \in \Sigma$. The scalar Green function was defined as the inverse of the scalar Laplacian on the torus on the space of functions orthogonal to constant functions in (5.5.12), solved in terms of Jacobi ϑ-functions in (5.5.16), and shown to be invariant under $\mathrm{SL}(2,\mathbb{Z})$ in (5.5.22). Here we shall use its expression in terms of a double Fourier series or Kronecker–Eisenstein series, given in (5.5.11),

$$G(z|\tau) = \sum_{(m,n)\neq(0,0)} \frac{\tau_2}{\pi |m + n\tau|^2} e^{2\pi i (nx - my)} \tag{9.2.1}$$

where $z = x + y\tau$ and x, y are real "co-moving coordinates" valued in $\mathbb{R}/\mathbb{Z}$. By concatenating the Green function repeatedly using convolution, we may construct an infinite family of modular invariant functions, which may be defined recursively as follows,

$$G_s(z|\tau) = \int_\Sigma \frac{d^2 u}{\tau_2} G(z - u|\tau) G_{s-1}(u|\tau) \tag{9.2.2}$$

where we set $G_1 = G$ and s is an arbitrary positive integer. To the Green function G and its generalizations G_s for positive integer s, it is natural to associate a graph, which physicists refer to as a Feynman graph,

$$G(z - w|\tau) = \quad \underset{z}{\circ}\!\!-\!\!-\!\!-\!\!-\!\!-\!\!\underset{w}{\circ}$$

$$G_s(z - w|\tau) = \quad \underset{z}{\circ}\!\!-\!\!-\!\!-\!\!\underset{u_1}{\bullet}\!\!-\!\!-\!\!\underset{u_2}{\bullet}\!\!-\!\!-\!\!\underset{u_3}{\bullet}\!\!\cdots\!\!\cdots\!\!\underset{u_{s-1}}{\bullet}\!\!-\!\!-\!\!\underset{w}{\circ} \qquad (9.2.3)$$

The black dots represent points $u_1, \cdots u_{s-1}$ that are integrated over Σ, while the white dots represent given points in Σ, that are not integrated. Upon Fourier transformation, convolution acts by multiplication, and we obtain the following representation for G_s,

$$G_s(z|\tau) = \sum_{(m,n)\neq(0,0)} \frac{\tau_2^s}{\pi^s |m + n\tau|^{2s}} e^{2\pi i(nx - my)} \qquad (9.2.4)$$

As long as $s \geq 2$, we may set the point $z = 0$ which graphically is equivalent to closing the open chain in (9.2.3) into the closed loop of (9.2.5),

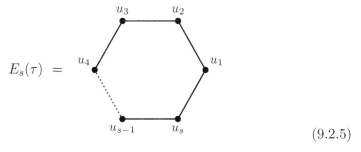

$$(9.2.5)$$

Comparing with the definition of non-holomorphic Eisenstein series in (4.2.1), we see that $G_s(0|\tau) = E_s(\tau)$. Actually, the double sum in (9.2.4) converges absolutely for any $s \in \mathbb{C}$ with $\mathrm{Re}\,(s) > 1$, but the graphical interpretation no longer makes sense when s is not a positive integer.

When s is integer and satisfies $s \geq 2$, the number s counts the number of edges in the graph (9.2.5) and is referred to as the *weight* of the graph (not to be confused with modular weight). In Section 12.8, we shall identify this definition of the weight with that of *transcendental weight*.

9.3 Maass operators and Laplacians

Before studying higher-loop modular graph functions in detail, we give a brief intermission on differential operators on the space $\mathcal{M}_{k,\ell}$. There exist several natural maps between the spaces $\mathcal{M}_{k,\ell}$ which we shall now exhibit.

One noteworthy element of $\mathcal{M}_{-1,-1}$ is the imaginary part τ_2 of the modulus τ, whose powers satisfy $\tau_2^s \in \mathcal{M}_{-s,-s}$. Multiplication by a suitable power of τ_2 therefore provides a map between the different spaces $\mathcal{M}_{k,\ell}$. Since a function $f(\tau)$ and its image $\tau_2^s f(\tau)$ under this map are essentially equivalent to one another, we may introduce the equivalence relations,

$$\mathcal{M}_{k,\ell} \approx \mathcal{M}_{k-\ell,0} \approx \mathcal{M}_{0,\ell-k} \tag{9.3.1}$$

This equivalence is particularly convenient when it comes to the action of partial differential operators. The Cauchy–Riemann operator $\partial_{\bar{\tau}}$ acts covariantly on the space $\mathcal{M}_{k,0}$ without the need for a connection and maps to the space $\mathcal{M}_{k,2}$. Similarly, the operator ∂_τ maps $\mathcal{M}_{0,\ell}$ covariantly to the space $\mathcal{M}_{2,\ell}$ without the need for a connection. One may, of course, include a k- and ℓ-dependent connection to obtain the Maass operators (see Appendix C.4). We shall circumvent doing so here by applying $\partial_{\bar{\tau}}$ to $\mathcal{M}_{k,0}$ and ∂_τ to $\mathcal{M}_{0,\ell}$ only, using the equivalence of (9.3.1). Actually, we can multiply the Cauchy–Riemann operators by suitable powers of τ_2 so that the operators $\nabla = 2i\tau_2^2\partial_\tau$ and $\bar{\nabla} = -2i\tau_2^2\partial_{\bar{\tau}}$ map the spaces as follows,

$$\bar{\nabla} : \mathcal{M}_{k,0} \to \mathcal{M}_{k-2,0}$$
$$\nabla : \mathcal{M}_{0,\ell} \to \mathcal{M}_{0,\ell-2} \tag{9.3.2}$$

Laplace–Beltrami operators, or simply Laplace operators, are defined by,

$$\Delta_k^+ : \mathcal{M}_{k,0} \to \mathcal{M}_{k,0} \qquad \Delta_k^+ = \tau_2^{-k} \nabla \tau_2^{k-2} \bar{\nabla}$$
$$\Delta_\ell^- : \mathcal{M}_{0,\ell} \to \mathcal{M}_{0,\ell} \qquad \Delta_\ell^- = \tau_2^{-\ell} \bar{\nabla} \tau_2^{\ell-2} \nabla \tag{9.3.3}$$

On modular forms of weight $(0,0)$ the Laplace operators Δ_0^+ and Δ_0^- coincide and are denoted by $\Delta = 4\tau_2^2\partial_\tau\partial_{\bar{\tau}}$. Famously, the Eisenstein series satisfies the Laplace eigenvalue equation,

$$\Delta E_s = s(s-1)E_s \tag{9.3.4}$$

The expansion of $E_s(\tau)$ at the cusp ∞ consists of two power behaved terms, given in (4.2.17), and which we recall here for convenience,

$$E_s(\tau) = 2\frac{\zeta(2s)}{\pi^s}\tau_2^s + 2\frac{\Gamma(s-\frac{1}{2})}{\pi^{s-\frac{1}{2}}\Gamma(s)}\zeta(2s-1)\tau_2^{1-s} + \mathcal{O}(e^{-2\pi\tau_2}) \tag{9.3.5}$$

plus an infinite number of exponentially suppressed terms. In terms of the independent variables $\tau, \bar{\tau}$ and the real co-moving coordinates x, y, the Green function $G_s(x + y\tau|\tau)$ is an eigenfunction of the Laplace operator Δ with eigenvalues that are independent of x, y,

$$\Delta G_s(x + y\tau|\tau) = s(s-1)G_s(x + y\tau|\tau) \tag{9.3.6}$$

Applying the operator ∇^k to E_s produces the following modular graph form of weight $(0, -2k)$,

$$\nabla^k E_s(\tau) = \frac{\Gamma(s+k)}{\Gamma(s)} \sum_{(m,n)\neq(0,0)} \frac{\tau_2^{s+k}}{\pi^s (m+n\tau)^{s+k} (m+n\bar{\tau})^{s-k}} \quad (9.3.7)$$

These functions coincide with Zagier's single-valued elliptic polylogarithm, up to an overall normalization.

9.4 Two-loop modular graph functions

We now investigate the special class of modular graph functions in $\mathcal{M}_{0,0}$ associated with arbitrary (decorated) two-loop graphs. A connected two-loop graph has two trivalent vertices, at points z and w, and an arbitrary number of bivalent vertices, distributed on the three concatenated edges connecting the trivalent vertices at z and w according to three positive integers a_1, a_2, and a_3. The most general such graph is depicted in (9.4.1),

$$C_{a_1,a_2,a_3}(\tau) = $$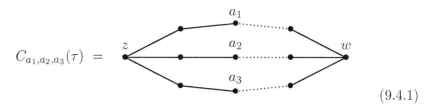

$$(9.4.1)$$

The associated modular graph function may be defined in terms of a double integral over Σ of a triple product of concatenated Green functions,

$$C_{a_1,a_2,a_3}(\tau) = \int_\Sigma \frac{d^2z}{\tau_2} \int_\Sigma \frac{d^2w}{\tau_2} G_{a_1}(z-w|\tau) G_{a_2}(z-w|\tau) G_{a_3}(z-w|\tau) \quad (9.4.2)$$

where G_s was defined in (9.2.2). Translation invariance on the torus renders one of the integrals trivial, giving the following equivalent expression,

$$C_{a_1,a_2,a_3}(\tau) = \int_\Sigma \frac{d^2z}{\tau_2} G_{a_1}(z|\tau) G_{a_2}(z|\tau) G_{a_3}(z|\tau) \quad (9.4.3)$$

Upon Fourier transforming, the corresponding modular graph function is given by the following Kronecker–Eisenstein sums,

$$C_{a_1,a_2,a_3}(\tau) = \sum_{\substack{m_r,n_r\in\mathbb{Z} \\ (m_r,n_r)\neq(0,0)}} \delta\left(\sum_{r=1}^3 m_r\right) \delta\left(\sum_{r=1}^3 n_r\right) \prod_{r=1}^3 \left(\frac{\tau_2}{\pi|m_r+n_r\tau|^2}\right)^{a_r} \quad (9.4.4)$$

Alternatively, we may express C_{a_1,a_2,a_3} as a Kronecker–Eisenstein sum over the lattice $\Lambda' = \Lambda \setminus \{0\}$ as follows,

$$C_{a_1,a_2,a_3}(\tau) = \sum_{p_m \in \Lambda'} \delta\Big(\sum_{r=1}^{3} p_r\Big) \prod_{r-1}^{3} \Big(\frac{\tau_2}{\pi |p_r|^2}\Big)^{a_r} \qquad (9.4.5)$$

where the lattice momenta $p_r = m_r + n_r\tau$ with $m_r, n_r \in \mathbb{Z}$ and $\delta(p) = 1$ when $p = 0$ and $\delta(p) = 0$ when $p \neq 0$. The *weight* of the modular graph function $C_{a_1,a_2,a_3}(\tau)$ is defined by $w = a_1 + a_2 + a_3$.

For integer $a_1, a_2,$ and a_3, the Kronecker–Eisenstein sums of (9.4.4) are absolutely convergent provided $a_r + a_s > 1$ for all pairs $r \neq s$, which requires the weight w to satisfy $w \geq 3$. The sums are absolutely convergent for complex values of $a_1, a_2,$ and a_3 as long as $\mathrm{Re}\,(a_1), \mathrm{Re}\,(a_2), \mathrm{Re}\,(a_3) > 1$. Contrarily to the case of the Eisenstein series, the problem of the existence of analytic continuations in these variables remains unexplored. Once convergent, $C_{a_1,a_2,a_3}(\tau)$ is readily verified to be modular invariant and thus belongs to $\mathcal{M}_{0,0}$.

Two key properties of the non-holomorphic Eisenstein series, namely, that they satisfy a differential equation and have a simple expansion at the cusp, have more complicated counterparts in two-loop modular graph functions.

Theorem 9.4.1. *Two-loop modular graph functions C_{a_1,a_2,a_3} of weight $w = a_1 + a_2 + a_3$ obey a system of differential equations of uniform weight w,*

$$\begin{aligned}
2\Delta C_{a_1,a_2,a_3} = {}& 2a_1 a_2\, C_{a_1+1,a_2-1,a_3} + a_1 a_2\, C_{a_1+1,a_2+1,a_3-2} \\
& - 4 a_1 a_2\, C_{a_1+1,a_2,a_3-1} + a_1(a_1 - 1)\, C_{a_1,a_2,a_3} \\
& + \text{ 5 permutations of } (a_1, a_2, a_3)
\end{aligned} \qquad (9.4.6)$$

where $\Delta = 4\tau_2^2 \partial_\tau \partial_{\bar\tau}$. When $a_1, a_2, a_3 \in \mathbb{N}$, the system of all the differential equations for a given weight w truncates to a finite-dimensional linear system of inhomogeneous equations where the inhomogeneous part consists of a linear combination of E_w and $E_{w_1} E_{w_2}$ with $w = w_1 + w_2$ and $w_1, w_2 \geq 2$, as a consequence of the following relations,

$$\begin{aligned}
C_{w_1,w_2,0} &= E_{w_1} E_{w_2} - E_w \\
C_{w_1+1,w_2,-1} &= E_{w_1} E_{w_2} + E_{w_1+1} E_{w_2-1}
\end{aligned} \qquad (9.4.7)$$

The eigenvalues of the linear system are given by $s(s-1)$ where $1 \leq s \leq w-2$ and $w - s \in 2\mathbb{N}$ and have multiplicity $[(s+2)/3]$.

The proof of (9.4.7) is the subject of Exercise 9.1. For example, up to weight 5, the relations of Theorem 9.4.1 are given by,

$$\Delta C_{1,1,1} = 6E_3$$
$$(\Delta - 2)C_{2,1,1} = 9E_4 - E_2^2$$
$$(\Delta - 6)C_{3,1,1} - 3C_{2,2,1} = 16E_5 - 4E_2 E_3$$
$$\Delta C_{2,2,1} = 8E_5 \qquad (9.4.8)$$

The first and last identities imply that $C_{1,1,1} - E_3$ and $5C_{2,2,1} - 2E_5$ must be constant. These constants may be determined from the asymptotic behavior of these functions at the cusp, a problem to which we now turn.

Theorem 9.4.2. *The expansion of $C_{a_1,a_2,a_3}(\tau)$ near the cusp $\tau \to i\infty$ is given by a Laurent polynomial in τ_2 of degree $(w, 1 - w)$*

$$C_{a_1,a_2,a_3}(\tau) = c_w(-4\pi\tau_2)^w + \frac{c_{2-w}}{(4\pi\tau_2)^{w-2}}$$

$$+ \sum_{k=1}^{w-1} \frac{c_{w-2k-1}\,\zeta(2k+1)}{(4\pi\tau_2)^{2k+1-w}} + \mathcal{O}(e^{-2\pi\tau_2}) \qquad (9.4.9)$$

where $c_w, c_{w-2k-1} \in \mathbb{Q}$ and c_{2-w} is a linear combination with integer coefficients of products of odd zeta-values,

$$c_{2-w} = \sum_{k=1}^{w-2} \gamma_k\,\zeta(2k+1)\zeta(2w - 2k - 3) \qquad \gamma_k \in \mathbb{Z} \quad (9.4.10)$$

A consequence of this theorem is that no multiple zeta-values occur in the Laurent polynomial of two-loop modular graph functions. Higher loop modular graph functions, however, do involve irreducible multiple zeta values.

At every odd weight, there is one linear combination of the above differential identities in the subspace of vanishing eigenvalue whose inhomogeneous part does not involve bilinears in the Eisenstein series. This property may be verified in the identities for $C_{1,1,1}$ and $C_{2,2,1}$ in (9.4.8) and holds for all odd weights. Each one of these identities may be integrated up to an additive constant which may be fixed by the asymptotics near the cusp, and we find,

$$E_3 + \zeta(3) = C_{1,1,1}$$

$$\frac{3}{7}E_7 + \frac{\zeta(7)}{252} = C_{3,3,1} + C_{3,2,2}$$

$$\frac{2}{5}E_5 + \frac{\zeta(5)}{30} = C_{2,2,1}$$

$$4E_7 + \frac{\zeta(9)}{240} = 9C_{4,4,1} + 18C_{4,3,2} + 4C_{3,3,3} \qquad (9.4.11)$$

The equation for $C_{1,1,1}$ was derived by Zagier in an unpublished note by direct summation of the Kronecker–Eisenstein series which defines $C_{1,1,1}$.

For the complete Fourier and Poincaré series representations of two-loop modular graph functions, we refer the reader to the works quoted in the bibliographical notes and to Exercises 9.2 and 9.3 for the Poincaré series.

9.5 General modular graph functions and forms

At higher loops, it is no longer true that the Laplace operator Δ (acting on the immediate generalization of two-loop modular graph functions to higher loops) closes onto functions defined by Kronecker–Eisenstein series in which the exponents of holomorphic momenta $m_r + n_r \tau$ and their complex conjugate coincide. To obtain a space of functions on which the action of the Laplacian closes, it is natural to lift the restriction to modular functions and consider more generally *modular graph forms*, in which the exponents of holomorphic momenta $p_r = m_r + n_r \tau$ and their complex conjugates no longer have to coincide, and which may have nontrivial modular weights. The required generalization may be defined as follows.

The connectivity matrix Γ of an arbitrary graph has components Γ_{vr}, where the index $v = 1, \cdots, V$ labels the vertices of the graph and the index $r = 1, \cdots, R$ labels its edges. No edge is allowed to begin and end on the same vertex, thereby excluding "tadpole graphs," in the terminology of quantum field theory. When the edge r contains vertex v we have $\Gamma_{vr} = \pm 1$, while otherwise we have $\Gamma_{vr} = 0$. To each edge r, we assign a pair of exponents (a_r, b_r), namely, a_r to the momentum $p_r = m_r + n_r \tau$ and b_r to its complex conjugate $\bar{p}_r$ through edge r, thereby providing a *decoration* of the graph. The decorated graph is then specified by the connectivity matrix Γ and the pair of arrays of exponents (A, B) defined by,

$$A = [a_1, \cdots, , a_R] \qquad \mathfrak{a} = a_1 + \cdots + a_R$$
$$B = [b_1, \cdots, , b_R] \qquad \mathfrak{b} = b_1 + \cdots + b_R \qquad (9.5.1)$$

where $a_r, b_r \in \mathbb{C}$ with $a_r - b_r \in \mathbb{Z}$ for all $r = 1, \cdots, R$. The pair of total exponents $(\mathfrak{a}, \mathfrak{b})$ generalizes the weight to the full decorated graph.

To a decorated graph (Γ, A, B), we associate a complex-valued function on $\mathcal{H}$, defined by the following Kronecker–Eisenstein sum, whenever this sum is absolutely convergent,

$$
\mathcal{C}_\Gamma \begin{bmatrix} A \\ B \end{bmatrix} (\tau) = \left(\frac{\tau_2}{\pi} \right)^{\frac{1}{2}\mathfrak{a}+\frac{1}{2}\mathfrak{b}} \sum_{p_1,\dots,p_R \in \Lambda'} \prod_{r=1}^{R} \frac{1}{(p_r)^{a_r} (\bar{p}_r)^{b_r}} \prod_{v=1}^{V} \delta \left(\sum_{s=1}^{R} \Gamma_{vs} p_s \right)
$$

$$(9.5.2)$$

where $p_r = m_r + n_r \tau \in \Lambda$ with $m_r, n_r \in \mathbb{Z}$, $\Lambda' = \Lambda \setminus \{0\}$, and $\delta(p) = 1$ for $p = 0$ and $\delta(p) = 0$ otherwise, generalizing the notation used in (9.4.5) for the case of two-loop modular graph functions. The Kronecker δ enforces the conservation of momentum at each vertex v in the graph.

The number of loops L is the number of independent momenta p_r and is given by $L = R - V + 1$. For a decorated graph (Γ, A, B), the domain of absolute convergence of the sums in (9.5.2) is given by a system of inequalities on the combinations $\mathrm{Re}\,(a_r + b_r)$. Outside the domain of convergence, it is an open question as to whether and how the functions $\mathcal{C}$ of (9.5.2) may be defined by analytic continuation in the variables $a_r + b_r$.

The function $\mathcal{C}$ in (9.5.2) vanishes whenever the integer $\mathfrak{a} - \mathfrak{b}$ is odd, or whenever it is associated with a graph Γ which becomes disconnected upon severing a single edge. A function $\mathcal{C}$ associated with a graph Γ which is the union of two graphs $\Gamma = \Gamma_1 \cup \Gamma_2$, such that the intersection of Γ_1 and Γ_2 consists of a single vertex, factorizes into the product of functions $\mathcal{C}$ for Γ_1 and Γ_2, with the corresponding partitions of the exponents.

9.5.1 Modular properties

Under $\gamma \in \mathrm{SL}(2, \mathbb{Z})$, the functions defined in (9.5.2) transform as follows,

$$
\mathcal{C}_\Gamma \begin{bmatrix} A \\ B \end{bmatrix} (\gamma\tau) = \left(\frac{c\tau + d}{c\bar{\tau} + d} \right)^{\frac{1}{2}\mathfrak{a}-\frac{1}{2}\mathfrak{b}} \mathcal{C}_\Gamma \begin{bmatrix} A \\ B \end{bmatrix} (\tau) \qquad\qquad \gamma = \begin{pmatrix} a & b \\ c & d \end{pmatrix} \quad (9.5.3)
$$

One defines the *modular weight* of $\mathcal{C}_\Gamma$ by the pair $(\frac{1}{2}(\mathfrak{a} - \mathfrak{b}), \frac{1}{2}(\mathfrak{b} - \mathfrak{a}))$, which is integer-valued for $\mathfrak{a} - \mathfrak{b}$ even (recall that $\mathcal{C}_\Gamma$ vanishes for odd $\mathfrak{a} - \mathfrak{b}$). As we have explained earlier, we refer to $\mathcal{C}_\Gamma$ as a *modular graph form* when $\mathfrak{a} \neq \mathfrak{b}$ and a *modular graph function* when $\mathfrak{a} = \mathfrak{b}$. For $\mathfrak{a} = \mathfrak{b}$, the exponent in the normalization factor τ_2 may be canonically chosen so that $\mathfrak{a} = \mathfrak{b} \to 0$. For $\mathfrak{a} \neq \mathfrak{b}$, however, the normalization factor of τ_2 is not canonical, and we introduce the forms $\mathcal{C}_\Gamma^+$ and $\mathcal{C}_\Gamma^-$ of modular weight $(0, \mathfrak{b} - \mathfrak{a})$ and $(\mathfrak{a} - \mathfrak{b}, 0)$, respectively as follows,

$$
\mathcal{C}_\Gamma^{\pm} \begin{bmatrix} A \\ B \end{bmatrix} = (\tau_2)^{\pm \frac{\mathfrak{a}-\mathfrak{b}}{2}} \mathcal{C}_\Gamma \begin{bmatrix} A \\ B \end{bmatrix} (\tau) \qquad\qquad (9.5.4)
$$

$\mathcal{C}_\Gamma^+$ and $\mathcal{C}_\Gamma^-$ transform with factors $(c\bar{\tau}+d)^{-a+b}$ and $(c\tau+d)^{a-b}$, respectively.

9.5.2 Examples: Eisenstein series and dihedral graphs

In this newfound notation, we may represent and generalize some of the examples of modular graph functions and forms encountered earlier. The Eisenstein series and its successive derivatives take the form,

$$
E_s = \mathcal{C}_\Gamma \begin{bmatrix} s & 0 \\ s & 0 \end{bmatrix}
\qquad
\nabla^k E_s = \frac{\Gamma(s+k)}{\Gamma(s)} \mathcal{C}_\Gamma^+ \begin{bmatrix} s+k & 0 \\ s-k & 0 \end{bmatrix}
\qquad (9.5.5)
$$

where Γ is a connected one-loop graph with only bivalent vertices. The two-loop modular graph functions C_{a_1,a_2,a_3} may be generalized to $(n-1)$-loop modular graph functions,

$$
C_{a_1,\cdots,a_n} = \mathcal{C}_\Gamma \begin{bmatrix} a_1 & \cdots & a_n \\ a_1 & \cdots & a_n \end{bmatrix}
\qquad (9.5.6)
$$

where the "dihedral" graph Γ contains two vertices of valence n in addition to the remaining vertices which are all bivalent. For $n = 3$, we recover the two-loop modular graph functions studied earlier.

9.5.3 Examples: holomorphic modular graph forms

When all anti-holomorphic exponents b_i vanish identically the corresponding modular graph form reduces to the following expression,

$$
\mathcal{C}_\Gamma \begin{bmatrix} a_1 & \cdots & a_n \\ 0 & \cdots & 0 \end{bmatrix} (\tau) = \sum_{p_1,\ldots,p_R \in \Lambda'} \prod_{r=1}^{R} \frac{1}{(p_r)^{a_r}} \prod_{v=1}^{V} \delta \left(\sum_{s=1}^{R} \Gamma_{vs} p_s \right)
\qquad (9.5.7)
$$

The multiple sum is absolutely convergent provided $a_r \geq 2$ for all $r = 1, \cdots, R \geq 2$. The result vanishes when $a = a_1 + \cdots + a_R$ is odd. When a is even, $\mathcal{C}_\Gamma^-$ produces a holomorphic modular form of weight $(a, 0)$ which may be expressed as a polynomial in the holomorphic modular forms E_4 and E_6 and therefore produces a map from graphs Γ to holomorphic modular forms, or equivalently to polynomials in E_4 and E_6. The nature of this map, which is much simpler than the map to non-holomorphic modular forms, remains to be studied. Under weaker assumptions of conditional convergence, the sum must be suitably regularized. Maintaining holomorphicity while regularizing leads to a violation of full modular invariance, such as is the case with $a_1 = 2, a_2 = 0$ for $R = 2$ which yields the quasi-modular form E_2.

9.5.4 Algebraic relations

Under complex conjugation, the arrays of holomorphic and anti-holomorphic exponents are swapped,

$$\overline{\mathcal{C}_\Gamma \begin{bmatrix} A \\ B \end{bmatrix} (\tau)} = \mathcal{C}_\Gamma \begin{bmatrix} B \\ A \end{bmatrix} (\tau) \tag{9.5.8}$$

In addition, modular graph forms obey *momentum conservation identities*,

$$\sum_{r=1}^{R} \Gamma_{kr} \, \mathcal{C}_\Gamma \begin{bmatrix} A - S_r \\ B \end{bmatrix} = \sum_{r=1}^{R} \Gamma_{kr} \, \mathcal{C}_\Gamma \begin{bmatrix} A \\ B - S_r \end{bmatrix} = 0 \tag{9.5.9}$$

where the R-dimensional row-vector S_r is defined to have zeros in all slots except for the r-th, which instead has value 1,

$$S_r = [\underbrace{0, \ldots, 0}_{r-1}, 1, \underbrace{0, \ldots, 0}_{R-r}] \tag{9.5.10}$$

These identities express the fact that the lattice momenta entering each vertex v add up to zero or, as physicists would state, that momenta is conserved at each vertex as a consequence of translation invariance on the torus.

9.5.5 Differential relations

The action of the Cauchy–Riemann operator $\nabla = 2i\tau_2^2 \partial_\tau$ on modular graph forms $\mathcal{C}_\Gamma^+$, which were defined and normalized in (9.5.4) for arbitrary arrays of exponents A, B, is given by,

$$\nabla \mathcal{C}_\Gamma^+ \begin{bmatrix} A \\ B \end{bmatrix} = \sum_{r=1}^{R} a_r \, \mathcal{C}^+ \begin{bmatrix} A + S_r \\ B - S_r \end{bmatrix} \tag{9.5.11}$$

where S_r was defined in (9.5.10) and the coefficients a_r are the entries of the array A given in (9.5.1). The action of the Laplace operator $\Delta = 4\tau_2^2 \partial_{\bar\tau} \partial_\tau$ on modular graph functions (for which $\mathfrak{a} = \mathfrak{b}$ and $\mathcal{C}_\Gamma^+ = \mathcal{C}_\Gamma$) is given by,

$$(\Delta + \mathfrak{a}) \, \mathcal{C}_\Gamma \begin{bmatrix} A \\ B \end{bmatrix} = \sum_{r,s=1}^{R} a_r b_s \, \mathcal{C} \begin{bmatrix} A + S_r - S_s \\ B - S_r + S_s \end{bmatrix} \tag{9.5.12}$$

One may also define the Laplace operator on modular graph forms of arbitrary modular weight with $\mathfrak{a} \neq \mathfrak{b}$, but we shall not need them here.

9.5.6 Higher loop algebraic relations

Higher-loop modular graph functions and forms satisfy systems of algebraic and differential relations that generalize those found for two-loops. For example, one has the following algebraic identities involving modular graph functions of weight 4 and 5,

$$C_{1,1,1,1} = 24\, C_{2,1,1} + 3\, E_2^2 - 18\, E_4$$
$$C_{1,1,1,1,1} = 60\, C_{3,1,1} + 10\, E_2\, C_{1,1,1} - 48\, E_5 + 16\, \zeta(5)$$
$$40\, C_{2,1,1,1} = 300\, C_{3,1,1} + 120\, E_2 E_3 - 276\, E_5 + 7\, \zeta(5) \qquad (9.5.13)$$

These identities were proven using *holomorphic subgraph reduction* and a *sieve algorithm*. We discuss the techniques used in the proof of these identities in Section 9.6 for the example of the first identity in (9.5.13). We refer to the bibliographical notes for the papers where all algebraic identities between modular graphs of weight up to 6, and some selected identities of weight 7 are proven, and where a Mathematica package is now available.

9.6 Relating modular graph functions to holomorphic forms

The Maass operator $\nabla = 2i\tau_2^2 \partial_\tau$ maps a modular graph form of modular weight $(0, \ell)$ to a modular graph form of modular weight $(0, \ell - 2)$, as was shown in Section 9.3. When the derivative is applied to a non-holomorphic Eisenstein series E_n, for example, and repeated n times, one obtains the following form, using (9.5.5),

$$\nabla^n E_n = \frac{\Gamma(2n)}{\Gamma(n)}\, \mathcal{C}^+ \begin{bmatrix} 2n & 0 \\ 0 & 0 \end{bmatrix} = \frac{\Gamma(2n)}{\Gamma(n)\pi^n}\, \tau_2^{2n}\, G_{2n} \qquad (9.6.1)$$

where $G_{2n}(\tau)$ is the holomorphic Eisenstein series of modular weight $(2n, 0)$, familiar from (3.4.3).

Applying the Maass operator ∇ to more complicated modular graph functions is not as simple as in the case of non-holomorphic Eisenstein series, but nonetheless a similar pattern emerges: ∇ decreases the anti-holomorphic exponents and increases the holomorphic exponents. Let's see how this works in the case of the modular graph function $C_{1,1,1,1}$. Using the rules of (9.5.11) for ∇, we obtain for the first derivative,

$$\nabla C_{1,1,1,1} = \nabla \mathcal{C} \begin{bmatrix} 1 & 1 & 1 & 1 \\ 1 & 1 & 1 & 1 \end{bmatrix} = 4\, \mathcal{C}^+ \begin{bmatrix} 2 & 1 & 1 & 1 \\ 0 & 1 & 1 & 1 \end{bmatrix} \qquad (9.6.2)$$

The form on the right has modular weight $(0, -2)$. Applying ∇ again gives,

$$\nabla^2 C_{1,1,1,1} = 12\,\mathcal{C}^+\begin{bmatrix} 2\ 2\ 1\ 1 \\ 0\ 0\ 1\ 1 \end{bmatrix} - 24\,\mathcal{C}^+\begin{bmatrix} 3\ 1\ 1\ 1 \\ 0\ 0\ 1\ 1 \end{bmatrix} \tag{9.6.3}$$

where the last term was obtained using the following momentum conservation identity to convert a negative entry on the lower row,

$$\mathcal{C}^+\begin{bmatrix} 3 & 1\ 1\ 1 \\ -1 & 1\ 1\ 1 \end{bmatrix} = -3\,\mathcal{C}^+\begin{bmatrix} 3\ 1\ 1\ 1 \\ 0\ 0\ 1\ 1 \end{bmatrix} \tag{9.6.4}$$

Further application of ∇ to (9.6.3), in its present form, would yield modular graph functions with some negative anti-holomorphic exponents that cannot be removed simply by using momentum conservation identities. However, the subgraph corresponding to the first two entries in the arrays of exponents in (9.6.3) is a closed loop with only holomorphic momenta. The momentum summations in such holomorphic subgraphs may be evaluated explicitly in terms of holomorphic modular forms. This procedure is dubbed *holomorphic subgraph reduction* and, in the present case, leads to the following identity,

$$\mathcal{C}^+\begin{bmatrix} 2\ 2\ 1\ 1 \\ 0\ 0\ 1\ 1 \end{bmatrix} - 2\,\mathcal{C}^+\begin{bmatrix} 3\ 1\ 1\ 1 \\ 0\ 0\ 1\ 1 \end{bmatrix} = 2\,\mathcal{C}^+\begin{bmatrix} 4\ 1\ 1 \\ 0\ 1\ 1 \end{bmatrix} + 3\,\mathcal{C}^+\begin{bmatrix} 4\ 0 \\ 0\ 0 \end{bmatrix}\mathcal{C}^+\begin{bmatrix} 2\ 0 \\ 2\ 0 \end{bmatrix} \tag{9.6.5}$$

Collecting all contributions to $\nabla^2 C_{1,1,1,1}$ and proceeding analogously for the entries $C_{2,1,1}$ and E_2^2 in the first identity in (9.5.13), we obtain,

$$\nabla^2 C_{1,1,1,1} = 24\,\mathcal{C}^+\begin{bmatrix} 4\ 1\ 1 \\ 0\ 1\ 1 \end{bmatrix} + 36\,\mathcal{C}^+\begin{bmatrix} 4\ 0 \\ 0\ 0 \end{bmatrix}\mathcal{C}^+\begin{bmatrix} 2\ 0 \\ 2\ 0 \end{bmatrix}$$

$$\nabla^2 C_{2,1,1} = 6\,\mathcal{C}^+\begin{bmatrix} 4\ 1\ 1 \\ 0\ 1\ 1 \end{bmatrix} + 6\,\mathcal{C}^+\begin{bmatrix} 3\ 2\ 1 \\ 1\ 0\ 1 \end{bmatrix} + 4\,\mathcal{C}^+\begin{bmatrix} 6\ 0 \\ 2\ 0 \end{bmatrix}$$

$$\nabla^2 E_2^2 = 8\,\mathcal{C}^+\begin{bmatrix} 3\ 0 \\ 1\ 0 \end{bmatrix}^2 + 12\,\mathcal{C}^+\begin{bmatrix} 4\ 0 \\ 0\ 0 \end{bmatrix}\mathcal{C}^+\begin{bmatrix} 2\ 0 \\ 2\ 0 \end{bmatrix} \tag{9.6.6}$$

Applying ∇ to either $\nabla^2 C_{1,1,1,1}$ or to E_2^2 produces modular graph functions with negative entries in the lower row that cannot be removed by using momentum conservation identities. This occurs when ∇ is applied to one of the modular graph functions that is proportional to a holomorphic modular form, such as the first factors in the last terms on the right of $\nabla^2 C_{1,1,1,1}$ and $\nabla^2 E_2^2$. However, $C_{2,1,1}$ and the linear combination $C_{1,1,1,1} - 3E_2^2$ (that cancels the last terms discussed earlier) produces the following ∇^3 derivatives,

$$\nabla^3(C_{1,1,1,1} - 3E_2^2) = 432\,\mathcal{C}^+ \begin{bmatrix} 7 & 0 \\ 1 & 0 \end{bmatrix} - 288\,\mathcal{C}^+ \begin{bmatrix} 4 & 0 \\ 0 & 0 \end{bmatrix} \mathcal{C}^+ \begin{bmatrix} 3 & 0 \\ 1 & 0 \end{bmatrix}$$

$$\nabla^3 C_{2,1,1} = 108\,\mathcal{C}^+ \begin{bmatrix} 7 & 0 \\ 1 & 0 \end{bmatrix} \quad 12\,\mathcal{C}^+ \begin{bmatrix} 4 & 0 \\ 0 & 0 \end{bmatrix} \mathcal{C}^+ \begin{bmatrix} 3 & 0 \\ 1 & 0 \end{bmatrix} \qquad (9.6.7)$$

Proceeding analogously for the fourth-order ∇-derivative, we obtain,

$$\nabla^4(C_{1,1,1,1} - 24C_{2,1,1} - 3E_2^2) = -15120\,\mathcal{C}^+ \begin{bmatrix} 8 & 0 \\ 0 & 0 \end{bmatrix} \qquad (9.6.8)$$

The right side is proportional to $\tau_2^8 G_8$, where G_8 is the holomorphic Eisenstein series of weight $(8,0)$. But this form is proportional to $\nabla^4 E_4$, so that, matching coefficients, we obtain,

$$\nabla^4\big(C_{1,1,1,1} - 24C_{2,1,1} - 3E_2^2 + 18E_4\big) = 0 \qquad (9.6.9)$$

Next, we invoke the following lemma.

Lemma 9.6.1. *Let F be a non-holomorphic modular function with polynomial growth near the cusp as $\tau_2 \to \infty$. If F satisfies the differential equation,*

$$\nabla^n F = 0 \qquad (9.6.10)$$

for an arbitrary integer $n \geq 1$, then F is constant as a function of τ.

A simple proof of the lemma proceeds by making use of the formula,

$$\bar{\nabla}^n(\tau_2)^{-2n}\nabla^n = \prod_{s=1}^{n}\Big(\Delta - s(s-1)\Big) \qquad (9.6.11)$$

which is proven in Exercise 9.4. This formula immediately implies that F must satisfy the weaker condition,

$$\prod_{s=1}^{n}\Big(\Delta - s(s-1)\Big)F = 0 \qquad (9.6.12)$$

Being a modular graph function, F has at most polynomial growth at the cusp. The solution to $\Delta f_s = s(s-1)f_s$ for positive integer s is either a constant for $s = 1$ or an Eisenstein series E_s for $s \geq 2$. This implies that F must be a linear combination of a constant and the Eisenstein series E_s for $s = 2, \cdots, n$. Returning to the original equation (9.6.10) in the formulation of the lemma and using the asymptotics of the Eisenstein series at the cusp of (4.2.17), we conclude that the coefficients of the Eisenstein series in F must all vanish so that F is constant. This completes the proof of the lemma.

Applying the lemma to (9.6.9), it follows that,

$$F_4 = C_{1,1,1,1} - 24C_{2,1,1} - 3E_2^2 + 18E_4 \qquad (9.6.13)$$

must be constant. To calculate the constant, we investigate the asymptotics at the cusp. Clearly, all τ-dependent terms must cancel on the right side as one may verify using the expansions of $C_{1,1,1,1}$ and $C_{2,1,1}$ in terms of the variable $y = \pi\tau_2$,

$$C_{1,1,1,1}(\tau) = \frac{y^4}{945} + \frac{2\zeta(3)y}{3} + \frac{10\zeta(5)}{y} - \frac{3\zeta(3)^2}{y^2} + \frac{9\zeta(7)}{4y^3} + \mathcal{O}(e^{-2y})$$

$$C_{2,1,1}(\tau) = \frac{2y^4}{14175} + \frac{\zeta(3)y}{45} + \frac{5\zeta(5)}{12y} - \frac{\zeta(3)^2}{4y^2} + \frac{9\zeta(7)}{16y^3} + \mathcal{O}(e^{-2y}) \quad (9.6.14)$$

and the expansion of the Eisenstein series in (4.2.17). The constant y^0 term has vanishing coefficient in each of the modular graph functions in F_4, so that $F_4 = 0$. This completes the proof of the first identity in (9.5.13). The other identities in (9.5.13) may be established with the same methods, and holomorphic subgraph reduction formulas that generalize (9.6.5). The proof of the second equation in (9.5.13) is carried out in Exercise 9.5.

9.7 Modular graph functions and iterated modular integrals

The Fourier expansion of nonholomorphic Eisenstein series $E_k(\tau)$ for integer $k \geq 2$, presented in (4.2.18), shows that the Fourier non-zero modes are of the form $(q^N + \bar{q}^N)$ times a polynomial in $1/\tau_2$. Such a simple form is no longer shared by modular graph functions and forms at higher loops since in that case terms of the form $q^M\bar{q}^N$ also occur. Nonetheless, modular graph forms are related to holomorphic modular forms, even at higher loop, as demonstrated by the fact that certain repeated derivatives ∇ map to holomorphic modular forms. Another way to exhibit this connection concretely is via iterated modular integrals.

The basic building blocks of iterated modular integrals are the functions $\mathcal{E}(k_1, \cdots, k_r|\tau)$ and $\mathcal{E}^0(k_1, \cdots, k_r|\tau)$ defined iteratively by the relations,

$$\mathcal{E}(k_1, \cdots, k_r|\tau) = 2\pi i \int_\tau^{i\infty} \frac{d\tau_r}{(2\pi i)^{k_r}} G_{k_r}(\tau_r)\, \mathcal{E}(k_1, \cdots, k_{r-1}|\tau)$$

$$\mathcal{E}^0(k_1, \cdots, k_r|\tau) = 2\pi i \int_\tau^{i\infty} \frac{d\tau_r}{(2\pi i)^{k_r}} G_{k_r}^0(\tau_r)\, \mathcal{E}^0(k_1, \cdots, k_{r-1}|\tau) \quad (9.7.1)$$

The holomorphic modular form $G_k(\tau)$ of weight k is defined for $k \geq 3$ in

(3.4.3) by the familiar absolutely convergent Kronecker–Eisenstein series,

$$G_k(\tau) = \sum_{\substack{m,n\in\mathbb{Z}\\(m,n)\neq(0,0)}} \frac{1}{(m+n\tau)^k} = 2\zeta(k) + G_k^0(\tau) \tag{9.7.2}$$

The function $G_k(\tau)$ vanishes for all odd $k \geq 3$ and evaluates to $2\zeta(k)$ at the cusp, while $G_k^0(\tau)$ consists of the nonconstant Fourier modes of $G_k(\tau)$, and decays exponentially at the cusp. We extend the definition to $k = 0$ and supply initial conditions for the iterated integrals,

$$G_0(\tau) = G_0^0(\tau) = -1$$
$$\mathcal{E}(\cdot|\tau) = \mathcal{E}^0(\cdot|\tau) = 1 \tag{9.7.3}$$

The exponential decay of $G_k^0(\tau)$ at the cusp for $k \geq 4$ renders the corresponding integrals convergent, while all other integrations are defined via the tangential base-point prescription at the cusp,

$$\int_\tau^{i\infty} d\tau_r \to -\tau \tag{9.7.4}$$

The iterated integrals satisfy the following differential equations,

$$\nabla\mathcal{E}(k_1,\cdots,k_r|\tau) = \frac{4\pi\tau_2^2}{(2\pi i)^{k_r}} G_{k_r}(\tau)\,\mathcal{E}(k_1,\cdots,k_{r-1}|\tau)$$

$$\nabla\mathcal{E}^0(k_1,\cdots,k_r|\tau) = \frac{4\pi\tau_2^2}{(2\pi i)^{k_r}} G_{k_r}^0(\tau)\,\mathcal{E}^0(k_1,\cdots,k_{r-1}|\tau) \tag{9.7.5}$$

For example, we have,

$$\nabla\mathcal{E}^0(2k|\tau) = \frac{4\pi\tau_2^2}{(2\pi i)^{2k}} G_{2k}^0(\tau)$$

$$\nabla\mathcal{E}^0(2k,0_n|\tau) = -4\pi\tau_2^2\,\mathcal{E}^0(2k,0_{n-1}|\tau) \tag{9.7.6}$$

where the notation 0_n stands for an array of n zeros.

We begin by showing how the simplest non-holomorphic Eisenstein series E_2 may be expressed in terms of these iterated integrals. The starting point is the differential equation obtained in (9.6.1),

$$\pi^2\nabla^2 E_2 = 6\tau_2^4\,G_4 = \frac{2\pi^4\tau_2^4}{15} + 6\tau_2^4\,G_4^0 \tag{9.7.7}$$

Using the relation $\nabla\tau_2^n = n\tau_2^{n+1}$ and eliminating G_4^0 in favor of $\nabla\mathcal{E}^0(4|\tau)$ using the first equation in (9.7.6), we obtain,

$$\pi^2\nabla^2 E_2 = \frac{2\pi^4}{45}\nabla\tau_2^3 + 24\pi^3\tau_2^2\,\nabla\mathcal{E}^0(4|\tau) \tag{9.7.8}$$

Successive use of the second relation in (9.7.6) for $n = 1, 2$ allows us to recast the right side in the form of a total derivative in terms of the iterated integrals $\mathcal{E}^0(4|\tau), \mathcal{E}^0(4, 0|\tau)$, and $\mathcal{E}^0(4, 0_2|\tau)$,

$$\nabla(\pi\nabla E_2) = \nabla\left(\frac{2\pi^3\tau_2^3}{45} + 24\pi^2\tau_2^2\mathcal{E}^0(4|\tau)\right.$$
$$\left. + 12\pi\tau_2\mathcal{E}^0(4, 0|\tau) + 3\mathcal{E}^0(4, 0_2|\tau)\right) \qquad (9.7.9)$$

The equation is readily solved, and the general solution is given as follows,

$$\pi\nabla E_2 = \frac{2\pi^3\tau_2^3}{45} + 24\pi^2\tau_2^2\,\mathcal{E}^0(4|\tau) + 12\pi\tau_2\,\mathcal{E}^0(4, 0|\tau)$$
$$+ 3\,\mathcal{E}^0(4, 0_2|\tau) + \overline{f_1(\tau)} \qquad (9.7.10)$$

where $f_1(\tau)$ is holomorphic. Making use anew of the equations (9.7.6), we can integrate this equation once more, and find,

$$E_2(\tau) = \frac{\pi^2\tau_2^2}{45} - 6\,\mathcal{E}^0(4, 0|\tau) - \frac{3}{\pi\tau_2}\mathcal{E}^0(4, 0_2|\tau) - \frac{\overline{f_1(\tau)}}{\pi\tau_2} + \overline{f_2(\tau)} \quad (9.7.11)$$

where $f_2(\tau)$ is another holomorphic function. The reality of E_2 determines f_1 and f_2 uniquely up to an additive constant in f_1, which may be fixed by the known asymptotics at the cusp. Using again $y = \pi\tau_2$, we have,

$$E_2(\tau) = \frac{y^2}{45} + \frac{\zeta(3)}{y} - 12\,\text{Re}\left[\mathcal{E}^0(4, 0|\tau)\right] - \frac{6}{y}\text{Re}\left[\mathcal{E}^0(4, 0_2|\tau)\right] \quad (9.7.12)$$

The above expression reflects the structure of the Fourier decomposition of the Eisenstein series mentioned at the beginning of this chapter.

Next, we provide an example of the decomposition into iterated integrals for the simplest modular graph function that is not an Eisenstein series, namely, $C_{2,1,1}$. The differential equations for $C_{2,1,1}$, given in the second line of (9.6.7), may be recast in the following form,

$$\pi^2\nabla^3\left(C_{2,1,1} - \tfrac{9}{10}E_4\right) = -6\tau_2^2 G_4(\tau)\nabla E_2 \qquad (9.7.13)$$

The subtraction of the term in E_4 is convenient as it readily cancels the first term on the right side of the second line in (9.6.7). One may begin by using the expression given in (9.7.10) to eliminate ∇E_2 from the above equation in favor of iterated integrals. Successive use of (9.7.5) again allows us to proceed by the same methods as used for E_2, and one obtains,

$$
\begin{aligned}
C_{2,1,1} = {}& -\frac{y^4}{20250} + \frac{\zeta(3)y}{45} + \frac{5\zeta(5)}{12y} - \frac{\zeta(3)^2}{4y^2} - \left(\frac{2y}{15} - \frac{3\zeta(3)}{y^2}\right) \mathrm{Re}\left[\mathcal{E}^0(4,0_2|\tau)\right] \\
& -\frac{9}{y^2}\,\mathrm{Re}\left[\mathcal{E}^0(4,0_2|\tau)\right]^2 - 72\,\mathrm{Re}\left[\mathcal{E}^0(4_2,0_2|\tau)\right] \\
& -\frac{1}{5}\,\mathrm{Re}\left[\mathcal{E}^0(4,0_3|\tau)\right] - \frac{3b}{y}\,\mathrm{Re}\left[\mathcal{E}^0(4,0,4,0_2|\tau)\right] \\
& -\frac{108}{y}\,\mathrm{Re}\left[\mathcal{E}^0(4_2,0_3|\tau)\right] - \frac{1}{10y}\,\mathrm{Re}\left[\mathcal{E}^0(4,0_4|\tau)\right]
\end{aligned}
\tag{9.7.14}
$$

In this formulation, the first term on the second line reveals the presence of terms of the form $q^M \bar{q}^N$ in the Fourier series of $C_{2,1,1}$, as expected.

• Bibliographical notes

Special cases of modular graph functions were encountered in explorations of the low-energy effective interactions produced by one-loop superstring amplitudes [GRV08], as will be described in Chapters 12 and 14. Systematic investigations of modular graph functions were initiated in [DGV15], [Zer16], and [DGGV17]. The proof of Theorem 9.4.1 was given in [DGV15], while the proof of Theorem 9.4.2 was given in [DD18], where explicit expressions for the coefficients in the Laurent expansion are provided. The relation between modular graph functions and single-valued elliptic polylogarithms was exhibited in [DGGV17]. Introductions to elliptic polylogarithms may be found in [Zag90], [BL11], [Enr16], [BDDT18], and [EZ23].

The systematic construction of the algebraic and differential identities satisfied by modular graph functions and forms was carried out in [DG18], [DK16], and [GK19] using the method of holomorphic subgraph reduction. A variety of other methods were used in [Bas16a], [Bas16b], [Bas16c], [KV17], and [Bas19] to obtain identities within special classes of modular graph functions. An efficient Mathematica package for algebraic and differential relations was developed in [Ger21] and [Ger20]. A class of related modular functions is presented in [Bro17a] and [Bro20].

Fourier series expansions were obtained in [DD18], [AK18], [DK19a], and [DK19b], while Poincaré series were derived in [AK18], [DK19a], and [DK19b]. The role of modular graph functions in genus-one amplitudes for open and Heterotic strings was elucidated in [BSZ19] and [GKS19]. The solutions to the differential identities obtained from holomorphic subgraph reduction may be obtained in terms of iterated elliptic integrals and systematically represented in terms of generating functions [BSZ19, GKS20a, GKS20b]. Additional discussions of the relation between modular graph functions and iterated Eisenstein integrals can be found in [DDD+22] and [HSV22].

The generalization of modular graph functions to higher genera was initiated in [DGP19b] and is motivated by the study of the low-energy expansion of superstring amplitudes, to be discussed in Chapter 12. The natural generalization of modular graph forms to higher genus is provided by modular graph tensors, which were introduced in [DS22]. Degenerations of the higher genus Riemann surfaces, studied in [DGP19a], naturally lead one to introduce elliptic modular graph functions, whose study was begun in [DKS21].

Exercises

9.1 Prove the formulas of (9.4.7), namely,

$$C_{w_1,w_2,0} = E_{w_1} E_{w_2} - E_{w_1+w_2}$$
$$C_{w_1+1,w_2,-1} = E_{w_1} E_{w_2} + E_{w_1+1} E_{w_2-1} \qquad (9.7.15)$$

9.2 Show that, for every $(m_1, n_1) \in \mathbb{Z}^2 \setminus \{(0,0)\}$, there exist unique $(0, n)$ and $\gamma \in \Gamma_\infty \backslash \mathrm{SL}(2, \mathbb{Z})$ such that $(m_1, n_1) = (0, n)\gamma$. Determine n, γ.

9.3 Use the result of Exercise 9.2 to obtain a Poincaré series representation for $C_{a_1,a_2,a_3}(\tau)$ with $a_1, a_2, a_3 \in \mathbb{N}$.

9.4 Prove the following relation between Maass operators and polynomials in the Laplace–Beltrami operator $\Delta = \Delta_1 = 4\tau_2^2 \partial_{\bar\tau} \partial_\tau$ on scalars,

$$\Delta_n = \prod_{m=1}^{n} \left(\Delta - m(m-1) \right) = \bar\nabla^n \tau_2^{-2n} \nabla^n = \nabla^n \tau_2^{-2n} \bar\nabla^n \qquad (9.7.16)$$

Show that $\Delta_n \tau_2^s = \lambda_n(s) \tau_2^s$, where $\lambda_n(s) = \prod_{\ell=1-n}^{n} (s - \ell)$.

9.5 Use holomorphic subgraph reduction and Lemma 9.6.1 to verify the second equation of (9.5.13).

Part II

Extensions and applications

In Part II, we discuss mathematical extensions and physical applications of the material discussed in Part I.

Chapter 10 is devoted to Hecke operators, including their action on holomorphic modular forms, Maass forms, and vector-valued modular forms. We illustrate their physical application to two-dimensional conformal field theory and the counting of the number of black holes in three-dimensional gravity. In Chapter 11, we introduce complex multiplication and singular moduli, prove that $j(\tau)$ is an algebraic integer at a complex multiplication point τ, and evaluate elliptic and ϑ-functions at these points.

We give an economical introduction to string amplitudes in Chapter 12, review their perturbative expansion, explain the role of conformal symmetry in decoupling negative norm states, and discuss how modular invariance ensures the good high-energy behavior of string theory. Amplitudes at tree-level, one-loop, and two-loop orders are presented without derivation, and modular graph functions, introduced in Chapter 9, are used to derive low energy effective interactions.

Toroidal compactifications are analyzed in Chapter 13, shown to give rise to higher arithmetic T-duality groups, and proven to exhibit special properties at complex multiplication points. S-duality in Type IIB string theory is discussed in Chapter 14, following a brief review of Type IIB supergravity and superstring theory. Wherever a comparison can be made, the modular properties of the low-energy effective interactions are found to match with the predictions derived from superstring amplitudes in Chapter 12.

In Chapter 15, we present a brief introduction to Yang–Mills theories with extended supersymmetry in four space-time dimensions. We review Montonen–Olive duality in $\mathcal{N} = 4$ theories and then discuss the Seiberg–Witten solution with emphasis on the role played by $\mathrm{SL}(2, \mathbb{Z})$. We close with a discussion of dualities of $\mathcal{N} = 2$ superconformal Yang–Mills theory.

In the last Chapter 16, we present a basic introduction to Galois theory, along with some applications to rational conformal field theory.

10

Hecke operators

A natural set of mutually commuting linear operators acting on the space of modular forms are the Hecke operators T_n labeled by a positive integer n. They map holomorphic functions to holomorphic functions, weight-k modular forms to weight-k modular forms, and weight-k cusp forms to weight-k cusp forms. For the full modular group $\mathrm{SL}(2, \mathbb{Z})$, the operators T_n map the space $\mathcal{M}_k$ of holomorphic modular forms into itself and map the subspace $\mathcal{S}_k$ of cusp forms into itself. For congruence subgroups, the operators T_n map weight-k modular forms of one congruence subgroup into those of another congruence subgroup. Hecke operators commute with the Laplace–Beltrami operator on the upper half plane so that Maass forms and cusp forms are simultaneous eigenfunctions of all Hecke operators. Finally, given a modular form with positive integer Fourier coefficients, the Hecke transforms also have positive integer Fourier coefficients. For this reason, Hecke operators are relevant in a number of physical problems, as we shall see shortly.

10.1 Definition of Hecke operators

Given a lattice $\Lambda = \mathbb{Z}\omega_1 + \mathbb{Z}\omega_2$ with modulus $\tau = \omega_2/\omega_1 \in \mathcal{H}$, a modular form $f(\tau)$ of weight k under $\mathrm{SL}(2, \mathbb{Z})$ may be represented by a degree k homogeneous function of the lattice Λ, denoted by the same letter, $f(\Lambda) = \omega_1^{-k} f(\omega_2/\omega_1)$, as was already explained in (3.4.3) for the special case of Eisenstein series. Modularity of $f(\tau)$ is then equivalent to the fact that $f(\Lambda)$ is intrinsic and depends only on the lattice Λ, but not on the particular basis vectors ω_1 and ω_2 used to represent Λ. The Hecke operator $\mathsf{T}_n : \mathcal{M}_k \to \mathcal{M}_k$ is defined on $f(\Lambda)$ as a sum over all sublattices $\Lambda_n \subset \Lambda$

of index $[\Lambda : \Lambda_n] = n$ by,

$$\mathsf{T}_n f(\Lambda) = n^{k-1} \sum_{\substack{\Lambda_n \subset \Lambda \\ [\Lambda:\Lambda_n]=n}} f(\Lambda_n) \qquad (10.1.1)$$

where the sublattice Λ_n is understood to be a subgroup of Λ. The definition of T_n given above is intrinsic, in that it depends only on the lattices Λ and Λ_n, and not on any particular choice of the basis vectors for each lattice. An arbitrary multiplicative normalization factor, which depends only on k and n, and is completely multiplicative in n, is allowed in the definition of T_n and has been chosen here to be n^{k-1} for later convenience.

To make contact with modular forms expressed as functions of $\tau \in \mathcal{H}$, we now choose specific basis vectors for both Λ and Λ_n. In particular, the set of index n sublattices Λ_n may be parametrized by choosing basis vectors $\omega_1^{(n)}$ and $\omega_2^{(n)}$ for Λ_n which are integer linear combinations of the basis vectors ω_1 and ω_2 of Λ,

$$\omega_2^{(n)} = a\,\omega_2 + b\,\omega_1 \qquad \omega_1^{(n)} = c\,\omega_2 + d\,\omega_1 \qquad (10.1.2)$$

where $a, b, c, d \in \mathbb{Z}$. Since Λ and Λ_n are Abelian groups, Λ_n is a normal subgroup of Λ and the quotient Λ/Λ_n is a finite group. One may represent Λ/Λ_n as the set of the points in Λ modulo Λ_n, equipped with addition modulo Λ_n. The cardinality of Λ/Λ_n is the index $[\Lambda : \Lambda_n] = n$, which is given by $n = ad - bc$, is positive when Λ_n inherits its orientation from Λ, and is equal to the number of points in Λ that belong in the fundamental parallelogram of Λ_n. The above parametrization of Λ_n motivates us to introduce the set M_n of all 2×2 matrices with integer coefficients of determinant $n > 0$,

$$M_n = \left\{ \begin{pmatrix} a & b \\ c & d \end{pmatrix} \,\middle|\, a, b, c, d \in \mathbb{Z};\ ad - bc = n > 0 \right\} \qquad (10.1.3)$$

The quotient $\mathrm{SL}(2, \mathbb{Z})\backslash M_n$ is the set of equivalence classes in M_n under the equivalence relation $A \sim B$ whenever $A, B \in M_n$ and there exists a matrix $\mu \in SL(2, \mathbb{Z})$ such that $A = \mu B$. This equivalence relation translates into an equivalence relation between sublattices Λ_n under $\mathrm{SL}(2, \mathbb{Z})$. Therefore, the cosets in $\mathrm{SL}(2, \mathbb{Z})\backslash M_n$ are in one-to-one correspondence with the lattices Λ_n and the Hecke operator T_n may equivalently be defined to act as follows,

$$\mathsf{T}_n f(\tau) = n^{k-1} \sum_{\gamma \in \mathrm{SL}(2,\mathbb{Z})\backslash M_n} (c\tau + d)^{-k} f(\gamma\tau) \qquad (10.1.4)$$

The above definition makes it clear that T_n maps holomorphic functions to holomorphic functions. Below we shall show that it maps modular forms of $\mathrm{SL}(2, \mathbb{Z})$ to modular forms of $\mathrm{SL}(2, \mathbb{Z})$.

10.2 Parametrization of equivalence classes SL(2, ℤ)\$\backslash M_n$

The action of the operators T_n can be made even more explicit by using the following explicit parametrization of the equivalence classes SL(2, ℤ)\$\backslash M_n$.

Lemma 10.2.1. *Every equivalence class in* SL(2, ℤ)\$\backslash M_n$ *contains an upper triangular representative of the form* $\left(\begin{smallmatrix} a & b \\ 0 & d \end{smallmatrix}\right) \in M_n$, *with* $d > 0$. *A complete set of inequivalent elements of* SL(2, ℤ)\$\backslash M_n$ *is given by* $b \in \{0, 1, \cdots, d-1\}$, *so that the coset* SL(2, ℤ)\$\backslash M_n$ *is isomorphic to* $\mathcal{A}_n$ *defined by,*

$$\mathcal{A}_n = \left\{ \begin{pmatrix} a & b \\ 0 & d \end{pmatrix} \text{ with } a, b, d \in \mathbb{Z}; ad = n; b \in \{0, 1, \cdots, d-1\} \right\} \quad (10.2.1)$$

To prove the first part of the Lemma, we show that for every matrix $A = \left(\begin{smallmatrix} a & b \\ c & d \end{smallmatrix}\right) \in M_n$, there exists a $\mu \in$ SL(2, ℤ) such that μA is of the form given in the Lemma. To do so we parameterize μ as follows,

$$\mu = \begin{pmatrix} p & q \\ r & s \end{pmatrix} \qquad \mu A = \begin{pmatrix} pa + qc & pb + qd \\ ra + sc & rb + sd \end{pmatrix} \quad (10.2.2)$$

If $c = 0$ then $d \neq 0$ in view of $ad - bc = n > 0$ since $\mu \in M_n$, and we set $r = q = 0$ and $p = s = \text{sign}(d)$ to obtain a representative of the desired form. For the remaining cases $c \neq 0$, we choose mutually prime r and s such that $s/r = -a/c$ to render the matrix μA upper triangular. By Bézout's lemma, there exist integers p and q such that $ps - qr = 1$ guarantees $\mu \in$ SL(2, ℤ), and either μ or $-\mu$ gives the desired matrix.

To prove the second part of the Lemma, we show that the set $\mathcal{A}_n$ is complete and that all such elements are inequivalent under SL(2, ℤ). To see that the set is complete, we take two matrices $A_1 = \left(\begin{smallmatrix} a & b_1 \\ 0 & d \end{smallmatrix}\right)$ and $A_2 = \left(\begin{smallmatrix} a & b_2 \\ 0 & d \end{smallmatrix}\right)$ with $b_1 \equiv b_2 \pmod{d}$. Then if $b_1 = b_2 + kd$, we define $\mu = \left(\begin{smallmatrix} 1 & k \\ 0 & 1 \end{smallmatrix}\right) \in$ SL(2, ℤ) so that $A_1 = \mu A_2$, and hence $A_1 \sim A_2$. Conversely, to see that all elements are inequivalent we proceed as follows. If A_1 and A_2 are such that $A_1 \sim A_2$, there exists $\mu = \left(\begin{smallmatrix} p & q \\ r & s \end{smallmatrix}\right) \in$ SL(2, ℤ) such that $A_1 = \mu A_2$, namely,

$$\begin{pmatrix} a_1 & b_1 \\ 0 & d_1 \end{pmatrix} = \begin{pmatrix} p & q \\ r & s \end{pmatrix} \begin{pmatrix} a_2 & b_2 \\ 0 & d_2 \end{pmatrix} = \begin{pmatrix} pa_2 & pb_2 + qd_2 \\ ra_2 & rb_2 + sd_2 \end{pmatrix} \quad (10.2.3)$$

Since $a_2 d_2 = n$, clearly $a_2 \neq 0$, and hence by equating the bottom-left entries on both sides, we conclude that $r = 0$. Then since $pq - rs = 1$, we have $pq = 1$ and thus $p = q = \pm 1$. We make the choice $p = q = +1$, which we can always do by replacing μ with $-\mu$ if needed. Then equating entries on both sides, we see that $a_1 = a_2$, $d_1 = d_2$, and $b_1 \equiv b_2 \pmod{d}$. Hence if we restrict to $b \in \{0, \ldots, d-1\}$, there are no equivalences.

We shall also make use of the following simple but important lemma.

Lemma 10.2.2. *For every pair (A_1, μ_1) with $A_1 \in M_n$ and $\mu_1 \in SL(2, \mathbb{Z})$, there exists a pair (A_2, μ_2) with $A_2 \in M_n$ and $\mu_2 \in SL(2, \mathbb{Z})$ such that $A_1 \mu_1 = \mu_2 A_2$. As a result, the left and right coset spaces $SL(2, \mathbb{Z}) \backslash M_n$ and $M_n / SL(2, \mathbb{Z})$ are isomorphic.*

To prove this lemma, we may choose $A_1, A_2 \in \mathcal{A}_n$ without loss of generality, and parametrize the matrices A_1, A_2, μ_1, and μ_2 involved as follows,

$$A_1 = \begin{pmatrix} a_1 & b_1 \\ 0 & d_1 \end{pmatrix} \qquad\qquad A_2 = \begin{pmatrix} a_2 & b_2 \\ 0 & d_2 \end{pmatrix}$$

$$\mu_1 = \begin{pmatrix} p_1 & q_1 \\ r_1 & s_1 \end{pmatrix} \qquad\qquad \mu_2 = \begin{pmatrix} p_2 & q_2 \\ r_2 & s_2 \end{pmatrix} \tag{10.2.4}$$

Given μ_1 and A_1, one solves for A_2 and μ_2 using $\mu_2 A_2 = A_1 \mu_1$,

$$p_2 a_2 = a_1 p_1 + b_1 r_1 \qquad\qquad p_2 b_2 + q_2 d_2 = a_1 q_1 + b_1 s_1$$

$$r_2 a_2 = d_1 r_1 \qquad\qquad r_2 b_2 + s_2 d_2 = d_1 s_1 \tag{10.2.5}$$

From the left equations, one obtains $p_2/r_2 = (a_1 p_1 + b_1 r_1)/(d_1 r_1)$, which determines (p_2, r_2) uniquely up to a common sign since they must be relatively prime and gives $a_2 = d_1 r_1 / r_2$. Bézout's lemma then guarantees the existence of a solution q_2, s_2 to $p_2 s_2 - q_2 r_2 = 1$. In terms of these solutions, we readily solve the remaining right equations to obtain,

$$b_2 = s_2 a_1 q_1 + s_2 b_1 s_1 - q_2 d_1 s_1$$

$$d_2 = p_2 d_1 s_1 - r_2 a_1 q_1 - r_2 b_1 s_1 \tag{10.2.6}$$

10.3 Hecke operators map $\mathcal{M}_k$ to $\mathcal{M}_k$

Lemma 10.2.1 allows us to recast the action of the Hecke operator T_n more explicitly as a summation over the set $\mathcal{A}_n$ of (10.2.1) as follows,

$$\mathsf{T}_n f(\tau) = n^{k-1} \sum_{A \in \mathcal{A}_n} \frac{1}{d^k} f(A\tau) \tag{10.3.1}$$

Parametrizing $\mathcal{A}_n$ using Lemma 10.2.1 gives even more explicit formulas,

$$\mathsf{T}_n f(\tau) = n^{k-1} \sum_{\substack{ad=n \\ d>0}} \frac{1}{d^k} \sum_{b=0}^{d-1} f\left(\frac{a\tau + b}{d}\right)$$

$$= n^{k-1} \sum_{\substack{d|n \\ d>0}} \frac{1}{d^k} \sum_{b=0}^{d-1} f\left(\frac{n\tau + bd}{d^2}\right) \tag{10.3.2}$$

where the second line is obtained from the first by solving for $a = n/d$. Using these expressions for T_n, one readily proves the modularity of $\mathsf{T}_n f(\tau)$. Applying a modular transformation $\mu_1 \in \mathrm{SL}(2,\mathbb{Z})$ to (10.3.1), we have,

$$
\mathsf{T}_n f(\mu_1 \tau) = n^{k-1} \sum_{A_1 \in \mathcal{A}_n} \frac{1}{d_1^k} f(A_1 \mu_1 \tau) = n^{k-1} \sum_{A_2 \in \mathcal{A}_n} \frac{1}{d_1^k} f(\mu_2 A_2 \tau)
$$

$$
= n^{k-1} \sum_{A_2 \in \mathcal{A}_n} \frac{1}{d_1^k} (r_2 A_2 \tau + s_2)^k f(A_2 \tau) \tag{10.3.3}
$$

where we have used Lemma 10.2.2 to relate $A_1 \mu_1 = \mu_2 A_2$ and the fact that if A_1 runs through all elements of $\mathcal{A}_n$, then so too does A_2. Substituting the two equations on the second line of (10.2.5), we then obtain,

$$
\mathsf{T}_n f(\mu_1 \tau) = n^{k-1} \sum_{A_2} \frac{1}{d_2^k} (r_1 \tau + s_1)^k f(A_2 \tau)
$$

$$
= (r_1 \tau + s_1)^k \, \mathsf{T}_n f(\tau) \tag{10.3.4}
$$

which shows that $\mathsf{T}_n f$ is a modular form of weight k, just as f was.

10.4 Fourier expansions

For many applications in physics and mathematics, we will often be interested in the Fourier expansions of Hecke transforms. Given a weight-k modular form $f(\tau)$ with Fourier expansion $f(\tau) = \sum_m f_m q^m$, the Fourier expansion of the Hecke transform can be computed explicitly using the second line in (10.3.2). We begin by noting that,

$$
\mathsf{T}_n f(\tau) = n^{k-1} \sum_{d|n} \frac{1}{d^k} \sum_{b=0}^{d-1} \sum_{m \in \mathbb{Z}} f_m \, q^{nm/d^2} e^{2\pi i b m/d}
$$

$$
= \sum_{m \in \mathbb{Z}} \sum_{d | \gcd(n,m)} \left(\frac{n}{d}\right)^{k-1} f_m \, q^{nm/d^2} \tag{10.4.1}
$$

The second equality follows from the first by using the fact that $\sum_{b=0}^{d-1} e^{2\pi i \frac{bm}{d}}$ equals d if $\frac{m}{d} \in \mathbb{Z}$ and vanishes otherwise. Representing $m = d\ell$ with $\ell \in \mathbb{Z}$, and changing summation variable $d \to n/d$, we obtain,

$$
\mathsf{T}_n f(\tau) = \sum_{\ell \in \mathbb{Z}} \sum_{d|n} \left(\frac{n}{d}\right)^{k-1} f_{\ell d} \, q^{\frac{n\ell}{d}} = \sum_{\ell \in \mathbb{Z}} \sum_{d|n} d^{k-1} f_{\ell n/d} \, q^{\ell d}
$$

$$
= \sum_{m \in \mathbb{Z}} \sum_{d | \gcd(m,n)} d^{k-1} f_{mn/d^2} \, q^m \tag{10.4.2}
$$

We conclude that the Fourier expansion of the Hecke transform is given by,

$$\mathsf{T}_n f(\tau) = \sum_{m \in \mathbb{Z}} g_m^{(n)} q^m \qquad g_m^{(n)} = \sum_{d | \gcd(m,n)} d^{k-1} f_{mn/d^2} \qquad (10.4.3)$$

If the Fourier coefficients f_m of $f(\tau)$ are integers then, by construction, so are the Fourier coefficients $g_m^{(n)}$ of the Hecke transform $\mathsf{T}_n f$ for all $n > 0$. Furthermore, we see from (10.4.3) that if $f_0 = 0$ then $g_0^{(n)} = 0$ as well, so that T_n acts as an endomorphism on the space $\mathcal{S}_k$ of weight-k cusp forms.

In the particular case of $n = p$ being prime, the Fourier coefficients $g_m^{(n)}$ of the Hecke transform take a particularly simple form,

$$g_m^{(p)} = \begin{cases} p^k f_{pm} & p \nmid m \\ p^{k-1}(p\, f_{pm} + f_{m/p}) & p \mid m \end{cases} \qquad (10.4.4)$$

If $k \geq 1$, we see that if $f_m \in \mathbb{Z}$, then $g_m^{(p)} \in \mathbb{Z}$ as well. For $k = 0$ though this is no longer the case, unless one drops the factor of n^{k-1} in (10.1.4).

10.5 Example: the Ramanujan tau function

We illustrate the utility of this machinery by means of a simple example. In particular, let us prove the product formulas (3.7.16) and (3.7.17) for the Ramanujan tau function, defined in (3.7.13) of Section 3.7.1, and repeated here for convenience,

$$\Delta(\tau) = (2\pi)^{12} \sum_{n=1}^{\infty} \tau(n) q^n \qquad (10.5.1)$$

so that $f_n = (2\pi)^{12} \tau(n)$ in the notation of the previous section. To do so, we apply the Hecke operators to the discriminant $\Delta(\tau)$ to obtain,

$$\mathsf{T}_n \Delta(\tau) = \sum_m g_m^{(n)} q^m \qquad (10.5.2)$$

where the Fourier coefficients $g_m^{(n)}$ of the Hecke transform $\mathsf{T}_n \Delta$ are given by,

$$g_m^{(n)} = (2\pi)^{12} \sum_{d | \gcd(m,n)} d^{11} \tau\left(\frac{nm}{d^2}\right) \qquad (10.5.3)$$

Since $\Delta(\tau)$ and $\mathsf{T}_n \Delta(\tau)$ are both holomorphic cusp forms of weight 12, and the space $\mathcal{S}_{12}$ is one dimensional, we conclude that we must have,

$$\mathsf{T}_n \Delta(\tau) = \alpha_n \Delta(\tau) \qquad (10.5.4)$$

for some sequence of constants $\alpha_n \in \mathbb{C}$. Equating Fourier coefficients using (10.4.3), we obtain the following relation,

$$\alpha_n \tau(m) = \sum_{d|\gcd(m,n)} d^{11}\, \tau\left(\frac{nm}{d^2}\right) \tag{10.5.5}$$

For the particular case of $m = 1$ we have $\tau(1) = 1$ and only the term with $d = 1$ contributes to the sum on the right side, so that $\alpha_n = \tau(n)$. Substituting this expression for α_n into (10.5.5) readily reproduces (3.7.17), which reduces to (3.7.16) when $\gcd(m,n) = 1$.

10.6 Multiplicative properties of Hecke operators

An important property of Hecke operators is their behavior under multiplication when acting on a modular form of weight k, and we have,

$$\mathsf{T}_n \mathsf{T}_m = \sum_{d|\gcd(n,m)} d^{k-1}\, \mathsf{T}_{mn/d^2} \tag{10.6.1}$$

The full proof of this property is standard (see, e.g., [Ser73]) but somewhat lengthy. Here we settle for a proof of a special case of this relation,

$$\mathsf{T}_n \mathsf{T}_m = \mathsf{T}_{nm} \qquad\qquad \gcd(m,n) = 1 \tag{10.6.2}$$

To prove this, we begin by explicitly applying the Hecke operators on a weight-k modular form $f(\tau)$ to obtain,

$$\mathsf{T}_n \mathsf{T}_m f(\tau) = \sum_{\substack{\alpha\delta=n \\ \delta>0}} \sum_{\substack{ad=m \\ d>0}} \sum_{\beta=0}^{\delta-1} \sum_{b=0}^{d-1} \frac{(nm)^{k-1}}{(d\delta)^k} f\left(\frac{\alpha a \tau + (\alpha b + \beta d)}{d\delta}\right) \tag{10.6.3}$$

Since we have $\gcd(m,n) = 1$ and d and δ run through positive divisors of n and m, respectively, $d\delta$ runs through the positive divisors of nm. Furthermore, since $b \pmod d$ runs through a complete set of representatives for $\mathrm{SL}(2,\mathbb{Z})\backslash M_m$ and $\beta \pmod \delta$ runs through a full set of representatives for $\mathrm{SL}(2,\mathbb{Z})\backslash M_n$, then $\alpha b + \beta d \pmod{d\delta}$ runs through a complete set of inequivalent representatives for $\mathrm{SL}(2,\mathbb{Z})\backslash M_{mn}$. Hence by the definition given in (10.1.4), the right side of (10.6.3) is exactly the action of T_{mn} on $f(\tau)$.

The multiplication formula (10.6.1) may be inverted to obtain a formula for T_{mn} established in Exercise 10.1. Formula (10.6.1) also allows one to decompose the evaluation of Hecke operators into those of powers of prime numbers T_{p^ν}. It is shown in Exercise 10.2 that the operators T_{p^ν} satisfy a two-step recursion relation and may be assembled into a simple generating function derived in Exercise 10.3.

10.7 Hecke eigenforms

The multiplication identity (10.6.1) implies that the Hecke operators mutually commute. As such, we may consider modular forms $f(\tau) \in \mathcal{M}_k$ which are simultaneous eigenfunctions of all Hecke operators,

$$\mathsf{T}_n f(\tau) = \lambda_n f(\tau) \qquad \text{for all } n \in \mathbb{N} \tag{10.7.1}$$

Such forms are referred to as *Hecke eigenforms*, or just *eigenforms* for short.[1] Comparing the $\mathcal{O}(q)$ terms in the Fourier expansions on both sides of (10.7.1), the Fourier coefficients of eigenforms are seen to satisfy,

$$g_m^{(n)} = \lambda_n f_m \tag{10.7.2}$$

using (10.4.3). Considering in particular $m = 1$, we see that $f_n = \lambda_n f_1$, and thus that any nontrivial eigenform must have $f_1 \neq 0$. One often finds it useful to normalize $f_1 = 1$ so that the Hecke eigenvalues coincide with the Fourier coefficients of $f(\tau)$,

$$\mathsf{T}_n f(\tau) = f_n f(\tau) \qquad \text{for } f_1 = 1 \tag{10.7.3}$$

Applying the multiplication relation (10.6.1) to f and using the above relation, we see that, with the normalization $f_1 = 1$, the Fourier coefficients f_n satisfy the multiplication rule,

$$f_m f_n = \sum_{d|\gcd(m,n)} d^{k-1} f_{mn/d^2} \qquad \text{for } f_1 = 1 \tag{10.7.4}$$

For certain weights, the existence of eigenforms is manifest. Indeed, for weights $k = 4, 6, 8, 10, 14$, Theorem 3.5.2 states that the space $\mathcal{M}_k$ is one-dimensional, being generated by the Eisenstein series $\mathsf{E}_k(\tau)$. Hence for such values of k, the Hecke transform $\mathsf{T}_n \mathsf{E}_k(\tau)$ must clearly be a multiple of $\mathsf{E}_k(\tau)$ for any n, making $\mathsf{E}_k(\tau)$ a Hecke eigenform. More surprisingly, the Eisenstein series are Hecke eigenforms even when $\dim \mathcal{M}_k > 1$, which is shown as follows. Consider a generic noncuspidal modular form f satisfying,

$$\mathsf{T}_n f = \lambda_n f \tag{10.7.5}$$

with f having even weight k. For noncuspidal forms f, the Fourier coefficient f_0 is nonzero, and by extension $g_0^{(n)}$ is nonzero as well. In particular, for

---

[1] In some literature, the term eigenform refers to eigenfunctions of only a single Hecke operator, whereas our definition of eigenforms would be referred to as *simultaneous eigenforms*.

$m = 0, 1$ we have from (10.4.3) that,

$$g_0^{(n)} = \sum_{d|n} d^{k-1} f_0 \;=\; f_0\, \sigma_{k-1}(n)$$

$$g_1^{(n)} = \sum_{d|\gcd(1,n)} d^{k-1} f_{n/d^2} \;-\; f_n \qquad (10.7.6)$$

where $\sigma_{k-1}(n)$ are the sum of divisors functions. Comparing the first equation to (10.7.2) allows us to conclude that $\lambda_n = \sigma_{k-1}(n)$, while the second gives $f_n = f_1\sigma_{k-1}(n)$. Choosing the normalization $f_0 = 1$ (which is different from, and incompatible with, the normalization $f_1 = 1$ chosen in (10.7.3) and (10.7.4), so that here $f_1 \neq 1$), the Fourier expansion of the Hecke eigenform $f(\tau)$ is required to take the following form,

$$f(\tau) = 1 + f_1 \sum_{N=1}^{\infty} \sigma_{k-1}(N)\, q^N \qquad (10.7.7)$$

The normalized Eisenstein series $\mathsf{E}_k(\tau)$ take exactly this form, as in (3.4.13). Incidentally, note that the divisor function indeed satisfies the multiplicative property (10.7.4) (see also part (b) of Exercise 3.3).

Besides Eisenstein series, all Hecke eigenforms are cusp forms. We have already encountered one example of a cuspidal eigenform in Section 10.5, where we saw that the modular discriminant $\Delta(\tau)$ is an eigenfunction of the Hecke operators, with the eigenvalues being precisely the values of the Ramanujan tau function. We also saw that the Ramanujan tau function satisfied a multiplicative property consistent with (10.7.4).

The fact that $\Delta(\tau)$ is an eigenform follows simply from the fact that $\dim \mathcal{S}_{12} = 1$. The same argument holds for cusp forms of weight $k = 16, 18, 20, 22$, and 26, where $\dim \mathcal{S}_k = 1$, which are generated by $\Delta(\tau)\mathsf{E}_k(\tau)$. However, it is a remarkable fact due to Hecke, that there exist cuspidal eigenforms for other k as well. In fact, Hecke showed that *eigenforms provide a basis of $\mathcal{M}_k$ for every k*. In order to prove this, he introduced an inner product on the space of cusp forms, as we now briefly review.

10.7.1 Petersson inner product

The space of weight-k cusp forms admits the Petersson inner product,

$$\langle f|g \rangle = \int_{\mathrm{SL}(2,\mathbb{Z})\backslash \mathcal{H}} \frac{d\tau_1 d\tau_2}{\tau_2^2} (\tau_2)^k \overline{f(\tau)}\, g(\tau) \qquad (10.7.8)$$

which is convergent due to the vanishing of f or g at the cusp. A key fact is that Hecke operators are self-adjoint under this inner product,

$$\langle \mathsf{T}_n f | g \rangle = \langle f | \mathsf{T}_n g \rangle \tag{10.7.9}$$

as proven in Exercise 10.4. This together with the mutual commutativity of the T_n implies that a generic $f \in \mathcal{S}_k$ can be decomposed into a basis of Hecke eigenforms and that the eigenvalues of T_n (and hence the Fourier coefficients of normalized eigenforms) are all real.

10.8 Hecke operators acting on Maass forms

Recall from Section 4.3 that a Maass form $f(\tau)$ is a complex-valued $\mathrm{SL}(2,\mathbb{Z})$-invariant function of $\tau \in \mathcal{H}$ that satisfies a Laplace eigenvalue equation,

$$\Delta f(\tau) = s(s-1) f(\tau) \tag{10.8.1}$$

and has at most polynomial growth at the cusp. The space of Maass forms for eigenvalue $s(s-1)$ is denoted $\mathcal{N}(s) = \mathcal{N}(\mathrm{SL}(2,\mathbb{Z}), s)$ with the understanding that it satisfies $\mathcal{N}(1-s) = \mathcal{N}(s)$. The Fourier series of an arbitrary Maass form $f(\tau)$ was obtained in (4.3.6) and is repeated below for convenience,

$$f(\tau) = a\,\tau_2^s + b\,\tau_2^{1-s} + \sum_{N \neq 0} \alpha_N \sqrt{\tau_2}\, K_{s-1/2}(2\pi|N|\tau_2)\, e^{2\pi i N \tau_1} \tag{10.8.2}$$

Non-holomorphic Eisenstein series $E_s(\tau)$, defined in (4.2.1), are Maass forms whose Fourier series were obtained in (4.2.17), and are given by,

$$a = \frac{2\zeta^*(2s)}{\Gamma(s)} \qquad b = \frac{2\zeta^*(2s-1)}{\Gamma(s)} \qquad \alpha_N = \frac{4\sigma_{2s-1}(|N|)}{\Gamma(s)|N|^{s-\frac{1}{2}}} \tag{10.8.3}$$

A Maass form $f(\tau)$ is a cusp form provided the constant Fourier mode in (10.8.2) vanishes, $a = b = 0$. Cusp forms decay exponentially at the cusp and have finite norm with respect to the Petersson inner product. The subspace of cusp forms $\mathcal{S}(s) \subset \mathcal{N}(s)$ is empty except for an infinite number of discrete eigenvalues of the form $s \in \frac{1}{2} + i\mathbb{R}$. Theorem 4.3.1 provides a basis for $\mathcal{N}(s)$ in terms of Eisenstein series and cusp forms.

Hecke operators T_n are linear operators whose action on Maass forms in $\mathcal{N}(s)$ is similar to their action on holomorphic modular forms,

$$\mathsf{T}_n f(\tau) = \frac{1}{\sqrt{n}} \sum_{\gamma \in \mathrm{SL}(2,\mathbb{Z}) \backslash M_n} f(\gamma \tau) \tag{10.8.4}$$

In contrast to (10.1.4), there is no power of $(c\tau + d)$ multiplying the summand, as the form $f(\tau)$ has weight 0, and the prefactor $1/\sqrt{n}$ has been chosen for later convenience. Using Lemma 10.2.2, and the fact that a Maass form is invariant under $\mathrm{SL}(2,\mathbb{Z})$, it follows immediately that $\mathsf{T}_n f(\tau)$ is also invariant under $\mathrm{SL}(2,\mathbb{Z})$. Furthermore, since $f(\tau)$ has at most polynomial growth at the cusp, so must $\mathsf{T}_n f(\tau)$. This is because it is defined to be a finite sum of terms, each of which has at most polynomial growth. By manipulations analogous to those used for the holomorphic case, one may use an explicit parametrization of the coset $\mathrm{SL}(2,\mathbb{Z})\backslash M_n$ to recast the action of the Hecke operators in the following form,

$$\mathsf{T}_n f(\tau) = \frac{1}{\sqrt{n}} \sum_{\substack{ad=n,d>0 \\ b=0,\cdots,d-1}} f\left(\frac{a\tau+b}{d}\right) \tag{10.8.5}$$

Finally, a fundamental ingredient in understanding the action of Hecke operators on Maass forms is the fact that the Laplacian $\Delta = 4\tau_2^2 \partial_\tau \partial_{\bar{\tau}}$ on $\mathcal{H}$ is invariant under T_n, and thus commutes with T_n, because the transformation $\tau \to (a\tau+b)/d$ implies $\tau_2 \to a/d\tau_2$ and $\partial_\tau \to d/a\,\partial_\tau$. The above considerations imply the following properties of Hecke operators acting on Maass forms, the proofs of which are left to the reader:

1. The Hecke operators T_n map $\mathcal{N}(s)$ to $\mathcal{N}(s)$ and $\mathcal{S}(s)$ to $\mathcal{S}(s)$.
2. The product of T_m and T_n on $\mathcal{N}(s)$ is commutative and given by,

$$\mathsf{T}_m \mathsf{T}_n = \sum_{d|\gcd(m,n)} \mathsf{T}_{mn/d^2} \tag{10.8.6}$$

3. The operators T_n are self-adjoint with respect to the Petersson inner product on $\mathcal{S}(s)$, and may be simultaneously diagonalized in view of 2.
4. If $f \in \mathcal{N}(s)$ is a simultaneous eigenfunction $\mathsf{T}_n f = \lambda_n f$ for all $n \geq 0$, then the coefficients α_n in its Fourier expansion (10.8.2) satisfy,

$$\alpha_{\pm n} = \lambda_n\, \alpha_{\perp 1} \qquad\qquad n > 0 \tag{10.8.7}$$

Therefore, up to the normalizations $\alpha_{\pm 1}$, the coefficients of the non-constant Fourier modes are the eigenvalues of the Hecke operators.

5. The Fourier series of a Maass form $f(\tau)$ in $\mathcal{N}(s)$, given in (10.8.2), and its Hecke image $\mathsf{T}_n f(\tau)$ expressed in the form given in (10.8.2) with $a \to a', b \to b', \alpha_N \to \alpha'_N$, are related as follows,

$$\begin{cases} a' = n^{1/2-s}\sigma_{2s-1}(n)\,a \\ b' = n^{s-1/2}\sigma_{1-2s}(n)\,b \end{cases} \qquad \alpha'_N = \sum_{d|\gcd(N,n)} \alpha_{Nn/d^2} \tag{10.8.8}$$

10.9 Hecke operators on vector-valued modular forms

In Chapter 8, we introduced modular differential equations (MDEs) and discussed how they were usefully organized according to the so-called Wronskian index ℓ, defined in (8.3.22), which gives a rough measure of the number of poles allowed in the coefficient functions. For the case of $\ell = 0$, the coefficient functions must be holomorphic, and for low orders the MDEs can be recast as hypergeometric equations and solved exactly, as was done in Subsections 8.2.2 and 8.2.3 for the cases of second and third orders, respectively.

Obtaining exact solutions to MDEs for general ℓ is more difficult. One way to obtain (a subset of) solutions to MDEs with non-zero ℓ is to make use of Hecke operators. Here we will give only a flavor of how this is done. Above we have defined the action of Hecke operators on modular forms for $SL(2,\mathbb{Z})$. In fact, there exists a straightforward generalization to modular forms for congruence subgroups. Invoking the integrality theorem of Section 8.3.2, this then leads to a component-wise action of the Hecke operators on *integral* vector-valued modular forms. For a vector-valued modular form $\mathbf{f}(\tau)$ with components $f_i(\tau)$ transforming covariantly under $\Gamma(N)$, the action of T_p for p prime such that $(p, N) = 1$ is given by

$$(\mathsf{T}_p f)_i(\tau) = \sum_j \rho_{ij}(\sigma_p) f_j(p\tau) + \sum_{b=0}^{p-1} f_i\left(\frac{\tau + bN}{p}\right) \qquad (10.9.1)$$

where $\rho(\sigma_p)$ is a matrix representation of

$$\sigma_p = T^{\bar{p}} S^{-1} T^p S T^{\bar{p}} S \qquad (10.9.2)$$

and $\bar{p}$ is the modular inverse of p, that is, $p\bar{p} \equiv 1 \pmod{N}$. This definition of the Hecke operators is chosen to lead to a simple action on the Fourier coefficients. Indeed, the Fourier expansions of the components f_i of a vector-valued modular form $\mathbf{f}$ and its transform $(\mathsf{T}_p \mathbf{f})_i$ are related by,

$$f_i(\tau) = \sum_n f_n^{(i)} q^{n/N} \qquad (\mathsf{T}_p \mathbf{f})_i(\tau) = \sum_n g_n^{(i)}(p) q^{n/N} \quad (10.9.3)$$

with

$$g_n^{(i)}(p) = \begin{cases} p f_{pn}^{(i)} & p \nmid n \\ p f_{pn}^{(i)} + \sum_j \rho_{ij}(\sigma_p) f_{n/p}^{(j)} & p \mid n \end{cases} \qquad (10.9.4)$$

This action of the Hecke operator gives a new vector-valued modular form with components that have the same weight k under $\Gamma(N)$ as before, but now with different leading exponents $n_0^{(i)}$ in the notation of (8.3.15). Because $\mathrm{ord}_{W_d}(i\infty) = \sum_i n_0^{(i)}$, we expect from (8.3.23) that the Hecke image $\mathsf{T}_p \mathbf{f}(\tau)$

generically satisfies an MDE with different Wronskian index ℓ. The Hecke operators can thus be used to transform solutions to MDEs with $\ell = 0$ to solutions to those with $\ell > 0$.

We present just a single example here, which is motivated by physics. Consider the three dimensional vector-valued modular function made up of characters of the Ising conformal field theory, given in (8.2.23). From (8.2.22), the leading exponents are seen to be,

$$n_0^{(1)} = \frac{23}{48} \qquad n_0^{(2)} = -\frac{1}{48} \qquad n_0^{(3)} = \frac{1}{24} \qquad (10.9.5)$$

in the notation of (8.3.15). Applying the Hecke operator T_p they change to,

$$\tilde{n}_0^{(1)} = \frac{23p \bmod 48}{48} \qquad \tilde{n}_0^{(2)} = -\frac{p}{48} \qquad \tilde{n}_0^{(3)} = \frac{2p \bmod 48}{48} \qquad (10.9.6)$$

from (10.9.4). Then using the formula (8.3.23), we conclude that, for example, $\mathsf{T}_{49}, \mathsf{T}_{53}$, and T_{55} we have $\ell = 6$, and hence application of these Hecke operators to the Ising characters gives solutions to an MDE of order 3 with meromorphic coefficient functions. These solutions are exact in the sense that their full Fourier expansions can be obtained, though they are not known to be expressible in terms of any standard functions.

Another interesting case is that of T_{47} acting on the Ising characters. In that case, we obtain

$$(\mathsf{T}_{47}\,\chi_1)(\tau) = q^{-\frac{47}{48}} \left(1 + 96256\,q^2 + 9646891\,q^3 + \dots\right) \qquad (10.9.7)$$
$$(\mathsf{T}_{47}\,\chi_\epsilon)(\tau) = q^{\frac{25}{48}} \left(4371 + 1143745\,q + 64680601\,q^2 + \dots\right)$$
$$(\mathsf{T}_{47}\,\chi_\sigma)(\tau) = q^{\frac{23}{24}} \left(96256 + 10602496\,q + 420831232\,q^2 + \dots\right)$$

These turn out to be solutions to an order 3 MDE with $\ell = 0$, and can be shown to be precisely the characters of the so-called Baby Monster conformal field theory, which has as its global symmetry group the Baby Monster $\mathbb{B}$, the second-largest finite sporadic group. Hence the Hecke operator T_{47} maps from the Ising conformal field theory to the Baby Monster conformal field theory, in some appropriate sense.

10.10 Further physics applications

We now give a further physical application of Hecke theory. Before doing so, we recall the basic notion of Virasoro characters. In Section A.7, we introduce characters for finite groups. Characters can be introduced for a Lie group

G as well, in which case, given a representation $\mathcal{R}$ of G, the character $\chi_\mathcal{R}$ is defined as a function on G by,

$$\chi_\mathcal{R}(g) = \text{Tr}_\mathcal{R}(g) \tag{10.10.1}$$

with the trace $\text{Tr}_\mathcal{R}$ taken in representation space of $\mathcal{R}$. The general theory of characters is rich, but largely unnecessary for our purposes. The particular case of importance to us here is that of the characters of the Virasoro group. As we have discussed in Subsection 5.1.1, the Virasoro group is an infinite-dimensional Lie group, which is the universal central extension of the group of diffeomorphisms of the circle. The corresponding Lie algebra is the Virasoro algebra, whose generators L_n satisfy the structure relations given in (5.1.3), which we repeat here for convenience,

$$[L_n, L_m] = (n - m)L_{n+m} + \frac{\mathfrak{c}}{12}(n^3 - n)\delta_{n,-m} \tag{10.10.2}$$

The parameter $\mathfrak{c}$ here is known as the central charge.

A quantum field theory with Virasoro symmetry is known as a conformal field theory. By "symmetry," one means that the state space of the theory, and the operators, are organized into Virasoro representations. In a unitary conformal field theory, the operators obey $L_m^\dagger = L_{-m}$ and its representations are unitary. The representations can be constructed starting from a highest weight state $|h\rangle$, where h is real in view of $L_0^\dagger = L_0$, and defined by,

$$L_0|h\rangle = h|h\rangle \qquad L_n|h\rangle = 0 \qquad n > 0 \tag{10.10.3}$$

and then generating other states in the representation by taking linear combinations of states of the form,

$$L_{-n_1}L_{-n_2}\ldots L_{-n_r}|h\rangle \qquad n_i > 0 \tag{10.10.4}$$

Such states are known as *conformal descendants* and will be collectively denoted by $|h, \{n\}\rangle$. The vector space of the states thus generated is referred to in mathematics as the *Verma module* $\mathcal{V}_h$ of weight h and in physics as the *conformal family* of weight h. The character associated to a representation labeled by h is then given by,

$$\chi_h(\tau) = q^{h - \frac{\mathfrak{c}}{24}} \sum_{\{n\}} q^N \tag{10.10.5}$$

with $q = e^{2\pi i \tau}$ and $N = \sum_i n_i$. This is sometimes also written as,

$$\chi_h(\tau) = \text{Tr}_{\mathcal{V}_h}\left(q^{L_0 - \frac{\mathfrak{c}}{24}}\right) \tag{10.10.6}$$

with the trace being over the weight-h Verma module $\mathcal{V}_h$.

Characters play an important role in the study of two-dimensional conformal field theories. In particular, the torus partition function of many conformal field theories can be decomposed into a sum over conformal families and a sum within each family, giving,

$$Z(\tau, \bar{\tau}) = \sum_{h, \tilde{h}}' \mathcal{N}_{h, \tilde{h}} \, \chi_h(\tau) \chi_{\tilde{h}}(\tau) \qquad (10.10.7)$$

for some Hermitian pairing matrix $\mathcal{N}$. Here we are assuming that the indices $h, \tilde{h}$ take discrete values, but not necessarily that the sums over conformal families are finite.

The partition function is, by definition, a weight-zero non-holomorphic modular function. But conversely, not every non-holomorphic modular function of weight 0 corresponds to a valid conformal field theory partition function. For example, since partition functions count degeneracies of states, any physically sensible partition function must have positive integer Fourier coefficients. The same is true for the constituent characters.

In this context, it will be useful to introduce a variant of the Hecke operators with alternative normalization,

$$\mathcal{T}_n = n \, \mathsf{T}_n \qquad (10.10.8)$$

which has the virtue that, if our original function $\chi(\tau)$ has positive integer Fourier coefficients $f_m \in \mathbb{Z}_{\geq 0}$, then the Hecke transform $\mathcal{T}_n \chi(\tau)$ has positive integer Fourier coefficients $n \, g_m^{(n)} \in \mathbb{Z}_{\geq 0}$ as well; c.f. the discussion after (10.4.4). The application of Hecke operators $\mathcal{T}_n$ to a conformal field theory character thus produces another function, which itself has some properties expected of a conformal field theory character. In this sense, we might hope that the Hecke operators provide a *map between conformal field theories*, as in the example of the Ising and Baby Monster conformal field theories mentioned above. We now study this phenomenon in the simplest examples of holomorphically factorizable conformal field theories, that is, those for which the partition function is given simply as a product $Z(\tau, \bar{\tau}) = Z(\tau)\overline{Z(\tau)}$.

Let us consider a class of two-dimensional conformal field theories arising as the holographic duals to three-dimensional gravity in anti-de Sitter space AdS$_3$. We will consider theories of pure gravity, which in three dimensions are known to not have any propagating gravitational waves. Indeed, three-dimensional gravity can be rewritten as a topological theory, namely, an SL$(2, \mathbb{R}) \times$ SL$(2, \mathbb{R})$ Chern–Simons theory specified by levels k_L, k_R. Here we work with the case of $k_L = k_R = \kappa$. By comparing the Chern–Simons formulation of the theory with the Einstein–Hilbert formulation, one shows that the curvature radius of AdS$_3$ is related to the level κ by $\ell = 16\kappa G_N$, where G_N is Newton's constant, so that ℓ is quantized in units of $1/(16G_N)$.

What do the tentative dual conformal field theories look like? By the Brown–Henneaux formula [BH86], they should have central charges,

$$\mathfrak{c} = \frac{3\ell}{2G_N} = 24\kappa \qquad (10.10.9)$$

The vacuum energy is given by $-\frac{\mathfrak{c}}{24} = -\kappa$. Additional states can be obtained from the vacuum state by application of the generators of the Virasoro algebra. In particular, the generators L_n for $n \geq -1$ annihilate the vacuum, while the generators L_{-n} for $n > 1$ act on the vacuum to give new states with energy $-\kappa + n$. For given energy $-\kappa + n$, the naive degeneracy would be given by $p(n)$, the number of integer partitions of n. For example, for $n = 3$ we have $p(3) = 3$, corresponding to the three states,

$$L_{-1}^3|0\rangle \qquad\qquad L_{-1}L_{-2}|0\rangle \qquad\qquad L_{-3}|0\rangle \qquad (10.10.10)$$

This would give total vacuum contribution to the chiral partition function,

$$Z_\kappa^{\text{vac}}(\tau) \stackrel{?}{=} q^{-\kappa} \sum_{n=0}^{\infty} p(n)q^n = q^{-\kappa} \prod_{n=1}^{\infty} \frac{1}{1 - q^n} \qquad (10.10.11)$$

as per the discussion in Subsection 5.6.2. However, in the current case, L_{-1} annihilates the vacuum, and the corresponding degeneracies must be removed. This is accomplished by including a factor of $(1 - q)$, giving the actual vacuum contribution,

$$Z_\kappa^{\text{vac}}(\tau) = q^{-\kappa} \prod_{n=2}^{\infty} \frac{1}{1 - q^n} \qquad (10.10.12)$$

The vacuum contribution to the partition function is not itself modular invariant, which means that there must be additional states beyond the vacuum and its descendants. Indeed, there must be some states in the conformal field theory which capture the known black holes states of the holographic dual. These black holes are expected on general grounds to appear at order $\mathcal{O}(q^1)$, giving a full partition function of the form,

$$Z_\kappa(\tau) = q^{-\kappa} \prod_{n=2}^{\infty} \frac{1}{1 - q^n} + \mathcal{O}(q) \qquad (10.10.13)$$

Remarkably, there is a unique such partition function for every κ.

For $\kappa = 1$, that is, pure AdS$_3$ gravity with the smallest value of the cosmological constant, the only possible candidate is

$$Z_1(\tau) = j(\tau) - 744 = q^{-1} + 196884\,q + \ldots \qquad (10.10.14)$$

with $j(\tau)$ the j-function introduced in Section 3.6. Physically, the 196884 states at level 1 can be interpreted as 1 descendant of the vacuum, and an additional 196883 black hole states. The conformal field theory with this partition function was first constructed by Frenkel, Lepowsky, and Merman and is known as the *Monster conformal field theory*, due to it having global symmetry group given by the Fischer–Greiss monster $\mathbb{M}$, the largest finite sporadic group. The smallest nontrivial representation of $\mathbb{M}$ has dimension 196883, and the black hole states indeed lie in this representation.

To obtain partition functions for higher $\kappa > 1$, we now make use of Hecke operators. Thus far in this section, we have focused on Hecke operators acting on holomorphic modular forms, but the extension to the meromorphic case is straightforward. The q-expansions of Hecke transforms can be obtained using (10.4.3), giving, for example,

$$\mathcal{T}_2 J(\tau) = q^{-2} + 42987520\,q + \ldots$$
$$\mathcal{T}_3 J(\tau) = q^{-3} + 2592899910\,q + \ldots \qquad (10.10.15)$$

where we have defined $J(\tau) = j(\tau) - 744$. For any κ, it is easy to show that,

$$\mathcal{T}_\kappa J(\tau) = q^{-\kappa} + \mathcal{O}(q) \qquad (10.10.16)$$

Using this result, we can now write down closed form expressions for $Z_\kappa(\tau)$ as follows. Say that,

$$Z_\kappa^{\mathrm{vac}}(\tau) = \sum_{r=-\kappa}^{\infty} c_r q^r \qquad (10.10.17)$$

Of course, using (10.10.12) all c_r are known. Then, the full chiral partition function is given simply by,

$$Z_\kappa(\tau) = \sum_{r=0}^{\kappa} c_{-r} \mathcal{T}_r J \qquad (10.10.18)$$

where we take $\mathcal{T}_0 J(\tau) = 1$. Using the formulas above, we get explicitly,

$$Z_2(\tau) = q^{-2} + 1 + 42987520\,q + \ldots$$
$$Z_3(\tau) = q^{-3} + q^{-1} + 1 + 2593096794\,q + \ldots \qquad (10.10.19)$$

and so on. Though no concrete construction of conformal field theories with these partition functions are known (assuming they actually exist), there exists much literature on the subject. Such tentative theories are referred to as *extremal conformal field theories*. We have thus seen that the Hecke operators act as maps between extremal conformal field theories.

Furthermore, by using Hecke operators, we have obtained the exact number of quantum black hole microstates in theories of pure gravity. In particular, this number is obtained by subtracting the number of vacuum descendants from the coefficient of the $\mathcal{O}(q^1)$ term. It is interesting to compare these results to those predicted by the Bekenstein–Hawking entropy,

$$S_{BH} = 4\pi\sqrt{\kappa L_0} \qquad (10.10.20)$$

For example, for $\kappa = L_0 = 1$ one finds $S_{BH} = 4\pi \approx 12.57$, to be compared to the exact result $S_{\text{exact}} = \ln 196883 \approx 12.19$. The agreement between the Bekenstein–Hawking entropy and the exact microstate count improves in the semiclassical limit $\kappa, L_0 \to \infty$ with L_0/k fixed. The origin of this agreement is in fact a classic result of Petersson and Rademacher. Writing the Fourier expansion of $j(\tau)$ as,

$$j(\tau) - 744 = \sum_{n=-1}^{\infty} c_n q^n \qquad (10.10.21)$$

with $c_0 = 0$, Petersson and Rademacher proved that as $n \to \infty$, one has,

$$\ln c_n \approx 4\pi\sqrt{n} - \frac{3}{4}\ln m - \frac{1}{2}\ln 2 + \dots \qquad (10.10.22)$$

We now combine this with the results for the Fourier coefficients of Hecke transforms. Since we are working in the limit $\kappa, L_0 \to \infty$, it suffices to consider only the $\mathcal{T}_\kappa J$ contribution to (10.10.18). The dominant contribution to the corresponding Fourier coefficients $b_n^{(\kappa)}$ are $b_n^{(\kappa)} \approx \kappa c_{\kappa n}$, which gives,

$$\ln b_n^{(\kappa)} \approx 4\pi\sqrt{\kappa n} + \frac{1}{4}\ln\kappa - \frac{3}{4}\ln n - \frac{1}{2}\ln 2 + \dots \qquad (10.10.23)$$

The first term reproduces the Bekenstein–Hawking result, while the others give the leading-order quantum corrections.

• Bibliographical notes

The formal theory of Hecke operators was developed by Hecke in [Hec37a] and [Hec37b], but the operators had been introduced two decades earlier when Mordell proved the multiplication law of the Ramanujan τ-function [Mor17], as reproduced in Section 10.5. Most of the mathematical material discussed here for meromorphic modular forms may be found in the books by

Apostol [Apo76b], Shimura [Shi71], Diamond and Shurman [DS05], and Serre [Ser73]. The theory of Hecke operators acting on Maass forms is given in the book by Terras [Ter85]. The double coset construction is described in the books by Diamond and Shurman [DS05] and Iwaniec [Iwa97]. The action of Hecke operators on vector-valued modular forms was developed recently in both physics [HW18, HHW20] and mathematics [BCJ19] literature.

The discussion of three-dimensional gravity given in Section 10.10 follows closely the work of [Wit07]. For standard introductions to two-dimensional quantum field theory and conformal field theory, we refer to [Gin88] and [DFMS97], already mentioned in Chapter 5. The basic tools of AdS/CFT are reviewed in [AGM⁺00] and [DF02] as well as in the book by Ammon and Erdmenger [AE15]. The nonexistence of several extremal conformal field theories was recently proven in [LP23], using the theory of topological modular forms. The asymptotic behavior of the coefficients of the j-function, quoted in (10.10.22), was obtained by Petersson in [Pet32], and by Rademacher [Rad38].

Exercises

10.1 Prove the inversion of the multiplication relation (10.6.1), given by,

$$
\mathsf{T}_{mn} = \sum_{d\mid \gcd(m,n)} \mu(d) d^{k-1} \mathsf{T}_{m/d} \mathsf{T}_{n/d} \tag{10.10.24}
$$

where $\mu(d)$ is the Möbius function.

10.2 Use Exercise 10.1 to derive a recursion relation for T_{p^ν} for p prime.

10.3 Use the result of Exercise 10.2 to compute the generating function,

$$
f_p(x) = \sum_{n=0}^{\infty} x^n \mathsf{T}_{p^n} \tag{10.10.25}
$$

10.4 Prove that the operators T_n are self-adjoint with respect to the Petersson inner product.

10.5 Taking $f(\tau)$ to be a weight-zero modular form with Fourier expansion $f(\tau) = \sum_{n=0}^{\infty} a_n q^n$, prove the following relation,

$$
\prod_{\substack{m>0 \\ n\in\mathbb{Z}}} (1 - p^m q^n)^{-a_{mn}} = \exp\left[\sum_{m>0} \mathsf{T}_m f(\tau) p^m \right]
$$

In the context of physics, this formula is crucial in the study of "symmetric product orbifolds" [DMVV97].

11

Singular moduli and complex multiplication

In Chapter 3, we introduced $SL(2, \mathbb{Z})$ as the automorphism group of an arbitrary lattice $\Lambda = \mathbb{Z}\omega_1 + \mathbb{Z}\omega_2$ with $\text{Im}\,(\omega_2/\omega_1) > 0$. For every value of the modulus $\tau = \omega_2/\omega_1$, the lattice also possesses a ring of endomorphisms which multiply the lattice $\Lambda \to n\Lambda$ by a nonvanishing integer $n \in \mathbb{Z}^*$ thereby producing a sublattice $n\Lambda \subset \Lambda$. On the other hand, multiplying the lattice Λ by an arbitrary complex number α gives a lattice $\alpha\Lambda$ that will generally not be a sublattice of Λ. However, for special values of the modulus τ, referred to as *singular moduli*, and associated special values of $\alpha \notin \mathbb{Z}$, the lattice possesses an enlarged ring of endomorphisms $\mathbb{Z}[\alpha]$ such that $\alpha\Lambda \subset \Lambda$. This phenomenon is referred to as *complex multiplication*. From a mathematics standpoint, various modular forms take on special values at singular moduli, as exemplified by the fact that the j-function is an algebraic integer. From a physics standpoint, the enlargement of the endomorphism ring has arithmetic consequences in conformal field theory, as exemplified by the fact that toroidal compactifications at singular moduli correspond to rational conformal field theories, as will be discussed in Chapter 13.

In the special cases where $\tau = i$ and $\tau = \rho = e^{2\pi i/3}$ (and $SL(2, \mathbb{Z})$ images thereof), the corresponding values are $\alpha = i$ and $\alpha = \rho$, respectively; they satisfy $|\alpha| = 1$ so that multiplication by α is actually an automorphism. The corresponding automorphism groups are $\{\pm I, \pm S\}$ of order 4 and $\{\pm I, \pm(ST), \pm(ST)^2\}$ of order 6, respectively.

11.1 Conditions for complex multiplication

We seek the lattices for which there exists a complex number $\alpha \in \mathbb{C} \setminus \mathbb{Z}$ such that $\alpha \Lambda \subset \Lambda$. The condition may be written out explicitly on the periods,

$$\alpha \omega_2 = a\omega_2 + b\omega_1$$
$$\alpha \omega_1 = c\omega_2 + d\omega_1 \qquad\qquad (11.1.1)$$

with $a, b, c, d \in \mathbb{Z}$ and $ad - bc \neq 0$. Note that we are not requiring $ad - bc = 1$ since we are only demanding that $\alpha \Lambda$ be contained in Λ, without necessarily being equal to Λ. Taking ratios gives a condition on τ, and a solution for α in terms of τ,

$$\tau = \frac{a\tau + b}{c\tau + d} \qquad\qquad \alpha = c\tau + d \qquad\qquad (11.1.2)$$

The condition on τ is given by a quadratic polynomial equation,

$$c\tau^2 + (d - a)\tau - b = 0 \qquad\qquad (11.1.3)$$

with integer coefficients. The discriminant of this equation is denoted by D,

$$D = (a - d)^2 + 4bc = (a + d)^2 - 4(ad - bc) \qquad\qquad (11.1.4)$$

The restriction $\text{Im}(\tau) \neq 0$ imposes the condition $D < 0$, which in turn requires $bc < 0$ and $ad - bc \geq 1$. Without loss of generality, we may assume that $c > 0$ and define $\sqrt{D}$ to be the square root of D satisfying $\text{Im}(\sqrt{D}) > 0$, so that we have $\alpha \in \mathbb{C} \setminus \mathbb{R}$ and,

$$\tau = \frac{a - d + \sqrt{D}}{2c} \qquad\qquad \alpha = \frac{a + d + \sqrt{D}}{2} \qquad\qquad |\alpha|^2 = ad - bc \quad (11.1.5)$$

Extending the field $\mathbb{Q}$ by τ produces the quadratic extension field $\mathbb{Q}(\tau)$, while the ring of homomorphisms is extended from $\mathbb{Z}$ to $\mathbb{Z}[\alpha]$.

11.2 Elliptic functions at complex multiplication points

When $\alpha \Lambda \subset \Lambda$ and $\alpha \in \mathbb{C} \setminus \mathbb{Z}$, the elliptic functions and modular forms at τ satisfy remarkable properties. A first key result is for elliptic functions, where it suffices to investigate the Weierstrass function $\wp(z) = \wp(z|\Lambda)$ for the lattice Λ. Since $\alpha \Lambda \subset \Lambda$, the function $\wp(\alpha z)$ is an even elliptic function for the original lattice Λ and may therefore be expressed as a rational function of $\wp(z)$. More specifically, we have,

$$\wp(\alpha z) = \frac{A(\wp(z))}{B(\wp(z))} \qquad\qquad (11.2.1)$$

with A and B relatively prime polynomials of degrees related by,

$$\deg A = 1 + \deg B = |\alpha|^2 = ad - bc \qquad (11.2.2)$$

The only assertion above that needs proof is the relation between the degrees. First, since $\wp(\alpha z)$ has a double pole at $z = 0$ and since A and B will have poles at $z = 0$ of respective degrees $2 \deg A$ and $2 \deg B$, we must have the first equality in (11.2.2) between the degrees. Second, the lattice $\alpha \Lambda$ has a fundamental parallelogram which is scaled up from Λ by a factor of $|\alpha|$, so that the area of the fundamental parallelogram is $|\alpha|^2$ times that of the fundamental parallelogram of Λ. Thus, the index of $\alpha \Lambda$ as a subgroup of Λ is given by $[\Lambda : \alpha \Lambda] = |\alpha|^2$. While $\wp(z)$ has one double pole in the fundamental parallelogram of Λ, the total number of poles of $\wp(\alpha z)$ in the fundamental parallelogram of Λ is $2|\alpha|^2$, which is also its number of zeros. It follows that $\deg A = |\alpha|^2$ giving the second equality and, by (11.1.5), gives the last equality in (11.2.2).

One may construct the polynomials A, B by matching the Laurent expansions of both sides at $z = 0$ using the Laurent series of $\wp(z)$ of (2.4.4),

$$\wp(z) = \frac{1}{z^2} + \sum_{m=1}^{\infty} (2m + 1)G_{2m+2}z^{2m} \qquad (11.2.3)$$

where we have taken into account the fact that $G_m = 0$ for m odd. We shall also use the fact that G_{2m} for $m \geq 4$ may be expressed as a polynomial in the generators G_4 and G_6 of the ring of modular forms of $\mathrm{SL}(2, \mathbb{Z})$ of (3.4.22),

$$G_{2m+4} = \sum_{k=1}^{m-1} \frac{3(2k + 1)(2m - 2k + 1)}{(m - 1)(2m + 3)(2m + 5)} G_{2k+2}G_{2m-2k+2} \qquad (11.2.4)$$

Of course, the relation between $\wp(\alpha z)$ and $\wp(z)$, and therefore the matching of the Laurent series, will only be possible at the singular modulus τ for which we have the corresponding complex conjugation factor α. Therefore, the matching will impose a relation between the Eisenstein series G_4 and G_6, and this will give a formula for the value of $j(\tau)$.

11.3 Examples

Consider the simple example $\alpha = i\sqrt{2}$ for which $a + d = 0$, the discriminant is given by $D = -8$, so that $|\alpha|^2 = 2$, and the corresponding singular modulus is $\tau = i\sqrt{2}$. In this case, A has degree 2 and B degree 1, so that we may express the relation (11.2.1) as,

$$\wp(i\sqrt{2}\,z) = c_1\wp(z) + c_2 + \frac{c_3}{\wp(z) + c_4} \tag{11.3.1}$$

Since $\wp(i\sqrt{2}\,z) = -1/(2z^2) + \mathcal{O}(z^2)$ we must have $c_1 = -\frac{1}{2}$ and $c_2 = 0$. Substituting the Laurent expansion we obtain the following relations,

$$
\begin{aligned}
\left(\wp(i\sqrt{2}\,z) + \tfrac{1}{2}\wp(z)\right)^{-1} &= -\frac{2}{9G_4}\left(\frac{1}{z^2} + 5\frac{G_6}{G_4} + 25\frac{G_6^2}{G_4^2}z^2 - 5G_4z^2 + \mathcal{O}(z^4)\right)\\
&= \frac{1}{c_3}\left(\wp(z) + c_4 + 25\frac{G_6^2}{G_4^2}z^2 - 8G_4z^2 + \mathcal{O}(z^4)\right)
\end{aligned}
\tag{11.3.2}
$$

To match the Laurent expansions of the first and second lines above to orders z^{-2} and z^0, we set,

$$c_3 = -\frac{9}{2}G_4 \qquad\qquad c_4 = 5\frac{G_6}{G_4} \tag{11.3.3}$$

But the relation must be exact to all orders in z by our global analysis, so the terms of order z^2 and higher must also match. This can be achieved only at the singular modulus associated with the complex multiplication factor α, and gives the relation $25\,G_6^2 = 8G_4^3$ between G_4 and G_6. One verifies that all higher-order terms indeed match. By the same token, we find that the j-invariant at the complex multiplication point is given by,

$$j(i\sqrt{2}) = \frac{(12)^3\,g_2^3}{g_2^3 - 27g_3^2} = (12)^3\left(1 - \frac{49\,G_6^2}{20G_4^3}\right)^{-1} = (20)^3 \tag{11.3.4}$$

Remarkably, $j(i\sqrt{2})$ is an integer, and more precisely the cube of an integer! Further examples of elliptic curves with complex multiplication are presented in Exercise 11.1. An alternative formulation of the multiplication formula for $\wp$ given in (11.2.1) is obtained in Exercise 11.2.

11.3.1 Heegner points

A related set of points are those labeled by the Heegner numbers, which are defined as integers d such that $j\left(\frac{1}{2}(1 + i\sqrt{d})\right)$ is an integer. There exist precisely eight such numbers d, given by, 1, 3, 7, 11, 19, 43, 67, and 163, as conjectured by Gauss and proven by Heegner, Baker, and Stark over a century later. Remarkably, the j-invariant evaluates to a cube on each one, as given in Table 11.1.

d	1	3	7	11	19	43	67	163
j	12^3	0	-15^3	-32^3	-96^3	-960^3	-5280^3	-640320^3

Table 11.1 Values of j at the Heegner points $\frac{1}{2}(1 + i\sqrt{d})$.

These are precisely the discriminants $D = -d$ for which the corresponding ring of integers has unique factorization, meaning that every nonzero non-unit element can be written as a product of irreducible elements in a unique way (up to ordering and multiplication of units). As a simple example without unique factorization, we may consider the quadratic integer ring defined by $D = -5$, in which the element 6 can be factored as both 2×3 and $(1 + i\sqrt{5}) \times (1 - i\sqrt{5})$, with all four factors being irreducible.

The Heegner numbers are related to some well-known examples of "almost integers" including the famous Ramanujan's constant $e^{\pi\sqrt{163}}$,

$$e^{\pi\sqrt{67}} = 147197952743.99999866\cdots$$
$$e^{\pi\sqrt{163}} = 262537412640768743.99999999999925\ldots \qquad (11.3.5)$$

The near-integrality is explained by the values of $j(\tau)$ at $\tau = \frac{1}{2}(1 + i\sqrt{67})$ and $\tau = \frac{1}{2}(1 + i\sqrt{163})$ given above, together with the q-expansion of $j(\tau)$,

$$e^{\pi\sqrt{67}} = 5280^3 + 744 + \mathcal{O}(e^{-\pi\sqrt{67}})$$
$$e^{\pi\sqrt{163}} = 640320^3 + 744 + \mathcal{O}(e^{-\pi\sqrt{163}}) \qquad (11.3.6)$$

11.4 $j(\tau)$ as an algebraic integer

An *algebraic integer* is a complex number that is a root of a monic polynomial with integer coefficients, namely, of a monic polynomial in $\mathbb{Z}[x]$. Recall that a *monic* polynomial is one whose leading monomial has coefficient 1. A general result regarding the arithmetic properties of $j(\tau)$ at a complex multiplication points τ is the following.

Theorem 11.4.1. *For a lattice $\Lambda = \mathbb{Z} + \mathbb{Z}\tau$ with complex multiplication, $j(\tau)$ is an algebraic integer.*

To prove this result, we define the monic polynomial in x for given τ,

$$P_n(x|\tau) = \prod_{\gamma \in SL(2,\mathbb{Z}) \backslash M_n} (x - j(\gamma\tau)) \qquad (11.4.1)$$

where M_n is the set of 2×2 matrices with determinant n, introduced in

(10.1.3). Since $j(\tau)$ is invariant under $\mathrm{SL}(2,\mathbb{Z})$ the product in γ is taken over the left cosets $\mathrm{SL}(2,\mathbb{Z})\backslash M_n$ discussed above (10.1.4). The function $P_n(x|\tau)$ is invariant under $\tau \to \gamma\tau$ for all $\gamma \in \mathrm{SL}(2,\mathbb{Z})$ because the right action of $\mathrm{SL}(2,\mathbb{Z})$ on the left cosets $\mathrm{SL}(2,\mathbb{Z})\backslash M_n$ simply permutes the cosets.

Because the set of matrices,

$$\left\{ \begin{pmatrix} a & b \\ 0 & d \end{pmatrix} \,\Big|\, ad = n, \, 0 \le b \le d-1 \right\} \tag{11.4.2}$$

provides a complete set of representatives of the quotient $\mathrm{SL}(2,\mathbb{Z})\backslash M_n$, as was explained in Section 10.1, we obtain the equivalent product formula,

$$P_n(x|\tau) = \prod_{ad=n} \prod_{b=0}^{d-1} \left(x - j\left(\frac{a\tau+b}{d}\right) \right) \tag{11.4.3}$$

This expression confirms that $P_n(x|\tau+1) = P_n(x|\tau)$ since the shift in τ amounts to a shift $b \to b+a \pmod{d}$ under which the product over b is invariant. The degree of P_n is given by the cardinality of $\mathrm{SL}(2,\mathbb{Z})\backslash M_n$,

$$\deg(P_n) = \left|\mathrm{SL}(2,\mathbb{Z})\backslash M_n\right| = \sigma_1(n) \tag{11.4.4}$$

where $\sigma_1(n) = \sum_{d|n} d$ is the divisor sum. The full structure of the τ-dependence of $P_n(x|\tau)$ is given by the following lemma.

Lemma 11.4.2. $P_n(x|\tau)$ *is a polynomial in x and $j(\tau)$ with integer coefficients, so that there exists a polynomial $Q_n(x,y) \in \mathbb{Z}[x,y]$ which satisfies $P_n(x|\tau) = Q_n(x, j(\tau))$.*

To prove the lemma, we begin by focusing on the product over b for given d. Writing the q-expansion of $j(\tau)$ as $j(\tau) = q^{-1} + 744 + \sum_{k\ge 1} c_k q^k$ with $c_k \in \mathbb{N}$, we have,

$$\prod_{0 \le b \le d-1} \left(x - j\left(\tfrac{a\tau+b}{d}\right) \right)$$

$$= (-1)^d \prod_{0 \le b \le d-1} \left(e^{-\frac{2\pi i b}{d}} q^{-\frac{a}{d}} + 744 + \sum_{k\ge 1} c_k e^{\frac{2\pi i k b}{d}} q^{\frac{ka}{d}} - x \right)$$

$$= (-1)^d q^{-a} + \sum_{\ell \ge 1-d} a_\ell(x) q^{\frac{\ell a}{d}} \tag{11.4.5}$$

By construction, the coefficients $a_\ell(x)$ are polynomials in x and the root of unity $e^{2\pi i/d}$, namely, $a_\ell(x) \in \mathbb{Z}\left[x, e^{2\pi i/d}\right]$. Actually, (11.4.5) implies that the $a_\ell(x)$ are invariant under all transformations $\sigma_r : e^{2\pi i/d} \to e^{2\pi i r/d}$ for $0 \le r \le d-1$; that is, the so-called *Galois automorphisms* of $\mathbb{Q}(e^{2\pi i/d})/\mathbb{Q}$, which will be further discussed in Chapter 16. As a result, $a_\ell(x)$ must be independent of $e^{2\pi i/d}$ and therefore belongs to $\mathbb{Z}[x]$. Furthermore, the left side of (11.4.5) is invariant under $\tau \mapsto \tau + 1$, so that we must have $a_\ell(x) = 0$ for all $d \nmid \ell$. The function $P_n(x|\tau)$ is obtained by taking the product of (11.4.5) over $ad = n$, which we reorganize as follows,

$$P_n(x|\tau) = \prod_{ad=n} \left[(-1)^d q^{-a} + \sum_{m \geq 0} a_{md}(x) q^{ma} \right] = \sum_{r=0}^{\deg(P_n)} h_r(\tau) x^r \quad (11.4.6)$$

where the coefficient functions $h_r(\tau)$ admit a Fourier expansion $h_r(\tau) = \sum_\ell a_\ell^{(r)} q^\ell$ with $a_\ell^{(r)} \in \mathbb{Z}$. Recall that any meromorphic modular function with integer Fourier coefficients can be written as a polynomial in $j(\tau)$ with integer coefficients; that is, we can write $h_r(\tau) = \sum_\ell \tilde{a}_\ell^{(r)} j(\tau)^\ell$ with $\tilde{a}_\ell^{(r)} \in \mathbb{Z}$. This allows us to conclude that the same is true of $P_n(x|\tau)$, thereby proving that $P_n(x|\tau)$ is a polynomial in x and $j(\tau)$ with integer coefficient. This result may be summarized by the statement that there exists a polynomial $Q_n(x, y)$ with integer coefficients such that,

$$P_n(x|\tau) = Q_n(x, j(\tau)) \quad (11.4.7)$$

Actually Q_n satisfies $Q_n(y, x) = \pm Q_n(x, y)$ and is of degree $\sigma_1(n)$ in x, y.

To complete the proof of Theorem 11.4.1, we now specialize to a point $\tau = \tau_{\text{CM}} \in \mathcal{H}$ of complex multiplication of order n, namely, satisfying $\gamma_0 \tau_{\text{CM}} = \tau_{\text{CM}}$ for a matrix $\gamma_0 = \left(\begin{smallmatrix} a & b \\ c & d \end{smallmatrix} \right) \in M_n$ with determinant $n = ad - bc$. The polynomial $P_n(x|\tau_{\text{CM}}) = Q_n(x, j(\tau_{\text{CM}}))$ must vanish at the point $x = j(\tau_{\text{CM}})$ since, up to an $SL(2, \mathbb{Z})$ transformation, γ_0 will coincide with one of the γ-matrices in the product in (11.4.1). Thus, the polynomial $Q_n(x, x) \in \mathbb{Z}[x]$ with integer coefficients vanishes at $x = j(\tau_{\text{CM}})$.

To complete the proof that $j(\tau_{\text{CM}})$ is an algebraic integer, it remains to show that it is a root of a monic polynomial with integer coefficients. To this end, we evaluate the leading behavior at the cusp of the polynomial $Q_n(x, x) \in \mathbb{Z}[x]$ from (11.4.3) with the help of the factorization formula,

$$\prod_{b=0}^{d-1} (x - e^{2\pi i b/d} y) = x^d - y^d \quad (11.4.8)$$

and we obtain,

$$Q_n(j(\tau), j(\tau)) = \prod_{ad=n} \left(q^{-d} - q^{-a} + \text{lower order terms} \right) \quad (11.4.9)$$

There are now two cases to be considered, depending on whether n is a perfect square or not. We first consider the case when n is not a perfect square. Every factor in the product has $a \neq d$ since otherwise $n = ad = a^2$ is a perfect square and the leading behavior at the cusp is given by the power of q obtained by summing the maxima of (d, a),

$$Q_n(j(\tau), j(\tau)) = \pm q^{-\sigma_1^+(n)} + \text{lower orders}$$

$$\sigma_1^+(n) = \sum_{ad=n} \max(a, d) \qquad (11.4.10)$$

Thus, the polynomial $Q_n(x, x)$ is nonvanishing and monic (up to a sign) when n is not a perfect square. When $n = \nu^2$ is a perfect square, then one element in $\mathrm{SL}(2, \mathbb{Z}) \backslash M_n$ is of the form νI so that $a = d = \nu$ and $Q_n(j(\tau), j(\tau)) = 0$. The vanishing is caused by the fact that $Q_n(x, y)$ has a factor $(x-y)$. Adapting the above arguments, one shows that the polynomial,

$$\tilde{Q}_n(x, y) = \frac{Q_n(x, y)}{x - y} \qquad (11.4.11)$$

evaluated at $x = y = j(\tau)$ is an integer multiple of a monic polynomial of degree $\sigma_1^+(n) - 1$. This completes the proof of the theorem.

11.4.1 The examples of $n = 2, 3, 4, 5$

Zagier worked out the example of $n = 2$ in detail. The cases $n = 3, 4, 5$ may be obtained analogously, and we find the following factorizations of $Q_n(x, x)$,

$$Q_2(x, x) = -(x - 12^3)(x - 20^3)(x + 15^3)^2$$
$$Q_3(x, x) = -x(x - 20^3)^2(x + 32^3)^2(x - 54000)$$
$$\tilde{Q}_4(x, x) = 2(x + 15^3)^2(x - 54000)^2(x - 66^3)$$
$$\times (x^2 + 191025x - 121287375)^2$$
$$Q_5(x, x) = (x - 12^3)^2(x + 32^3)^2(x - 66^3)^2(x + 96^3)^2$$
$$\times (x^2 - 1264000x - 681472000) \qquad (11.4.12)$$

The second-order polynomials factor as follows,

$$x^2 + 191025x - 121287375$$
$$= \left(x - \tfrac{1}{2}(-191025 + 85995\sqrt{5})\right)\left(x - \tfrac{1}{2}(-191025 - 85995\sqrt{5})\right)$$
$$x^2 - 1264000x - 681472000$$
$$= \left(x - 632000 - 282889\sqrt{5}\right)\left(x - 632000 + 282889\sqrt{5}\right) \qquad (11.4.13)$$

The expressions for $Q_3(x, y)$ and $Q_3(x, x)$ are obtained in Exercise 11.3. In addition to the entries listed in Table 11.1, we also recognize,

$$j(i) = 12^3 \qquad\qquad j(i\sqrt{3}) = 2 \cdot 30^3$$

$$j(2i) = 66^3 \qquad\qquad j(i\sqrt{5}) = (50 + 26\sqrt{5})^3$$

$$j(i\sqrt{2}) = 20^3 \qquad j\left(\tfrac{1}{2} + \tfrac{i}{2}\sqrt{15}\right) = -\tfrac{1}{2}(191025 + 85995\sqrt{5}) \qquad (11.4.14)$$

11.4.2 $j(\tau)$ as a rational integer

$-D$	τ	$j(\tau)$
3	$\frac{1}{2}(1 + i\sqrt{3})$	0
	$i\sqrt{3}$	$2 \cdot 30^3$
	$\frac{1}{2}(1 + i3\sqrt{3})$	$-3 \cdot 160^3$
4	i	12^3
	$2i$	66^3
7	$\frac{1}{2}(1 + i\sqrt{7})$	-15^3
	$i\sqrt{7}$	255^3
8	$i\sqrt{2}$	20^3
11	$\frac{1}{2}(1 + i\sqrt{11})$	-32^3
19	$\frac{1}{2}(1 + i\sqrt{19})$	-96^3
43	$\frac{1}{2}(1 + i\sqrt{43})$	-960^3
67	$\frac{1}{2}(1 + i\sqrt{67})$	-5280^3
163	$\frac{1}{2}(1 + i\sqrt{163})$	-640320^3

Table 11.2   The full list of complex multiplication points for which $j(\tau)$ is a rational (i.e., standard) integer.

Above, we have encountered several examples of complex multiplication points τ at which $j(\tau)$ is an integer. In fact, among complex multiplication points, there exist precisely 13 values of τ for which this is the case. The corresponding values of τ and $j(\tau)$, together with the corresponding discriminant D, are given in Table 11.2. In addition to the Heegner points in Table 11.1 and the values in (11.4.14), there are only two other points, namely, $\tau = \frac{1}{2}(1 + i3\sqrt{3})$ and $\tau = i\sqrt{7}$. Note the intimate relation of the above values of D with positive binary integer-valued quadratic forms of fundamental class number 1, discussed in Exercise 3.2.

From this fact, it is easy to see that the converse of Theorem 11.4.1 does not hold – namely that there exist noncomplex multiplication points such that $j(\tau)$ is an algebraic, and in fact rational, integer. Indeed, as will be shown in Exercise 11.4, $j(\tau)$ is a surjective function onto $\mathbb{C}$, which in particular means that for any integer $n \in \mathbb{Z}$ there exists a value of τ such that $j(\tau) = n$. By contrast, the only integers n which can be obtained from complex multiplication points are those in the rightmost column of Table 11.2. It follows that there are an infinite number of noncomplex multiplication points for which $j(\tau)$ is nevertheless an algebraic integer.

11.5 ϑ-functions at complex multiplication points

We now examine Jacobi theta functions at some arguments $\tau = i\sqrt{n}$ with $n \in \mathbb{N}$. To do so, we begin by defining the function,

$$K(k) = \int_0^{\frac{\pi}{2}} \frac{dx}{\sqrt{1 - k^2 \sin^2 x}} = \frac{\pi}{2} \, {}_2F_1\left(\tfrac{1}{2}, \tfrac{1}{2}; 1; k^2\right) \qquad (11.5.1)$$

known as a *complete elliptic integral of the first kind*. Valuable relations between elliptic integrals and Jacobi ϑ-functions are (see p. 362 of [Erd81a]),

$$K(k) = \tfrac{\pi}{2}\vartheta_3(0|\tau)^2 \qquad\qquad K(\sqrt{1-k^2}) = -i\tau\tfrac{\pi}{2}\vartheta_3(0|\tau)^2 \quad (11.5.2)$$

where the parameter τ is given in terms of the elliptic integrals as,

$$\tau = iF(k), \quad F(k) = \frac{K(\sqrt{1-k^2})}{K(k)} = \frac{{}_2F_1(\tfrac{1}{2},\tfrac{1}{2};1;1-k^2)}{{}_2F_1(\tfrac{1}{2},\tfrac{1}{2};1;k^2)} \quad (11.5.3)$$

In this presentation, the generators T and S of $\mathrm{SL}(2,\mathbb{Z})$ act as follows,

$$\begin{aligned} T: &\qquad \tau \to \tau + 1 &\qquad k^2 \to -k^2/(1-k^2) \\ S: &\qquad \tau \to -1/\tau &\qquad k^2 \to 1 - k^2 \end{aligned} \qquad (11.5.4)$$

which may be readily verified using the transformation formulas for the hypergeometric function (see p. 318 of [Erd81a]).

As we will now describe, the above relations between ϑ-functions and hypergeometric functions can be used to evaluate $\vartheta_3(0|\tau)$ at special values of τ. Having obtained $\vartheta_3(0|\tau)$, one can then derive the values of other theta functions by means of $\vartheta_2(0|\tau) = \sqrt{k}\,\vartheta_3(0|\tau)$ and $\vartheta_4(0|\tau) = (1-k^2)^{1/4}\vartheta_3(0|\tau)$.

11.5.1 The point $\tau = i$

To evaluate $\vartheta_3(0|i)$, Eq. (11.5.3) tells us to consider $F(k) = 1$, namely,

$$K(\sqrt{1-k^2}) = K(k) \qquad (11.5.5)$$

This is solved by $k = \frac{1}{\sqrt{2}}$. Using this in (11.5.2) then gives,

$$\vartheta_3(0|i) = \sqrt{\frac{2}{\pi} K\left(\frac{1}{\sqrt{2}}\right)} \tag{11.5.6}$$

In order to evaluate the remaining elliptic integral, we convert it to a hypergeometric function as per the second equality of (11.5.1) and make use of the following standard result,

$$_2F_1\left(a, b; \tfrac{1}{2}(a+b+1); \tfrac{1}{2}\right) = \frac{\sqrt{\pi}\,\Gamma(\tfrac{1}{2}(a+b+1))}{\Gamma(\tfrac{1}{2}(1+a))\Gamma(\tfrac{1}{2}(1+b))} \tag{11.5.7}$$

valid for arbitrary a and b. Using this, we conclude that,

$$\vartheta_3(0|i) = \frac{\pi^{\frac{1}{4}}}{\Gamma(\tfrac{3}{4})} \tag{11.5.8}$$

11.5.2 The point $\tau = 2i$

Next consider $\vartheta_3(0|2i)$. We begin by searching for k such that $F(k) = 2$. To do so, we make use of the elliptic integral identity,

$$F(k) = 2F\left(\frac{2\sqrt{k}}{1+k}\right) \tag{11.5.9}$$

which follows from a similar functional identity involving hypergeometric functions (see p. 110 of [Erd81b]),

$$_2F_1\left(\tfrac{1}{2}, \tfrac{1}{2}; 1; \frac{4x}{(1+x)^2}\right) = (1+x)\,_2F_1\left(\tfrac{1}{2}, \tfrac{1}{2}; 1; x^2\right) \tag{11.5.10}$$

Using this identity, we may recast our equation for k as follows,

$$F\left(\frac{2\sqrt{k}}{1+k}\right) = 1 \tag{11.5.11}$$

Its solution was identified in Section 11.5.1 – we require,

$$\frac{2\sqrt{k}}{1+k} = \frac{1}{\sqrt{2}} \qquad \Rightarrow \qquad k = (\sqrt{2} - 1)^2 \tag{11.5.12}$$

and thus we have

$$\vartheta_3(0|2i) = \sqrt{_2F_1\left(\tfrac{1}{2}, \tfrac{1}{2}; 1; (\sqrt{2} - 1)^4\right)} \tag{11.5.13}$$

Making use of duplication and reflection identities of hypergeometric functions, one shows,

$$_2F_1\left(\tfrac{1}{2},\tfrac{1}{2};1;(\sqrt{2}-1)^4\right) = \frac{1}{4-2\sqrt{2}} \, _2F_1\left(\tfrac{1}{2},\tfrac{1}{2};1;\tfrac{1}{2}\right) \qquad (11.5.14)$$

and then using (11.5.7), we obtain,

$$\vartheta_3(0|2i) = \frac{\pi^{\frac{1}{4}}}{2\Gamma(\tfrac{3}{4})}\sqrt{2+\sqrt{2}} \qquad (11.5.15)$$

11.5.3 The point $\tau = \sqrt{2}\,i$

To obtain $\vartheta_3(0|\sqrt{2}i)$, we would like to solve $F(k) = \sqrt{2}$, or equivalently,

$$\sqrt{2}\,_2F_1\left(\tfrac{1}{2},\tfrac{1}{2};1;k^2\right) = \,_2F_1\left(\tfrac{1}{2},\tfrac{1}{2};1;1-k^2\right) \qquad (11.5.16)$$

Comparison with (11.5.10) shows that $k = \sqrt{2}-1$ is a solution, so that,

$$\theta_3(0|\sqrt{2}i) = \sqrt{_2F_1\left(\tfrac{1}{2},\tfrac{1}{2};1;(\sqrt{2}-1)^2\right)} \qquad (11.5.17)$$

To proceed, we again use duplication and reflection identities of hypergeometric functions to show that,

$$_2F_1\left(\tfrac{1}{2},\tfrac{1}{2};1;(\sqrt{2}-1)^2\right) = \frac{1}{\sqrt{4-2\sqrt{2}}}\,_2F_1\left(\tfrac{1}{4},\tfrac{3}{4};1;\tfrac{1}{2}\right) \qquad (11.5.18)$$

from which we can evaluate,

$$\vartheta_3(0|\sqrt{2}i) = \frac{\Gamma(\tfrac{9}{8})}{\Gamma(\tfrac{5}{4})}\sqrt{\frac{\Gamma(\tfrac{1}{4})}{2^{1/4}\pi}} \qquad (11.5.19)$$

11.5.4 The points $\tau = in$ and $\tau = i/n$ with $n \in \mathbb{N}$

To close, we quote without proof the results for ϑ-constants at imaginary complex multiplication points $\vartheta_3(0|in)$ and $\vartheta_3(0|i/n)$ with $n \in \mathbb{N}$. We have the following general result,

$$\vartheta_3(0|in) = n^{-\frac{1}{2}}\vartheta_3(0|i/n) = \frac{\vartheta_3(0|i)}{n^{1/4}h_n} \qquad (11.5.20)$$

with $\vartheta_3(0|i)$ given in (11.5.8). The first few values of h_n are found to be,

$$h_1 = 1$$
$$h_2 = (2\sqrt{2} - 2)^{1/2}$$
$$h_3 = (2\sqrt{3} - 3)^{1/4}$$
$$h_4 = 2^{3/4}(2^{1/4} + 1)^{-1/2}$$
$$h_5 = (5 - 2\sqrt{5})^{1/2}$$
$$h_6 = \frac{2^{3/4}3^{1/8}((\sqrt{2} - 1)(\sqrt{3} - 1))^{1/6}}{(-4 + 3\sqrt{2} + 3^{5/4} + 2\sqrt{3} - 3^{3/4} + 2^{3/2}3^{3/4})^{1/3}} \qquad (11.5.21)$$

The evaluation of $\vartheta_3\left(0 \,\middle|\, \frac{1+i}{2}\right)$ is the subject of Exercise 11.5.

11.6 The values of E_2, E_4, and E_6 at the points $\tau = i, \rho$

The quasi-modular transformation property of $\mathsf{E}_2(\tau)$,

$$\mathsf{E}_2(\gamma\tau) = (c\tau + d)^2\mathsf{E}_2(\tau) + \frac{12}{2\pi i}c(c\tau + d) \qquad (11.6.1)$$

allows one to evaluate $\mathsf{E}_2(\tau)$ at the points $\tau = i$ and $\tau = \rho = e^{2\pi i/3}$. These are fixed points of the modular transformations $\gamma = S$ for which $(c, d) = (1, 0)$, and $\gamma = ST$ for which $(c, d) = (1, 1)$, respectively. Substituting these expressions into the transformation law (11.6.1), we readily obtain,

$$\mathsf{E}_2(i) = \frac{3}{\pi} \qquad\qquad \mathsf{E}_2(\rho) = \frac{2\sqrt{3}}{\pi} \qquad (11.6.2)$$

The modular forms E_4 and E_6 vanish at the points ρ and i, respectively,

$$\mathsf{E}_4(\rho) = 0 \qquad\qquad \mathsf{E}_6(i) = 0 \qquad (11.6.3)$$

as was already shown in (3.5.4) in terms of the closely related modular forms G_4 and G_6. Evaluating E_4 and E_6 at the points i and ρ, respectively, requires expressing elliptic integrals in terms of hypergeometric functions, and we shall quote here the result without proof,

$$\mathsf{E}_4(i) = \frac{48\,\Gamma(\frac{5}{4})^4}{\pi^2\,\Gamma(\frac{3}{4})^4} \qquad\qquad \mathsf{E}_6(\rho) = \frac{729\,\Gamma(\frac{4}{3})^6}{2\pi^3\,\Gamma(\frac{5}{6})^6} \qquad (11.6.4)$$

Derivations of these relations may be found in [DS05].

• Bibliographical notes

The algebraic nature of $j(\tau)$ at complex multiplication points is a classic result; the discussion given in Section 11.4 closely follows the one given by

Zagier in Section 6.1 of [Zag08] and in the beautiful book by Cox [Cox11b]. Singular moduli are discussed in [GZ85].

The method used to evaluate ϑ-functions at special complex multiplication points in Section 11.5 was pioneered by Ramanujan and can be found in his third notebook [Ber12], see Chapter 17, Entry 6. It was systematized by Chowla and Selberg in [CS49, SC67]. The explicit values of h_n listed in (11.5.21) were obtained in [Yi04].

Points of enhanced symmetry generally tend to play a special role in physics, and singular moduli are precisely the values of the moduli where such enhancements occur. As such, singular moduli and complex multiplication have entered physics in various places. Concrete examples can be found in Seiberg–Witten theory [SW94a, SW94b, AD95, MN96, MN97], where points on the Coulomb branch corresponding to singular moduli have extra massless degrees of freedom, as well as in string theory where the additional discrete symmetries at singular values of the axion-dilaton can be gauged to give so-called S-folds [GER16, AT16, GER17, Evt20, KMZ22]. In Section 13.6, we will introduce another physical application, namely, to the fact that a nonlinear sigma model conformal field theory with $\mathbb{T}^2$ target space is rational if and only if the torus admits complex multiplication [GV04].

We close by noting that another context where singular moduli appear is the study of orbifolds, in which one quotients by the emergent symmetry at such points. In the context of perturbative string theory on a target space torus, orbifolds were introduced in [DHVW85] and [DHVW86] were further developed in [DVVV89] among many other publications. An extensive bibliography, with emphasis on string phenomenology, may be found in [CHST22]. The study of singular moduli at genus-2 was carried out in the 1960s by Gottschling in [Got61a], [Got61b], and [Got67], though the application to genus-2 orbifolds is fairly recent [NRSTV21]. Some more general comments at arbitrary genus, from a two-dimensional conformal field theory perspective, can be found in [RV20].

Exercises

11.1 By explicitly solving for τ, show that the following elliptic curves have complex multiplication and identify the imaginary quadratic fields in which τ takes values: (a) $y^2 = x^3 - 15x + 22$, (b) $y^2 + y = x^3 - 38x + 90$, and (c) $y^2 + y = x^3 - 2174420x + 1234136692$. This may require computer assistance at intermediate steps.

11.2 Prove the following multiplication formula,

$$N^2 \wp(Nz|\tau) = \sum_{\alpha,\beta \in \mathbb{Z}_N} \wp\left(z + \tfrac{\alpha+\beta\tau}{N}\Big|\tau\right)$$

11.3 Prove the expression for $Q_3(x, j(\tau))$ of (11.4.7) and $Q_3(x, x)$ of (11.4.12).

11.4 Show that $j(\tau)$ is a surjective function onto $\mathbb{C}$. To do so, assume conversely that there exists a value $\alpha \in \mathbb{C}$ such that $j^{-1}(\alpha)$ is empty, and then by integrating $\frac{j'(\tau)}{j(\tau)-\alpha}$ along the contour shown in Figure 3.3, arrive at a contradiction.

11.5 Use the method described in the text to show that,

$$\vartheta_3\left(0 \Big| \tfrac{-1+i}{2}\right) = \frac{\Gamma\left(\tfrac{1}{4}\right)}{\sqrt{2}\,\pi^{3/4}}\, e^{-\tfrac{i\pi}{8}} \qquad\qquad (11.6.5)$$

12

String amplitudes

The starting point for string theory is the idea that the elementary constituents of the theory, which in quantum field theory are assumed to be point-like, are in fact one-dimensional objects, namely, strings. As time evolves, a string sweeps out a Riemann surface whose topology governs the interactions that result from joining and splitting strings. The Feynman–Polyakov prescription for quantum mechanical string amplitudes reduces to summing over all topologies of the Riemann surface, for each topology integrating over the moduli of the Riemann surface, and for each value of the moduli solving a conformal field theory. Modular invariance plays a key role in the reduction of the integral over moduli to an integral over a single copy of moduli space and, in particular, is responsible for rendering string amplitudes well behaved at short distances. In this chapter, we shall present a highly condensed introduction to key ingredients of string theory and string amplitudes, relegating the important aspects of toroidal compactification and T-duality to Chapter 13 and a discussion of S-duality in Type IIB string theory to Chapter 14.

12.1 Overview

Being a one-dimensional object, a connected string may have two different topologies: that of a circle or *closed string*, or that of an interval or *open string*. The time evolution of a closed string sweeps out a two-dimensional surface with the topology of a cylinder, represented in the left panel of Figure 12.1, while an open string sweeps out a rectangle. The surface swept out by a string is referred to as the *worldsheet*. One of the most striking features of string theory is that all interactions are governed by the joining and splitting of strings, as represented in the right panel of Figure 12.1 for closed strings, without the need for the point-like interactions required in quantum field theory. The interaction by joining and splitting is unique once the free propagation of the string is known, since locally on the worldsheet there is no way to distinguish free propagation from interaction.

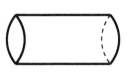   

Figure 12.1 Time evolution of a free closed string is represented in the left panel. Interaction by joining and splitting of closed strings is represented in the right panel.

String theory may be promoted to a consistent quantum theory provided certain conditions are met on the space-time in which strings propagate and interact. Superstrings can propagate in flat space-time provided the dimension of the space-time is 10. The spectrum of superstring theory then automatically contains a graviton, and string theory is automatically a theory of quantum gravity, albeit in space-time dimension greater than 4. Supersymmetry, which swaps boson and fermion states, is a key ingredient in making superstring theory mathematically and physically consistent. The focus in this book is modular invariance, which is responsible for making superstring theory well behaved at short distances, a property referred to as *ultra-violet finiteness* (or UV finiteness for short).

The ultimate goal of string theory is the unification of gravity with the Standard Model of Particle Physics into a single theory that is consistent with quantum mechanics and general relativity. Successful unification will require principles and mechanisms for selecting our four-dimensional universe from all the solutions of string theory, for breaking space-time supersymmetry and for maintaining a small cosmological constant in the process. Satisfactory answers to these questions remain largely outstanding today.

In this chapter, we shall review some of the string amplitudes derived from *superstring perturbation theory* and use these to obtain corrections to supergravity and super Yang–Mills theory in the form of local *effective interactions*. Superstring perturbation theory uses a series expansion in powers of the string coupling g_s to evaluate *quantum mechanical probability amplitudes*, or simply *amplitudes*. The absolute value squared of an amplitude gives the quantum mechanical probability for a given process to occur. A process is specified by the data of the incoming and the outgoing string states, and the amplitude for the process is schematically given by the Feynman functional integral prescription of summing over all possible configurations of the string worldsheet given by the data of the initial and final states.

The only topological information that remains, once the initial and final string data have been fixed by specifying the process, is the genus $g \geq 0$ of the surface, namely, the number of handles on the worldsheet. The string coupling g_s assigns a weight given to each topology, so that the perturbative string amplitude $\mathcal{A}$ for a given process is provided by the following topological expansion in powers of g_s,

$$\mathcal{A} = \sum_{g=0}^{\infty} g_s^{2g-2} \mathcal{A}^{(g)} + \text{nonperturbative effects} \qquad (12.1.1)$$

where $\mathcal{A}^{(g)}$ is the *genus g* (or *g*-loop) contribution to the string amplitude $\mathcal{A}$. The expansion is schematically represented in Figure 12.2. The prescription of the initial and final data for the incoming and/or outgoing string states will be discussed in Section 12.5.

Figure 12.2 Schematic representation of the perturbative expansion in the string coupling g_s of the string amplitude $\mathcal{A}$ for a process of four incoming and/or outgoing string states.

The construction of string amplitudes that we shall sketch below contains only bosonic states, and the corresponding string theory is referred to as the *closed bosonic string*. Including fermions introduces many technical complications and has led to the elaboration of several different formulations, which are mostly beyond the scope of this book. Instead we shall focus on the bosonic string, and some mild extensions thereof, to illustrate the questions and results that relate to modular invariance of string perturbation theory in this chapter, to T-duality for toroidal compactifications in Chapter 13, and to S-duality in Type IIB superstring theory in Chapter 14.

12.2 The Polyakov formulation

Space-time is assumed to be a manifold M (or an orbifold) with a space-time metric G. Physically, the signature of G is Minkowskian. However, as is familiar from quantum field theory, it will often be convenient to construct string amplitudes in terms of a metric with Euclidean signature, so that M is a Riemannian manifold, and then analytically continue to Minkowski signature. String amplitudes are formulated in terms of two-dimensional orientable surfaces Σ and continuous maps $X : \Sigma \to M$. The Riemannian metric G on M induces a Riemannian metric $X^*(G)$ on Σ under the pull-back X^* of the map X. Since Σ is orientable and carries a Riemannian metric, it is automatically a Riemann surface (see Appendix B).

The Polyakov formulation of string theory also invokes an intrinsic Riemannian metric $\mathfrak{g}$ on Σ, which is independent of the induced metric $X^*(G)$. The map X and the intrinsic metric $\mathfrak{g}$ on Σ are governed by a worldsheet action. String amplitudes are required to be independent of the coordinates used to parametrize Σ and M, which leads us to require invariance under the diffeomorphism groups $\mathrm{Diff}(\Sigma)$ and $\mathrm{Diff}(M)$. A natural candidate satisfying these conditions is given by the following action,[1]

$$I_G[X, \mathfrak{g}] = \frac{1}{4\pi\alpha'} \int_\Sigma d\mu_\mathfrak{g}\, \mathfrak{g}^{mn} \partial_m X^\mu \partial_n X^\nu G_{\mu\nu}(X) \qquad (12.2.1)$$

Assuming the coordinates X^μ on M to have dimensions of length and the coordinates ξ^m on Σ, the metrics $\mathfrak{g}$ and G, and the action I_G to be dimensionless, the parameter α' has dimensions of length squared, while $\hbar^2/(c^2\alpha')$ has dimensions of mass squared and plays the role of a string tension.

Any relativistic quantum theory of gravity has a built-in length scale, referred to as the Planck length ℓ_P, and given by $\ell_P^2 = G_N\hbar/c^3$ in terms of the speed of light c, Planck's constant $\hbar$, and Newton's gravitational constant G_N. The Planck length evaluates to $\ell_P \approx 10^{-33}\mathrm{cm}$, namely, 10^{-19} times the size of a proton. If string theory is to be a quantum theory of gravity, then the dimensionless ratio α'/ℓ_P^2 should be a function of the string coupling and the specific compactification from 10 to 4 dimensions. Depending on the details of the coupling and compactification, the typical size of strings may be of order the Planck length, or may differ significantly from it.

Feynman's functional integral prescription for probability amplitudes gives string amplitudes as integrals over the spaces $\mathrm{Map}(\Sigma)$ of maps X and $\mathrm{Met}(\Sigma)$ of metrics $\mathfrak{g}$. In this section, we investigate the two-dimensional quantum field theory defined by $I_G[X, \mathfrak{g}]$, for fixed metrics $\mathfrak{g}$ and G, which is referred

---

[1] Throughout, ξ^m with $m = 1, 2$ are real local coordinates on Σ and $\partial_m = \partial/\partial\xi^m$ are the derivatives with respect to ξ^m. On Σ, the Riemannian metric $\mathfrak{g} = \mathfrak{g}_{mn}d\xi^m \otimes d\xi^n$ induces the volume form $d\mu_\mathfrak{g} = \frac{1}{2}\sqrt{\det \mathfrak{g}}\, \epsilon_{mn}d\xi^m \wedge d\xi^n$ and $\epsilon_{mn} = -\epsilon_{nm}$ specifies the orientation $\epsilon_{12} = 1$. The Einstein summation convention on a pair of matching upper and lower indices is implied, both for the worldsheet indices $m, n = 1, 2$ and the space-time indices $\mu, \nu = 1, \cdots, \dim(M)$.

to as a *nonlinear sigma model*, and relegate the integration over the metrics $\mathfrak{g} \in \mathrm{Met}(\Sigma)$ to Section 12.3.

The functional measure of integration DX may be constructed from a suitable L^2 norm on the tangent space to $\mathrm{Map}(\Sigma)$. Denoting the tangent vectors by δX, the norm is dictated by $\mathrm{Diff}(\Sigma)$ and $\mathrm{Diff}(M)$ invariance,

$$\|\delta X\|^2 = \int_\Sigma d\mu_\mathfrak{g} \, G_{\mu\nu}(X) \, \delta X^\mu \delta X^\nu \qquad (12.2.2)$$

The functional integral over $\mathrm{Map}(\Sigma)$ with measure DX is given as follows,

$$e^{-W_G[\mathfrak{g}]} = \int_{\mathrm{Map}(\Sigma)} DX \, e^{-I_G[X,\mathfrak{g}]} \qquad (12.2.3)$$

Note that α' in the functional integral over the field X plays the role of $\hbar$ in the functional integrals of standard quantum field theory.

Ordinary quantum field theories, such as the Standard Model, depend on a finite number of coupling parameters. The nonlinear sigma model is no ordinary quantum field theory, as the dependence of $I_G[X, \mathfrak{g}]$ on a space-time metric $G_{\mu\nu}(X)$ generally introduces an infinite number of couplings. This may be seen, for example, by Taylor expanding $G_{\mu\nu}(X)$ in powers of the field X in the neighborhood of a flat metric $G_{\mu\nu}^{(0)}$,

$$G_{\mu\nu}(X) = G_{\mu\nu}^{(0)} + G_{\mu\nu;\rho}^{(1)} X^\rho + \frac{1}{2} G_{\mu\nu;\rho\sigma}^{(2)} X^\rho X^\sigma + \mathcal{O}(X^3) \qquad (12.2.4)$$

The coefficients $G_{\mu\nu}^{(0)}$, $G_{\mu\nu;\rho}^{(1)}$, $G_{\mu\nu;\rho\sigma}^{(2)}, \cdots$ are independent of X and may be viewed as independent coupling parameters of the nonlinear sigma model. Quantization of a field theory requires regularization of short-distance singularities and renormalization. We refer to the bibliographical notes for standard references on the subject.

What is important for us here is that renormalization produces a *renormalization group flow* in the space of coupling parameters of the quantum field theory. This flow depends on a single parameter t referred to as the *renormalization scale*. For the nonlinear sigma model, the effect of renormalization is to render each one of the couplings $G_{\mu\nu}^{(0)}$, $G_{\mu\nu;\rho}^{(1)}$, $G_{\mu\nu;\rho\sigma}^{(2)}, \cdots$ dependent on the parameter t. Alternatively, the flow promotes the metric $G_{\mu\nu}(X)$ to a one-parameter family of metrics $G_{\mu\nu}(X;t)$. The flow is governed by the so-called *renormalization group equation*, namely, a first-order differential equation in t of the form,

$$t\frac{d}{dt} G_{\mu\nu}(X;t) = \beta_{\mu\nu}\big(G(X;t)\big) \qquad (12.2.5)$$

Ordinarily, one would view a theory with an infinite number of coupling parameters as *non-renormalizable* because it would require an infinite number of input data. Friedan has shown, however, that the theory is renormalizable provided the renormalization group equation (12.2.5) is invariant under $\mathrm{Diff}(M)$, which requires that the function $\beta_{\mu\nu}(G(X;t))$ is a proper $\mathrm{Diff}(M)$ tensor built out of the metric $G_{\mu\nu}$ (up to diffeomorphisms of the metric). The $\beta_{\mu\nu}$ function may be calculated perturbatively in an expansion in powers of α' and, to lowest order, gives the following renormalization group equation,

$$t\frac{d}{dt}G_{\mu\nu}(X;t) = \tfrac{1}{2}R_{\mu\nu} + \mathcal{O}(\alpha') \tag{12.2.6}$$

where $R_{\mu\nu}$ is the Ricci tensor for the metric $G_{\mu\nu}(X;t)$. We note that this equation may be viewed as a nonlinear generalization of the diffusion equation for the metric. In mathematics, it is also referred to as *Ricci flow* and was used by Hamilton [Ham82] and generalized by Perelman [Per02] in the proof of the Poincaré conjecture for the three sphere [Per03].

12.3 String amplitudes as integrals over moduli space

Returning to string amplitudes, we shall now define the functional integral over metrics $\mathfrak{g} \in \mathrm{Met}(\Sigma)$ using the following integration measure constructed to be invariant under $\mathrm{Diff}(\Sigma)$,

$$\|\delta\mathfrak{g}\|^2 = \int_\Sigma d\mu_\mathfrak{g}\, \mathfrak{g}^{mn}\, \mathfrak{g}^{pq}\, \delta\mathfrak{g}_{mp}\, \delta\mathfrak{g}_{nq} \tag{12.3.1}$$

The genus g contribution $\mathcal{A}^{(g)}$ to the string amplitude for metric G at genus g is *formally* given by the functional integral over $\mathrm{Met}(\Sigma)$,

$$\mathcal{A}^{(g)} \sim \int_{\mathrm{Met}(\Sigma)} D\mathfrak{g}\, e^{-W_G[\mathfrak{g}]} \tag{12.3.2}$$

where the symbol $\sim$ indicates that this relation is formal. Taken literally, the integral over $\mathfrak{g}$ would be divergent as the measure and the integrand are both invariant under $\mathrm{Diff}(\Sigma)$ so that $\mathrm{Met}(\Sigma)$ contains an infinite number of images of each metric $\mathfrak{g}$ under $\mathrm{Diff}(\Sigma)$. Factoring out $\mathrm{Diff}(\Sigma)$ amounts to a reduction of the integration to the quotient $\mathrm{Met}(\Sigma)/\mathrm{Diff}(\Sigma)$ of orbits, and may be carried out using the *Faddeev–Popov gauge fixing procedure* familiar from Yang–Mills gauge theory.

An infinitesimal diffeomorphism in $\mathrm{Diff}(\Sigma)$ is represented by a vector field v^m that acts on the coordinates of Σ by $\xi^m \to \xi^m + v^m$ and on the metric by $\mathfrak{g}_{mn} \to \mathfrak{g}_{mn} + \nabla_m v_n + \nabla_n v_m$, up to quadratic terms in v^m, where ∇_m

is the covariant derivative with respect to the Christoffel connection for the metric $\mathfrak{g}$. It is convenient to gauge fix $\mathrm{Diff}(\Sigma)$ by choosing local complex coordinates $z, \bar{z}$ in terms of which the metric $\mathfrak{g}$ takes the form,

$$\mathfrak{g}_{mn} d\xi^m d\xi^n = \mathfrak{g}_{z\bar{z}} |dz|^2 \tag{12.3.3}$$

On a surface Σ of genus $g \geq 1$, the metric cannot be expressed as such globally on Σ as its global form will depend on orbit in $\mathrm{Met}(\Sigma)/\mathrm{Diff}(\Sigma)$. The Faddeev-Popov ghost system for the operator ∇_m acting on vector fields v^n may be decomposed in complex coordinates and consists of ghost fields b and c of conformal weight $(2,0)$ and $(-1,0)$, respectively, together with their complex conjugates. The action $I_{\mathrm{gh}}[b, c, \mathfrak{g}]$,

$$I_{\mathrm{gh}}[b, c, \mathfrak{g}] = \frac{1}{2\pi} \int_\Sigma d^2 z \left(b\partial_{\bar{z}} c + \bar{b}\partial_z \bar{c} \right) \tag{12.3.4}$$

is familiar from (5.2.1) in our study of the bc system at genus one in Chapter 5, where it was denoted simply $I[b, c]$, and at higher genus in Section D.7 of Appendix D. Functional integral quantization proceeds as follows,

$$e^{-W_{\mathrm{gh}}[\mathfrak{g}]} = \int D(bc) e^{-I_{\mathrm{gh}}[b, c, \mathfrak{g}]} \tag{12.3.5}$$

The resulting string amplitude is then given by,

$$\mathcal{A}^{(g)} = \int_{\mathrm{Met}(\Sigma)/\mathrm{Diff}(\Sigma)} D\mathfrak{g} \, e^{-W_G[\mathfrak{g}] - W_{\mathrm{gh}}[\mathfrak{g}]} \tag{12.3.6}$$

The reduction to $\mathrm{Met}(\Sigma)/\mathrm{Diff}(\Sigma)$ removes two of the three functional degrees of freedom of the metric $\mathfrak{g}_{mn}$, leaving a single real field $\mathfrak{g}_{z\bar{z}}$. Such theories are referred to as *noncritical strings*.

Critical string theories are obtained by requiring that the combination of the measure $D\mathfrak{g}$ and the integrand are invariant also under Weyl transformations. The group of Weyl transformations $\mathrm{Weyl}(\Sigma)$ leaves the space-time metric G and the map X invariant, and multiplies the metric $\mathfrak{g}$ by a positive scalar function, $\mathfrak{g}_{mn}(\xi) = e^{2\sigma(\xi)} \hat{\mathfrak{g}}_{mn}(\xi)$, where $\sigma : \Sigma \to \mathbb{R}$. The quotient of the space $\mathrm{Met}(\Sigma)$ of metrics by the semi-direct product $\mathrm{Diff}(\Sigma) \ltimes \mathrm{Weyl}(\Sigma)$ may be identified with the moduli space $\mathcal{M}_h$ of *conformal structures*, or equivalently *complex structures*, on a Riemann surface Σ of genus g,

$$\mathcal{M}_g = \mathrm{Met}(\Sigma)/\{\mathrm{Diff}(\Sigma) \ltimes \mathrm{Weyl}(\Sigma)\} \tag{12.3.7}$$

Thus, the genus g contribution $\mathcal{A}^{(g)}$ to the amplitude for the critical string is given by a finite-dimensional integral over moduli space,

$$\mathcal{A}^{(g)} = \int_{\mathcal{M}_g} D\hat{\mathfrak{g}} \, e^{-W_G[\hat{\mathfrak{g}}] - W_{\mathrm{gh}}[\hat{\mathfrak{g}}]} \tag{12.3.8}$$

where $\hat{\mathfrak{g}}$ is the worldsheet metric parametrized by $\mathcal{M}_g$, and $D\hat{\mathfrak{g}}$ is the integration measure induced by $D\mathfrak{g}$ on $\mathcal{M}_g$. The resulting amplitude $\mathcal{A}^{(g)}$ is a function of the metric G on the space-time M.

12.4 Conformal invariance and decoupling negative norm states

The classical actions $I_G[X, \mathfrak{g}]$ and $I_{\mathrm{gh}}[b, c, \mathfrak{g}]$ are Weyl invariant, but the measures DX, $D\mathfrak{g}$, and $D(bc)$ are not Weyl invariant, thereby producing Weyl anomalies. In particular, the functional $W_G[\mathfrak{g}]$, defined in (12.2.1), is not invariant under $\mathrm{Weyl}(\Sigma)$. For general metric G, the Weyl transformation law of W_G is complicated and not known explicitly. However, when the transformation law of W_G takes the following special form,

$$\delta W_G[\mathfrak{g}] = \frac{\mathfrak{c}}{24\pi} \int_\Sigma d\mu_\mathfrak{g} \, R_\mathfrak{g} \, \delta\sigma \tag{12.4.1}$$

where $R_\mathfrak{g}$ is the scalar curvature of the metric $\mathfrak{g}$ and $\mathfrak{c}$ is a constant, then I_G produces a *conformal field theory with central charge* $\mathfrak{c}$. For example, when G is the flat metric on $\mathbb{R}^D$, then the action I_G is quadratic in X, and defines a conformal field theory with central charge $\mathfrak{c} = D$, and I_{gh} is a conformal field theory with central charge $\mathfrak{c} = -26$. In the critical dimension $D = 26$, the combined field theory of X, b, and c is therefore Weyl invariant.

When the metric G is not flat, the Weyl transformation of W_G may be evaluated in an expansion in powers of α', which is equivalent to an expansion in powers of the Riemann tensor of the metric G and its derivatives. A constant Weyl transformation amounts to a change in the renormalization scale t, whose effect on G is captured by the renormalization group equation of (12.2.5). Weyl invariance thus requires the vanishing of the $\beta_{\mu\nu}$ function which, to leading order in α', requires the vanishing of the Ricci tensor $R_{\mu\nu} = 0$ for the metric G. Remarkably, the requirement of Weyl symmetry on the worldsheet metric $\mathfrak{g}$ imposes Einstein's equations $R_{\mu\nu} = 0$ on the space-time metric G, to leading order in α'. Although this result was obtained by considering only constant Weyl transformations, one proves that the condition is sufficient for arbitrary Weyl transformations (see the bibliographical notes for references where these results were obtained).

Conformal symmetry plays a fundamental role in decoupling the negative norm and null states which arise in the Lorentz-covariant formulation of string theory. For simplicity, we restrict attention to the case of flat Minkowski space-time $M = \mathbb{R}^{1,25}$ with metric $\eta_{\mu\nu} = \mathrm{diag}(-+\cdots+)_{\mu\nu}$. The functions X^μ satisfy the Laplace equation $\partial_z \partial_{\bar{z}} X^\mu = 0$ in the local complex coordinates $z, \bar{z}$ introduced in (12.3.3), so that $\partial_z X^\mu$ is holomorphic, while $\partial_{\bar{z}} X^\mu$ is anti-holomorphic. Both fields may be expanded in Laurent series,[2]

$$i\partial_z X^\mu = \sum_{m \in \mathbb{Z}} z^{-m-1} X_m^\mu \qquad -i\partial_{\bar{z}} X^\mu = \sum_{m \in \mathbb{Z}} \bar{z}^{-m-1} \tilde{X}_m^\mu \qquad (12.4.2)$$

which upon canonical quantization satisfy,

$$\begin{aligned} (X_m^\mu)^\dagger &= X_{-m}^\mu & [X_m^\mu, X_n^\nu] &= m\, \eta^{\mu\nu}\, \delta_{m+n,0} \\ (\tilde{X}_m^\mu)^\dagger &= \tilde{X}_{-m}^\mu & [\tilde{X}_m^\mu, \tilde{X}_n^\nu] &= m\, \eta^{\mu\nu}\, \delta_{m+n,0} \end{aligned} \qquad (12.4.3)$$

while $[X_m^\mu, \tilde{X}_n^\nu] = 0$ for all m, n. The operators X_0^μ and $\tilde{X}_0^\mu$ are self-adjoint, commute with one another and with all other modes, and correspond to left- and right-moving momentum operators. They will play a key role in toroidal compactification and will be discussed in greater detail in Chapter 13.

The Fock space of the closed bosonic string in flat Minkowski space-time $\mathbb{R}^{1,25}$ is constructed as follows. The operators X_0^μ and $\tilde{X}_0^\mu$ may be diagonalized simultaneously; their eigenvalues are real and are denoted by k^μ and $\tilde{k}^\mu$; their eigenvectors $|0; k\rangle$ and $|0; \tilde{k}\rangle$ are ground states of the string excitation spectrum of momenta k^μ and $\tilde{k}^\mu$, respectively. In flat Minkowski space-time, the left and right momenta must be equal $\tilde{k}^\mu = k^\mu$. (This condition will be relaxed in Chapter 13 for toroidal compactifications in order to accommodate winding modes.) The Fock space is the sum over string momenta $k \in \mathbb{R}^{1,25}$ of the tensor product of the left and right chiral Fock spaces,

$$\mathfrak{F}_{\text{closed}} = \bigoplus_{k \in \mathbb{R}^{26}} \mathfrak{F}_k \otimes \tilde{\mathfrak{F}}_k \qquad (12.4.4)$$

The *chiral Fock space* $\mathfrak{F}_k$ is the infinite tensor product of the Fock spaces for the operators X_n^μ with $n \neq 0$. The ground state $|0; k\rangle$ is defined as follows,

$$X_0^\mu |0; k\rangle = k^\mu |0; k\rangle \qquad X_n^\mu |0; k\rangle = 0 \qquad n > 0 \qquad (12.4.5)$$

The chiral Fock space $\mathfrak{F}_k$ is obtained by linear combinations of all possible monomials in the operators X_{-n}^μ for $n > 0$ applied to the ground state,

$$\mathfrak{F}_k = \bigoplus_{n_1, n_2, \cdots, n_\ell > 0;\, \ell \geq 0} \left\{ X_{-n_1}^{\mu_1} \cdots X_{-n_\ell}^{\mu_\ell} |0; k\rangle \right\} \qquad (12.4.6)$$

with the analogous construction for $\tilde{\mathfrak{F}}_k$. We define an inner product on these

---

[2] Henceforth, we shall set $c = \hbar = 1$ and $\alpha' = 2$. The fields $\partial_z X^\mu$ and $\partial_{\bar{z}} X^\mu$ are treated as independent of one another, and not as complex conjugates of one another, a property that is inherited from the worldsheet with Minkowskian signature where the fields correspond to the data on the two light-cone directions.

spaces by normalizing the string ground state with continuous momenta,

$$\langle 0; k'|0; k \rangle = \delta(k' - k) \tag{12.4.7}$$

The lowest excited state in $\mathfrak{F}_k$ with polarization vector $\varepsilon^\mu(k)$ is given by $|\varepsilon; k\rangle = \varepsilon_\mu(k) X^\mu_{-1}|0; k\rangle$ and has the following inner product,

$$\langle \varepsilon; k'|\varepsilon; k \rangle = \varepsilon_\mu(k) \varepsilon^\mu(k) \delta(k - k') \tag{12.4.8}$$

Since the signature of $M = \mathbb{R}^{1,25}$ is given by the Minkowski metric, the "norm" $\varepsilon_\mu(k)\varepsilon^\mu(k)$ can be positive, null, or negative, in which cases $\varepsilon^\mu(k)$ is referred to as *space-like*, *null*, or *time-like*. States in the Fock space that have *negative norm* are inconsistent with the principles of quantum mechanics, but they are unavoidable byproducts of Lorentz-covariant quantization, just as they are unavoidable in the covariant quantization of gauge theories.

The negative norm states and null states may be decoupled with the help of conformal symmetry, whose Virasoro generators take the form,

$$L_m = \sum_{n \in \mathbb{Z}} \frac{1}{2} X_{m-n} \cdot X_n \qquad L_0 = \frac{1}{2} X_0^2 + \sum_{n \in \mathbb{N}} X_{-n} \cdot X_n \tag{12.4.9}$$

which was derived for a single scalar field in (5.4.8). The conformal transformations by the Virasoro generators act as follows on the oscillators,

$$[L_m, X^\mu_n] = -n X^\mu_{m+n} \tag{12.4.10}$$

and similarly for tilde operators. One defines the subspace of *physical states* $\mathfrak{F}^{\text{phys}}_k \subset \mathfrak{F}_k$ by imposing the following conditions on the states $|\psi\rangle \in \mathfrak{F}_k$,

$$(L_0 - 1)|\psi\rangle = L_m|\psi\rangle = 0 \qquad\qquad m > 0 \tag{12.4.11}$$

All negative norm states are eliminated by these physical state conditions, and all null states decouple, just as is the case when imposing Gauss's law in gauge theory. The ground state is physical provided $k^2 = 2$, which makes it a tachyon. The tachyon renders the bosonic string vacuum unstable, but it is eliminated in the superstring, and we shall omit its discussion in the sequel. Returning to the example of the first excited state, $|\varepsilon; k\rangle = \varepsilon_\mu(k) X^\mu_{-1}|0; k\rangle$, we see that the physical state conditions $L_m|\psi\rangle = 0$ for $m \geq 2$ are automatically satisfied, while those for $m = 1, 0$ impose the constraints $k \cdot \varepsilon(k) = 0$ and the massless condition $k^2 = 0$ on the momentum k. The resulting physical states constitute the spectrum of the open bosonic string and include a massless vector particle, such as the photon.

The above construction of the Fock space also shows that the closed bosonic string spectrum contains the massless states $|\varepsilon; k\rangle \otimes |\tilde{\varepsilon}; k\rangle$, where

$k^2 = 0$ and $k \cdot \varepsilon(k) = k \cdot \tilde{\varepsilon}(k) = 0$. The traceless symmetric part of $\varepsilon_\mu(k)\tilde{\varepsilon}_\nu(k)$ corresponds to the graviton state as expected from the presence of the metric G in the worldsheet action. However, there is also a trace-part that corresponds to the scalar dilaton state and an antisymmetric part that corresponds to a rank-2 antisymmetric state. These additional states and particles are not required by general relativity, but their presence in the spectrum is, in fact, mandated by supergravity, as we shall see in Chapter 14.

In view of the presence of the extra massless states of the dilaton and rank 2 antisymmetric tensor field, we may generalize the nonlinear sigma model action of (12.2.1) to include a field $\Phi(X)$ for the dilaton as well as a rank 2 antisymmetric field $B_{\mu\nu}(X)$,[3]

$$I_{G,\Phi,B}[X,\mathfrak{g}] = \frac{1}{8\pi} \int_\Sigma d\mu_{\mathfrak{g}} \left\{ \left(\mathfrak{g}^{mn} G_{\mu\nu}(X) - i\varepsilon^{mn} B_{\mu\nu}(X) \right) \partial_m X^\mu \partial_n X^\nu \right.$$

$$\left. + 4 R_{\mathfrak{g}} \, \Phi(X) \right\} \tag{12.4.12}$$

where $R_{\mathfrak{g}}$ is the scalar curvature of the metric $\mathfrak{g}$, already encountered in (12.4.1). In the presence of the fields G, Φ, B, conformal invariance requires, up to lowest nontrivial order in the α'-expansion, a generalization of Einstein's equations that includes the dynamics of the fields Φ and $B_{\mu\nu}$. Those field equations may be derived from the following space-time action,

$$S[G,\Phi,B] = \frac{1}{2\kappa^2} \int d\mu_G \, e^{-2\Phi} \left(R_G + 4D_\mu \Phi D^\mu \Phi - \frac{1}{12} H_{\mu\nu\rho} H^{\mu\nu\rho} \right) \tag{12.4.13}$$

where κ is a constant, D_μ is the covariant derivative, and,

$$H_{\mu\nu\rho} = D_\mu B_{\nu\rho} + D_\nu B_{\rho\mu} + D_\rho B_{\mu\nu} \tag{12.4.14}$$

In summary, conformal symmetry is the local gauge symmetry that eliminates negative norm states and decouples null states. In this chapter, we have illustrated this mechanism when space-time is flat, but the result is expected to hold in space-times that are asymptotically flat or anti-de Sitter.

12.5 String amplitudes in terms of vertex operators

Conformal symmetry also allows us to simplify the formulation of string scattering processes. Instead of representing each incoming and outgoing string in a scattering process by a boundary of the string worldsheet, as was presented in Figure 12.2, we can encode all the physical information of the

---

[3] The factor of i multiplying the $B_{\mu\nu}$ term is present for a worldsheet metric $\mathfrak{g}$ of Euclidean signature but is absent when Minkowski signature is used.

incoming or outgoing strings in terms of vertex operators. The data for these vertex operators is specified at vertex insertion points on the worldsheet, as illustrated in Figure 12.3. One may view these insertion points as the limit in which the radii of the boundary discs of Figure 12.2 are shrunk to zero by a conformal transformation.

Figure 12.3 Schematic representation of the perturbative expansion in the string coupling g_s formulated in terms of vertex operators on the world-sheet.

Consider the scattering of N on-shell gravitons in flat space-time $\mathbb{R}^{1,25}$, with momenta k_i and polarization tensors $\varepsilon_{i;\mu\nu}(k)$ for $i = 1, \cdots, N$. A graviton of momentum k_i introduces a small ripple in the metric of space-time, given in terms of a plane monochromatic wave. Adding up the contributions of the N gravitons, the space-time metric is perturbed as follows to lowest order in the amplitude of the wave given by the polarizations tensors ε_i,

$$G_{\mu\nu}(X) = \eta_{\mu\nu} + \sum_{i=1}^{N} \varepsilon_{i;\mu\nu}(k)\, e^{ik_i \cdot X} + \mathcal{O}(\varepsilon_i^2) \qquad (12.5.1)$$

Conformal invariance requires G to satisfy Einstein's equations, as was discussed in the second paragraph of Section 12.4. Einstein's equations apply in linearized form, because we consider the metric only to first order in each ε_i, and impose the conditions $k_i^2 = 0$ and $k_i^\mu \varepsilon_{i;\mu\nu}(k) = 0$. Substituting the expression (12.5.1) for the metric fluctuations into the action $I_G[X, \mathfrak{g}]$ of (12.2.1) and the functional integral of (12.3.8), and retaining the contribution which is linear in each ε_i gives the genus g contribution to the amplitude for the scattering of N gravitons in terms of vertex operators,

$$\mathcal{A}^{(g)} = \int_{\mathcal{M}_g} D\hat{\mathfrak{g}} \int_{\mathrm{Map}(\Sigma)} DX\, \mathcal{V}_1[X, \hat{\mathfrak{g}}] \cdots \mathcal{V}_N[X, \hat{\mathfrak{g}}]\, e^{-I_G[X,\hat{\mathfrak{g}}]-W_{\mathrm{gh}}[\hat{\mathfrak{g}}]} \qquad (12.5.2)$$

The vertex operators are read off from the expansion in ε_i to linear order and are given by,

$$\mathcal{V}_i[X, \hat{\mathfrak{g}}] = \varepsilon_{i;\mu\nu}(k) \int_\Sigma d\mu_{\hat{\mathfrak{g}}}\, \hat{\mathfrak{g}}^{mn} \partial_m X^\mu \partial_n X^\nu\, e^{ik_i \cdot X} \qquad (12.5.3)$$

The physical state conditions for the graviton, namely, its massless condition

$k_i^2 = 0$, the transversality $k_i^\mu \varepsilon_{i;\mu\nu}(k) = 0$, and tracelessness $\varepsilon_{i;\mu\nu}\eta^{\mu\nu} = 0$ of its polarization tensor, guarantee that the unintegrated vertex operator $\mathfrak{g}^{mn}\partial_m X^\mu \partial_n X^\nu e^{ik_i \cdot X}$ has conformal dimension $(1,1)$ and may be integrated over Σ consistently with conformal symmetry. One may proceed analogously for the vertex operators of the fields Φ and $B_{\mu\nu}$.

12.6 Superstring amplitudes

Scattering amplitudes for the superstrings may similarly be formulated in terms of functional integrals over the string degrees of freedom, such as X^μ, b, and c, with appropriate insertions of vertex operators. Unlike for the case of the bosonic string, several different formulations have been developed over time for the case of the superstring, each with certain advantages and disadvantages. Of central concern to us here is not so much how the amplitudes have been obtained but rather what their structure is. Thus, we shall proceed below by simply quoting the final expressions for the amplitudes we need, and refer to the literature for their detailed derivations, which are often quite involved. We shall concentrate on Type II theories here, but a similar analysis may be carried out for Type I and Heterotic superstrings.

For definiteness, we shall consider the scattering process that involves four gravitons, whose momenta we denote by k_i and whose polarization tensors are chosen in factorized form $\varepsilon_i^\mu(k_i)\tilde{\varepsilon}_i^\nu(k_i)$ with $k_i^2 = k_i \cdot \varepsilon_i(k_i) = k_i \cdot \tilde{\varepsilon}_i(k_i) = 0$ for $i = 1, 2, 3, 4$. It will be useful to introduce the dimensionless kinematic Lorentz-invariants defined by (recall that we have set $\alpha' = 2$),

$$s_{ij} = -\tfrac{1}{2}(k_i + k_j)^2 = -k_i \cdot k_j \qquad (12.6.1)$$

These variables satisfy a number of kinematic relations as a result of momentum conservation and the massless conditions $k_i^2 = 0$, which may be solved as follows,

$$s = s_{12} = s_{34}$$
$$t = s_{14} = s_{23}$$
$$u = s_{13} = s_{24} \qquad\qquad s + t + u = 0 \qquad (12.6.2)$$

Closed superstring perturbation theory produces the following on-shell four-graviton amplitude in Type II superstring theory,

$$\mathcal{A}(\varepsilon_i; k_i) = \kappa_{10}^2 \, \mathcal{R}^4 \sum_{g=0}^{\infty} g_s^{2g-2} \mathcal{A}^{(g)}(s_{ij}) \qquad (12.6.3)$$

where $\mathcal{R}$ stands for the on-shell linearized Riemann tensor, $\mathcal{R}^4$ for a particular scalar invariant constructed out of $\mathcal{R}$, and $\mathcal{A}^{(h)}(s_{ij})$ is a Lorentz scalar function that depends only on s, t, u. The fact that a single kinematic combination $\mathcal{R}^4$ is shared by the amplitudes at all loop order is a consequence of the space-time supersymmetry of Type II superstrings and does not hold for other theories such as Type I or Heterotic strings. The particular structure of the Lorentz contractions needed to form the scalar $\mathcal{R}^4$ is also a direct consequence of supersymmetry. In the factorized basis for the polarization tensors for the gravitons, the combination $\mathcal{R}^4$ is itself factorized,

$$2^6 \mathcal{R}^4 = \mathcal{K}\tilde{\mathcal{K}} \tag{12.6.4}$$

where $\mathcal{K}$ depends only on the polarization vectors ε_i^μ and $\tilde{\mathcal{K}}$ depends only on the polarization vectors $\tilde{\varepsilon}_i^\mu$. In terms of the linearized field strengths,

$$\begin{aligned} f_i^{\mu\nu} &= k_i^\mu \varepsilon_i^\nu - k_i^\nu \varepsilon_i^\mu \\ \tilde{f}_i^{\mu\nu} &= k_i^\mu \tilde{\varepsilon}_i^\nu - k_i^\nu \tilde{\varepsilon}_i^\mu \end{aligned} \tag{12.6.5}$$

the factor $\mathcal{K}$ is given as follows,

$$\begin{aligned} \mathcal{K} = (f_1 f_2)(f_3 f_4) + (f_1 f_4)(f_2 f_3) + (f_1 f_3)(f_2 f_4) \\ -4(f_1 f_2 f_3 f_4) - 4(f_1 f_2 f_4 f_3) - 4(f_1 f_3 f_2 f_4) \end{aligned} \tag{12.6.6}$$

The parentheses are defined by,

$$\begin{aligned} (f_i f_j) &= f_i^{\mu\nu} f_j^{\nu\mu} \\ (f_i f_j f_k f_\ell) &= f_i^{\mu\nu} f_j^{\nu\rho} f_k^{\rho\sigma} f_\ell^{\sigma\mu} \end{aligned} \tag{12.6.7}$$

The expression for $\tilde{\mathcal{K}}$ is given by the above expression with $f \to \tilde{f}$.

Explicit formulas for $\mathcal{A}^{(g)}(s_{ij})$ have been established from first principles for tree-level $g = 0$, one-loop $g = 1$, and two loops $g = 2$ and are given by,[4]

$$\begin{aligned} \mathcal{A}^{(0)}(s_{ij}) &= \frac{1}{stu} \frac{\Gamma(1-s)\Gamma(1-t)\Gamma(1-u)}{\Gamma(1+s)\Gamma(1+t)\Gamma(1+u)} \\ \mathcal{A}^{(1)}(s_{ij}) &= \frac{\pi}{8} \int_{\mathcal{M}_1} \frac{d^2\tau}{(\operatorname{Im}\tau)^2} \mathcal{B}^{(1)}(s_{ij}|\tau) \\ \mathcal{A}^{(2)}(s_{ij}) &= \frac{\pi}{8} \int_{\mathcal{M}_2} \frac{d^6\Omega}{(\det Y)^3} \mathcal{B}^{(2)}(s_{ij}|\Omega) \end{aligned} \tag{12.6.8}$$

---

[4] The normalizations used here are $d^2z = \frac{i}{2}dz \wedge d\bar{z}$ and $d^2\tau = \frac{i}{2}d\tau \wedge d\bar{\tau}$ and differ from those used in [DGP05] and [DG14] by a factor of 2. The normalization $d^6\Omega = d^2\Omega_{11} d^2\Omega_{12} d^2\Omega_{22}$ where $d^2\Omega_{IJ} = \frac{i}{2}d\Omega_{IJ} \wedge d\bar{\Omega}_{IJ}$ differs from the volume form $|d\Omega|^2$ used in [DGP05] and [DG14] by a factor of 8. These various factors of 2 account for the differences in prefactors in the expressions for $\mathcal{A}^{(g)}(s_{ij})$, $\mathcal{B}^{(1)}(s_{ij}|\tau)$, and $\mathcal{B}^{(2)}(s_{ij}|\Omega)$ compared to [DGP05] and [DG14].

where $Y = \operatorname{Im}\Omega$. The integrands $\mathcal{B}^{(1)}$ and $\mathcal{B}^{(2)}$ are dimensionless Lorentz scalar functions of s, t, and u given by,

$$\mathcal{B}^{(1)}(s_{ij}|\tau) = 16 \int_{\Sigma^4} \prod_{i=1}^{4} \frac{d^2 z_i}{\operatorname{Im}\tau} \exp\left\{ \sum_{i<j} s_{ij}\, G(z_i, z_j|\tau) \right\}$$

$$\mathcal{B}^{(2)}(s_{ij}|\Omega) = \int_{\Sigma^4} \frac{\mathcal{Y} \wedge \bar{\mathcal{Y}}}{(\det Y)^2} \exp\left\{ \sum_{i<j} s_{ij}\, \mathcal{G}(z_i, z_j|\Omega) \right\} \quad (12.6.9)$$

Here, z_i are the vertex insertion points on the surface Σ, τ is the modulus used as a local complex coordinate for the genus-one moduli space $\mathcal{M}_1$, and Ω is the period matrix used as a set of local coordinates for genus-two moduli space $\mathcal{M}_2$, defined in Appendix B. Furthermore, in the genus-two amplitude, $\mathcal{Y}$ is a holomorphic $(1,0)$ form in each one of the vertex points z_i given by,

$$3\mathcal{Y} = (t - u)\Delta(1, 2)\Delta(3, 4) + (s - t)\Delta(1, 3)\Delta(4, 2)$$
$$+ (u - s)\Delta(1, 4)\Delta(2, 3) \quad (12.6.10)$$

where Δ is a holomorphic $(1,0)$ form in its entries defined by,

$$\Delta(i, j) = \omega_1(z_i)\omega_2(z_j) - \omega_2(z_i)\omega_1(z_j) \quad (12.6.11)$$

and $\omega_I(z)$ are the normalized holomorphic Abelian differentials on Σ (see Appendix B). In $\mathcal{B}^{(1)}(s_{ij}|\tau)$, the scalar Green function $G(x_i, z_j|\tau)$ on the torus of modulus τ was defined in (5.5.12) and given explicitly in (5.5.11) and (5.5.16). In $\mathcal{B}^{(2)}(s_{ij}|\Omega)$, the Arakelov Green function $\mathcal{G}(z_i, z_j|\Omega)$ on the genus-two Riemann surface Σ with period matrix Ω will be constructed next.

Analogous formulas hold for amplitudes with more than four external states, as well as for Type I and Heterotic strings.

12.6.1 The Arakelov–Green function

The Arakelov Green function on a Riemann surface Σ of genus g is defined via the canonical Kähler form κ on Σ, which is the pull-back under the Abel map of the canonical Kähler form on the Jacobian $J(\Sigma)$ (see Appendix D),

$$\kappa = \frac{i}{2g} \sum_{I,J=1}^{g} (Y^{-1})^{IJ} \omega_I \wedge \bar{\omega}_J \qquad \int_{\Sigma} \kappa = 1 \qquad (12.6.12)$$

where $Y = \text{Im}\,\Omega$. In terms of local complex coordinates $z, \bar{z}$ on Σ, we have $\kappa = \frac{i}{2}\kappa_{z\bar{z}}dz \wedge d\bar{z}$, and the Arakelov Green function $\mathcal{G}(z, w|\Omega)$ is defined by,

$$\partial_{\bar{z}}\partial_z \mathcal{G}(z, w|\Omega) = -\pi\delta(z, w) + \pi\kappa_{z\bar{z}}(z)$$

$$\int_\Sigma \kappa(z)\mathcal{G}(z, w|\Omega) = 0 \qquad\qquad (12.6.13)$$

It is constructed in terms of the prime form $E(z, w|\Omega)$, which is defined for z, w in a simply connected fundamental domain F_Σ for Σ (see Appendix B), and Abelian integrals. One first introduces the following function,

$$G(z, w|\Omega) = -\ln|E(z, w|\Omega)|^2 + 2\pi\sum_{I,J}\text{Im}\left(\int_w^z \omega_I\right)Y_{IJ}^{-1}\left(\text{Im}\int_w^z \omega_J\right)$$

which is also well defined on F_Σ. The Arakelov Green function is then obtained by ensuring that its integral against κ vanishes, and we find,

$$\mathcal{G}(z, w|\Omega) = G(z, w|\Omega) - \gamma(z|\Omega) - \gamma(w|\Omega) + \int_{F_\Sigma} \kappa(z)\gamma(z|\Omega)$$

$$\gamma(z|\Omega) = \int_{F_\Sigma} \kappa(w)G(z, w|\Omega) \qquad\qquad (12.6.14)$$

In the special case of genus $g = 1$, the functions $\mathcal{G}(z, w|\Omega)$ and $G(z, w|\Omega)$ coincide with the Green function $G(z, w|\tau)$ of (5.5.11) and (5.5.16).

12.6.2 Physical singularity structure of amplitudes

The singularity structure of the genus-one and genus-two amplitudes is as follows. For fixed moduli of the compact Riemann surface Σ, the singularities of $\mathcal{B}^{(g)}$ as a function of s, t, and u are governed by the operator product expansion of the graviton vertex operators, namely, the conformal field theory of free scalar fields on Σ. The Green functions $G(z, w|\tau)$ and $\mathcal{G}(z, w|\Omega)$ are smooth throughout Σ, except for a logarithmic singularity at $z = w$, where they both behave as $-\ln|z - w|^2$ plus regular terms. As a result, the integrations which define $\mathcal{B}^{(g)}$ are absolutely convergent for $\text{Re}\,(s_{ij}) < 1$. The analytic continuation of $\mathcal{B}^{(g)}$ may be carried out to the complex plane in each variable and produces simple poles at positive integers $s, t, u \in \mathbb{N}$.

The further integration over the moduli of Σ, which is required to obtain the physical amplitude $\mathcal{A}^{(g)}$ for $g = 1, 2$, is absolutely convergent only when $\text{Re}\,(s_{ij}) = 0$. Analytic continuation in s_{ij} to the complex plane may be carried out using a decomposition of moduli space into subregions in each of which the domain of convergence is enlarged. This procedure has been performed explicitly only for genus one. The resulting analytic continuation has branch cuts in s_{ij} starting at any nonnegative integer, which signals two-particle intermediate states. The analytic continuation process will be illustrated in detail in Section 12.11 using the low-energy expansion near the massless branch cut starting at $s_{ij} = 0$.

12.7 Ultraviolet finiteness from modular invariance

An alternative presentation of the genus-one and genus-two amplitudes may be given by introducing internal loop momenta. We shall illustrate this here for the case of genus one. One begins by choosing a canonical homology basis of cycles $\mathfrak{A}, \mathfrak{B}$ on the torus Σ and takes the loop momentum p to flow through the cycle $\mathfrak{A}$. The amplitude is then given in terms of a Hermitian pairing between left and right chiral amplitudes,

$$\mathcal{A}^{(1)}(s_{ij}) = 2\pi i \int_{\mathbb{R}^{10}} dp \int_{\mathcal{M}_1} \int_{\Sigma^4} \mathcal{F}(z_i, k_i, p|\tau) \wedge \overline{\mathcal{F}(z_i, -k_i, p|\tau)} \quad (12.7.1)$$

where we have suppressed the dependence on the polarization tensors. The chiral amplitude $\mathcal{F}(z_i, k_i, p|\tau)$ is locally holomorphic in τ and in z_i for $i = 1, \ldots, 4$. Its explicit form is given by the following top holomorphic differential form on $\mathcal{M}_1 \times \Sigma^4$,

$$\mathcal{F}(z_i, k_i, p|\tau) = e^{i\pi\tau p^2 + 2\pi i \sum_i p \cdot k_i z_i} \prod_{i<j} \vartheta_1(z_i - z_j|\tau)^{-s_{ij}} d\tau \bigwedge_{i=1}^{4} dz_i \quad (12.7.2)$$

The price to pay for the local holomorphicity is that $\mathcal{F}$ has nontrivial monodromy when a point z_ℓ is taken around one of the homology cycles of the surface, and we have,

$$\mathcal{F}(z_i + \delta_{i,\ell} \mathfrak{A}, k_i, p|\tau) = e^{2\pi i k_\ell \cdot p} \mathcal{F}(z_i, k_i, p|\tau)$$
$$\mathcal{F}(z_i + \delta_{i,\ell} \mathfrak{B}, k_i, p|\tau) = \mathcal{F}(z_i, k_i, p + k_\ell|\tau) \quad (12.7.3)$$

When the point z_ℓ is taken around the point z_i with $\ell \neq i$, the chiral amplitude $\mathcal{F}$ is multiplied by a phase factor $e^{-2\pi i s_{ij}}$. Modular transformations of $\mathcal{F}$ involve a change of homology basis $\mathfrak{A}, \mathfrak{B}$ and thus a change of momentum routing through the torus. The Hermitian pairing of $\mathcal{F}$ and $\bar{\mathcal{F}}$ is familiar from two-dimensional conformal field theory where the loop momentum p labels the conformal blocks of 10 copies of a $\mathfrak{c} = 1$ theory.

Thanks to modular invariance, all string amplitudes are ultraviolet (UV) finite. This result was obtained by Shapiro for the bosonic string at genus one, but holds for all modular invariant superstrings to all genera.

We can illustrate the mechanism by which the genus-one string amplitude is UV finite by inspection of the chiral amplitude derived in the previous paragraph. For any string theory, the chiral amplitude, with loop momentum p chosen to flow through the cycle $\mathfrak{A}$, has the following universal prefactor which is Gaussian in the loop momentum, as in (12.7.2),

$$\mathcal{F}(z_i, \varepsilon_i, k_i, p|\tau) = e^{i\pi\tau p^2} \times \text{ exponentials and powers in } p \qquad (12.7.4)$$

The Gaussian factor is universal in the sense that it is independent of the momenta k_i and polarization tensors ε_i of the external states and actually independent of the specific string theory under consideration. Modular invariance allows one to choose the fundamental domain $\mathcal{M}_1$ for $\text{SL}(2, \mathbb{Z})\backslash\mathcal{H}$ that is adapted to the particular choice for the flow of loop momentum. For our choice of loop momentum traversing the $\mathfrak{A}$-cycle, the choice of fundamental domain is the standard one, $\mathcal{M}_1 = \{\tau \in \mathcal{H}, |\tau| \geq 1, |\text{Re}(\tau)| \leq \frac{1}{2}\}$. The high-energy behavior of the loop momentum is governed by the magnitude of $\text{Im}(\tau)$, which is uniformly bounded away from zero by our choice of fundamental domain. The uniform Gaussian suppression at large loop momenta implies UV finiteness as all other factors are either only exponential or polynomial in p. Higher genus amplitudes have a generalization of the above factor that involves the period matrix of the corresponding higher genus surface, but the structure is otherwise analogous to the one-loop case.

12.8 Effective interactions from the four-graviton amplitude

The tree-level, one-loop, and two-loop string amplitudes listed in (12.6.8) for four gravitons have important applications to the evaluation of string theory corrections to supergravity in the low-energy limit. Type IIB supergravity and its relation to Type IIB string theory will be reviewed in Chapter 14. In preparation for those discussions and the comparison of perturbative results with the implications from supersymmetry and $\text{SL}(2, \mathbb{Z})$-duality of Type IIB string theory, we shall evaluate the behavior of the amplitudes in (12.6.8) in the low-energy limit.

To illustrate the emergence of low-energy effective interactions from superstring amplitudes, we recall the expansion in powers of the string coupling g_s of the four-graviton amplitude, given in (12.6.3),

$$\mathcal{A}(\varepsilon_i; k_i) = \kappa_{10}^2 \mathcal{R}^4 \sum_{g=0}^{\infty} g_s^{2g-2} \mathcal{A}^{(g)}(s_{ij}) \qquad (12.8.1)$$

where $\mathcal{R}$ stands for the on-shell linearized Weyl tensor, whose expression was given in (12.6.4) and (12.6.6), while expressions for the integral representations for $\mathcal{A}^{(0)}, \mathcal{A}^{(1)}$, and $\mathcal{A}^{(2)}$ were given in (12.6.8).

12.8.1 Low-energy expansion at tree level

We concentrate first on the tree-level contribution $\mathcal{A}^{(0)}(s_{ij})$ and use the following series expansion for the ratio of Γ-functions (see Exercise 2.1),

$$\frac{\Gamma(1-s)\Gamma(1-t)\Gamma(1-u)}{\Gamma(1+s)\Gamma(1+t)\Gamma(1+u)} \tag{12.8.2}$$

$$= \exp\left\{ \sum_{n=1}^{\infty} \frac{2\zeta(2n+1)}{2n+1}(s^{2n+1} + t^{2n+1} + u^{2n+1}) \right\}$$

The argument of the exponential must clearly be odd in s, t, and u, since the left side is mapped to its inverse under simultaneous sign reversal of s, t, and u. The linear term cancels by the relation $s + t + u = 0$ of (12.6.2). The amplitude $\mathcal{A}^{(0)}$ is a symmetric function of s, t, u and may therefore be expanded in powers of the two remaining symmetric polynomials in s, t, u,

$$\sigma_2 = s^2 + t^2 + u^2 \qquad\qquad \sigma_3 = s^3 + t^3 + u^3 = 3stu \tag{12.8.3}$$

The first few terms in the expansion in powers of σ_2 and σ_3 are given by,

$$\mathcal{A}^{(0)}(s_{ij}) = \frac{1}{stu} + 2\zeta(3) + \sigma_2\,\zeta(5) + \frac{2\sigma_3}{3}\zeta(3)^2 + \frac{\sigma_2^2}{2}\zeta(7) + \mathcal{O}(s_{ij})^5 \tag{12.8.4}$$

The first term on the right side in the expansion is nonanalytic at low energy and corresponds to the exchange of massless gravitons and other massless bosons in Type IIB supergravity. The remaining terms on the right side are all local and correspond to effective interactions that are schematically of the form $D^{2k}\mathcal{R}^4$ for $k = 0, 2, 3$, and 4. The effective interaction $D^2\mathcal{R}^4$ is missing in this list in view of the relation $s + t + u = 0$.

12.8.2 Transcendental weight

An important notion that has arisen over the past few years in the study of quantum field theory amplitudes is that of *transcendental weight*. A natural starting point for its definition is with the Riemann ζ-function at integer argument $\zeta(k)$ for $k \geq 2$, referred to as *zeta values*, and *multiple zeta-values*,

$$\zeta(k_1, \cdots, k_p) = \sum_{n_1 \geq n_2 \geq \cdots \geq n_p \geq 1} \frac{1}{n_1^{k_1} \cdots n_p^{k_p}} \tag{12.8.5}$$

for integer $k_1 \geq 2$ and $k_2, \cdots, k_p \geq 1$. When k is even, $\zeta(k) \in \pi^k \mathbb{Q}$. Transcendental weight provides a grading upon multiplication and one assigns weight 0 to any algebraic number, weight 1 to π, and thus weight k to $\zeta(k)$ for k even. By extension, one assigns weight k to $\zeta(k)$ for odd k and weight $k = k_1 + \cdots + k_p$ to the multiple zeta value $\zeta(k_1, \cdots, k_p)$.

With these assignments of transcendental weight, we observe that the argument of the exponential on the right of (12.8.2) has transcendental weight 0 provided we assign weight -1 to s, t, and u. Each term in the expansion of the tree-level amplitude in (12.8.4) thus has weight 3. The amplitude is said to exhibit *uniform transcendentality*.

Remarkably, string theory has nontrivial transcendentality properties already at tree level. This is in contrast with the case of quantum field theory where tree-level amplitudes have trivial transcendentality, and it is only through the expansion of the Γ-functions that arise in dimensional regularization at loop level that nontrivial transcendentality arises. Uniform transcendentality, observed here for tree-level Type II string amplitudes, is shared in quantum field theory by the maximally supersymmetric $\mathcal{N} = 4$ Yang–Mills theory in four dimensions, to be discussed in Chapter 15. A full understanding of how this property arises, and for precisely which correlators it holds, remains an open problem.

12.9 Genus one in terms of modular graph functions

The genus-one amplitude $\mathcal{A}^{(1)}(s_{ij})$ is obtained in (12.6.8) by integrating the partial amplitude $\mathcal{B}^{(1)}(s_{ij}|\tau)$, which is a modular function in τ, over the genus-one moduli space $\mathcal{M}_1$. The partial amplitude in turn is given by the integration over four copies of the torus Σ with modulus τ in (12.6.9). For fixed $\tau \in \mathcal{H}$, the scalar Green function $G(z|\tau)$ on the torus is smooth throughout the torus except for the logarithmic singularity at $z = 0$ given by $G(z|\tau) \approx -\ln|z|^2$. As a result, for fixed τ, the integrals that define $\mathcal{B}^{(1)}(s_{ij}|\tau)$ as a function of s, t, and u are absolutely convergent for $\operatorname{Re}(s), \operatorname{Re}(t), \operatorname{Re}(u) < 1$, and may be expanded in a Taylor series in powers of s, t, and u with radius of convergence $|s|, |t|, |u| < 1$.

Since the function $\mathcal{B}^{(1)}(s_{ij}|\tau)$ is symmetric in s, t, and u, we may organize this expansion in powers of the symmetric polynomials σ_2 and σ_3 in s, t, and u as we did for the tree-level amplitude,

$$\mathcal{B}^{(1)}(s_{ij}|\tau) = \sum_{p,q=0}^{\infty} \mathcal{B}^{(1)}_{(p,q)}(\tau) \frac{\sigma_2^p \sigma_3^q}{p!\, q!} \tag{12.9.1}$$

Since $\mathcal{B}^{(1)}(s_{ij}|\tau)$ is a modular function in τ, each coefficient $\mathcal{B}^{(1)}_{(p,q)}(\tau)$ is a modular function of τ. The coefficients may be computed by expanding the exponential in the integrand of (12.6.9) in powers of its argument to a given order w, which is the overall degree in the variables s, t, and u, given by $w = 2p + 3q$, so that we have,

$$\sum_{\substack{p,q \geq 0 \\ 2p+3q=w}} \mathcal{B}^{(1)}_{(p,q)}(\tau) \frac{\sigma_2^p \sigma_3^q}{p!\,q!} = \frac{1}{w!} \prod_{i=1}^{4} \int_{\Sigma} \frac{d^2 z_i}{\operatorname{Im}\tau} \left(\sum_{i<j} s_{ij} G(z_i - z_j|\tau) \right)^w \quad (12.9.2)$$

The coefficients $\mathcal{B}^{(1)}_{(p,q)}(\tau)$ for different values of p and q satisfying $w = 2p+3q$ may be sorted by further expanding the wth power on the right side.

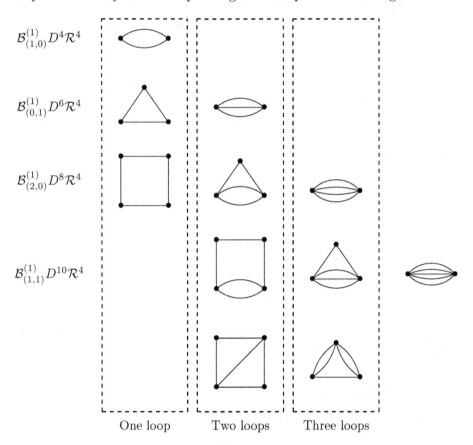

Figure 12.4 Modular graph functions contributing to the genus-one four-graviton amplitude, up to weight $w = 5$ included.

The expansion may be presented graphically, and the modular function corresponding to a given graph is precisely one of the *modular graph func-*

tions introduced in Chapter 9. Each graph has four vertices, namely, the points z_i for $i = 1, 2, 3, 4$, and w edges. An edge may connect any pair of distinct vertices but is not allowed to begin and end on the same vertex. The expansion is simplified by the following observations. Any modular graph function corresponding to a graph that

- contains a vertex on which only a single edge begins or ends vanishes;
- becomes disconnected upon removing one edge vanishes; and
- becomes disconnected upon removing a single vertex factorizes into the modular graph functions of the resulting connected components.

The graphs corresponding to the modular graph functions which contribute to $\mathcal{B}_{(p,q)}(\tau)$ up to weight $w = 5$ are presented in Figure 12.4. Instead of organizing the graphs by weight w, they may alternatively be organized by loop order, which is indicated in Figure 12.4 by the three rectangular boxes corresponding to 1, 2, and 3 loop orders, respectively.

The weight w of the graphs contributing to $\mathcal{B}_{(p,q)}(\tau)$ for $w = 2p + 3q$ may now be seen to coincide with the definition of *transcendental weight* given in Section 12.8.2, provided we assign weight -1 to s, t, and u and weight 0 to τ.

Upon expanding (12.9.2) and using identities such as (9.4.11), one finds,

$$\mathcal{B}_{(0,0)}^{(1)} = 1$$

$$\mathcal{B}_{(1,0)}^{(1)} = E_2$$

$$\mathcal{B}_{(0,1)}^{(1)} = \tfrac{5}{3}E_3 + \tfrac{1}{3}\zeta(3)$$

$$\mathcal{B}_{(2,0)}^{(1)} = 2C_{2,1,1} + E_2^2 - E_4$$

$$\mathcal{B}_{(1,1)}^{(1)} = \tfrac{7}{3}C_{3,1,1} + \tfrac{5}{3}E_2 E_3 + \tfrac{1}{3}\zeta(3)E_3 - \tfrac{34}{15}E_5 + \tfrac{1}{5}\zeta(5) \qquad (12.9.3)$$

up to weight $w = 5$ included. The graphs that contribute to each partial amplitude are shown in Figure 12.4.

12.10 Genus two in terms of modular graph functions

The genus-two amplitude $\mathcal{A}^{(2)}(s_{ij})$ is obtained in (12.6.8) by integrating the partial amplitude $\mathcal{B}^{(2)}(s_{ij}|\Omega)$ over the genus-two moduli space $\mathcal{M}_2$, while $\mathcal{B}^{(2)}(s_{ij}|\Omega)$ in turn is given by the integration over four copies on the genus-two surface Σ with moduli Ω in (12.6.9). For fixed $\Omega \in \mathcal{H}_2$, the scalar Arakelov Green function $\mathcal{G}(z, w|\Omega)$ is smooth throughout Σ except for the logarithmic singularity at $z = w$ given by $\mathcal{G}(z, w|\Omega) \approx -\ln|z - w|^2$. As a result, for fixed Ω, the integrals that define $\mathcal{B}^{(2)}(s_{ij}|\Omega)$ as a function of

s, t, and u are absolutely convergent for $\mathrm{Re}\,(s), \mathrm{Re}\,(t), \mathrm{Re}\,(u) < 1$ and may be expanded in a Taylor series in powers of s, t, and u with radius of convergence $|s|, |t|, |u| < 1$. This situation is completely analogous to the genus-one case. The presence of the additional measure factor $|\mathcal{Y}|^2$ is inconsequential for these results to hold as $\mathcal{Y}$ is holomorphic on Σ.

Since the function $\mathcal{B}^{(2)}(s_{ij}|\Omega)$ is symmetric in s, t, and u, we may organize its Taylor expansion in powers of s, t, and u via the symmetric polynomials σ_2 and σ_3, as we did for tree-level and genus-one amplitudes,

$$\mathcal{B}^{(2)}(s_{ij}|\Omega) = \sum_{p,q=0}^{\infty} \mathcal{B}^{(2)}_{(p,q)}(\Omega)\,\frac{\sigma_2^p\,\sigma_3^q}{p!\,q!} \qquad (12.10.1)$$

Since $\mathcal{B}^{(2)}(s_{ij}|\Omega)$ is a modular function of Ω, namely, it is invariant under the genus-two modular group $\mathrm{Sp}(4, \mathbb{Z})$, each coefficient $\mathcal{B}^{(2)}_{(p,q)}(\Omega)$ is itself a modular function of Ω. The coefficients may be computed by expanding the exponential in the integrand of (12.6.9) in powers of its argument to a given order w, which is related to the overall degree in the variables s, t, and u, given by $w + 2 = 2p + 3q$, so that we have,

$$\sum_{\substack{p,q\geq 0 \\ 2p+3q=w+2}} \mathcal{B}^{(2)}_{(p,q)}(\Omega)\,\frac{\sigma_2^p\,\sigma_3^q}{p!\,q!} = \frac{1}{w!}\int_{\Sigma^4} \frac{\mathcal{Y}\wedge\bar{\mathcal{Y}}}{(\det Y)^2}\left(\sum_{i<j} s_{ij}\mathcal{G}(z_i, z_j|\Omega)\right)^{\!w} \qquad (12.10.2)$$

where $Y = \mathrm{Im}\,\Omega$ and $\mathcal{Y}$ was given in (12.6.10). Note that the shift to $w + 2$ included in the range of summation on the left side is due to the fact that $\mathcal{Y}$ is linear in the variables s, t, and u.

12.10.1 Contributions to low weight

Linearity of $\mathcal{Y}$ in s, t, and u immediately implies that the genus-two contribution to the $\mathcal{R}^4$ interaction without derivatives must vanish, $\mathcal{B}^{(2)}_{(0,0)} = 0$. As was already mentioned earlier, the $D^2\mathcal{R}^4$ contribution vanishes in view of the momentum conservation identity $s + t + u = 0$.

The leading low-energy contribution is the $D^4\mathcal{R}^4$ interaction obtained from the zeroth order expansion term of the exponential in (12.9.2). The remaining integrations required to evaluate $\mathcal{B}^{(2)}_{(1,0)}$ may be carried out using the Riemann bilinear relations, reviewed in Appendix B, and yield an Ω-independent contribution.

The coefficient $\mathcal{B}^{(2)}_{(0,1)}$ has nontrivial dependence on Ω, denoted by $64\,\varphi(\Omega)$. Assembling these results to order $D^6\mathcal{R}^4$, we have,

$$\mathcal{B}^{(2)}(s_{ij}|\Omega) = 32(s^2 + t^2 + u^2) + 192\,stu\,\varphi(\Omega) + \mathcal{O}(s_{ij}^4) \qquad (12.10.3)$$

The genus-two modular function $\varphi(\Omega)$ is given in terms of the Arakelov Green function $\mathcal{G}(x,y|\Omega)$ by,

$$\varphi(\Omega) = \frac{1}{8} \int_{\Sigma^2} \omega_I(x)\omega_J(y)\left(\bar{\omega}^I(x)\bar{\omega}^J(y) - 2\bar{\omega}^J(x)\bar{\omega}^I(y)\right)\mathcal{G}(x,y|\Omega) \quad (12.10.4)$$

where $\bar{\omega}^I = (Y^{-1})^{IJ}\overline{\omega_J}$. This object was identified with the spectral invariant introduced by Kawazumi and Zhang. Higher order contributions to $\mathcal{B}^{(2)}(s_{ij}|\Omega)$ may similarly be expressed in terms of integrals of products of Arakelov Green functions and may be viewed as higher genus generalizations of the genus-one modular graph functions investigated in Chapter 9. Their general structure has, however, been less well studied and is much less well understood than that of their genus-one counterparts.

12.11 Integration over the genus-one moduli space

When considering one-loop or two-loop amplitudes we have, so far, focused attention on the structure of the partial amplitudes $\mathcal{B}^{(1)}(s_{ij}|\tau)$ and $\mathcal{B}^{(2)}(s_{ij}|\Omega)$, for which we have obtained a systematic low-energy expansion in terms of modular graph functions, both at genus one and genus two. For fixed τ and Ω, this expansion is absolutely convergent for $\mathrm{Re}\,(s), \mathrm{Re}\,(t), \mathrm{Re}\,(u) < 1$ and admits an analytic continuation in s, t, and u whose sole singularities are simple poles at positive integers in s, t, and u, corresponding to the exchanges of massive string states, as may be seen from the operator product expansion.

Obtaining concrete expressions for the corresponding physical string amplitudes $\mathcal{A}^{(1)}(s_{ij})$ and $\mathcal{A}^{(2)}(s_{ij})$ further requires integrating over the moduli spaces of genus-one and genus-two Riemann surfaces in (12.6.8). These integrations are absolutely convergent only when s, t, and u are purely imaginary, in which case $\mathcal{B}^{(1)}(s_{ij}|\tau)$ and $\mathcal{B}^{(2)}(s_{ij}|\Omega)$ may be uniformly bounded on their respective moduli spaces.

For the case of genus one, the lack of convergence originates from the integration in the region near the cusp. This may be seen by expressing the genus-one Green function $G(z|\tau)$ in the co-moving coordinates $z = x + y\tau$ for $x, y \in \mathbb{R}/\mathbb{Z}$, first introduced in Section 5.5.1, and given by (5.5.16),

$$G(x+y\tau|\tau) = -\ln\left|\frac{\vartheta_1(x+y\tau|\tau)}{\eta(\tau)}\right|^2 + 2\pi\tau_2 y^2 \quad (12.11.1)$$

For $\tau_2 \to \infty$ and fixed x and $|y| < \frac{1}{2}$, the Green function behaves as follows,

$$G(x+y\tau|\tau) \approx 2\pi\tau_2(y^2 - |y|) \quad (12.11.2)$$

which diverges as $\tau_2 \to \infty$.

The analytic continuation in s, t, and u of the integrals over moduli now produces double and additional simple poles at all positive integers, as well as branch cuts in s, t, and u that originate at every nonnegative integer, including zero. These singularities are fully expected on physical grounds. The poles produce mass renormalization and nonzero decay widths of massive string states, while the branch cuts correspond to the decay of a single string into a two-string state. The presence of the branch cuts implies that the string amplitudes $\mathcal{A}$ do not admit a convergent low-energy Taylor expansion. We now review the derivation of the branch cuts that arise in the low-energy expansion for the case of one-loop amplitudes.

12.11.1 *Partitioning the genus-one moduli space*

To proceed with the analysis of the branch cuts produced in the low-energy limit, we partition the fundamental domain F for the moduli space $\mathcal{M}_1$ of genus-one Riemann surfaces into a region F_R near the cusp and its complement F_L, as illustrated in the Figure 12.5, and given for $L > 1$ by,

$$F_R = F \cap \{\tau_2 > L\} \qquad\qquad F_L \cap F_R = \emptyset$$
$$F_L = F \cap \{\tau_2 \leq L\} \qquad\qquad F_L \cup F_R = F \qquad (12.11.3)$$

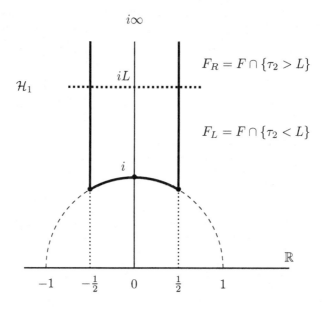

Figure 12.5 Partition of the genus-one moduli space F into F_R and F_L.

While $L > 1$ can be an arbitrary free parameter, in practice we shall take $L \gg 1$ so that F_R is a small neighborhood of the cusp. The amplitude $\mathcal{A}^{(1)}(s_{ij})$ is a sum of the corresponding integrals,

$$\mathcal{A}^{(1)}(s_{ij}) = \mathcal{A}_L^{(1)}(s_{ij}) + \mathcal{A}_R^{(1)}(s_{ij}) \tag{12.11.4}$$

where

$$\mathcal{A}_{L,R}^{(1)}(s_{ij}) = \frac{\pi}{8} \int_{F_{L,R}} \frac{|d\tau|^2}{(\mathrm{Im}\,\tau)^2} \mathcal{B}^{(1)}(s_{ij}|\tau) \tag{12.11.5}$$

The region F_L is compact and the integrand $\mathcal{B}^{(1)}(s_{ij}|\tau)$ admits a uniform low-energy expansion in the region of convergence $|s|, |t|, |u| < 1$. Therefore, the contribution $\mathcal{A}_L^{(1)}(s_{ij})$ will be analytic in s, t, and u but dependent on L.

The region F_R is not compact and $\mathcal{B}^{(1)}(s_{ij}|\tau)$ cannot be handled with a uniformly convergent low-energy expansion. Instead we need a treatment of this contribution that is exact up to contributions exponentially suppressed in L. The resulting $\mathcal{A}_R^{(1)}(s_{ij})$ will exhibit branch cuts in s, t, and u, and its L-dependence will cancel the L-dependence of $\mathcal{A}_L^{(1)}(s_{ij})$.

12.11.2 Integrals involving Eisenstein series

To carry out the integrations over the fundamental domain of the partial amplitudes $\mathcal{B}_{(p,q)}^{(1)}(\tau)$, one invariably encounters integrals of Eisenstein series, products of Eisenstein series, and other modular graph functions, over the truncated fundamental domain F_L. These integrals are of interest in their own right, and we dedicate this section to deriving some general results and their application to the simplest cases and then close by quoting the integral of the product of three Eisenstein series from Zagier's work. We begin with the following straightforward application of Stokes's theorem to an arbitrary modular graph function $\mathcal{C}$ of weight $(0,0)$ or to an arbitrary modular graph form $\mathcal{C}^+$ of weight $(0,2)$,

$$\int_{F_L} \frac{d^2\tau}{\tau_2^2} \Delta\mathcal{C} = \int_0^1 d\tau_1\, \partial_{\tau_2}\mathcal{C}(\tau_1,\tau_2)\Big|_{\tau_2=L}$$

$$\int_{F_L} \frac{d^2\tau}{\tau_2^2} \nabla\mathcal{C}^+ = \int_0^1 d\tau_1\, \mathcal{C}^+(\tau_1,\tau_2)\Big|_{\tau_2=L} \tag{12.11.6}$$

where we recall that $d^2\tau = \frac{i}{2} d\tau \wedge d\bar{\tau}$. The remaining integrals on the right side may be read off from the constant Fourier mode part of $\mathcal{C}$ and $\mathcal{C}^+$. The result may be readily used to evaluate the integral of a non-holomorphic Eisenstein series or products thereof. The simplest expressions are obtained in terms of the normalized Riemann zeta function $\zeta^*(s)$, introduced long ago in (2.1.16), and the normalized non-holomorphic Eisenstein series $E_s^*(\tau)$, defined by,[5]

$$\zeta^*(s) = \frac{\Gamma(s/2)\zeta(s)}{\pi^{\frac{s}{2}}} \qquad E_s^*(\tau) = \tfrac{1}{2}\Gamma(s)E_s(\tau) \qquad (12.11.7)$$

The normalization guarantees the functional equations $\zeta^*(1-s) = \zeta^*(s)$ and $E_{1-s}^*(\tau) = E_s^*(\tau)$, in addition to the following simple Fourier expansion,

$$E_s^*(\tau) = \zeta^*(2s)\,\tau_2^s + \zeta^*(2s-1)\,\tau_2^{1-s} + \mathcal{O}(e^{-2\pi\tau_2}) \qquad (12.11.8)$$

In terms of these functions, we have the integrals below.

Lemma 12.11.1. *The following integrals involving Eisenstein series are given, up to exponentially decaying terms $\mathcal{O}(e^{-2\pi\tau_2})$ at the cusp, by,*

$$\int_{F_L} \frac{d^2\tau}{\tau_2^2}\, E_s^* = \sum_{x=s,1-s} \zeta^*(2x)\frac{L^{x-1}}{x-1} \qquad (12.11.9)$$

$$\int_{F_L} \frac{d^2\tau}{\tau_2^2}\, E_s^* E_t^* = \sum_{\substack{x=s,1-s \\ y=t,1-t}} \zeta^*(2x)\zeta^*(2y)\frac{L^{x+y-1}}{x+y-1}$$

$$\int_{F_L} \frac{d^2\tau}{\tau_2^2}\, E_s^* E_t^* E_u^* = \zeta^*(w-1)\zeta^*(w-2s)\zeta^*(w-2t)\zeta^*(w-2u)$$

$$+ \sum_{\substack{x=s,1-s \\ y=t,1-t \\ z=u,1-u}} \zeta^*(2x)\zeta^*(2y)\zeta^*(2z)\frac{L^{x+y+z-1}}{x+y+z-1}$$

where $w = s+t+u$ in the first line on the right side of the last integral.

The derivatives with respect to L on both sides readily match upon using the Laurent expansions of E_s^*. Thus the nontrivial part of the above relations resides entirely in the L-independent terms. The proof of the first two integrals is provided in Exercise 12.1. The third line in Lemma 12.11.1 was proven by Zagier using different methods, as will be indicated in the bibliographical notes. An example of an integration involving two-loop modular graph functions is presented in Exercise 12.2.

12.11.3 The nonanalytic contribution $\mathcal{A}_R^{(1)}$

To evaluate $\mathcal{B}^{(1)}(s_{ij}|\tau)$ for $\tau \in F_R$, we cannot use (12.9.1) since this expansion is not uniformly convergent throughout F_R. Instead, our starting point is the original expression in (12.6.8), which we shall evaluate exactly up to

---

[5] Note that the parameters s, t, and u used in this subsection and in Theorem 12.11.1 are, of course, unrelated to the kinematical variables of (12.6.2) used elsewhere.

terms that are exponentially suppressed in L. To do so, we use co-moving coordinates, $z_i = x_i + y_i \tau$ with $i = 1, 2, 3, 4$ and partition the integration region Σ^4 into the six possible orderings of y_1, y_2, y_3, and y_4 up to cyclic permutations. Doing so allows us to decompose $\mathcal{A}_R^{(1)}(s_{ij})$ into a sum of six more elementary contributions, which are pairwise equal to one another,

$$\mathcal{A}_R^{(1)}(s_{ij}) = 2\mathcal{A}_*(L; s, t) + 2\mathcal{A}_*(L; t, u) + 2\mathcal{A}_*(L; u, s) \qquad (12.11.10)$$

where as usual we have $s + t + u = 0$. The function $\mathcal{A}_*(L; s, t)$ may be computed exactly, up to exponentially suppressed terms,

$$\mathcal{A}_*(L; s, t) = \mathfrak{C}(s, t; 0) \ln(-4\pi L s) + \left. \frac{\partial}{\partial \varepsilon} \mathfrak{C}(s, t; \varepsilon) \right|_{\varepsilon = 0} + \mathcal{O}(e^{-2\pi L}) \quad (12.11.11)$$

where $\mathfrak{C}$ is conveniently expressed as the sum of two integral representations,

$$\mathfrak{C}(s, t; \varepsilon) = -\frac{8\pi s^4}{\Gamma(3 + \varepsilon)} \int_0^1 dx \int_0^1 dy \, x \left[x^2 y (1 - y) \right]^{2+\varepsilon} W(s, -sxy + t(1 - x))$$

$$-2\pi s^7 \sum_{k=0}^{\infty} s^k t^k \int_0^1 dx \, \frac{[x(1 - x)]^{k+3+\varepsilon}}{\Gamma(k + 4 + \varepsilon) \, k!} \left[\left. \frac{\partial^k}{\partial \mu^k} W(s, \mu - sx) \right|_{\mu = 0} \right]^2$$

$$(12.11.12)$$

The function W is closely related to the Virasoro–Shapiro amplitude,

$$W(s, t) = \frac{1}{stu} \left(\frac{\Gamma(1 - s)\Gamma(1 - t)\Gamma(1 - u)}{\Gamma(1 + s)\Gamma(1 + t)\Gamma(1 + u)} - 1 \right) \qquad (12.11.13)$$

where we set $u = -s - t$. Although the above expression may look daunting at first, its analytical structure is manifest. The function $W(s, t)$ is mero-morphic throughout $s, t \in \mathbb{C}$, and holomorphic in discs of unit radius near the origin. It has only simple poles at positive integer values of s, t, and u and thus admits a Taylor expansion in s and t. Substituting this Taylor expansion into the integrals defining $\mathfrak{C}$, we see that the integrals preserve the holomorphicity in s, t, and u near the origin, so that $\mathfrak{C}$ has a convergent Taylor expansion near the origin, which starts at order s^4. To this lowest order, we have,

$$\mathfrak{C}(s, t; \varepsilon) = -16\pi \zeta(3) s^4 \frac{\Gamma(3 + \varepsilon)}{\Gamma(7 + 2\varepsilon)} + \mathcal{O}(s^5, s^4 t) \qquad (12.11.14)$$

The coefficients of the expansion to all orders inherit contributions linear and bilinear in odd zeta values from the expansion of the Virasoro–Shapiro am-plitude, with rational coefficients in the coefficient $\mathfrak{C}(s, t; 0)$ of the $\ln(-4\pi L s)$ term, and additional harmonic sums in the $\partial_\varepsilon \mathfrak{C}(s, t; \varepsilon)|_{\varepsilon=0}$ analytic contribu-tion to $\mathcal{A}_*(L; s, t)$. The discontinuity of $\ln(-4\pi L s)$ near the origin reproduces the square of the tree-level amplitude by unitarity.

12.11.4 The analytic contribution $\mathcal{A}_L^{(1)}$

The contribution $\mathcal{A}_L^{(1)}(s_{ij})$ is analytic in s, t, and u near the origin, and we may use the expansion of $\mathcal{B}^{(1)}(s_{ij}|\tau)$ in terms of modular graph functions of (12.9.1) to evaluate $\mathcal{A}_L^{(1)}(s_{ij})$ order by order in powers of s, t, and u,

$$\mathcal{A}_L^{(1)}(s_{ij}) = \sum_{p,q=0}^{\infty} \mathcal{A}_{(p,q)}^{(1)} \frac{\sigma_2^p \sigma_3^q}{p!\,q!} \tag{12.11.15}$$

Since the nonanalytic part $\mathcal{A}_R^{(1)}(s_{ij})$ starts contributing to $\mathcal{A}^{(1)}(s_{ij})$ only at order s^4, all its lower orders are given entirely by $\mathcal{A}_L^{(1)}(s_{ij})$. The corresponding amplitudes are readily evaluated, and we find,

$$\mathcal{A}_{(0,0)}^{(1)} = \frac{\pi}{3} \quad \mathcal{A}_{(0,1)}^{(1)} = \frac{\pi}{9}\zeta(3) \quad \mathcal{A}_{(1,0)}^{(1)} = 0 \quad \mathcal{A}_{(1,1)}^{(1)} = \frac{29\pi}{540}\zeta(5) \tag{12.11.16}$$

The remaining coefficients at weight 4 are given as follows,

$$\mathcal{A}_{(2,0)}^{(1)} = \frac{4\pi\zeta(3)}{45}\left(\ln(2L) + \frac{\zeta'(4)}{\zeta(4)} - \frac{\zeta'(3)}{\zeta(3)} - \frac{1}{4}\right) \tag{12.11.17}$$

Some of the intermediate results needed to establish the above expressions are obtained in Exercise 12.3.

12.11.5 Assembling analytic and nonanalytic parts

Assembling the contributions to $\mathcal{A}_L^{(1)}(s_{ij})$ in (12.11.17) and to $\mathcal{A}_R^{(1)}(s_{ij})$ in (12.11.14) gives $\mathcal{A}^{(1)}(s_{ij})$, the total amplitude, to order $\mathcal{O}(s_{ij}^4)$ included. We see explicitly that all L-dependence cancels as required for the consistency of the calculation. It is instructive to rearrange the total amplitude as the sum of "analytic" and "nonanalytic" pieces,

$$\mathcal{A}^{(1)}(s_{ij}) = 2\pi\left(\mathcal{A}_L^{(1)}(L; s_{ij}) + \mathcal{A}_R^{(1)}(L; s_{ij})\right)$$
$$= \mathcal{A}_{\text{an}}(s_{ij}) + \mathcal{A}_{\text{non-an}}(s_{ij}) \tag{12.11.18}$$

The reason for the quotation marks on analytic and nonanalytic is that the non-analytic piece actually contains also analytic contributions, so that the nomenclature is natural and suggestive but not entirely precise. The "analytic" piece $\mathcal{A}_{\text{an}}(s_{ij})$ is given by,

$$\mathcal{A}_{\text{an}}(s_{ij}) = \frac{2\pi^2}{3}\left(1 + \frac{\zeta(3)\sigma_3}{3} + \mathcal{O}(s_{ij}^5)\right) \tag{12.11.19}$$

and the "nonanalytic" piece $\mathcal{A}_{\text{non-an}}(s_{ij})$ has the form,

$$\mathcal{A}_{\text{non-an}}(s_{ij}) = \frac{2\pi^2}{3}\left(\mathcal{A}_{\text{sugra}} + \mathcal{A}_4 + \mathcal{O}(s_{ij}^5)\right) \qquad (12.11.20)$$

The lowest-order term $\mathcal{A}_{\text{sugra}}$ is a regularized version of the 10-dimensional one-loop supergravity amplitude. The contribution $\mathcal{A}_4$ is given by,

$$\mathcal{A}_4 = \frac{4s^4}{15}\zeta(3)\left(-\ln(-2\pi s) + \frac{\zeta'(4)}{\zeta(4)} - \frac{\zeta'(3)}{\zeta(3)} - \gamma + \frac{63}{20}\right)$$
$$+ \text{cycl}\,(s, t, u) \qquad (12.11.21)$$

where γ is the Euler constant. The discontinuity of $\mathcal{A}_{\text{non-an}}(s_{ij})$, namely, the coefficients of the $\ln(-2\pi s)$ terms in (12.11.21), reproduce those of [GRV08].

12.12 Integration over the genus-two moduli space

The function $\mathcal{B}_{(1,0)}^{(2)}(\Omega)$ needed to obtain the coefficient of the $D^4\mathcal{R}^4$ inter-action was found to be independent of moduli Ω in Section 12.10.1. Thus, its integral over moduli space $\mathcal{M}_2$ is given by the volume of $\mathcal{M}_2$ with the Siegel metric of constant negative curvature on the Siegel upper half-space,

$$ds^2 = \sum_{I,J,K,L} Y_{IJ}^{-1} Y_{KL}^{-1}\, d\bar{\Omega}_{JK}\, d\Omega_{IL} \qquad (12.12.1)$$

The invariant volume form is $d^6\Omega/(\det Y)^3$, where $d^6\Omega$ is normalized by,

$$d^6\Omega = d^2\Omega_{11} d^2\Omega_{12} d^2\Omega_{22} \qquad d^2\Omega_{IJ} = \tfrac{i}{2} d\Omega_{IJ} \wedge d\bar{\Omega}_{IJ} \qquad (12.12.2)$$

and the value of the volume, computed by Siegel, is given by,

$$\int_{\mathcal{M}_2} \frac{d^6\Omega}{(\det Y)^3} = \frac{\pi^3}{270} \qquad (12.12.3)$$

As a result, the low-energy effective interaction to this order is given by,

$$\mathcal{A}_{(1,0)}^{(2)} = \frac{4\pi^4}{270}(s^2 + t^2 + u^2)\mathcal{R}^4 \qquad (12.12.4)$$

The function $\mathcal{B}_{(0,1)}^{(2)}(\Omega)$ is proportional to the Kawazumi–Zhang invari-ant φ. Its integral over $\mathcal{M}_2$ is convergent and yields the contribution $\mathcal{A}_{(0,1)}^{(2)}$ to the low-energy effective action. Evaluation of the integral is made possible by the fact that φ satisfies an inhomogeneous Laplace eigenvalue equation,

$$(\Delta - 5)\varphi = -2\pi\delta_{SN} \qquad (12.12.5)$$

where Δ is the Laplacian on the Siegel upper half-space for the Siegel metric,

$$\Delta = \sum_{I,J,K,L} 4 Y_{IK} Y_{JL} \bar{\partial}_{IJ} \partial_{KL} \qquad \partial_{IJ} = \frac{1}{2}(1 + \delta_{IJ}) \frac{\partial}{\partial \Omega_{IJ}} \qquad (12.12.6)$$

and δ_{SN} is the Dirac δ-function supported on the separating degeneration node. The coefficient $\mathcal{B}^{(2)}_{(0,1)}$ is then given by the integral of the Kawazumi–Zhang invariant over $\mathcal{M}_2$,

$$\mathcal{B}^{(2)}_{(0,1)} = \frac{\pi}{8} \int_{\mathcal{M}_2} \frac{d^6 \Omega}{(\det Y)^3} \varphi$$

$$= \frac{\pi}{40} \int_{\mathcal{M}_2} \frac{d^6 \Omega}{(\det Y)^3} \left(\Delta \varphi + 2\pi \delta_{SN} \right) = \frac{2\pi^3}{45} \qquad (12.12.7)$$

The integral may be evaluated using Stokes's theorem and the asymptotic behavior for the Kawazumi–Zhang invariant, and we find,

$$\mathcal{A}^{(2)}_{(0,1)} = \frac{2\pi^3}{45}(s^3 + t^3 + u^3)\mathcal{R}^4 \qquad (12.12.8)$$

Genus-three studies are cited in the bibliographical notes of this chapter.

12.13 The Selberg trace formula and functional determinants

Key ingredients in the construction of string amplitudes as integrals over moduli space are the Green functions of the quantum fields on the world-sheet, as well as the functional determinants of Laplace and Cauchy–Riemann operators acting on these fields. In this last section of the chapter, we use functional determinants of certain combinations of the Laplace operator on scalars to segue into a discussion of the *Selberg trace formula*. The Selberg trace formula is an essential tool in modern analytic number theory that generalizes the Poisson resummation formula to non-Abelian groups and may be formulated for arbitrary trace-class integral operators. Here, we shall specialize to the case of functional determinants where the operator is the heat-kernel $\mathcal{K}^t_\Sigma(z, z')$ with $z, z' \in \Sigma$, namely, the integral operator $e^{t\Delta}$ of the scalar Laplacian Δ on a Riemann surface Σ endowed with the Poincaré metric. We begin with some setup.

The Riemann surface Σ may be represented by the quotient $\Sigma = \Gamma \backslash \mathcal{H}$ of the Poincaré upper half plane $\mathcal{H}$ by a Fuchsian group Γ, namely, a subgroup of $\mathrm{PSL}(2, \mathbb{R})$ whose elements $\gamma \neq I$ are hyperbolic so that they satisfy the inequality $\mathrm{tr}(\gamma) > 2$. The heat-kernel $\mathcal{K}^t_\Sigma(z, z')$ on Σ may be obtained from the heat-kernel $\mathcal{K}^t_\mathcal{H}(z, z')$ on $\mathcal{H}$ (which was evaluated in Exercise 6.5) by using the Poincaré series with respect to the Fuchsian group Γ,

$$\mathcal{K}_\Sigma^t(z, z') = \sum_{\gamma \in \Gamma} \mathcal{K}_\mathcal{H}^t(z, \gamma z') \qquad (12.13.1)$$

For a compact Riemann surface Σ, the Poincaré series is absolutely convergent, and the operator $\mathcal{K}_\Sigma^t$ admits a finite trace given by,

$$\text{Tr}(\mathcal{K}_\Sigma^t) = \int_{\Gamma \backslash \mathcal{H}} d\mu(z)\, \mathcal{K}_\Sigma^t(z, z) \qquad d\mu(z) = \frac{d^2 z}{(\text{Im } z)^2} \qquad (12.13.2)$$

The Selberg trace formula gives an explicit expression for this trace by parametrizing the elements of Γ via the conjugacy classes $\{\gamma\}$ defined by,

$$\{\gamma\} = \{\sigma^{-1}\gamma\sigma \text{ for } \sigma \in \Gamma\} \qquad \gamma \in \Gamma \qquad (12.13.3)$$

The centralizer Γ_γ of an element $\gamma \in \Gamma$ is a cyclic group defined by,

$$\Gamma_\gamma = \{\delta \in \Gamma \text{ such that } \gamma\delta = \delta\gamma\} \qquad (12.13.4)$$

Denoting the *primitive element* of Γ_γ by β, we have $\Gamma_\gamma = \{\beta^p, \, p \in \mathbb{Z}\}$. The Selberg trace formula is given by the following theorem.

Theorem 12.13.1 (Selberg Trace Formula). *The quantity* $\text{Tr}(\mathcal{K}_\Sigma^t)$ *can be written as a sum over the primitive elements β of Γ as follows,*

$$\text{Tr}(\mathcal{K}_\Sigma^t) = \frac{e^{-t/4}}{4\sqrt{\pi t}} \sum_{\beta \text{ primitive}} \sum_{p=1}^{\infty} \frac{\ell(\beta)}{\sinh(\frac{1}{2}p\ell(\beta))} e^{-p^2 \ell(\beta)^2/4t} \qquad (12.13.5)$$

where $2\cosh\left(\ell(\beta)/2\right) = \text{tr}(\beta)$. *A trivial additional term proportional to* $\mathcal{K}_\mathcal{H}^t(z, z)$, *which only depends on* area(Σ), *is not exhibited above.*

The proof of Theorem 12.13.1 is relegated to Exercise 12.4. The functional determinant of the operator $\mathcal{O}_s = -\Delta + s(s-1)$ for $s \in \mathbb{C}$ is evaluated in terms of the Selberg zeta function in Exercise 12.5.

• Bibliographical notes

The books by Green, Schwarz, and Witten [GSW12a, GSW12b], Polchinski [Pol07a, Pol07b], and Blumenhagen, Lüst, and Theisen [BLT13] present comprehensive perspectives on string theory. A pedagogical introduction, aimed at undergraduate and graduate students, is given in the book by Zwiebach [Zwi04]. Introductions to D-branes dualities, and compactifications are in the books by Johnson [Joh06], Becker, Becker, and Schwarz [BBS06], and Kiritsis [Kir19]. The Polyakov formulation of string theory, along with many other key topics in modern theoretical physics, is presented in the book by Polyakov [Pol87]. A detailed introduction to the connection between automorphic forms and string amplitudes can be found in the book by Fleig, Gustafsson, Kleinschmidt, and Persson [FGKP18].

There are many standard textbooks on quantum field theory, where regularization and renormalization are discussed in detail, including the books by Zinn-Justin [ZJ89], Weinberg [Wei95, Wei96], and Srednicki [Sre07].

The early development of string theory was dominated by the study of string amplitudes. The Ramond and Neveu–Schwarz sectors of the RNS formulation were introduced in [Ram71] and [NS71], respectively. The different spin structure sectors and the Gliozzi–Scherk–Olive (GSO) projection onto supersymmetric string spectra were introduced in [GSO77] and formulated more generally in [SW86]. Friedan's renormalization of the non-linear σ-model and the relation between Einstein's equations and conformal invariance may be found in [Fri85]. Generalizations with worldsheet fermions are given in [AGFM81], including torsion in [BCZ85], and the dilaton in [Osb91]. The α' expansion of the effective action including the metric, antisymmetric tensor field, and dilaton is presented in detail in the book of Princeton Institute for Advanced Study lecture notes [D'H99].

The standard treatment of the decoupling of negative norm states for both the bosonic and superstrings is reviewed in [GSW12a]. The modern BRST-based approach to the covariant quantization of superstrings in the RNS formulation was developed in [FMS86], and reviewed in [GSW12a, GSW12b] and [Pol07a, Pol07b]. A perspective from the point of view of partial wave decompositions was advanced recently in [AHEHM22]. The structure of gauge and gravitational anomalies, relevant to supergravity and string theories in various dimensions, was obtained in [AGW84]. The cancellation of anomalies in the Type I theory with gauge algebra $\mathfrak{so}(32)$ was famously discovered in [GS84]. Useful lecture notes on anomalies may be found in [Har05].

Reviews of superstring perturbation theory may be found in [DP88] and [DP02c] for the RNS formulation and in [Ber00] and [Ber04] for the pure spinor formulation. Lectures accessible to a more mathematically oriented readership are in [D'H99], while an elaboration of the mathematical underpinnings of the RNS formulation is provided in [Wit19], [Wit12], and [SW15]. A broad overview of the current status of superstring perturbation theory, and open problems, is given in [BDG$^+$22], which contains an extensive bibliography.

The tree-level and one-loop four graviton amplitudes were derived in the original papers in which the Type II [GS82] and Heterotic strings [GHMR85, GHMR86] were discovered. The existence of the analytic continuation of the one-loop four-graviton amplitude was shown in [DP93] and [DP95]. Analytic continuation of the amplitudes by contour deformation was developed in [Ber94] and [Wit15a], while an explicit evaluation of the one-loop open superstring amplitude with four bosons was obtained using Rademacher sums in [EM23].

The fact that the low-energy expansion of string theory leads to $\mathcal{N} = 4$ super Yang–Mills theory and $\mathcal{N} = 8$ supergravity in four space-time dimensions was discovered in [GSB82]. The systematic analysis of the low-energy effective interactions induced by the tree-level four-graviton amplitude was initiated in [GW86] and [GvdVZ86] and extended to the Heterotic string in [GS87], while the effects at genus one were obtained in [GGV97], [GV00], and [GRV08]. The low-energy expansion, up to order $D^8\mathcal{R}^4$, including the logarithmic branch cuts and the transcendentality properties of the amplitude discussed in Section 12.8.2 were obtained in [DG19].

The original Rankin–Selberg method for integrating cusp forms over the fundamental domain for $SL(2, \mathbb{Z})$ was developed in [Ran39] and [Sel40], and was extended to automorphic functions of nonrapid decay in [Zag82]. It was applied to string amplitudes in [FP17] building on the earlier works [AFP12a, AFP12b]. The proof of the last integral identity in Lemma 12.11.1 is in [Zag82], while methods to carry out integrals of modular graph functions may be found in [DG22] and references therein.

Considerable progress has been made recently toward evaluating tree-level and genus-one string amplitudes with arbitrary numbers of external states in [MSS13], [ACJS18], [MS19a], [MS19b], [MS19c], and [MS23] and references therein. The role of motivic and multiple zeta values played in the form of the amplitudes was examined in [BSS13], [Sti14], and [SS13]. Genus-one amplitudes have also recently been reproduced from $\mathcal{N} = 4$ supersymmetric Yang–Mills theory by considering a flat-space limit of $AdS_5 \times S^5$ [ABP19].

There is an intimate and fruitful relation between string amplitudes and quantum field theory amplitudes which is, unfortunately, beyond the scope of this book. For comprehensive presentations of modern methods in the study of quantum field theory amplitudes, we refer to the books by Elvang and Huang [EH15] and by Arkani-Hamed, Bourjaily, Cachazo, Goncharov, Postnikov, and Trnka [AHBC+16], as well as to the recent SAGEX summaries [T+22, BCC+22], and references therein.

For genus-two amplitudes, the measure on supermoduli space was derived in the RNS formulation in [DP02a] and [DP02b] and reproduced using algebraic-geometric methods in [Wit13]. The full amplitude for four gravitons was obtained in [DP05] in the RNS formulation, and extended to include external fermion states in [Ber06] and [BM06] using the pure spinor formulation. The overall normalization of the amplitude was obtained in [DGP05] and [GM10]. The genus-two amplitudes with five massless states were constructed in [DMPS20] and [DMPS21] using an amalgam of methods from the RNS and pure spinor formulations. For external NS states and even spin structure, these results were confirmed from first principles in the RNS formulation in [DS21].

The generalization of modular graph functions to genus two and higher genus was developed in [DGP19b], while the relation between the coefficient of the $D^6\mathcal{R}^4$ term and the invariant of Kawazumi [Kaw08] and Zhang [Zha08] was identified in [DG14] and used to derive the corresponding differential equation in [DGPR15]. For more recent developments, see also [Kaw16]. Degenerations of genus-two modular forms naturally produce non-holomorphic versions of Jacobi forms, the systematic treatment of which may be found in the book by Eichler and Zagier [EZ85]. A ϑ-lift representation of the genus-two Kawazumi–Zhang invariant was obtained in [Pio16]. For the derivation of the results in Section 12.12, we refer to [DGPR15].

For amplitudes at genus three and beyond, no first-principle derivations are available at the time of this writing. A proposal for the measure on supermoduli space in the RNS formulation was advanced in [CDPvG08b]. The proposal of [CDPvG08b] was critiqued, however, in the appendix of [Wit15b]. The leading low-energy effective interaction in Type II string theory was obtained using the pure spinor formulation in [GM13]. An explicit expression for the genus-three four-graviton amplitude has been conjectured in [GMSMa21]. Proposals for the measure on supermoduli space in the RNS formulation at genus four may be found in [CDPvG08a] and [Gru09].

Finally, the Selberg trace formula was formulated in [Sel56] and is reviewed in the books by Gel'fand, Graev, and Pyatetskii-Shapiro [GGPS69], Iwaniec [Iwa02], Lang [Lan75], and in [McK72]. Many of its applications are discussed, for example, in [Zag81a] and [Hej83]. For the calculation of functional determinants using the heat-kernel in terms of the Selberg Zeta function discussed here, we refer to [Fri86], [DP86b], [Sar87], and [JL93]. Its application to the bosonic string may be found in [DP86a], whose results were used to estimate the behavior of string amplitudes to large orders in the string coupling in [GP88]. A path integral derivation of the Selberg trace formula in supersymmetric quantum mechanics can be found in [CT21] and [CT23].

Exercises

12.1 To prove the first two integrals in Lemma 12.11.1, evaluate the following integrals of non-holomorphic Eisenstein series,

$$G_1(s; L) = \int_{F_L} \frac{d^2\tau}{\tau_2^2} E_s^*(\tau)$$

$$G_2(s_1, s_2; L) = \int_{F_L} \frac{d^2\tau}{\tau_2^2} E_{s_1}^*(\tau) E_{s_2}^*(\tau) \qquad (12.13.6)$$

where $F_L = F \cap \{\tau_2 < L\}$ for $L > 1$. Evaluate the limit $G_2(s, s; L)$.

12.2 Use the result of Exercise 9.1 to show that $\Delta_n C_{a,b,c}$ for $n = a + b + c$ is a cusp form. Using the equation $(\Delta - 2)C_{2,1,1} = 9E_4 - E_2^2$ of (9.4.8), calculate the function $f(s)$ defined by the integral,

$$f(s) = \int_F \frac{d^2\tau}{\tau_2^2} E_s(\tau) \Delta_4 C_{2,1,1}(\tau) \qquad (12.13.7)$$

12.3 The weight 6 portion of the partial amplitude $\mathcal{B}^{(1)}(s_{ij}|\tau)$ in (12.9.2) has two kinematic contributions, $\mathcal{B}_{(3,0)}(\tau)$ and $\mathcal{B}_{(0,2)}(\tau)$. By expanding (12.10.2), obtain the explicit form of $\mathcal{B}_{(3,0)}(\tau)$ in terms of modular graph functions. Then make use of the integrals in Exercises 12.1 and 12.2 to compute its contribution to the analytic part of the amplitude $\mathcal{A}_L^{(1)}(s_{ij})$.

12.4 The purpose of this exercise is to prove Theorem 12.13.1 for the Selberg trace formula using the results of Exercises 3.1 and 6.5.

(a) Using the fact that $\mathcal{K}_{\mathcal{H}}^t(z, z')$ is a function only of the hyperbolic distance $\ell(z, z')$, introduced in Exercise 3.1, show that the trace may be decomposed into a sum over primitive elements β as follows,

$$\text{Tr}(\mathcal{K}_{\Sigma}^t) = \sum_{\beta \text{ primitive}} \sum_{p=1}^{\infty} \int_{\mathbb{R}} dx \int_1^{e^{\ell(\beta)}} \frac{dy}{y^2} \mathcal{K}_{\mathcal{H}}^t\left(z, e^{p\ell(\beta)} z\right) \qquad (12.13.8)$$

where $2\cosh\left(\frac{1}{2}\ell(\beta)\right) = \text{tr}(\beta)$, and we have dropped a trivial term, proportional to $\mathcal{K}_{\mathcal{H}}^t(z, z)$, which only depends on the area of Σ.

(b) Use the expression for the heat-kernel $\mathcal{K}_{\mathcal{H}}^t(z, z')$ on $\mathcal{H}$, obtained in Exercise 6.5, to complete the proof of Theorem 12.13.1.

12.5 Using the results of Exercise 12.4, compute the functional determinant of the operator $-\Delta + s(s-1)$ in terms of the Selberg zeta function, which is defined by,

$$Z(s) = \prod_{\beta \text{ primitive}} \prod_{k=0}^{\infty} \left(1 - e^{(s+k)\ell(\beta)}\right) \qquad (12.13.9)$$

for $\text{Re}(s) > 1$ and may be analytically continued in s throughout $\mathbb{C}$.

13

Toroidal compactification

In this chapter, we discuss the modular properties of quantum field theory of scalar fields that take values in a d-dimensional torus with a flat metric and constant antisymmetric tensor. This problem is of great interest in quantum field theory and string theory in view of the fact that such *toroidal compact-ifications* admit solutions using free-field theory methods on the worldsheet, preserve Poincaré supersymmetries, and may be used to relate different perturbative string theories via *T-duality*. Toroidal compactifications produce large duality groups, to which we have already alluded in the Introduction, that generalize the modular group $\mathrm{SL}(2,\mathbb{Z})$ considered thus far.

13.1 Conformal field theory on flat manifolds

In Chapter 5, we considered the functional integral for a single free scalar field φ that takes values in $\mathbb{R}$. The situation is readily generalized to the case of d free scalar fields taking values in a flat Euclidean $\mathbb{R}^d$. As discussed in Chapter 12, the generalization to d scalar fields taking values in (the local coordinate systems of) an arbitrary space-time manifold in general leads to fully interacting quantum field theories on the worldsheet that cannot be solved exactly.

There is one class of theories, however, where an exact quantization may be obtained. This is the case when the metric G, the dilaton Φ, and the anti-symmetric field B, introduced in (12.4.12), are constant, and the manifold M is flat. Although flat, M may have nontrivial topology as in the case of a d-dimensional torus $M = \mathbb{T}^d$, which is the case we shall consider here. Constant Φ in the worldsheet action $I_{G,\Phi,B}[X,\mathfrak{g}]$ of (12.4.12) gives rise to a term proportional to the *Euler number* (also referred to as the *Euler characteristic*) of Σ and weighs the functional integral by a power of the string coupling which is independent of the quantum field X. The term involving B is topological when B is constant. It does not contribute when the topology of M is trivial, but when M is a torus, it does have interesting physical effects that we shall now examine. For constant G, B, Φ, the X-dependent part of the action $I_{G,\Phi,B}[X, \mathfrak{g}]$ of (12.4.12) thus reduces to,

$$I_{G,\Phi,B}[X, \mathfrak{g}] = \frac{1}{4\pi\alpha'} \int_\Sigma d\mu_\mathfrak{g} \left(\mathfrak{g}^{mn} G_{\mu\nu} - i\,\varepsilon^{mn} B_{\mu\nu} \right) \partial_m X^\mu \partial_n X^\nu \quad (13.1.1)$$

Henceforth, we shall restrict to the case where the worldsheet is a torus of modulus τ, so that $\Sigma = \mathbb{C}/\Lambda$ with $\Lambda = \mathbb{Z} + \mathbb{Z}\tau$. We shall choose a system of local complex coordinates $z, \bar{z}$ on Σ in which the metric $\mathfrak{g}$ takes the form,

$$\mathfrak{g}_{mn} d\xi^m d\xi^n = \frac{|dz|^2}{\tau_2} \quad (13.1.2)$$

or in components we have $\mathfrak{g}_{zz} = \mathfrak{g}_{\bar{z}\bar{z}} = 0$ and $\mathfrak{g}_{z\bar{z}} = 1/(2\tau_2)$. Furthermore, the space-time torus $\mathbb{T}^d$ is given by $\mathbb{R}^d/L$ for some d-dimensional lattice L.

13.2 Lattices and tori of dimension d

A lattice L of rank d, also referred to as a lattice of dimension d, may be represented in terms of a basis of d vectors $v_1, \cdots, v_d$ in $\mathbb{R}^d$,

$$L = \left\{ \ell = \sum_{i=1}^d n_i v_i \quad n_i \in \mathbb{Z} \right\} \quad (13.2.1)$$

It will be useful to view the assignment of basis vectors as obtained via a $\mathrm{GL}(d, \mathbb{R})$ transformation applied to a standard basis of d vectors, such as, $v_1^0 = (1, 0, \cdots, 0)$, $v_2^0 = (0, 1, 0, \cdots, 0)$, and $v_n^0 = (0, \cdots, 0, 1)$. Decomposing $\mathrm{GL}(d, \mathbb{R}) = \mathrm{SL}(d, \mathbb{R}) \times \mathbb{R}^+$, the factor $\mathbb{R}^+$ corresponds to overall rescaling of the lattice. An arbitrary set of d linearly independent vectors $v_i' \in L$ for $i = 1, \cdots, d$ may be decomposed in the basis v_i,

$$v_i' = \sum_{j=1}^d M_{ij} v_j \qquad\qquad M_{ij} \in \mathbb{Z} \quad (13.2.2)$$

where $\det M \neq 0$ and generates a sublattice $L' \subset L$ of rank d. Lattices may be viewed as Abelian groups so that L' is a subgroup of L. The index of L' as a subgroup of L is given as follows,

$$[L : L'] = |\det M| \quad (13.2.3)$$

Equivalently, the unit lattice cell of L' contains $|\det M|$ copies of the unit lattice cell of L. In the special case where $\det M = 1$, we have $L' = L$ and therefore M is an automorphism of L. The group of all automorphisms of

L is $SL(d, \mathbb{Z})$, thereby generalizing to d dimensions the result familiar from the two-dimensional torus. When $\det M = -1$, the lattice L' is the mirror image of L, reversing the orientation of the lattice.

Introducing the equivalence relation between two lattices that are related to one another by an overall rescaling in $\mathbb{R}^+$, an overall rotation in $SO(d)$, and a lattice automorphism in $SL(d, \mathbb{Z})$, we may identify the space of all inequivalent lattices as follows,

$$\text{Space of inequivalent lattices} = SL(d, \mathbb{Z})\backslash SL(d, \mathbb{R})/SO(d, \mathbb{R}) \quad (13.2.4)$$

In the case of a two-dimensional lattice L with $d = 2$, we may identify $SL(2, \mathbb{R})/SO(2)$ with the Poincaré upper half plane $\mathcal{H}$ and the entire coset $SL(2, \mathbb{Z})\backslash SL(2, \mathbb{R})/SO(2, \mathbb{R})$ with the moduli space of toroidal Riemann surfaces. The dynamics of point particles propagating on the torus $\mathbb{T}^d = \mathbb{R}^d/L$ is sensitive to the lattice geometry of this quotient. We shall see below that the dynamics of strings propagating on the torus $\mathbb{R}^d/L$ is sensitive to a different equivalence class of lattices, which is "smaller" than $SL(d, \mathbb{Z})\backslash SL(d, \mathbb{R})$ thanks to the quintessentially string-theoretic phenomenon of T-duality.

13.3 Fields taking values in a torus

We begin by reviewing the construction of the space of d scalar fields $X(z)$ that take values in the torus $\mathbb{R}^d/L$ and are functions on a worldsheet Σ with the topology of a two-dimensional torus. The complex structure of $\Sigma = \mathbb{C}/\Lambda$ with $\Lambda = \mathbb{Z} \oplus \mathbb{Z}\tau$ is given by the modulus $\tau \in \mathcal{H}$. The space of maps,

$$X : \Sigma = \mathbb{C}/\Lambda \rightarrow \mathbb{T}^d = \mathbb{R}^d/L \quad (13.3.1)$$

is disconnected and may be decomposed into the union of an infinite number of connected components that are labeled by two copies of the lattice L. In a sector labeled by a pair $(\ell_A, \ell_B) \in L \times L$, the field X^μ satisfies the following monodromy conditions for $z \in \Sigma$,

$$\begin{aligned} X(z + 1) &= X(z) + \ell_A \\ X(z + \tau) &= X(z) + \ell_B \end{aligned} \qquad \ell_A, \ell_B \in L \qquad (13.3.2)$$

The field X may be decomposed into a purely periodic part Y satisfying,

$$Y(z + 1) = Y(z + \tau) = Y(z) \quad (13.3.3)$$

and a special solution to the monodromy conditions. The latter is arbitrary, but it will be convenient to choose it linear in z and $\bar{z}$. Since the field X and the vectors ℓ_A, ℓ_B are real valued, we have the following decomposition,

$$X_{\ell_A,\ell_B}(z) = Y(z) + z\,\frac{\ell_B - \bar\tau\ell_A}{\tau - \bar\tau} - \bar z\,\frac{\ell_B - \tau\ell_A}{\tau - \bar\tau} \tag{13.3.4}$$

Such a configuration is sometimes referred to as a *worldsheet instanton*. The lattice elements ℓ_A and ℓ_B are related to the momentum and the winding modes of strings on $\mathbb{T}^d$, as we shall clarify below. The integration over all $X : \Sigma \to \mathbb{T}^d$ decomposes into a sum over sectors labeled by $L \times L$. The functional integral takes the following form,

$$\int_{\mathrm{Map}(\Sigma\to\mathbb{T}^d)} DX\, e^{-I[X,\mathfrak{g}]}$$

$$= \mathrm{Vol}(\mathbb{T}^d) \sum_{\ell_A,\ell_B\in L} \int_{\mathrm{Map}(\Sigma\to\mathbb{R}^d)} D'Y\, e^{-I[X_{\ell_A,\ell_B},\mathfrak{g}]} \tag{13.3.5}$$

For given worldsheet metric $\mathfrak{g}$, the measures DX and $D'Y$ are evaluated using the norm of (12.2.2). The factor $\mathrm{Vol}(\mathbb{T}^d)$ arises from the constant normalizable zero mode of the field Y, and the integration $D'Y$ is over all nonzero modes, that is, all functions Y orthogonal to constants. The contribution of the momentum and winding modes may be evaluated explicitly,[1]

$$I[X_{\ell_A,\ell_B},\mathfrak{g}] = I[Y,\mathfrak{g}] + \frac{G_{\mu\nu} + B_{\mu\nu}}{4\pi\alpha'\tau_2}(\ell_B^\mu - \bar\tau\ell_A^\mu)(\ell_B^\nu - \tau\ell_A^\nu) \tag{13.3.6}$$

The integral over Y factorizes, the dependence on $B_{\mu\nu}$ cancels out in this integral and reduces to the functional integrals computed earlier in Chapter 5. The remaining factor is given by the following lattice sum,

$$\mathrm{Vol}(\mathbb{T}^d) \sum_{\ell_A,\ell_B\in L} \exp\left\{-\frac{G_{\mu\nu} + B_{\mu\nu}}{4\pi\alpha'\tau_2}(\ell_B^\mu - \bar\tau\ell_A^\mu)(\ell_B^\nu - \tau\ell_A^\nu)\right\} \tag{13.3.7}$$

The dependence on $B_{\mu\nu}$ reduces to $-iB_{\mu\nu}\,\ell_A^\mu\,\ell_B^\nu/(2\pi\alpha')$ and is independent of τ, as is expected for a topological term. Before investigating the properties for a general torus $\mathbb{T}^d$, we first look at the much simpler case of $d = 1$.

13.4 T-duality on a circle

In this section, we study the simplest case of toroidal compactification when $d = 1$ and exhibit the phenomenon of T-duality for a circle. We take $G = 1$ and introduce the lattice L associated with a circle of radius R, namely,

---

[1] In this chapter, it will be illuminating to render the dependence on α' explicit again.

$L = 2\pi R\, \mathbb{Z}$. We label the monodromy vectors ℓ_A and ℓ_B, which are one dimensional in this case, as follows,

$$\ell_A = 2\pi R\, n$$
$$\ell_B = 2\pi R\, m \qquad\qquad m, n \in \mathbb{Z} \qquad (13.4.1)$$

The partition function then simplifies to,

$$Z(R|\tau) = \frac{\tau_2^{-1/2}}{|\eta(\tau)|^2} \sqrt{\frac{R^2}{\alpha'}} \sum_{m,n \in \mathbb{Z}} \exp\left\{ -\frac{\pi R^2}{\alpha' \tau_2} |m - \tau n|^2 \right\} \qquad (13.4.2)$$

The prefactor in $\sqrt{R^2/\alpha'}$ arises from the zero mode of the field Y which, for a circle of radius R, gives a finite contribution. The above sum $Z(R|\tau)$ is a rather familiar looking object. To exhibit T-duality, we perform a Poisson resummation on both m and n. To do so, we compute the Fourier transform in both variables,

$$\int_{\mathbb{R}^2} dm\, dn\, e^{-2\pi i(mx+ny)} \exp\left\{ -\frac{\pi R^2}{\alpha' \tau_2} |m - \tau n|^2 \right\}$$
$$= \frac{\alpha'}{R^2} \exp\left\{ -\frac{\pi \alpha'}{R^2 \tau_2} |y + \tau x|^2 \right\} \qquad (13.4.3)$$

so that Poisson resummation gives us,

$$Z(R|\tau) = \frac{\tau_2^{-1/2}}{|\eta(\tau)|^2} \sqrt{\frac{\alpha'}{R^2}} \sum_{m,n \in \mathbb{Z}} \exp\left\{ -\frac{\pi \alpha'}{R^2 \tau_2} |m + \tau n|^2 \right\} \qquad (13.4.4)$$

Noticing that Poisson resummation has effectively inverted the dependence on R^2/α', we have,

$$Z(R|\tau) = Z\left(\frac{\alpha'}{R}\Big|\tau\right) \qquad (13.4.5)$$

Thus, a free bosonic string on a circle of radius R has exactly the same partition function as a free bosonic string on a circle of radius α'/R. While we have proven this above for the free string on a genus-one worldsheet only, the effect actually is valid on surfaces of arbitrary genus and thus holds for fully interacting string theory. The effect is referred to as *T-duality*.

The double sum may be organized in a different way that makes the decomposition into momentum and winding modes transparent. Performing a Poisson resummation on (13.4.2) in m only, we obtain the representation,

$$Z(R|\tau) = \frac{1}{|\eta(\tau)|^2} \sum_{m,n \in \mathbb{Z}} e^{2\pi i \tau p_L^2}\, e^{-2\pi i \bar{\tau} p_R^2} \qquad (13.4.6)$$

where the left and right momenta are given in terms of the integers m and n by,

$$p_L = \frac{1}{2}\left(\frac{\sqrt{\alpha'}}{R}m + \frac{R}{\sqrt{\alpha'}}n\right) \qquad p_L + p_R = \frac{\sqrt{\alpha'}}{R}m$$

$$p_R = \frac{1}{2}\left(\frac{\sqrt{\alpha'}}{R}m - \frac{R}{\sqrt{\alpha'}}n\right) \qquad p_L - p_R = \frac{R}{\sqrt{\alpha'}}n \qquad (13.4.7)$$

The lattice points $p_L + p_R \in \sqrt{\alpha'}/R\,\mathbb{Z}$ are clearly associated with the total momentum of the string, while the lattice points $p_L - p_R \in R/\sqrt{\alpha'}\,\mathbb{Z}$ are associated with the winding modes.

The space of all possible compactifications on the circle is labeled by the radius R modulo the identification $R \equiv \alpha'/R$, a space we may denote by $\mathbb{R}^+/\mathbb{Z}_2$. Clearly, the lattices $L = \sqrt{\alpha'}/R\,\mathbb{Z}$ and $L^+ = R/\sqrt{\alpha'}\,\mathbb{Z}$ are dual to one another, a fact that is reflected in the relation $(p_L + p_R)(p_L - p_R) = mn \in \mathbb{Z}$. Thus, the lattice sum may be written alternatively as,

$$Z(R|\tau) = \frac{1}{|\eta(\tau)|^2} \sum_{p \in L}\sum_{w \in L^+} e^{\pi i \tau (p+w)^2/2}\, e^{-\pi i \bar{\tau}(p-w)^2/2} \qquad (13.4.8)$$

At the radius $R = \sqrt{\alpha'}$, the conformal field theory is self-dual and enjoys an enhanced symmetry to the SU(2) Kac–Moody algebra. The radius $R = \sqrt{2\alpha'}$ is also special in a different sense: It is at this point that the $\mathfrak{c} = 1$ Virasoro algebra representation of the conformal field theory becomes reducible to two $\mathfrak{c} = \frac{1}{2}$ fermions.

The presence of T-duality has several important consequences for string theory, which include the following.

1. The fact that (closed bosonic) string theory on a circle of radius R is physically equivalent to a string theory on a circle of radius α'/R means that one can never probe distances smaller than $\sqrt{\alpha'}$ with perturbative strings. Analogous arguments exist for the energy of physical scattering processes, namely, probing a string with energy E and with energy $1/(\alpha'E)$ are physically equivalent. This means that perturbative string theory has built in its dynamics a smallest length scale $\sqrt{\alpha'}$ (though nonperturbatively there is structure at shorter distances).

2. Actually, the above picture is modified in the case of superstring theories. Compactification of Type IIA string theory on a circle of radius R is physically equivalent to Type IIB string theory on a circle of radius α'/R. The Heterotic $E_8 \times E_8$ string theory is invariant under $R \leftrightarrow \alpha'/R$, while the Heterotic Spin(32)/$\mathbb{Z}_2$ string theory is exchanged with the Type I

theory of open and closed strings. Thus, T-duality provides a mechanism for a partial unification of the five 10-dimensional superstring theories.

3. T-duality extends to string theories on curved manifolds under certain conditions. For example, on Calabi–Yau spaces, T-duality is promoted to *mirror symmetry*.

13.5 T-duality on a torus $\mathbb{T}^d$

For a general torus $\mathbb{T}^d = \mathbb{R}^d/L$ with $d \geq 2$ new effects arise. We consider the lattice sum of (13.3.7) and introduce a basis $\{v_1^\mu, \cdots, v_d^\mu\}$ with $\mu = 1, \cdots, d$ for the lattice L. Generalizing the conventions of (13.4.1) used for the circle, the vectors ℓ_A^μ and ℓ_B^μ may be parametrized as follows,

$$\ell_A^\mu = 2\pi \sum_{i=1}^{d} n^i\, v_i^\mu \qquad\qquad n^i \in \mathbb{Z}$$

$$\ell_B^\mu = 2\pi \sum_{i=1}^{d} m^i\, v_i^\mu \qquad\qquad m^i \in \mathbb{Z} \qquad (13.5.1)$$

What enters $Z(L|\tau)$ is the combination,

$$\frac{1}{\alpha'}(G_{\mu\nu} + B_{\mu\nu})v_i^\mu v_j^\nu = \mathcal{G}_{ij} + \mathcal{B}_{ij} \qquad (13.5.2)$$

where $\mathcal{G}_{ij}$ and $\mathcal{B}_{ij}$ are, respectively, symmetric and antisymmetric under the interchange of the indices i and j. Using matrix notation for $m, n, \mathcal{G}$, and $\mathcal{B}$, we have,

$$Z(L|\tau) = \frac{\tau_2^{-d/2}}{|\eta(\tau)|^{2d}}\mathrm{Vol}(\mathbb{T}^d) \qquad (13.5.3)$$

$$\times \sum_{m,n\in\mathbb{Z}^d} \exp\left\{-\frac{\pi}{\tau_2}(m+\tau n)^\dagger(\mathcal{G}+\mathcal{B})(m+\tau n)\right\}$$

Since v_i forms a basis of L, the volume of the torus is given by,

$$\mathrm{Vol}(\mathbb{T}^d) = (\det \mathcal{G})^{\frac{1}{2}} \qquad (13.5.4)$$

up to some factor of 2π which will be of no interest here.

13.5.1 A first look at T-duality

The argument of the exponential may be decomposed as follows,

$$-\frac{\pi}{\tau_2}(m+\tau n)^\dagger(\mathcal{G}+\mathcal{B})(m+\tau n) \qquad (13.5.5)$$

$$= -\frac{\pi}{\tau_2}(m+\tau_1 n)^t\mathcal{G}(m+\tau_1 n) - \pi\tau_2 n^t\mathcal{G}n - 2\pi i m^t\mathcal{B}n$$

As a result, each term in the exponential is invariant under a shift in $\mathcal{B}$ by an arbitrary antisymmetric matrix with integer entries, or in components,

$$\mathcal{B}_{ij} \to \mathcal{B}_{ij} + b_{ij} \qquad\qquad b_{ij} = -b_{ji} \in \mathbb{Z} \qquad (13.5.6)$$

This invariance is an immediate consequence of the fact that, for constant B, the B-term is topological and its values on the field configurations for the torus $\mathbb{T}^d$ are quantized. The corresponding symmetry is the analogue of the T-transformation $\tau_1 \to \tau_1 + 1$ for $\mathrm{SL}(2,\mathbb{Z})$ and will provide some of the generators of the full T-duality group for $\mathbb{T}^d$.

Next, we wish to obtain the analogue of the S-transformation. To do so, we Poisson resum in both m and n. It is helpful to express the quadratic form combining m and n into a single column matrix, so that,

$$-\frac{\pi}{\tau_2}(m + \tau n)^\dagger (\mathcal{G} + \mathcal{B})(m + \tau n) = -\pi \begin{pmatrix} m \\ n \end{pmatrix}^t \mathcal{M} \begin{pmatrix} m \\ n \end{pmatrix} \qquad (13.5.7)$$

where $\mathcal{M}$ is given by,

$$\mathcal{M} = \frac{1}{\tau_2} \begin{pmatrix} \mathcal{G} & \tau_1 \mathcal{G} + i\tau_2 \mathcal{B} \\ \tau_1 \mathcal{G} - i\tau_2 \mathcal{B} & (\tau_1^2 + \tau_2^2)\mathcal{G} \end{pmatrix} \qquad (13.5.8)$$

The inverse matrix may be expressed as follows,

$$\mathcal{M}^{-1} = \frac{1}{\tau_2} \begin{pmatrix} 0 & I \\ -I & 0 \end{pmatrix}^t \begin{pmatrix} \tilde{\mathcal{G}} & \tau_1 \tilde{\mathcal{G}} + i\tau_2 \tilde{\mathcal{B}} \\ \tau_1 \tilde{\mathcal{G}} - i\tau_2 \tilde{\mathcal{B}} & (\tau_1^2 + \tau_2^2)\tilde{\mathcal{G}} \end{pmatrix} \begin{pmatrix} 0 & I \\ -I & 0 \end{pmatrix}$$

$$(13.5.9)$$

where $\tilde{\mathcal{G}}$ and $\tilde{\mathcal{B}}$ are given by the S-transformation,

$$\tilde{\mathcal{G}} = \left(\mathcal{G} + \mathcal{B}^t \mathcal{G}^{-1} \mathcal{B}\right)^{-1} \qquad\qquad \tilde{\mathcal{B}} = -\mathcal{G}^{-1}\mathcal{B}\tilde{\mathcal{G}} \qquad (13.5.10)$$

The action of S^2 is obtained by repeating the transformation,

$$\hat{\mathcal{G}} = \left(\tilde{\mathcal{G}} + \tilde{\mathcal{B}}^t \tilde{\mathcal{G}}^{-1} \tilde{\mathcal{B}}\right)^{-1} \qquad\qquad \hat{\mathcal{B}} = -\tilde{\mathcal{G}}^{-1}\tilde{\mathcal{B}}\hat{\mathcal{G}} \qquad (13.5.11)$$

and eliminating $\tilde{\mathcal{B}}$ and $\tilde{\mathcal{G}}$. The result is $\hat{\mathcal{B}} = \mathcal{B}$ and $\hat{\mathcal{G}} = \mathcal{G}$, so that the action of S^2 on $\mathcal{G}, \mathcal{B}$ is the identity.[2]

---

[2] Although in $\mathrm{SL}(2,\mathbb{Z})$ the matrix S squares to $S^2 = -I$, its action on τ reduces to the identity; the above statement is the direct analogue thereof.

13.5.2 Holomorphic block decomposition

As in the case of the scalar field theory valued in the circle, it is instructive to perform a Poisson resummation only on the integers m and expose the holomorphic structure in τ. To do so, we start from the decomposition of the quadratic form given earlier, and we need the following Fourier transform,

$$\int_{\mathbb{R}^d} d^d x \, e^{-2\pi i x^t m} \exp\left\{ -\frac{\pi}{\tau_2}(x + \tau_1 n)^t \mathcal{G}(x + \tau_1 n) - 2\pi i x^t \mathcal{B} n \right\}$$

$$= \frac{(\tau_2)^{\frac{d}{2}}}{(\det \mathcal{G})^{\frac{1}{2}}} \exp\left\{ 2\pi i \tau p_L^2 - 2\pi i \bar{\tau} p_R^2 \right\} \tag{13.5.12}$$

where we obtain after minimal simplifications,

$$4p_L^2 = (m + \mathcal{B}n)^t \mathcal{G}^{-1}(m + \mathcal{B}n) + n^t \mathcal{G}n + 2m^t n$$
$$4p_R^2 = (m + \mathcal{B}n)^t \mathcal{G}^{-1}(m + \mathcal{B}n) + n^t \mathcal{G}n - 2m^t n \tag{13.5.13}$$

Note that while the integers n were defined with an upper index, as were the old m, by contrast the Poisson resummation variable m have lower indices. So, it is instructive to write out the above relations in terms of indices,

$$4p_L^2 = \left(m_i + \mathcal{B}_{ii'}n^{i'}\right)\mathcal{G}^{ij}\left(m_j + \mathcal{B}_{jj'}n^{j'}\right) + n^i \mathcal{G}_{ij}n^j + 2m_i n^i$$
$$4p_R^2 = \left(m_i + \mathcal{B}_{ii'}n^{i'}\right)\mathcal{G}^{ij}\left(m_j + \mathcal{B}_{jj'}n^{j'}\right) + n^i \mathcal{G}_{ij}n^j - 2m_i n^i \tag{13.5.14}$$

where $\mathcal{G}^{ij}$ are the components of $\mathcal{G}^{-1}$. In evaluating the partition function (13.3.7), the volume factor $\text{Vol}(\mathbb{T}^d)$ cancels out, and we find,

$$Z(L|\tau) = \frac{1}{|\eta(\tau)|^{2d}} \sum_{m,n \in \mathbb{Z}^d} \exp\left\{ 2\pi i \tau \, p_L^2 - 2\pi i \bar{\tau} \, p_R^2 \right\} \tag{13.5.15}$$

with p_L^2 and p_R^2 functions of m, n as defined in (13.5.13). The invariance under integer shifts of the matrix $\mathcal{B}$ is now manifested by the fact that such a shift may be compensated in the sum by a shift in the integers m.

13.5.3 T-duality in terms of the lattice

The description of the torus with both an arbitrary lattice L and an arbitrary metric G is in fact redundant. By general linear transformation, we may always make the metric equal to the identity. Geometrically, this is equivalent to expressing the metric $\mathcal{G}_{ij}$ in terms of an orthonormal frame $e_i{}^a$ and its inverse $e_a{}^i$ by,

$$\mathcal{G}_{ij} = e_i{}^a e_j{}^b \delta_{ab} \tag{13.5.16}$$

where i are *Einstein indices* and a are *frame indices* and the inverses are related by $e_i{}^a e_a{}^j = \delta_i{}^j$ and $e_a{}^i e_i{}^b = \delta_a{}^b$. The lattice L with identity metric is then generated by the vector $e_a{}^i$, while the dual lattice L^+ is generated by the dual vector $e_i{}^a$. Defining also the B-field in frame indices,

$$\mathcal{B}_{ab} = \mathcal{B}_{ij}\, e_a{}^i e_b{}^j \tag{13.5.17}$$

and defining the lattice and dual lattice vectors,

$$\begin{aligned}
\mu_a &= e_a{}^i\, m_i \\
\nu^a &= n^i\, e_i{}^a
\end{aligned} \tag{13.5.18}$$

we find that p_L^2 and p_R^2 may be interpreted as squares of vectors, given by,

$$\begin{aligned}
p_L &= \frac{1}{2}\left(\mu + \mathcal{B}\,\nu + \nu\right) & \mu &\in L^+ \\
p_R &= \frac{1}{2}\left(\mu + \mathcal{B}\,\nu - \nu\right) & \nu &\in L
\end{aligned} \tag{13.5.19}$$

respectively, with,

$$p_L \cdot p_L - p_R \cdot p_R = \mu \cdot \nu = m^t n \in \mathbb{Z} \tag{13.5.20}$$

To parametrize the space of all lattices, we proceed as follows. A given lattice L may be deformed continuously provided we maintain the condition $p_L \cdot p_L - p_R \cdot p_R \in \mathbb{Z}$ which is invariant under the group $O(d, d; \mathbb{R})$. But rotations under $O(d)$ on both p_L and p_R produce lattices that are physically indistinguishable from the original lattice. Therefore, any lattice and its dual may be parametrized by a point in the coset space,

$$O(d, d; \mathbb{R}) / \big(O(d; \mathbb{R}) \times O(d; \mathbb{R})\big) \tag{13.5.21}$$

The dimension of this space is,

$$\tfrac{1}{2}(2d)(2d - 1) - 2 \times \tfrac{1}{2}d(d - 1) = d^2 \tag{13.5.22}$$

This dimension is easily understood in terms of the original formulation with arbitrary metric G and B-field in a square torus: together G and B have d^2 independent components.

Two different points in $O(d, d; \mathbb{R})/(O(d; \mathbb{R}) \times O(d; \mathbb{R}))$ may correspond to the same lattice. To understand this, it is convenient to represent the elements $\gamma \in O(d, d; \mathbb{R})$ by $2d \times 2d$ matrices,

$$\gamma = \begin{pmatrix} A & B \\ C & D \end{pmatrix} \tag{13.5.23}$$

where $A, B, C,$ and D are real $d \times d$ matrices subject to the constraint,

$$\gamma^T J \gamma = J \qquad J = \begin{pmatrix} 0 & \mathbb{1}_{d \times d} \\ \mathbb{1}_{d \times d} & 0 \end{pmatrix} \tag{13.5.24}$$

In this presentation, the action of $O(d, d; \mathbb{R})$ on $\mathcal{E}_{ij} = \mathcal{G}_{ij} + \mathcal{B}_{ij}$ is given succinctly by,

$$\gamma(\mathcal{E}) = (A\mathcal{E} + B)(C\mathcal{E} + D)^{-1} \tag{13.5.25}$$

We may now list, without proof, the set of transformations leaving the lattice invariant. First, there are transformations,

$$\gamma(K) = \begin{pmatrix} \mathbb{1}_{d \times d} & K \\ 0 & \mathbb{1}_{d \times d} \end{pmatrix} \tag{13.5.26}$$

with K an antisymmetric $d \times d$ matrix with integer entries. These are clearly elements of $O(d, d; \mathbb{R})$, and correspond to shifting $\mathcal{B}_{ij}$ by integers K_{ij}. This, we already know, leaves the lattice invariant.

Second, given any $M \in GL(d, \mathbb{Z})$, the following transformations,

$$\gamma(M) = \begin{pmatrix} M & 0 \\ 0 & (M^T)^{-1} \end{pmatrix} \tag{13.5.27}$$

can be shown to leave the lattice invariant. These correspond to modular invariances of the space-time torus.

Finally, there are a finite set of d transformations given by,

$$\gamma_i = \begin{pmatrix} \mathbb{1}_{d \times d} - \delta_i & \delta_i \\ \delta_i & \mathbb{1}_{d \times d} - \delta_i \end{pmatrix} \tag{13.5.28}$$

where δ_i is a matrix whose components are zero except for $(\delta_i)_{ii} = 1$. These transformations give the generalization of inversions of the radii $R \to 1/R$ along each of the 1-cycles of T^d. Note that the S transformation defined in earlier subsections mapped $\mathcal{E} \to \mathcal{E}^{-1}$, and hence in the current notation is given by,

$$\gamma_S = \begin{pmatrix} 0 & \mathbb{1}_{d \times d} \\ \mathbb{1}_{d \times d} & 0 \end{pmatrix} \tag{13.5.29}$$

It is easy to verify that $\gamma_S = \prod_{i=1}^{d} \gamma_i$, that is, that S can be thought of as a simultaneous inversion of all directions of the space-time torus.

The transformations given above turn out to generate all of the invariances of the lattice, and in total form the discrete group $O(d, d; \mathbb{Z})$. The space of inequivalent tori $\mathbb{T}^d$ from the string point of view is thus parametrized by the double coset space,

$$O(d, d; \mathbb{Z}) \backslash O(d, d; \mathbb{R}) / \left(O(d; \mathbb{R}) \times O(d; \mathbb{R}) \right) \qquad (13.5.30)$$

In the case of Heterotic string, the result extends to include the left-moving 16-dimensional torus, and the duality group is $O(d, d + 16; \mathbb{Z})$.

13.6 Rationality and complex multiplication

In Section 10.10, we introduced the torus partition function for a large class of two-dimensional conformal field theories, for example, in terms of a sum over characters in (10.10.7). A conformal field theory is said to be *rational* if the sum over characters is finite. In a sense, a rational conformal field theory is one which is exactly solvable, and great effort has been expended toward obtaining a classification of such theories. In this chapter, we ask when a $\mathbb{T}^2$ nonlinear sigma model is rational.

13.6.1 The case $\mathbb{T}^1$

To begin, let us consider the simpler case of an S^1 nonlinear sigma model, with the circle having radius R. In this case, the partition function may be computed to be,

$$Z(R|\tau) = \frac{1}{|\eta(\tau)|^2} \sum_{m,n \in \mathbb{Z}} q^{\frac{1}{2}\left(\frac{m}{R} + \frac{nR}{2}\right)^2} \bar{q}^{\frac{1}{2}\left(\frac{m}{R} - \frac{nR}{2}\right)^2} \qquad (13.6.1)$$

This result may be obtained from (13.4.6) by choosing $\alpha' = 2$ and dressing with a factor of $|\eta(\tau)|^2$ to account for Virasoro descendants. To address the question of rationality, we must ask when $Z(R|\tau)$ can be written as a sum over a finite number of characters. It turns out that this is possible if $R^2 \in \mathbb{Q}$. Indeed, say that $R^2 = \frac{2k}{\ell}$ for $k, \ell \in \mathbb{Z}$. In this case, we can write the partition function as,

$$Z\left(\sqrt{\frac{2k}{\ell}} \middle| \tau\right) = \frac{1}{|\eta(\tau)|^2} \sum_{m,n \in \mathbb{Z}} q^{\frac{1}{4k\ell}(m\ell + nk)^2} \bar{q}^{\frac{1}{4k\ell}(m\ell - nk)^2} \qquad (13.6.2)$$

It will be useful to write,

$$m = \mu + 2kr \qquad\qquad n = \nu + 2\ell s \qquad (13.6.3)$$

where $\mu \in \{0, \ldots, 2k-1\}$, $\nu \in \{0, \ldots, 2\ell-1\}$, and $r, s \in \mathbb{Z}$. We may then split the sums as,

$$Z\left(\sqrt{\frac{2k}{\ell}} \,\Big|\, \tau\right) = \frac{1}{|\eta(\tau)|^2} \sum_{\mu=0}^{2k-1} \sum_{\nu-0}^{2\ell-1} \sum_{r,s \in \mathbb{Z}} q^{\frac{1}{4k\ell}(2k\ell(r+s)+\mu\ell+\nu k)^2}$$

$$\times \bar{q}^{\frac{1}{4k\ell}(2k\ell(r-s)+\mu\ell-\nu k)^2}$$

$$(13.6.4)$$

We now again change summation variables to,

$$\tilde{\mu} = \mu\ell + \nu k \qquad \tilde{\nu} = \mu\ell - \nu k \qquad \rho = r+s \qquad \sigma = r-s \quad (13.6.5)$$

By definition we have,

$$\tilde{\mu} + \tilde{\nu} \equiv 0 \pmod{2\ell} \qquad\qquad \tilde{\mu} - \tilde{\nu} \equiv 0 \pmod{2k} \qquad (13.6.6)$$

Hence we can rewrite the partition function as a finite sum,

$$Z\left(\sqrt{\frac{2k}{\ell}} \,\Big|\, \tau\right) = \sideset{}{'}\sum_{\tilde{\mu},\tilde{\nu}=0,\ldots,2k\ell-1} \chi_{\tilde{\mu}}(\tau)\overline{\chi}_{\tilde{\nu}}(\bar{\tau}) \qquad (13.6.7)$$

where the primed sum denotes a sum subject to restrictions (13.6.6). The result has been written in terms of the characters,

$$\chi_{\tilde{\mu}}(\tau) = \frac{1}{\eta(\tau)} \sum_{\rho \in \mathbb{Z}} q^{k\ell\left(\rho + \frac{\tilde{\mu}}{2k\ell}\right)^2} \qquad (13.6.8)$$

That these are the correct characters for $\mathbb{T}^1$ is confirmed by noting that the compact boson at radius $R^2 = \frac{2k}{\ell}$ is equivalent to a $\mathbb{Z}_\ell$ orbifold of the $U(1)_k$ current algebra theory.

13.6.2 The case $\mathbb{T}^2$

We now proceed to the case of $M = \mathbb{T}^2$. For the moment we will consider only tori that are trivial products of circles, with radii R_1 and R_2. Then the conformal field theory factorizes into two S^1 CFTs studied above, and the total CFT is rational if and only if the two components conformal field theories are. In other words, we must take $R_{1,2}^2 \in \mathbb{Q}$. Say concretely that,

$$R_1^2 = \frac{2k_1}{\ell_1} \qquad\qquad R_2^2 = \frac{2k_2}{\ell_2} \qquad k_i, \ell_i \in \mathbb{Z} \qquad (13.6.9)$$

The complex structure of the space-time torus, which we will denote as σ to avoid confusion with the complex structure τ of the worldsheet torus, is given by $\sigma = iR_1/R_2$ up to $\mathrm{SL}(2,\mathbb{Z})$ transformations. Hence σ satisfies a quadratic equation with integer coefficients,

$$\ell_1 k_2 \sigma^2 + \ell_2 k_1 = 0 \qquad (13.6.10)$$

This is precisely the condition that the space-time $\mathbb{T}^2$ admit complex multiplication. Though we have restricted ourselves to tori which are trivial products of circles here, one can argue by similar methods that for generic elliptic curves $\mathbb{C}/L$, rationality of the conformal field theory is tantamount to the presence of complex multiplication on $\mathbb{C}/L$. Details are relegated to Exercise 13.5.

• Bibliographical notes

Toroidal compactification goes back to [SS79] and was generalized to the Heterotic string in [Nar86]and [NSW87]. It was used to map out the moduli space of $\mathfrak{c} = 1$ unitary conformal field theories in [DVV88]. Comprehensive overviews of toroidal compactification of the bosonic string may be found in the first volume of Polchinski's book [Pol07a], while toroidal compactification of the superstring, T-duality, and related topics are discussed in detail in the second volume of Polchinski's book [Pol07b] as well as in the book by Kiritsis [Kir19]. Further useful references on correlation functions for compact scalars, representations of current and Kac–Moody algebras, and related topics may be found in [DFSZ87] and [DFSZ88]. A particularly lucid discussion of the $\mathrm{O}(d,d;\mathbb{Z})$ T-duality group of the torus $\mathbb{T}^d$ can be found in [GPR94]. For the contents of Section 13.6, we refer again to [GV04], and to Chapter 11 on complex multiplication. For discussions of the structure of the space of all rational conformal field theories, we refer to [MS88], [MS89a], [MS89b], and [MS89c].

Exercises

13.1 Given the constant metric G and B-field of a $\mathbb{T}^2$ sigma model, we define a pair of complex variables $\sigma = \sigma_1 + i\sigma_2$ and $\rho = \rho_1 + i\rho_2$ such that,

$$G = \frac{\rho_2}{\sigma_2}\begin{pmatrix} |\sigma|^2 & \sigma_1 \\ \sigma_1 & 1 \end{pmatrix} \qquad B = \begin{pmatrix} 0 & \rho_1 \\ -\rho_1 & 0 \end{pmatrix} \qquad (13.6.11)$$

Show that the T-duality group can be realized on these variables as $\mathrm{PSL}(2,\mathbb{Z})_\sigma \times \mathrm{PSL}(2,\mathbb{Z})_\rho \rtimes \mathbb{Z}_2^2$, where $\mathrm{PSL}(2,\mathbb{Z})_{\sigma,\rho}$ acts on σ, ρ via linear fractional transformations, and the two $\mathbb{Z}_2$ transformations correspond to $(\sigma, \rho) \to (\rho, \sigma)$ and $(\sigma, \rho) \to (-\bar{\sigma}, -\bar{\rho})$. In terms of these variables, write the expressions for p_L^2 and p_R^2.

13.2 Show that any $\mathbb{T}^2$ sigma model with $\sigma = i$ and $|\rho|^2 = 1$ in the notation of Exercise 13.1 has an additional discrete symmetry coming from a combination worldsheet parity and T-duality.

13.3 At the level of the path integral, T-duality can be shown by defining a field $f^\mu = dX^\mu$, enforcing the Bianchi identity for f^μ via Lagrange multipliers $\widetilde{X}_\mu$, and integrating out f^μ to leave a theory involving only $\widetilde{X}_\mu$. Carry out this procedure, and show that it reproduces (13.5.10).

13.4 It is useful to consider the momenta and winding lattices L and L^+ as forming a single lattice L of signature (d, d), equipped with the product

$$p \circ p' = p_L \cdot p'_L - p_R \cdot p'_R \in \mathbb{Z} \qquad (13.6.12)$$

Show that modular invariance of the partition function requires L to be even and self-dual, that is,

$$p \circ p \in 2\mathbb{Z} \qquad \mathsf{L} = \mathsf{L}^* \qquad (13.6.13)$$

13.5 Using the results of Exercise 13.1, prove the claim of Subsection 13.6.2, namely that for a generic elliptic curve rationality of the corresponding conformal field theory is tantamount to the presence of complex multiplication.

14

S-duality of Type IIB superstrings

In this chapter, we shall draw together a number of different strands of inquiry addressed in earlier chapters. We shall study the interplay between superstring amplitudes, their low-energy effective interactions, Type IIB supergravity, and the S-duality symmetry of Type IIB superstring theory. We begin with a brief review of Type IIB supergravity which, in particular, provides the massless sector of Type IIB superstring theory. We then discuss how the $SL(2,\mathbb{R})$ symmetry of Type IIB supergravity is reduced to the $SL(2,\mathbb{Z})$ symmetry of Type IIB superstring theory. We conclude with a discussion of how the low-energy effective interactions induced by string theory on supergravity may be organized in terms of modular functions and forms under this $SL(2,\mathbb{Z})$ symmetry and match the predictions provided by perturbative calculations of Chapter 12.

14.1 Type IIB supergravity

While the expansion of string amplitudes in powers of the string coupling g_s, discussed in Chapter 12, was given only by an asymptotic series in g_s, it has the advantage of being valid for all energy ranges or, equivalently, for all distance scales. A different and complementary expansion of string theory, referred to as the *low-energy expansion*, in powers of the string scale α' was used for string amplitudes in Section 12.8. To leading order in this expansion, one omits the effects of all the massive string states, whose mass-squares are set by the string tension $1/\alpha'$. The remaining massless states make up precisely the particle contents of a corresponding supergravity.

To each one of the five 10-dimensional perturbative string theories, namely, the Type I theory of open and closed strings, and the four theories of purely closed strings Type IIA, Type IIB, Heterotic $E_8 \times E_8$, and $\mathrm{Spin}(32)/\mathbb{Z}_2$, there is a corresponding supergravity. The Type II theories have the maximal number of 32 supersymmetries, while the other three have 16 supersymmetries. We shall focus here on Type IIB supergravity as its corresponding Type IIB superstring theory possesses the famous $\mathrm{SL}(2,\mathbb{Z})$ duality symmetry, which is our principal interest in this book.

14.1.1 Fields of Type IIB supergravity

The content of Type IIB supergravity consists of the following fields,

G	the metric, whose excitations are the graviton
C_2	a complex 2-form potential field
τ	a complex scalar axion–dilaton field
C_4	a real 4-form potential with self-dual field strength
ψ	a chirality $-$ Weyl spinor gravitino field
λ	a chirality $+$ Weyl spinor dilatino field $\qquad$ (14.1.1)

The real fields G and C_4 and the complex fields C_2 and τ are bosonic fields transforming under tensor (or single-valued) representations of the Lorentz group $\mathrm{SO}(1,9)$, while ψ and λ are fermions transforming under Weyl spinor (or double-valued) representations of $\mathrm{SO}(1,9)$. The complex-valued field τ is related to the real-valued axion χ and dilaton Φ as follows,

$$\tau = \chi + i e^{-\Phi} \tag{14.1.2}$$

Since Φ is real-valued, we have $\mathrm{Im}\,\tau > 0$ so that τ takes values in the Poincaré upper half plane. The metric G discussed in Chapter 12, and which enters the nonlinear sigma model in (12.2.1), is referred to as the *string-frame metric*. In the presence of the dilaton field Φ, the string-frame metric G may be Weyl-rescaled by a factor involving Φ to the *Einstein-frame metric*, which is denoted by G_E and related to G by,

$$G_{E\mu\nu} = e^{-\Phi/2}\, G_{\mu\nu} \tag{14.1.3}$$

One may formulate Type IIB supergravity either in terms of G or G_E. The advantage of the Einstein-frame metric G_E is that it will be invariant under the groups $\mathrm{SL}(2,\mathbb{R})$ and $\mathrm{SL}(2,\mathbb{Z})$ mentioned earlier and to be discussed in detail in Section 14.2, while the string-frame metric G is not invariant. The full set of field equations is a complicated system of nonlinear partial differential equations. Concentrating on the bosonic fields simplifies matters greatly. One may set the fermionic fields to zero consistently with the fermion field equations since they all involve a sum of monomials of odd degree in the fermion fields and their derivatives.

14.1.2 Field equations of Type IIB supergravity

We proceed here for vanishing fermion fields. The field strength of C_2 is $F_3 = dC_2$, while the field strengths of τ and C_4 are defined as follows,

$$
\begin{aligned}
P &= f^2 \, dB & B &= (1 + i\tau)/(1 - i\tau) \\
Q &= f^2 \operatorname{Im} (B d\bar{B}) & f^2 &= (1 - |B|^2)^{-1} \\
F_5 &= dC_4 + \tfrac{i}{16}(C_2 \wedge \bar{F}_3 - \bar{C}_2 \wedge F_3) & \star F_5 &= F_5
\end{aligned}
\tag{14.1.4}
$$

The field equation for F_5 coincides with its Bianchi identity $dF_5 = \tfrac{i}{8} F_3 \wedge \bar{F}_3$ in view of its self-duality relation. The remaining Bianchi identities are,

$$
\begin{aligned}
dQ + iP \wedge \bar{P} &= 0 & dF_5 - \tfrac{i}{8} K \wedge \bar{K} &= 0 \\
dP - 2iQ \wedge P &= 0 & dK - iQ \wedge K + P \wedge \bar{K} &= 0
\end{aligned}
\tag{14.1.5}
$$

where $K = f(F_3 - B\bar{F}_3)$. In terms of the component fields of $P = P_\mu \, dx^\mu$, $Q = Q_\mu \, dx^\mu$, as well as F_5 and K given by,

$$
\begin{aligned}
K &= \tfrac{1}{3!} K_{\mu\nu\rho} \, dx^\mu \wedge dx^\nu \wedge dx^\rho \\
F_5 &= \tfrac{1}{5!} F_{5\mu\nu\rho\sigma\tau} \, dx^\mu \wedge dx^\nu \wedge dx^\rho \wedge dx^\sigma \wedge dx^\tau
\end{aligned}
\tag{14.1.6}
$$

the remaining field equations are conveniently expressed as follows,

$$
\begin{aligned}
0 &= \nabla_E^\mu P_\mu - 2iQ^\mu P_\mu + \tfrac{1}{24} K_{\mu\nu\rho} K^{\mu\nu\rho} \\
0 &= \nabla_E^\rho K_{\mu\nu\rho} - iQ^\rho K_{\mu\nu\rho} - P^\rho \bar{K}_{\mu\nu\rho} + \tfrac{2i}{3} F_{5\mu\nu\rho\sigma\tau} K^{\rho\sigma\tau} \\
0 &= R_{E\mu\nu} - P_\mu \bar{P}_\nu - \bar{P}_\mu P_\nu - \tfrac{1}{6}(F_5^2)_{\mu\nu} - \tfrac{1}{4}\operatorname{Re}(K_\mu{}^{\rho\sigma} \bar{K}_{\nu\rho\sigma}) \\
&\quad + \tfrac{1}{48} G_{E\mu\nu} K^{\rho\tau\sigma} \bar{K}_{\rho\tau\sigma}
\end{aligned}
\tag{14.1.7}
$$

where $R_{E\mu\nu}$ is the Ricci tensor for the Einstein-frame metric $G_{E\mu\nu}$ and ∇_E^μ is the covariant derivative with respect to the affine connection for G_E.

In view of the self-duality condition on F_5, there does not exist a Lorentz-covariant action from which the field equations of Type IIB supergravity may be deduced using the standard variational principle. Instead one defines an action S_{IIB} for Type IIB supergravity on an orientable manifold M by,[1]

$$
\begin{aligned}
S_{\mathrm{IIB}} = \frac{1}{2\kappa_{10}^2} \int_M d\mu_{G_E} \bigg(& R_E - \frac{\partial_\mu \tau \partial^\mu \bar{\tau}}{2\tau_2^2} - \frac{(F_3^2 - \tau F_3^1) \cdot (F_3^2 - \bar{\tau} F_3^1)}{2\,\tau_2} \\
& - \frac{1}{4} F_5 \cdot F_5 \bigg) - \frac{2}{\kappa_{10}^2} \int_M C_4 \wedge F_3^1 \wedge F_3^2
\end{aligned}
\tag{14.1.8}
$$

Here κ_{10}^2 is related to the 10-dimensional Newton constant, and we

---

[1] Here, $d\mu_{G_E} = d^{10}x \sqrt{-\det G_{E\mu\nu}}$ is the invariant volume form for the metric G_E; all contractions of Einstein indices are carried out with the help of the metric G_E; and the contraction of rank n antisymmetric tensors is denoted by $F^1 \cdot F^2 = \tfrac{1}{n!} F^1_{\mu_1\cdots\mu_n} F^{2\,\mu_1\cdots\mu_n}$.

decompose the complex fields $F_3 = F_3^1 + iF_3^2$ and $\tau = \tau_1 + i\tau_2$ into the real-field components F_3^1, F_3^2, τ_1, and τ_2. Importantly, the action S_{IIB} should be considered for a field F_5 that is *unconstrained by the self-duality condition*. The field equations for Type IIB supergravity may be obtained from this action by varying the *unconstrained field C_4*, along with all the other fields, and only subsequently imposing the self-duality condition $\star F_5 = F_5$.

Finally, for the sake of completeness and later use in Section 14.7, we also include the supersymmetry transformations of the fermion dilatino λ and gravitino ψ_μ fields, evaluated at vanishing fermion fields. They are given as follows for an infinitesimal supersymmetry transformation spinor field ε,[2]

$$\delta\psi_\mu = D_\mu\varepsilon + \tfrac{i}{4}(\gamma \cdot F_5)\gamma_\mu\varepsilon - \tfrac{1}{16}\Big(\gamma_\mu(\gamma \cdot K) + 2(\gamma \cdot K)\gamma_\mu\Big)\mathcal{B}^{-1}\varepsilon^*$$

$$\delta\lambda = i(\gamma \cdot P)\mathcal{B}^{-1}\varepsilon^* - \tfrac{i}{4}(\gamma \cdot K)\varepsilon \qquad\qquad (14.1.9)$$

where ε^* is the complex conjugate of ε.

14.2 The SL(2, ℝ) symmetry of classical Type IIB supergravity

The Type IIB supergravity field equations are invariant under transformation by elements γ of the group SL(2, ℝ), under which the metric G_E and the antisymmetric tensor fields C_4 and F_5 are invariant, while the fields P, Q, K, τ, and $F_3 = F_3^1 + iF_3^2$ transform as follows,

$$\begin{aligned} \gamma P &= e^{2i\theta} P & \gamma\tau &= (a\tau + b)/(c\tau + d) \\ \gamma Q &= Q + d\theta & \gamma F_3^2 &= aF_3^2 + bF_3^1 \\ \gamma K &= e^{i\theta} K & \gamma F_3^1 &= cF_3^2 + dF_3^1 \end{aligned} \qquad (14.2.1)$$

where $\gamma = \begin{pmatrix} a & b \\ c & d \end{pmatrix} \in$ SL(2, ℝ) and the function θ is given by,

$$\theta = \phi + \theta_\tau - \theta_{\gamma\tau} \qquad e^{-i\phi} = \frac{c\tau + d}{|c\tau + d|}, \qquad e^{i\theta_\tau} = \frac{\tau + i}{|\tau + i|} \quad (14.2.2)$$

The axion–dilaton takes values in the coset SL(2, ℝ)/U(1), and the phase $e^{i\theta}$ performs a field-dependent U(1) rotation. The gravitino field ψ and the dilatino field λ transform with U(1) phase factors under SL(2, ℝ), given by,

$$\gamma\psi = e^{i\theta/2}\psi \qquad\qquad \gamma\lambda = e^{3i\theta/2}\lambda \qquad (14.2.3)$$

The group SL(2, ℝ) is a symmetry of the classical supergravity field equations, of the action S_{IIB}, and of the self-duality condition $\star F_5 = F_5$.

---

[2] Our conventions for the Dirac matrices γ_μ are as follows: they satisfy $\{\gamma_\mu, \gamma_\nu\} = 2\eta_{\mu\nu}$, the matrix $\mathcal{B}$ represents charge conjugation $(\gamma_\mu)^* = \mathcal{B}\gamma_\mu\mathcal{B}^{-1}$ and $\mathcal{B}\mathcal{B}^* = I$, while $\gamma \cdot P = \gamma_\mu P^\mu$, $\gamma \cdot K = \frac{1}{3!}\gamma_\mu\gamma_\nu\gamma_\rho K^{\mu\nu\rho}$, and $\gamma \cdot F_5 = \frac{1}{5!}\gamma_\mu\gamma_\nu\gamma_\rho\gamma_\sigma\gamma_\tau F_5^{\mu\nu\rho\sigma\tau}$.

14.3 $SL(2, \mathbb{R})$ to $SL(2, \mathbb{Z})$ via an anomaly mechanism

A classical symmetry cannot always be promoted to a symmetry of the corresponding quantum theory, in which case the symmetry is said to suffer an *anomaly*. Phase rotations on chiral fermions, such as those of (14.2.3), often suffer anomalies. In the case at hand, the field-dependent $U(1)$ symmetry does not suffer a local anomaly – such an anomaly would invalidate the decoupling of the spurious θ-field – but a subtle related effect reduces the classical continuous $SL(2, \mathbb{R})$ symmetry of supergravity to the discrete $SL(2, \mathbb{Z})$ symmetry of Type IIB string theory. A thorough discussion of perturbative anomalies in quantum field theory is beyond the scope of this book. Instead, as our starting point, we shall adopt without derivation the results of the anomaly calculations from the literature [AGW84] and [GG98].

Noether's theorem implies that the $U(1)$ transformations of (14.2.3) on the gravitino and dilatino fields produce a current,

$$J^\mu = \frac{1}{2} \bar{\psi}_\nu \gamma^{\nu\mu\rho} \psi_\rho + \frac{3}{2} \bar{\lambda} \gamma^\mu \lambda \tag{14.3.1}$$

which is conserved, namely, it obeys $\partial_\mu J^\mu = 0$ using the classical field equations of Type IIB supergravity. Upon quantizing Type IIB supergravity the current J^μ is no longer conserved and the classical action (of which S_{IIB} is the part for vanishing fermion fields) receives quantum correction that may be summarized in terms of an *effective action S_{eff}*. The only contribution that is of interest here is the one that results from carrying out an $SL(2, \mathbb{R})$ transformation on the fields (14.2.1) and (14.2.3), which is accompanied by the phase rotation of (14.2.2). The change in the effective action S_{eff} under an infinitesimal $SL(2, \mathbb{R})$ transformation, and the accompanying infinitesimal $U(1)$ transformation by the phase $\delta\theta$, is given by,

$$\delta S_{\text{eff}} = -\frac{1}{4\pi} \int_M \delta\theta \, dQ \wedge X_8(R_E) \tag{14.3.2}$$

Here, $6X_8(R_E)$ is the Euler characteristic class: a closed 8-form that depends only on the curvature 2-form R_E in the Einstein frame and satisfies $dX_8(R_E) = 0$. Since both $dQ = -iP \wedge \bar{P}$ and R_E are $SL(2, \mathbb{R})$-invariant, the change in the action under an infinitesimal transformation (14.3.2) may be integrated uniquely to a finite $SL(2, \mathbb{R})$ transformation,

$$\Delta S_{\text{eff}} = -\frac{1}{4\pi} \int_M \theta \, dQ \wedge X_8(R_E) \tag{14.3.3}$$

where Δ stands for the change in S_{eff} under the finite $SL(2, \mathbb{R})$ transformation γ of (14.2.1) and (14.2.2). As the classical Type IIB supergravity action is $SL(2, \mathbb{R})$ invariant, the quantum theory is invariant provided,

$$e^{i\Delta S_{\text{eff}}} = 1 \tag{14.3.4}$$

Since the supergravity fields are unchanged as $\theta \to \theta + 2\pi$, we must have,

$$\nu = \frac{1}{4\pi} \int_M dQ \wedge X_8(R_E) \in \mathbb{Z} \tag{14.3.5}$$

The presence of a noninvariant effective action does not necessarily imply that the corresponding symmetry transformation is anomalous. Indeed, if a *local counter-term*, namely, a local expression in the fields, can be found to restore invariance, then a quantization scheme may be adopted where the counter-term is added and the symmetry survives quantization. The existence of such a counter-term is a question in cohomology. Decomposing the angle $\theta = \phi + \theta_\tau - \theta_{\gamma\tau}$ as given in (14.2.2), the dependence on $\theta_\tau - \theta_{\gamma\tau}$ may clearly be cancelled by the following local counter-term,

$$\Delta S_{\text{c-t}}^{(1)} = \frac{1}{4\pi} \int_M (\theta_\tau - \theta_{\gamma\tau}) \, dQ \wedge X_8(R_E) \tag{14.3.6}$$

Inclusion of this counter-term reduces the problem to finding a local counter-term for the ϕ term in the expression for θ in (14.2.2), of the form,

$$S_{\text{c-t}}^{(2)} = -\frac{i}{4\pi} \int_M \ln f(\tau) \, dQ \wedge X_8(R_E) \tag{14.3.7}$$

such that,

$$\Delta S_{\text{eff}} + \Delta S_{\text{c-t}}^{(1)} + \Delta S_{\text{c-t}}^{(2)} \in 2\pi\mathbb{Z} \tag{14.3.8}$$

The requirement that this condition imposes on $f(\tau)$ for any value of ν is,

$$\frac{f(\gamma\tau)}{f(\tau)} = U_\gamma \frac{(c\tau + d)}{|c\tau + d|} \tag{14.3.9}$$

where U_γ is a τ-independent factor satisfying $|U_\gamma| = 1$. While the product of the factors $(c\tau + d)$ and $(c\bar{\tau} + d)$ is given by the ratio of the function $\text{Im}\,(\tau)$ evaluated at τ and $\gamma\tau$, namely, $|c\tau + d|^2 = \text{Im}\,(\tau)/\text{Im}\,(\gamma\tau)$, the ratio $(c\tau + d)/(c\bar{\tau} + d)$ of these factors cannot be expressed as a function of τ and $\gamma\tau$ alone, because this phase takes values in the $SO(2)$ fiber of the bundle $SL(2, \mathbb{R}) \mapsto \mathcal{H}$ and is not just a function of the base. As a result, the relation (14.3.9), imposed for all $\gamma \in SL(2, \mathbb{R})$, admits no solutions $f(\tau)$. Therefore, no counter-term of the form (14.3.7) exists for $\gamma \in SL(2, \mathbb{R})$, so that this classical symmetry fails to be a symmetry of the quantum theory.

While no solutions exist for the condition (14.3.9) valid for all $\gamma \in SL(2, \mathbb{R})$, solutions for $f(\tau)$ that are valid for all γ in a discrete subgroup of $SL(2, \mathbb{R})$

do exist. We shall limit our study here to $\gamma \in \mathrm{SL}(2,\mathbb{Z})$ and leave open the problem of proving that no larger subgroup of $\mathrm{SL}(2,\mathbb{R})$ works. The corresponding solutions may be constructed in two steps: first for the continuous τ-dependent part in a single fundamental domain F for $\mathrm{SL}(2,\mathbb{Z})$ and then for the discrete dependence through the fundamental domain that contains τ. The continuous τ-dependent part can always be solved with the help of the Dedekind η function, thereby reducing the problem as follows,

$$f(\tau) = \frac{\eta(\tau)^2}{|\eta(\tau)|^2} \times \text{ discrete phase} \tag{14.3.10}$$

When $\nu = 0$, the discrete phase is immaterial and cancels out, so that the full $\mathrm{SL}(2,\mathbb{Z})$ symmetry is realized.

We now address the case where $\nu \neq 0$ and note that ν can be nonzero only if dQ generates nontrivial cohomology – that is, if we are in the presence of a singular axion–dilaton profile. This requires the insertion of so-called 7-branes, introduced in Exercises 14.1–14.3, in which case the cohomology generator $dQ/4\pi$ integrates to integer values. In this sense, the discrete anomaly that we shall discuss below is not an anomaly in Type IIB supergravity proper, but rather in Type IIB supergravity embellished with such nonperturbative objects as 7-branes.[3]

The requirement imposed on $f(\tau)$ for $\nu \neq 0$ including the discrete phases sharpens the condition of (14.3.9) and is given as follows,

$$\frac{f(\gamma\tau)^\nu}{f(\tau)^\nu} = \frac{(c\tau + d)^\nu}{|c\tau + d|^\nu} \tag{14.3.11}$$

For $\gamma \in \Gamma \subset \mathrm{SL}(2,\mathbb{Z})$ where Γ is a subgroup of $\mathrm{SL}(2,\mathbb{Z})$, the relation (14.3.11) is one on the existence of holomorphic modular forms of weight ν. Expressing the function as $f(\tau) = \varphi(\tau)/|\varphi(\tau)|$, the solutions are,

- For $12|\nu$ the solution $\varphi(\tau) = \eta(\tau)^2$ preserves $\Gamma = \mathrm{SL}(2,\mathbb{Z})$;
- For $4|\nu$ the solution $\varphi(\tau) = \mathsf{E}_4(\tau)^{\frac{1}{4}}$ preserves $\Gamma = \mathrm{SL}(2,\mathbb{Z})$;
- For $6|\nu$ the solution $\varphi(\tau) = \mathsf{E}_6(\tau)^{\frac{1}{6}}$ preserves $\Gamma = \mathrm{SL}(2,\mathbb{Z})$;
- For $2|\nu$ the solution $\varphi(\tau) = \left(\mathsf{E}_2(\tau) - 2\mathsf{E}_2(2\tau)\right)^{\frac{1}{2}}$ preserves $\Gamma = \Gamma_0(2)$;
- For ν odd, the solution $\varphi(\tau) = \vartheta_3(0|2\tau)^{2\nu}$ preserving $\Gamma_0(4)$.

The fact that $\mathsf{E}_2(\tau) - 2\mathsf{E}_2(2\tau)$ is a modular form of weight 2 for $\Gamma_0(2)$ may be established by combining the transformation laws of (4.1.3) and (7.8.11)

---

[3] Equivalently, it is an anomaly in Type IIB supergravity with gauged $\mathrm{SL}(2,\mathbb{Z})$ symmetry. At the level of supergravity alone this gauging is optional and may in fact be obstructed by an entirely different sort of anomaly. String theory demands that these more subtle anomalies cancel and that $\mathrm{SL}(2,\mathbb{Z})$ be gauged. The cancellation of these anomalies requires the addition of subtle topological terms to the supergravity action (14.1.8), which we will not discuss here.

for $n = N = 2$. Furthermore, $\vartheta_3(0|2\tau)^2$ is a modular form of weight 1 under $\Gamma_0(4)$ by the arguments used to establish (7.8.6).

The relation between the above solutions for $4|\nu$ and $6|\nu$ and the partial solution of (14.3.9) is made manifest by the following identities,

$$E_4(\tau)^{\frac{1}{4}} = \eta(\tau)^2 j(\tau)^{\frac{1}{12}}$$
$$E_6(\tau)^{\frac{1}{6}} = \eta(\tau)^2 \left(j(\tau) - 1728\right)^{\frac{1}{12}} \qquad (14.3.12)$$

As an aside, we note that the coefficients of the following q-expansions,

$$E_4(\tau)^{\frac{1}{4}} = 1 + 60q - 4860q^2 + 660480q^3 - 105063420q^4 + \cdots \qquad (14.3.13)$$
$$E_6(\tau)^{\frac{1}{6}} = 1 - 84q - 20412q^2 - 6617856q^3 - 2505409788q^4 + \cdots$$

are all integers.

To summarize, $SL(2, \mathbb{Z})$ is preserved whenever $\nu \equiv 0, \pm 4, 6 \pmod{12}$, while $\Gamma_0(2)$ is preserved for $\nu \equiv \pm 2 \pmod{12}$, and $\Gamma_0(4)$ is preserved for ν odd. As we have mentioned earlier, at the current stage it is possible that additional elements survive which are not contained in $SL(2, \mathbb{Z})$, but we will argue that this is not the case in Section 14.4.

Note that our results agree with those obtained in [GG98] for ν divisible by 4. While the case of ν divisible by 6 seems admissible from the point of view of this anomaly, there are indications that this case is incompatible with a sensible compactification to two dimensions [Muk98]. Finally, we note that the above solutions for $f(\tau)$ to (14.3.9) are not unique, as one may multiply $f(\tau)$ by a rational function of $j(\tau)$ to obtain another solution to (14.3.9). Fixing this ambiguity would require analyzing the scattering amplitudes for four gravitons and at least two axion–dilatons.

14.4 SL(2, ℝ) to SL(2, ℤ) via the Dirac quantization condition

The arguments given in Section 14.3 for the reduction of the classical $SL(2, \mathbb{R})$ symmetry to the quantum $SL(2, \mathbb{Z})$ symmetry assumed the existence of singular profiles for the axion–dilaton, or in other words on the existence of 7-branes. Furthermore, it did not prove that $SL(2, \mathbb{Z})$ is the maximum possible symmetry preserved. One may invoke a different reasoning, without assuming the existence of 7-branes, to conclude that the maximum symmetry is indeed $SL(2, \mathbb{Z})$. This comes at the cost of assuming the existence of certain 5-branes.

The field $C_2^1 = \frac{1}{2} B_{\mu\nu} dx^\mu \wedge dx^\nu$ couples directly to the fundamental string in Type IIB string theory, essentially via the nonlinear sigma model coupling presented in (12.4.12). But the Type IIB superstring contains, in addition to the fundamental string, also a D1-brane to which the field C_2^2 couples directly. Strings that carry both charges are referred to as (p, q) strings; they couple directly to the complex field C_2 whose field strength is the 3-form F_3. Its dual $\star F_3$ is a 7-form whose 6-form potential couples to a (p, q) 5-brane (generalizing NS- and D5-branes). These (p, q) 5-branes may be viewed as the magnetic duals to the (p, q) strings and therefore obey a generalized Dirac quantization condition. Upon making the assumption that such five-branes exist (not just as semiclassical solutions to Type IIB supergravity, but as dynamical objects in the full theory), then the charges of the field F_3 must be quantized. If the field F_3 were to transform under $\mathrm{SL}(2, \mathbb{R})$ as given in (14.2.1), then it would be impossible to maintain integer quantized charges. Instead, the transformations that maintain this condition for all (p, q) five branes are precisely those with integer entries, namely, the elements of $\mathrm{SL}(2, \mathbb{Z})$. The original arguments along these lines were given in [HT95], and used in [Pol07b] and [BBS06].

Some further properties of (p, q) strings, as well as (p, q) 7-branes, are given in Exercises 14.4 and 14.5.

14.5 Low-energy effective interactions

Supergravity, as a classical theory, is valid for all values of the string coupling. Thus, it permits investigations into certain strong coupling phenomena such as soliton states, NS-branes, and D-branes. Supergravity has been an invaluable tool in the search for semi-realistic compactifications of superstring theory, such as on tori, orbifolds, and Calabi–Yau spaces.

However, supergravity alone is not a consistent quantum theory. In 10 space-time dimensions, it exhibits ultraviolet (UV) divergences starting at one loop. The UV convergence situation is somewhat improved by lowering the dimension of space-time. In four space-time dimensions, the four-graviton amplitude is UV convergent up to five loops, but it is likely that non-renormalizable UV divergences start occurring at seven loops, rendering the quantum theory inconsistent, or more accurately, incomplete. The UV completion of supergravity is precisely superstring theory. From the vantage point of superstring theory, supergravity should be thought of as an *effective low-energy field theory* that captures the dynamics of the massless states of Type IIB string theory, to leading order in the α' expansion.

The validity of an effective field theory may be extended beyond its leading order contributions. Viewed as an effective field theory of Type IIB superstring theory, the corresponding supergravity may be supplemented with contributions of higher order in α'. Such contributions are due to the effects of massive string states whose mass, we recall, is of order $1/\sqrt{\alpha'}$ and are referred to as *effective interactions*. In the approximation where the momenta used to probe string amplitudes are small compared to $1/\sqrt{\alpha'}$ the effective interactions are local, as is illustrated in Figure 14.1 for the exchange of a massive string state. Although these string-induced effective interactions are highly suppressed at low energies, they provide systematic corrections to supergravity in an expansion in powers of α'.

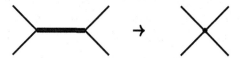

Figure 14.1 The exchange of a massive string state, indicated by the thick line in the left panel, induces a local effective interaction indicated by a thick dot in the right panel.

The space-time we observe has dimension 4 and is approximately flat when its curvature is measured in units of the Planck scale. Therefore, if the space-time of superstring theory is to be 10, then the six extra dimensions must have radii that are smaller than the smallest length scales accessible to experiment today. Physically realistic superstring theories are often based on space-times of the form $\mathbb{R}^{10-d} \times M_d$, where M_d is a compact manifold or orbifold and the physically observed dimension of space-time corresponds to $d = 6$. For the case where M_d is a flat torus, superstring perturbation theory continues to be well understood. When M_d is an orbifold of a torus, the space M_d is flat away from isolated point-like singularities, and string theory on toroidal orbifold spaces still lends itself to reasonably explicit solutions. However, when M_d is an arbitrary curved manifold, the predictions of superstring theory are much more difficult to obtain. This is the case even when space-time is a curved maximally symmetric space such as $AdS_5 \times S^5$, for which the quantization of the superstring is still largely an unresolved problem. An important exception is when the curvature of M_d is uniformly small compared to the Planck scale, a case that is referred to as the large radius expansion, as we shall discuss next.

Restricting, for example, to purely gravitational effects, the supergravity Lagrangian reduces to the Einstein–Hilbert term given by the Ricci scalar for the space-time metric. In Type II superstring theory, for example, the lowest order α' correction to the Einstein–Hilbert action is given by a term which we symbolically represent by $\mathcal{R}^4$, where $\mathcal{R}$ stands for the Riemann tensor, and the contribution to the action is obtained via a special contraction of the four factors, consistent with space-time supersymmetry, as will be explained later in this chapter. This effective interaction arises at order $(\alpha')^3$. Higher order effective interactions involve more derivatives and higher powers of α' and are symbolically represented by $D^{2k}\mathcal{R}^4$, where D represents a covariant derivative, again suitably contracted.

14.6 SL(2, $\mathbb{Z}$) duality in Type IIB superstring theory

Type IIB superstring theory exhibits SL(2, $\mathbb{Z}$) duality. This symmetry leaves the space-time metric G_E in Einstein frame and the self-dual 5-form invariant and transforms the 3-form flux field strengths linearly into one another,

$$\begin{pmatrix} F_3^2 \\ F_3^1 \end{pmatrix} \rightarrow \begin{pmatrix} a & b \\ c & d \end{pmatrix}\begin{pmatrix} F_3^2 \\ F_3^1 \end{pmatrix} \qquad \begin{pmatrix} a & b \\ c & d \end{pmatrix} \in \mathrm{SL}(2,\mathbb{Z}) \qquad (14.6.1)$$

As for the axion–dilaton field obtained by combining the dilaton field Φ and the axion field χ into a complex scalar field $\tau = \chi + ie^{-\Phi}$, transformations under SL(2, $\mathbb{Z}$) occur as follows,

$$\tau \rightarrow \frac{a\tau + b}{c\tau + d} \qquad (14.6.2)$$

The transformation properties of the fermionic fields are more complicated and will not be reproduced here. In fact, when one accounts for the fermionic fields, the true duality symmetry of Type IIB is *not* SL(2, $\mathbb{Z}$), but rather an extension of it given by the Pin$^+$ double cover of the meta-linear group denoted by either $ML^+(2,\mathbb{Z})$ or $GL^+(2,\mathbb{Z})$; see [TY19] and [DDHM21] for details. This subtlety will not affect the purely bosonic analysis below.

Being a symmetry of Type IIB string theory, SL(2, $\mathbb{Z}$) must be a symmetry of the low-energy effective theory. The effective interactions accessible from the four-graviton amplitude are of the form $D^{2k}\mathcal{R}^4$, as explained in Section 14.5, but they have coefficients that depend upon the vacuum expectation value of the dilaton $g_s = e^{\Phi}$ and the axion field χ. Expressing the effective action in terms of the Einstein frame metric G_E, with associated Riemann tensor $\mathcal{R}_E$, and the vacuum expectation value τ, we find to order $(\alpha')^7$,

$$\int d\mu_G \Big(\mathcal{E}_0(\tau)\mathcal{R}_E^4 + \mathcal{E}_4(\tau)D^4\mathcal{R}_E^4 + \mathcal{E}_6(\tau)D^6\mathcal{R}_E^4$$

$$+ \mathcal{E}_8(\tau)D^8\mathcal{R}_E^4 + \mathcal{O}((\alpha')^8)\Big) \qquad (14.6.3)$$

The coefficients $\mathcal{E}_{2k}(\tau)$ are real-valued scalar functions of τ. Derivatives of τ may occur as well but will be systematically omitted here for simplicity. Since the action must be SL(2, ℤ)-invariant and, in Einstein frame, G_E and $\mathcal{R}_E$ are invariant as well, we see that each coefficient $\mathcal{E}_{2k}(\tau)$ must be invariant,

$$\mathcal{E}_{2k}\left(\frac{a\tau + b}{c\tau + d}\right) = \mathcal{E}_{2k}(\tau) \qquad \begin{pmatrix} a & b \\ c & d \end{pmatrix} \in \text{SL}(2, \mathbb{Z}) \qquad (14.6.4)$$

a property that makes them into *non-holomorphic modular functions*.

String perturbation theory calculations are carried out in the string frame metric that we shall denote by $G_{\mu\nu}$ and that is related to the Einstein frame metric by $G_{\mu\nu} = e^{\Phi/2}G_{E\mu\nu}$. Converting the expression for the effective action of (14.6.3) to string frame may be done by using the following relations (for constant values of τ),

$$\sqrt{G_E}\,\mathcal{E}_{2k}(\tau)\,D_E^{2k}\mathcal{R}_E^4 = e^{(k-1)\Phi/2}\,\sqrt{G}\,\mathcal{E}_{2k}(\tau)\,D^{2k}\mathcal{R}^4 \qquad (14.6.5)$$

where D and $\mathcal{R}$ now stand, respectively, for the covariant derivative and Riemann tensor in the string frame. Now let us carry out the following exercise: express combinations of the non-holomorphic Eisenstein series with effective interactions $D_E^{2k}\mathcal{R}_E^4$ in the Einstein frame such that the leading perturbative contribution in string frame is tree level,

$$\tfrac{1}{2}\pi^{3/2}\sqrt{G_E}\,E_{\frac{3}{2}}(\tau)\mathcal{R}_E^4 = e^{-2\Phi}\zeta(3)\mathcal{R}^4 + \frac{\pi^2}{3}\mathcal{R}^4 + \cdots \qquad (14.6.6)$$

$$\tfrac{1}{2}\pi^{5/2}\sqrt{G_E}\,E_{\frac{5}{2}}(\tau)D_E^4\mathcal{R}_E^4 = e^{-2\Phi}\zeta(5)D^4\mathcal{R}^4 + \frac{2\pi^4}{135}e^{2\Phi}D^4\mathcal{R}^4 + \cdots$$

Comparing the terms in (14.6.6) which have the Φ-dependence of tree-level perturbative contributions with the known result of (12.11.16) at genus one and (12.12.4) and (12.12.8) at genus two, we see that their dependence on odd zeta-values is precisely reproduced. This observation would suggest that the coefficients $\mathcal{E}_{2k}(\tau)$ in (14.6.3) may be given by the non-holomorphic Eisenstein series.

14.7 Eisenstein series from supersymmetry and S-duality

We will now show how supersymmetry and S-duality can be used to prove the identifications between $\mathcal{E}_{2k}(\tau)$ and the appropriate Eisenstein series. We will begin with the case of $\mathcal{E}_0(\tau)$, namely, the coefficient of $\mathcal{R}^4$. Because the metric G is in the same supermultiplet as all of the other supergravity fields reviewed in Section 14.1, supersymmetry requires that the $\mathcal{R}^4$ term in the low-energy effective action be accompanied by a number of analogous terms related by supersymmetry. To make this more concrete, we first introduce some notation. As we have discussed earlier, the complex 2-form potential

C_2, and consequently the corresponding field strength F_3, transforms as a doublet under $\mathrm{SL}(2,\mathbb{Z})$. It is useful to package these field strengths into an $\mathrm{SL}(2,\mathbb{Z})$ singlet via,

$$M = \frac{e^{i\phi}}{\sqrt{-2i\tau_2}}(F_3^2 - \tau F_3^1) \tag{14.7.1}$$

where the superscript on F_3 is a doublet index. We further define the "super covariantized" combination,

$$\widehat{M}_{\mu\nu\rho} = M_{\mu\nu\rho} - 3\overline{\psi}_{[\mu}\gamma_{\nu\rho]}\lambda - 6i\overline{\psi}^*_{[\mu}\gamma_\nu\psi_{\rho]} \tag{14.7.2}$$

obtained by dressing M with appropriate combinations of the gravitino and dilatino fields ψ and λ. The combination $\widehat{M}$ has the property that under supersymmetry transformations, it does not contain derivatives of the transformation parameter.

In terms of this supercovariant field strength, the supersymmetric completion to the $\mathcal{R}^4$ term can be written as [GS99],

$$S^{(3)} = \int d\mu_G \left(\mathcal{E}_0^{(12,-12)}\lambda^{16} + \mathcal{E}_0^{(11,-11)}\widehat{M}\lambda^{14} + \cdots + \mathcal{E}_0^{(8,-8)}\widehat{M}^8 \right.$$
$$\left. + \cdots + \mathcal{E}_0^{(0,0)}\mathcal{R}^4 + \cdots + \mathcal{E}_0^{(-12,12)}(\lambda^*)^{16} \right) \tag{14.7.3}$$

where λ^{16} and $\widehat{M}\lambda^{14}$ (which will be the main terms of interest to us below) are short for the following contractions,

$$\lambda^{16} = \frac{1}{16!}\epsilon_{a_1\ldots a_{16}}\lambda^{a_1}\ldots\lambda^{a_{16}}$$
$$\widehat{M}\lambda^{14} = \frac{1}{14!}\widehat{M}_{\mu\nu\rho}\epsilon_{a_1\ldots a_{16}}\lambda^{a_1}\ldots\lambda^{a_{14}}(\gamma^{\mu\nu\rho}\gamma^0)^{a_{15}a_{16}} \tag{14.7.4}$$

The $\mathrm{SL}(2,\mathbb{Z})$ symmetry of Type IIB demands that the coefficient functions $\mathcal{E}_0^{(w,-w)}(\tau)$ be modular forms of holomorphic weight w and anti-holomorphic weight $-w$. The coefficient function $\mathcal{E}_0^{(0,0)}(\tau) = \mathcal{E}_0(\tau)$ is the original modular invariant coefficient function of $\mathcal{R}^4$. The coefficient functions are furthermore required by supersymmetry to satisfy,

$$\mathcal{E}_0^{(w,-w)}(\tau) = D_{w-1}\ldots D_0\mathcal{E}_0(\tau) \tag{14.7.5}$$

with $D_w = i\tau_2\left(\frac{\partial}{\partial\tau} - i\frac{w}{2\tau_2}\right)$ the covariant derivative mapping weight (w, w') modular forms to weight $(w+1, w'-1)$ modular forms. Because of these relations, constraints on $\mathcal{E}_0^{(w,-w)}$ for any w can imply nontrivial constraints on $\mathcal{E}_0$. We will now see how to use supersymmetry to obtain differential constraints on $\mathcal{E}_0^{(12,-12)}$ and $\mathcal{E}_0^{(11,-11)}$ in particular, which will suffice to prove that $\mathcal{E}_0$ is, up to normalization, equivalent to $E_{\frac{3}{2}}(\tau)$.

Let us begin by schematically writing the low-energy effective action as,

$$S = \frac{1}{(\alpha')^4}\left[S^{(0)} + (\alpha')^3 S^{(3)} + (\alpha')^4 S^{(4)} + (\alpha')^5 S^{(5)} + \ldots\right] \quad (14.7.6)$$

with $S^{(3)}$ in particular given in (14.7.3) above. Likewise, a generic SUSY transformation δ can be expanded in powers of α' as,

$$\delta = \delta^{(0)} + (\alpha')^3 \delta^{(3)} + (\alpha')^4 \delta^{(4)} + (\alpha')^5 \delta^{(5)} + \ldots \quad (14.7.7)$$

Requiring that the action has supersymmetry implies that we must have,

$$\delta^{(0)} S^{(0)} = \delta^{(0)} S^{(3)} + \delta^{(3)} S^{(0)} = \delta^{(0)} S^{(5)} + \delta^{(5)} S^{(0)} = \cdots = 0 \quad (14.7.8)$$

on the equations of motion.

We now consider the first two terms in the action,

$$\mathcal{L}^{(3)} = \mathcal{E}_0^{(12,-12)} \lambda^{16} + \mathcal{E}_0^{(11,-11)} \widehat{M} \lambda^{14} \quad (14.7.9)$$

A crucial feature of these two terms is that they are related by a subset of the SUSY transformations that do not mix with any of the other terms at this order, though we will have to keep track of variations of the lowest order action $S^{(0)}$ by $\delta^{(3)}$. It is a straightforward exercise to find the lowest-order $\delta^{(0)}$ supersymmetry variation of $\mathcal{L}^{(3)}$. Denoting the supersymmetry parameter by ϵ and keeping only terms proportional to $\lambda^{16} \psi_\mu^* \epsilon$ gives,

$$\delta^{(0)} \mathcal{L}^{(3)}|_{\lambda^{16}\psi_\mu^*\epsilon} = -8i\left(\mathcal{E}_0^{(12,-12)}(\tau) + 108 D_{11}\mathcal{E}_0^{(11,-11)}(\tau)\right) \quad (14.7.10)$$

On the other hand, the $\delta^{(3)}$ variation of $S^{(0)}$ cannot produce any such term. Hence we require $\delta^{(0)} \mathcal{L}^{(3)}|_{\lambda^{16}\psi_\mu^*\epsilon} = 0$ alone, which demands,

$$D_{11}\mathcal{E}_0^{(11,-11)}(\tau) = -\frac{1}{108}\mathcal{E}_0^{(12,-12)}(\tau) \quad (14.7.11)$$

Further constraints can be obtained by considering the term in the supersymmetry variation proportional to $\lambda^{16}\lambda^*\epsilon^*$. In this case, the computation is slightly more involved since now there *is* a contribution from the $\delta^{(3)}$ variation of $S^{(0)}$. The analysis will not be reproduced here, but the result is the following constraint,

$$\overline{D}_{-12}\, \mathcal{E}_0^{(12,-12)}(\tau) + 3240\, \mathcal{E}_0^{(11,-11)}(\tau) - 90g = 0 \quad (14.7.12)$$

where $g(\tau,\overline{\tau})$ is an unknown function. A final constraint may be obtained by demanding closure of the SUSY algebra on λ^*, giving,

$$32 D_{11} g = \mathcal{E}_0^{(12,-12)}(\tau) \tag{14.7.13}$$

We have now obtained three constraints (14.7.11), (14.7.12), and (14.7.13). Combining them, we may derive a Laplace eigenvalue equation for $\mathcal{E}_0^{(12,-12)}$,

$$\Delta \, \mathcal{E}_0^{(12,-12)}(\tau) = \left(-132 + \frac{3}{4} \right) \mathcal{E}_0^{(12,-12)}(\tau) \tag{14.7.14}$$

where we have noted that, when acting on weight-$(12,-12)$ modular functions, the Laplacian is given by $\Delta = 4 D_{11} \overline{D}_{-12}$.

It remains only to translate this differential constraint on $\mathcal{E}_0^{(12,-12)}(\tau)$ to a constraint on $\mathcal{E}_0(\tau)$, which can be done using the relation (14.7.5) between the two. This gives the following Laplace eigenvalue equation for $\mathcal{E}_0(\tau)$,

$$\Delta \mathcal{E}_0(\tau) = \tfrac{3}{4} \mathcal{E}_0(\tau) \tag{14.7.15}$$

Together with the asymptotic expansion obtained from string perturbation theory, this equation uniquely fixes $\mathcal{E}_0(\tau)$ to be the non-holomorphic Eisenstein series $E_{\frac{3}{2}}(\tau)$. Indeed, we saw already in (4.2.5) that the Eisenstein series satisfy precisely such eigenvalue equations.

Next, we briefly outline the analogous calculation for the $D^4 \mathcal{R}^4$ term. As before, the first step is to identify a subset of the terms appearing in $S^{(5)}$ which mix with each other, but not with other terms, under some subset of the supersymmetry transformations. In the current case, a judicious choice is the following,

$$\mathcal{L}^{(5)} = \lambda^{16} \widehat{M}^4 \mathcal{E}_4^{(14,-14)} + \lambda^{15} \gamma^\mu \psi_\mu^* \widehat{M}^4 \mathcal{E}_4^{(13,-13)}$$
$$+ \lambda^{16} \widehat{M}^2 \widehat{M}_{\rho_1 \rho_2 \rho_3} \widehat{M}^{\rho_1 \rho_2 \rho_3} \widetilde{\mathcal{E}}_4^{(13,-13)} \tag{14.7.16}$$

As before, modular invariance demands that the coefficient functions $\mathcal{E}_4^{(w,-w)}$ be modular forms of weight $(w,-w)$, with $\mathcal{E}_4^{(0,0)} = \mathcal{E}_4$ the coefficient of $D^4 \mathcal{R}^4$ itself. Note that there are two coefficient functions of weight-$(13,-13)$ above, which we have denoted by $\mathcal{E}_4^{(13,-13)}$ and $\widetilde{\mathcal{E}}_4^{(13,-13)}$. We now proceed by again checking invariance under supersymmetry. Considering the term proportional to $\lambda^{16} \psi_\mu^* \epsilon$, we find only the contribution from $\delta^{(0)} \mathcal{L}^{(5)}$, and demanding that this vanishes gives the constraint,

$$2 D_{13} \, \mathcal{E}_4^{(13,-13)} - 11 \, \mathcal{E}_4^{(14,-14)} = 0 \tag{14.7.17}$$

Considering instead the term proportional to $\lambda^{16} \lambda^* \epsilon^*$, we obtain both a contribution from the $\delta^{(0)}$ variation of $\mathcal{L}^{(5)}$ as well as the $\delta^{(5)}$ variation of $S^{(0)}$. Demanding the vanishing of the combination gives the constraint,

$$2\overline{D}_{-14}\,\mathcal{E}_4^{(14,-14)} + 15\,\mathcal{E}_4^{(13,-13)} - \frac{9i}{16}\,\widetilde{\mathcal{E}}_4^{(13,-13)} - 1080i\,g_1 - \frac{3}{4}i\,g_2 = 0$$

$$(14.7.18)$$

with $g_1(\tau,\overline{\tau})$ and $g_2(\tau,\overline{\tau})$ undetermined functions. Closure of the supersymmetry algebra on λ^* further gives rise to the following three equations,

$$-192i\,D_{13}g_1 = \mathcal{E}_4^{(14,-14)} \qquad\qquad ig_2 + 191ig_1 = \tfrac{1}{2}\mathcal{E}_4^{(13,-13)}$$

$$-108\,g_1 = \widetilde{\mathcal{E}}_4^{(13,-13)} \qquad\qquad\qquad\qquad (14.7.19)$$

These constraints can then be combined to obtain a Laplace eigenvalue equation for $\mathcal{E}_4^{(14,-14)}$, namely,

$$\Delta\mathcal{E}_4^{(14,-14)} = \left(-182 + \tfrac{15}{4}\right)\mathcal{E}_4^{(14,-14)} \qquad\qquad (14.7.20)$$

Recalling that $\mathcal{E}_4^{(14,-14)} = D_{13}\ldots D_0\mathcal{E}_4$, we conclude that $\mathcal{E}_4$ satisfies,

$$\Delta\mathcal{E}_4(\tau) = \tfrac{15}{4}\,\mathcal{E}_4(\tau) \qquad\qquad (14.7.21)$$

This eigenvalue equation, together with the asymptotics at the cusp, fix $\mathcal{E}_4$ to be equal to $E_{5/2}(\tau)$ up to an overall constant factor.

Finally, for the coefficient of $D^6\mathcal{R}^4$, analogous steps can again be carried out, though one now obtains an *inhomogeneous* Laplace eigenvalue equation,

$$(\Delta - 12)\mathcal{E}_6(\tau) = -6\pi^3 E_{3/2}(\tau)^2 \qquad\qquad (14.7.22)$$

the solution to which is no longer an Eisenstein series. No solution is known in closed form, but it is possible to get an explicit expression for the Laurent polynomial $\mathcal{E}_6^{(0)}$ of $\mathcal{E}_6$, namely, the portions of $\mathcal{E}_6$ which are power-law in τ_2 and independent of τ_1. Fortunately, the Laurent polynomial pieces are the most interesting for the present purposes, since they are the ones that can be compared to calculations in string perturbation theory.

To obtain the Laurent polynomial $\mathcal{E}_6^{(0)}(\tau)$, we begin by inserting the Fourier series of $E_{3/2}(\tau)$, given in (4.2.17), into (14.7.22) and keeping only the Laurent polynomial pieces,

$$(\Delta - 12)\mathcal{E}_6^{(0)}(\tau) = -6\left(2\zeta(3)\tau_2^{3/2} + \frac{2}{3}\pi^2\tau_2^{-1/2}\right)^2 \qquad\qquad (14.7.23)$$

$$-384\pi^2\tau_2\sum_{N\neq 0}|N|^{-2}\sigma_2(|N|)^2 K_1(2\pi\tau_2|N|)^2$$

Because the Laplacian is given by $\Delta = \tau_2^2(\partial_{\tau_1}^2 + \partial_{\tau_2}^2)$, and hence the derivatives $\partial_{\tau_2}^2$ are dressed with factors of τ_2^2, the function $\mathcal{E}_6^{(0)}$ must be a polynomial in τ_2 with powers being those appearing on the right side, namely, τ_2^3, τ_2, and τ_2^{-1}, together with the powers solving the homogeneous equation $(\Delta - 12)f = 0$, namely, τ_2^{-3} and τ_2^4. The term τ_2^4 can be ruled out on physical grounds, since there should be no term more singular than the tree-level τ_2^3 term. On the other hand, the τ_2^{-3} contribution corresponds to a three-loop contribution which can be (and indeed is) nonzero.

The general Ansatz for $\mathcal{E}_6^{(0)}(\tau)$ is then,

$$\mathcal{E}_6^{(0)}(\tau) = a_1\tau_2^3 + a_2\tau_2 + a_3\tau_2^{-1} + a_4\tau_2^{-3} \tag{14.7.24}$$

By plugging into (14.7.22), we may uniquely fix a_1, a_2, and a_3, giving,

$$\mathcal{E}_6^{(0)}(\tau) = 4\zeta(3)^2\tau_2^3 + \frac{4}{3}\pi^2\zeta(3)\tau_2 + \frac{4\pi^4}{15}\tau_2^{-1} + a_4\tau_2^{-3} \tag{14.7.25}$$

These match with the expectations from string perturbation given in (12.8.4), (12.11.16), and (12.12.8) at tree-level, one-loop, and two-loops, respectively. The three-loop coefficient a_4 cannot be fixed by simply inserting the Ansatz into (14.7.22), since this term already solves the homogenous equation. Instead, this coefficient was fixed in [GV06] by multiplying both sides of the Laplace equation by $E_4(\tau)$ and integrating over the fundamental domain, using the results presented in Lemma 12.11.1. The final result is,

$$a_4 = \frac{8\pi^6}{8505} \tag{14.7.26}$$

We note in closing that beyond $D^6\mathcal{R}_4$, the terms in the low-energy effective action are no longer protected by supersymmetry, and thus we do not expect to be able to extend the above analysis to study them.

14.7.1 Non-renormalization theorems

Perhaps the most striking implications of the modular structure of the $\mathcal{R}^4$, $D^4\mathcal{R}^4$, and $D^6\mathcal{R}^4$ effective interactions is that they make predictions for the amplitudes at all orders of perturbation theory, that is, all genera. From the Fourier series expansion of the functions $E_{\frac{3}{2}}(\tau)$ we see that, aside from the leading perturbative contribution, there is only one other perturbative contribution, given by the second term, and behaving as $\tau_2^{-\frac{1}{2}}$. In string frame, this gives a contribution of order τ_2^0 which corresponds to genus one. This means that from superstring perturbation theory, the $\mathcal{R}^4$ effective interaction receives contributions from genus zero and genus one only. There are no contributions to genus 2 or higher. Similarly, for the $D^4\mathcal{R}^4$ effective interaction, there is a genus zero contribution and the only other contribution behaves as $\tau_2^{-\frac{3}{2}}$ in the Einstein frame, or τ_2^{-2} in the string frame. Hence there is no genus-one correction to the $D^4\mathcal{R}^4$ effective interaction, but there is a nonzero genus-two contribution. There are no higher contributions.

• Bibliographical notes

The existence of a web of dualities between different supergravities and string theories was proposed in [HT95] and led to the discovery of M-theory in [Wit95b]. The action of $\mathrm{SL}(2,\mathbb{Z})$ on branes in Type IIB was discovered in [Sch95] and geometrized in [Vaf96]. Evidence for the duality between Type I and Heterotic string theory was given in [PW96]. Useful lecture notes may be found in [Sch97], and an early account of the role played by automorphic functions in dualities may be found in [OP99] and [OP00]. The books by Polchinski [Pol07a, Pol07b], Johnson [Joh06], and Becker, Becker and Schwarz [BBS06] provide excellent and comprehensive accounts of string dualities.

The anomaly involving $\mathrm{SL}(2,\mathbb{Z})$ discussed in Section 14.3 was first described in [GG98], with important follow-ups in [Muk98] and [MSS17]. A more subtle anomaly in $\mathrm{SL}(2,\mathbb{Z})$, which poses an obstruction to its gauging in Type IIB supergravity in the absence of extra topological terms, was described in [DDHM21].

The use of S-duality to constrain the low-energy effective action goes back to [GG97], where the coefficient of the $\mathcal{R}^4$ term was related to the non-holomorphic Eisenstein series $E_{3/2}(\tau)$ using results from tree-level string theory and D-instantons. A derivation using space-time supersymmetry, as was reviewed above, was provided in [GS99], extended in [Pio98], and applied to the $D^4\mathcal{R}^4$ term in [GS99] and [Sin02]. The Laplace eigenvalue equation satisfied by the coefficient of the $D^6\mathcal{R}^4$ term, and related properties, were discussed in [GV06], [GRV10], [Pio15], and [GW19]. An implicit expression for the $D^6\mathcal{R}^4$ term may be found in [BKP20]. Implications for the presence of higher arithmetic groups were treated in [GMRV10], [GRV10], [GMV15b], and [GMV15a].

As mentioned in Subsection 14.7.1, the results quoted in the previous paragraph imply a number of non-renormalization theorems, some of which were anticipated from superstring perturbation theory in [Mar86] and [Ber04]. For the $\mathcal{R}^4$ term, the S-duality and supersymmetry analysis predict the absence of perturbative corrections beyond one-loop, and indeed the vanishing at genus two was proven in [DP05]. Conversely, for the $D^4\mathcal{R}^4$ term, there is expected to be no genus-one correction, as was verified in [GV00], but there is a nonzero genus-two contribution which was computed and found to match the predictions of S-duality and space-time supersymmetry in [DGP05].

The study of genus-one contributions to effective interactions of the form $D^{2\ell}\mathcal{R}^4$ with $\ell \geq 3$ was initiated in [GRV08] and [DGV15] and systematically developed in [DGGV17] and [DG18] using modular graph functions, which were introduced earlier in Chapter 9.

One of the hallmarks of superstring theory is the infinite tower of BPS states, which includes black hole micro-states. If a string background contains a torus, then the functions which count these micro-states often enjoy nice modular or automorphic properties. A well-studied setup is string theory compactified to four dimensions with $\mathcal{N} = 4$ supersymmetry which can be engineered via Type II strings on $K3 \times T^2$ or Heterotic strings on T^6. Either way, the theory is expected to have an $\mathrm{SL}(2,\mathbb{Z}) \times \mathrm{SO}(22,6,\mathbb{Z})$ duality symmetry, with the corresponding counting functions expressible as Siegel modular forms [DVV97]. The contribution of multicentered black holes was studied in [DKST17], while the single-centered black holes were shown to have interesting connections with mock modular and Jacobi forms in [DMZ12]. Recent connections have been made with Hurwitz class numbers [KT17] and class groups [BKOR18]. The tools of Rademacher sums and Farey tail expansions have been applied to the study of microstate counting in [DMMV00], [dBCD$^+$06], and [MM10]. An introduction to the relation between blackhole microstate counting and automorphic forms is in [Pio06].

Exercises

14.1 The next three exercises deal with the existence and structure of D7 brane solutions in Type IIB supergravity. A flat D7 brane has an 8-dimensional world-volume parametrized by $x^0, x^1, \cdots, x^7$ that is invariant under the Poincaré group $\mathrm{ISO}(1,7)$. In the space transverse to the D7 brane, which we parametrize by $z = x^8 + ix^9$, the D7 brane corresponds to a point, say $z = 0$, around which the axion–dilaton field τ is allowed to have $\mathrm{SL}(2,\mathbb{Z})$-valued monodromy. The antisymmetric tensor fields C_2, F_3, C_4, and F_5 all vanish, leaving only the metric G and τ governed by the action S_{IIB} of (14.1.8). Show that the general metric G_E satisfying these assumptions – in Einstein frame – takes the following form for a real-valued function $A(z)$,

$$G_{E\mu\nu}dx^\mu dx^\nu = -(dx^0)^2 + (dx^1)^2 + \cdots + (dx^7)^2 + e^{A(z)}|dz|^2$$

Show that the field equations for τ admit solutions for which τ is locally holomorphic in z and has SL$(2, \mathbb{Z})$-valued monodromy.

14.2 Obtain the equation for the function A of Exercise 14.1 from Einstein's equations; show that its solution gives the metric G_E for a cone asymptotically far away from the brane; compute the deficit angle of this cone.

14.3 Generalize your solutions to Exercises 14.1 and 14.2 to the case of multiple D7 branes. Show that a globally well-defined solution requires 24 D7 branes.

14.4 Type IIB string theory has two real 2-form fields (C_2^1, C_2^2), which are the components of the complex 2-form C_2, and we can consider strings carrying charges (p, q) under them. The tension of a charge (p, q) string is given (in Einstein frame) by

$$T_{(p,q)} = \frac{1}{2\pi\alpha'} \frac{1}{\sqrt{\tau_2}} |p + \tau q| \qquad (14.7.27)$$

Deduce from this expression the transformation of charge (p, q) strings under SL$(2, \mathbb{Z})$. Check that the transformation rule is consistent with the one predicted from the worldsheet coupling $p \int C_2^1 + q \int C_2^2$.

14.5 Type IIB string theory also has two scalar fields (χ, Φ), which can be used to define charge (p, q) 7-branes. As seen in Exercise 14.1, a D7-brane leads to monodromy $M_{(1,0)} = T$ acting on $\tau = \chi + ie^{-\Phi}$ by $\tau \to \tau + 1$. It can be shown that (p, q) strings can end on (p, q) 7-branes, and *only* on charge (p, q) 7-branes. Use this fact to identify the monodromy matrix $M_{(p,q)}$ for a (p, q) 7-brane.

15

Dualities in $\mathcal{N} = 2$ super Yang–Mills theories

In this penultimate chapter, we shall discuss dualities in Yang–Mills theories with extended supersymmetry in four-dimensional Minkowski spacetime. We briefly review supersymmetry multiplets of states and fields and the construction of supersymmetric Lagrangian theories with $\mathcal{N} = 1, 2, 4$ Poincaré supersymmetries. We then discuss the $\mathrm{SL}(2, \mathbb{Z})$ Montonen–Olive duality properties of the maximally supersymmetric $\mathcal{N} = 4$ theory and the low energy effective Lagrangians for $\mathcal{N} = 2$ theories via the Seiberg–Witten solution. We close the chapter with a discussion of dualities of $\mathcal{N} = 2$ superconformal gauge theories, which possess interesting spaces of marginal gauge couplings. In some cases, these spaces of couplings can be identified with the moduli spaces for Riemann surfaces of various genera.

15.1 Super Yang–Mills: states and fields

The quantum field theories we shall consider here are invariant under the Poincaré algebra $\mathrm{ISO}(1, 3)$, whose generators are translations P_μ and Lorentz transformations $L_{\mu\nu}$ satisfying the following structure relations,[1]

$$[L_{\mu\nu}, L_{\rho\sigma}] = i\left(\eta_{\nu\rho}L_{\mu\sigma} + \eta_{\mu\sigma}L_{\nu\rho} - \eta_{\mu\rho}L_{\nu\sigma} - \eta_{\nu\sigma}L_{\mu\rho}\right)$$
$$[L_{\mu\nu}, P_\rho] = i\left(\eta_{\nu\rho}P_\mu - \eta_{\mu\rho}P_\nu\right)$$
$$[P_\mu, P_\nu] = 0 \qquad\qquad (15.1.1)$$

In supersymmetric theories, the Poincaré algebra is extended to a super Poincaré algebra by supplementing P_μ and $L_{\mu\nu}$ with $\mathcal{N}$ supercharges Q_α^I

---

[1] Our conventions follow [WB92]. The nonzero components of the Minkowski metric $\eta_{\mu\nu}$ are $-\eta_{00} = \eta_{11} = \eta_{22} = \eta_{33} = 1$. The two-component Dirac matrices satisfy $\{\sigma^\mu, \bar\sigma^\nu\} = -2\eta^{\mu\nu}I_2$, where $\sigma^0 = \bar\sigma^0 = -I_2$, I_2 is the identity matrix, and $\sigma^i = -\bar\sigma^i$ are the Pauli matrices for $i = 1, 2, 3$. The spinor indices $\alpha = 1, 2, \dot\alpha = \dot{1}, \dot{2}$ are raised and lowered with the help of the antisymmetric symbols $\epsilon^{\alpha\beta}, \epsilon^{\dot\alpha\dot\beta}, \epsilon_{\alpha\beta}$, and $\epsilon_{\dot\alpha\dot\beta}$ normalized to $\epsilon^{12} = \epsilon^{\dot{1}\dot{2}} = 1$ and $\epsilon_{12} = \epsilon_{\dot{1}\dot{2}} = -11$. We use the notation $\psi\chi = \psi^\alpha\chi_\alpha$, $\bar\psi\bar\chi = \bar\psi_{\dot\alpha}\bar\chi^{\dot\alpha}$ and $\chi\sigma^\mu\bar\psi = \chi^\alpha\sigma^\mu_{\alpha\dot\alpha}\bar\psi^{\dot\alpha}$.

and their adjoints $\bar{Q}_{\dot{\alpha}I} = (Q_\alpha^I)^\dagger$ for spinor index $\alpha = 1, 2$ and $I = 1, \cdots, \mathcal{N}$. The supercharges commute with P_μ, are Weyl spinors under the Lorentz generators $L_{\mu\nu}$ and obey the following anticommutation relations,

$$\{Q_\alpha^I, \bar{Q}_{\dot{\beta}J}\} = 2\,\sigma^\mu_{\alpha\dot{\beta}}\,P_\mu\,\delta^I_J \qquad\qquad I, J = 1, \cdots, \mathcal{N}$$

$$\{Q_\alpha^I, Q_\beta^J\} = 2\,\epsilon_{\alpha\beta}\,Z^{IJ} \qquad\qquad\qquad (15.1.2)$$

The generators $Z^{IJ} = -Z^{JI}$ can arise only when $\mathcal{N} \geq 2$. They are central generators in the Poincaré superalgebra since they commute with one another and with all other generators. When $Z^{IJ} = 0$, the supersymmetry algebra is invariant under a $U(\mathcal{N})_R$ automorphism group which, in physics, is referred to as the R-symmetry group. The mass squared operator $M^2 = -P^\mu P_\mu$ is a Casimir operator for the super Poincaré algebra and commutes with $P_\mu, L_{\mu\nu}, Q_\alpha^I$, and $\bar{Q}_{\dot{\beta}J}$.

15.1.1 States

The spectrum of states in a supersymmetric theory exhibits a characteristic pattern. All states have positive or zero energy. If the state of lowest energy in the spectrum – namely the ground state – has exactly zero energy then supersymmetry is said to be *unbroken* or *manifest*. In this case, all positive energy states occur in boson–fermion pairs of equal mass, momentum, and internal quantum numbers. If the ground state has strictly positive energy, however, the ground state is not supersymmetric and supersymmetry is said to be *spontaneously broken*. Bosons and fermions of identical internal quantum numbers do not generally have the same mass.

In this chapter, we shall consider only the case where supersymmetry is manifest. Applying a supercharge to a state changes the spin by $\pm\frac{1}{2}$ so that the values of the spin in the multiplets of the Poincaré superalgebra must increase with $\mathcal{N}$. When $\mathcal{N} > 4$, the multiplets necessarily involve states of spin greater than 1, which cannot be accommodated in Yang–Mills theory. For $\mathcal{N} \leq 8$ they can be accommodated in supergravity, but this requires states of spin 2. Thus $\mathcal{N} = 4$ is the maximal number of Poincaré supersymmetries allowed for supersymmetric Yang–Mills theories in four dimensions. As soon as $\mathcal{N} \geq 2$, there is an important interplay between the mass of the state and its central charge which we shall now discuss.

• For $\mathcal{N} = 2$, there is a single complex-valued central charge $Z = Z^{12}$, which we decompose into its absolute value and phase, $Z = e^{i\varphi}|Z|$ for $\varphi \in \mathbb{R}$. We consider a state $|M, Z\rangle$ with mass $M > 0$ and central charge Z and use a Poincaré transformation to the rest frame of the state with momentum

$P^\mu = (M, 0, 0, 0)$. In terms of the combinations $\mathcal{Q}_\alpha^\pm = \frac{1}{2}(Q_\alpha^1 \mp e^{i\varphi} \sigma_{\alpha\dot{\beta}}^0 \bar{Q}_2^{\dot{\beta}})$, the anticommutation relations of (15.1.2), acting on the state $|M, Z\rangle$, are equivalent to the following relations,

$$\{\mathcal{Q}_\alpha^\pm, (\mathcal{Q}_\beta^\pm)^\dagger\} = \delta_\alpha^\beta (M \pm |Z|) \tag{15.1.3}$$

while all other anticommutators vanish identically. Since the operators $\{\mathcal{Q}_\alpha^\pm, (\mathcal{Q}_\alpha^\pm)^\dagger\}$ are self-adjoint and positive in a unitary theory, we obtain the following so-called *BPS bound* between the mass M and the central charge Z of an arbitrary state $|M, Z\rangle$,

$$M \geq |Z| \tag{15.1.4}$$

In the limit where the state becomes massless, its central charge must vanish. Of particular interest are the massive states for which the BPS bound is saturated, say by $M = |Z|$. In such a state, which is referred to as a BPS state, the supersymmetry algebra reduces to,

$$\{\mathcal{Q}_\alpha^+, (\mathcal{Q}_\beta^+)^\dagger\} = 2M\delta_\alpha^\beta \qquad \{\mathcal{Q}_\alpha^-, (\mathcal{Q}_\beta^-)^\dagger\} = 0 \tag{15.1.5}$$

while all other anticommutators vanish. The operators $\mathcal{Q}_\alpha^-$ and $(\mathcal{Q}_\alpha^-)^\dagger$ both annihilate the BPS state which means that $\mathcal{Q}_\alpha^- |M, Z\rangle = (\mathcal{Q}_\alpha^-)^\dagger |M, Z\rangle = 0$ so that BPS states are invariant under half of the total number of Poincaré supersymmetries. More specifically, such states are referred to as $\frac{1}{2}$-BPS states, and the action of the remaining supercharges $\mathcal{Q}_\alpha^+$ produces a supermultiplet that is half as long as the non-BPS supermultiplets.

• For $\mathcal{N} = 4$, the matrix of central charges $Z^{IJ} = -Z^{JI}$ has two complex eigenvalues Z_1 and Z_2. The BPS bound is now $M \geq |Z_1|$ and $M \geq |Z_2|$. When $|Z_1| \neq |Z_2|$, one obtains $\frac{1}{2}$-BPS states by setting M equal to the maximum of $|Z_1|$ and $|Z_2|$. But when $|Z_1| = |Z_2|$, a further shortening of the supersymmetry multiplet takes place to $\frac{1}{4}$-BPS states. CPT invariance in four space-time dimensions implies that Lagrangian supersymmetric Yang–Mills theories with $\mathcal{N} = 3$ automatically enjoy the full $\mathcal{N} = 4$ supersymmetry, so that the case $\mathcal{N} = 3$ need not be considered separately (though this is no longer true if we give up the assumption of existence of a Lagrangian).

15.1.2 Fields

A super Yang–Mills theory with gauge group G and associated Lie algebra $\mathfrak{g}$ is constructed in terms of the customary fields of four-dimensional quantum field theory absent gravity, namely, spin-1 gauge fields A_μ in the adjoint representation of G, together with spin-$\frac{1}{2}$ left-handed Weyl fermion fields

$\psi_\alpha, \lambda_\alpha$ and spin-0 scalar fields ϕ and h in various representations of G. Recall that, in even dimensions, a Dirac spinor is the direct sum of two Weyl spinors of opposite chirality and that, in four dimensions, a Majorana spinor is equivalent to a Weyl spinor. The field multiplets for $\mathcal{N} = 1, 2, 4$ are as follows:

- $\mathcal{N} = 1$ *Gauge multiplet* (A_μ, λ_α) in the adjoint representation of G, where A_μ is the gauge field and λ_α is the *gaugino* left Weyl fermion;
  $\mathcal{N} = 1$ *Chiral multiplet* (ψ_α, ϕ) in a representation $\mathcal{R}$ of G, where ψ_α is a left Weyl fermion and ϕ is a complex scalar field;

- $\mathcal{N} = 2$ *Gauge multiplet* $(A_\mu, \lambda_\alpha, \tilde{\lambda}_\alpha, \phi)$ in the adjoint representation of G, where λ and $\tilde{\lambda}$ are left Weyl gaugino fermions in a doublet of $\mathrm{SU}(2)_R$, while A_μ and ϕ are singlets under $\mathrm{SU}(2)_R$;
  $\mathcal{N} = 2$ *Hypermultiplet* $(h, \tilde{h}, \psi_\alpha, \tilde{\psi}_\alpha)$ in a representation $\mathcal{R}$ of G, where $(h, \tilde{h})$ is a doublet of $\mathrm{SU}(2)_R$, while the Weyl fermions $\psi_\alpha, \tilde{\psi}_\alpha$ are singlets;

- $\mathcal{N} = 4$ *Gauge multiplet* $(A_\mu, \lambda_\alpha^A, \phi^I)$ in the adjoint representation of G, where λ_α^A with $A = 1, \cdots, 4$ are left Weyl gaugino fermions in the **4** of $\mathrm{SU}(4)_R$, and ϕ^I with $I = 1, \cdots, 6$ are complex scalars in the **6** of $\mathrm{SU}(4)_R$, while A_μ is a singlet of $\mathrm{SU}(4)_R$.

The fields of an $\mathcal{N} = 1$ theory may be collected in fully off-shell *superfields* with linear transformations under the action of the super Poincaré algebra, as will be made explicit in Section 15.2.

The construction of superfields for higher numbers of supersymmetries is, however, more complicated and will not be addressed here. For this reason, it will be useful to decompose the field content of the $\mathcal{N} = 2$ and $\mathcal{N} = 4$ theories in terms of $\mathcal{N} = 1$ superfields. In this spirit, the $\mathcal{N} = 2$ gauge multiplet is the direct sum of an $\mathcal{N} = 1$ gauge multiplet and an $\mathcal{N} = 1$ chiral multiplet, both in the adjoint representation of G. The $\mathcal{N} = 2$ hypermultiplet is a sum of two $\mathcal{N} = 1$ chiral multiplets in complex conjugate representations of one another $\mathcal{R} \oplus \bar{\mathcal{R}}$. Similarly, the $\mathcal{N} = 4$ gauge multiplet is the direct sum of an $\mathcal{N} = 2$ gauge multiplet and an $\mathcal{N} = 2$ hypermultiplet, both in the adjoint representation of G. Note that for $\mathcal{N} = 4$ theories, the $\mathrm{U}(1)_R$ factor of the automorphism group $\mathrm{U}(\mathcal{N})_R$ of the supersymmetry algebra is anomalous and fails to be a quantum symmetry. The charge assignments of the $\mathrm{U}(1)_R$ factor in the case of $\mathcal{N} = 1$ and $\mathcal{N} = 2$ will not be important in the sequel, and we shall not spell them out here.

Before moving on, we remark that when $\mathcal{R}$ is a pseudo-real representation, and hence admits an antisymmetric invariant tensor ϵ_{ij}, the fields h_i and $\tilde{h}^j$ in the $\mathcal{N} = 2$ hypermultiplet (with indices $i, j = 1, \cdots, \dim \mathcal{R}$) can be constrained to satisfy,

$$h_i = \epsilon_{ij} \tilde{h}^j \tag{15.1.6}$$

This constraint is compatible with $\mathcal{N} = 2$ supersymmetry, and halves the number of degrees of freedom in the multiplet. The resulting multiplet is referred to as a *half-hypermultiplet*. More generally, given a complex representation $\mathcal{R}$ there is a standard pseudo-real structure on $\mathcal{R} \oplus \overline{\mathcal{R}}$, and hence we may define half-hypermultiplets in the representation $\mathcal{R} \oplus \overline{\mathcal{R}}$. These are precisely the same as full hypermultiplets in the complex representation $\mathcal{R}$.

15.2 Super Yang–Mills: Lagrangians

Two types of supersymmetric Lagrangians will be of interest to us: those that correspond to renormalizable quantum field theories, and those that collect the low energy effective interactions of some arbitrary Lagrangian or non-Lagrangian quantum field theory. In either case, we shall restrict attention to Lagrangians in which bosonic fields have a total of at most two derivatives, while fermions have a total of at most one derivative acting on the fields. Furthermore, the Lagrangians for theories with $\mathcal{N} = 2$ and $\mathcal{N} = 4$ supersymmetries will be expressed in terms of $\mathcal{N} = 1$ chiral and vector superfields, using the decomposition of the corresponding multiplets given in the penultimate paragraph of the preceding section. Throughout we shall assume that the gauge group G is compact with Lie algebra[2] $\mathfrak{g}$ and may be written as the direct product of a semi-simple group times a certain number of $U(1)$ factors times a finite discrete group.

To introduce $\mathcal{N} = 1$ superfields, we supplement the coordinates x^μ of Minkowski space-time with Grassmann spinor variables θ^α and $\bar{\theta}^{\dot\alpha}$ of odd grading. The superfield Φ for the $\mathcal{N} = 1$ chiral multiplet containing the fields (ϕ, ψ_α) is given by,

$$\Phi = \phi + \sqrt{2} \theta \psi + \theta \theta F \tag{15.2.1}$$

where F is a nondynamical complex-valued scalar *auxiliary field*. All fields are evaluated at $y_\mu = x_\mu + i \theta \sigma_\mu \bar{\theta}$. The chiral superfield Φ is generally matrix-valued with components Φ_i and transforms under an arbitrary representation $\mathcal{R}$ of the gauge algebra $\mathfrak{g}$ with $i = 1, \cdots, \dim \mathcal{R}$. The superfield V for the $\mathcal{N} = 1$ gauge multiplet satisfies a reality condition $V^\dagger = V$ and transforms

---

[2] The gauge algebra $\mathfrak{g}$ should not be confused with the string worldsheet metric, which was denoted by the same symbol in Chapters 5 and 12.

in the adjoint representation of the gauge algebra $\mathfrak{g}$. The general decomposition of the superfield V is more complicated than that of the chiral multiplet, and we present it here in *Wess–Zumino gauge* only,

$$V = -\theta\sigma^\mu\bar\theta A_\mu + i\theta\theta\bar\theta\bar\lambda - i\bar\theta\bar\theta\theta\lambda + \frac{1}{2}\theta\theta\bar\theta\bar\theta\,(D - i\partial_\mu A^\mu) \qquad (15.2.2)$$

where D is a real-valued scalar auxiliary field. Once again, all fields are evaluated at $y_\mu = x_\mu + i\theta\sigma_\mu\bar\theta$. The vector superfield V is generally matrix-valued and may be decomposed $V = \sum_a T^a V^a$ onto the generators T^a of $\mathfrak{g}$, with $a = 1, \cdots, \dim\mathfrak{g}$, in the adjoint representation of the gauge algebra $\mathfrak{g}$.

15.2.1 $\mathcal{N} = 1$ Lagrangians

In terms of the $\mathcal{N} = 1$ gauge superfield V and chiral superfield Φ, the most general supersymmetric Lagrangian density, under the assumptions spelled out earlier, is of the following form,

$$\mathcal{L} = \int d^4\theta\, K(e^V\Phi, \Phi^\dagger) + \mathrm{Re}\int d^2\theta\Big(U(\Phi) + \tau_{ab}(\Phi)W^a W^b\Big) \qquad (15.2.3)$$

in superspace notation. Note that K is the real-valued *Kähler potential*, U is the *superpotential*, and τ_{ab} is the matrix of couplings and mixings of the field strengths W^a_α of the gauge fields V^a. The functions $U(\Phi)$ and $\tau_{ab}(\Phi)$ are locally holomorphic in Φ. The superfield $W_\alpha = \sum_a W^a_\alpha T^a$ is obtained from the vector superfield V by taking,

$$W_\alpha = -\frac{1}{4}\bar D\bar D\left(e^{-V}D_\alpha e^V\right) \qquad\qquad D_\alpha = \frac{\partial}{\partial\theta^\alpha} + i\sigma^\mu_{\alpha\dot\alpha}\bar\theta^{\dot\alpha}\partial_\mu \qquad (15.2.4)$$

The field W_α is a chiral superfield since $\bar D^{\dot\alpha}W_\beta = 0$ by construction.

15.2.2 Renormalizable $\mathcal{N} = 2$ Lagrangians

A general four-dimensional $\mathcal{N} = 2$ Lagrangian theory is specified by a gauge algebra $\mathfrak{g}$ which determines the vector multiplets, as well as a representation $\mathcal{R}$ of $\mathfrak{g}$ for the hypermultiplets, or half-hypermultiplets if $\mathcal{R}$ is pseudo-real.[3] The $\mathcal{N} = 2$ gauge multiplet consists of an $\mathcal{N} = 1$ gauge multiplet V and an $\mathcal{N} = 1$ chiral multiplet Φ in the adjoint representation of $\mathfrak{g}$, while the $\mathcal{N} = 2$

---

[3] This data is not quite complete: in general, additional global data such as the spectrum of line operators is necessary to fully specify the theory. We will discuss such additional data in the context of $\mathcal{N} = 4$ Yang–Mills in Section 15.5 but, for the time being, will ignore this subtlety. As long as one is concerned only with the local operator spectrum of the theory, no harm arises from this omission.

hypermultiplet consists of two $\mathcal{N} = 1$ chiral multiplets $H, \tilde{H}$ in the representations $\mathcal{R}$ and $\bar{\mathcal{R}}$ of $\mathfrak{g}$, respectively. The $\mathcal{N} = 2$ half-hypermultiplet consists of a single $\mathcal{N} = 1$ chiral multiplet H in a pseudo-real representation $\mathcal{R}$.

Let us consider first a renormalizable $\mathcal{N} = 2$ theory without hypermultiplets. We begin by arguing that the superpotential $U(\Phi)$ must vanish. Indeed, for the theory to be renormalizable, one requires that τ_{ab} be independent of Φ and that the superpotential $U(\Phi)$ be at most cubic in the chiral superfield Φ. Upon integrating over superspace, the final two terms of the Lagrangian (15.2.3) are the real parts of,

$$\int d^2\theta \, U(\Phi) = F^a \partial_a U(\phi) + \psi^a \psi^b \partial_a \partial_b U(\phi) \tag{15.2.5}$$

$$\int d^2\theta \, \tau_{ab} \, W^a W^b = -\tau_{ab} \left(\tfrac{i}{2} F^a_{\mu\nu} F^{b\mu\nu} - \tfrac{1}{2} F^a_{\mu\nu} \tilde{F}^{b\mu\nu} + \overline{\lambda}^a \overline{\sigma}^\mu D_\mu \lambda^b \right)$$

where $\partial_a = \partial/\partial\phi^a$. To obtain a Lagrangian with $\mathcal{N} = 2$ supersymmetry, the combined contributions of the $\mathcal{N} = 1$ gauge and chiral multiplets must be invariant under the $\mathrm{SU}(2)_R$ symmetry rotating ψ_α into λ_α, which is possible only if $U(\Phi)$ is linear in Φ. Upon integrating out the auxiliary field F, a linear $U(\Phi)$ would add a constant $|\partial U|^2$ to the energy density and spontaneously break supersymmetry. Therefore, in a theory with a supersymmetric vacuum, the superpotential U is a constant that may be set to zero.

What this means is that the only potential for the scalar fields ϕ^a comes from the Kähler potential. To proceed, we shall assume for simplicity that the gauge group G is a simple Lie group, so that there is a single $\mathcal{N} = 1$ vector superfield V transforming under the adjoint representation of $\mathfrak{g}$. For a renormalizable theory, the Kähler potential must be *canonical*, that is, $K(e^V \Phi, \Phi^\dagger) = \mathrm{Tr} \left(\Phi^\dagger e^V \Phi \right)$. In that case, the auxiliary field D may be integrated out using its field equations and the resulting scalar potential takes the form,

$$V_{\mathrm{gauge}}(\phi, \bar{\phi}) = \mathrm{Tr} \left([\phi, \bar{\phi}] \right)^2 \tag{15.2.6}$$

The structure of this potential will be of central importance in disentangling the vacuum structure of the $\mathcal{N} = 2$ theories.

Next, we include an $\mathcal{N} = 2$ hypermultiplet, which consists of $\mathcal{N} = 1$ chiral superfields $H = (h, \psi_\alpha)$ and $\tilde{H} = (\tilde{h}, \tilde{\psi}_\alpha)$ in representations $\mathcal{R}$ and $\bar{\mathcal{R}}$ of $\mathfrak{g}$, respectively. The representation $\mathcal{R} \oplus \bar{\mathcal{R}}$ may be reducible into a direct sum of irreducible representations $\mathcal{R}_f$ and $\bar{\mathcal{R}}_f$ of $\mathfrak{g}$,

$$\mathcal{R} = \bigoplus_f n_f \, \mathcal{R}_f \oplus \bar{\mathcal{R}}_f \tag{15.2.7}$$

where n_f is the number of hypermultiplets in representation $\mathcal{R}_f$. We decompose the hypermultiplet accordingly into hypermultiplets H_f and $\tilde{H}_f$ in the representations $\mathcal{R}_f$ and $\bar{\mathcal{R}}_f$, respectively. The scalar fields h_f and $\tilde{h}_f$ in the hypermultiplets produce the following addition to the scalar potential,

$$V_{\text{hyper}}(h, \tilde{h}, \phi) = \sum_{a=1}^{\dim \mathfrak{g}} \phi_a \tilde{h}^t T^a h + \sum_f \sum_{s,t=1}^{n_f} M_{f,s,t} h_{f,s} \tilde{h}_{f,t} \qquad (15.2.8)$$

Recall that ϕ is the scalar component of the superfield Φ; the T^a are the representation matrices of the representation $\mathcal{R}$ of the gauge algebra $\mathfrak{g}$; the index $a = 1, \cdots, \dim \mathfrak{g}$ labels the adjoint representation of $\mathfrak{g}$; and the indices $s, t = 1, \ldots, n_f$ label the flavors in each irreducible representation $\mathcal{R}_f$. The matrix $M_{f,s,t}$ contains parameters that can be interpreted as masses and mixings of the hypermultiplets in a given representation $\mathcal{R}_f$ of $\mathfrak{g}$. Gauge invariance precludes mixings between hypermultiplets $\mathcal{R}_f$ and $\mathcal{R}_{f'}$ with $f' \neq f$. Combining this scalar potential with that in (15.2.6) allows for interesting structure for the moduli space of vacua.

15.3 Low-energy Lagrangian on the Coulomb branch

In a supersymmetric Yang–Mills theory, the energy of the vacuum state is zero and the energy of any other state is strictly positive. In the sequel, we shall assume that the gauge group of the UV theory is a compact semi-simple Lie group G with Lie algebra $\mathfrak{g}$. In particular, the potential energy V_{gauge} of (15.2.6) is nonnegative since the Cartan–Killing form of $\mathfrak{g}$ is negative definite. Any supersymmetric vacuum state has exactly zero energy, which requires vanishing fermion fields, $D_\mu \phi = F_{\mu\nu} = 0$ and, for the solutions to these equations in the gauge $A_\mu = 0$, a constant field ϕ satisfying,

$$[\phi, \bar{\phi}] = 0 \qquad (15.3.1)$$

Decomposing $\phi = \phi_1 + i\phi_2$ into Hermitian matrices ϕ_1, ϕ_2 gives the equivalent condition $[\phi_1, \phi_2] = 0$, which guarantees that ϕ_1 and ϕ_2 may be simultaneously diagonalized. As a result, the general solution to (15.3.1) for the vacuum expectation value of the field ϕ may be parametrized as follows,

$$\phi = \sum_{I=1}^{r} a_I \mathfrak{h}_I \qquad\qquad r = \text{rank}(\mathfrak{g}) \qquad (15.3.2)$$

where $\mathfrak{h}_I$ for $I = 1, \cdots, r = \text{rank}(\mathfrak{g})$ are the mutually commuting generators of the Cartan subalgebra of $\mathfrak{g}$. For generic values of $a_I \neq 0$, the vacuum is referred to as the *Coulomb branch*. The standard Higgs mechanism breaks the gauge symmetry G to the Cartan subgroup $\mathrm{U}(1)^r$. Actually, the Cartan subgroup is invariant under the Weyl group $\mathcal{W}(\mathfrak{g})$, which is the residual symmetry remaining from the non-Abelian part of G. Thus, the precise symmetry breaking pattern is as follows,

$$G \longrightarrow \mathrm{U}(1)^r/\mathcal{W}(\mathfrak{g}) \tag{15.3.3}$$

Therefore, the low-energy effective theory consists of r copies of $\mathcal{N} = 2$ electromagnetism with compact $U(1)$ gauge groups, up to identification under the Weyl group $\mathcal{W}(\mathfrak{g})$.

The field content of the low-energy theory consists of $\mathcal{N} = 2$ gauge multiplets with component fields $(A^I_\mu, \lambda^I_\alpha, \tilde{\lambda}^I_\alpha, \phi^I)$ for gauge group $U(1)^r$ and $I = 1, \cdots, r$, which may be decomposed into $\mathcal{N} = 1$ gauge superfields V^I and chiral superfields Φ^I. The theory is governed by the Lagrangian constructed in (15.2.3) for gauge group $U(1)^r$,

$$\mathcal{L} = \frac{1}{4\pi} \int d^4\theta \, K\left(\Phi, \Phi^\dagger\right) + \frac{1}{4\pi} \mathrm{Re} \int d^2\theta \, \tau_{IJ} \, W^I W^J \tag{15.3.4}$$

where we have inserted a conventional factor of 4π. The field strength superfields W^I of the Abelian gauge multiplets simplifies,

$$W^I_\alpha = -\frac{1}{4}\bar{D}\bar{D}D_\alpha V^I \tag{15.3.5}$$

and $\tau_{IJ}(\Phi)$ is locally holomorphic in the fields Φ^I. Since the chiral fields Φ^I are neutral under the gauge group $U(1)^r$ the factor e^V multiplying Φ in (15.2.3) is absent here. No superpotential $U(\Phi^I)$ is allowed in view of $\mathcal{N} = 2$ supersymmetry, as may be shown by the same argument used earlier in the case of renormalizable Lagrangians. Expressed in terms of component fields, the Lagrangian takes the form,[4]

$$4\pi\mathcal{L} = -g_{I\bar{J}}(\phi, \bar{\phi})\left(D_\mu\phi^I D^\mu\phi^{\bar{J}} + i\tilde{\lambda}^{\bar{J}}\bar{\sigma}^\mu D_\mu\tilde{\lambda}^I\right)$$
$$-\mathrm{Re}\left\{\tau_{IJ}(\phi)\left(\frac{i}{2}F^I_{\mu\nu}F^{J\mu\nu} - \frac{1}{2}F^I_{\mu\nu}\tilde{F}^{J\mu\nu} + \bar{\lambda}^I\bar{\sigma}^\mu D_\mu\lambda^J\right)\right\} \tag{15.3.6}$$

The $SU(2)_R$ symmetry of $\mathcal{N} = 2$ rotates λ^I into $\tilde{\lambda}^I$ so that invariance of $\mathcal{L}$ under $SU(2)_R$ requires the following relation between their coefficients τ_{IJ}

---

[4] We use the standard notation for coordinates on Kähler manifolds, where $\phi^{\bar{I}} = \bar{\phi}^I$ so that the metric is $g_{I\bar{J}} \, d\phi^I \otimes d\phi^{\bar{J}}$. The Levi–Civita connection Γ for the metric $g_{I\bar{J}}$ is compatible with the complex structure so that its mixed components $\Gamma^I_{\bar{J}K}$ and $\Gamma^{\bar{I}}_{J\bar{K}}$ and their complex conjugates vanish. The corresponding covariant derivatives are given by $D_\mu f^I = \partial_\mu f^I + \Gamma^I_{JK}\partial_\mu\phi^J \, f^K$ for the fields $f^I = \phi^I, \lambda^I, \tilde{\lambda}^I$.

and the Kähler metric $g_{I\bar{J}}$,

$$\operatorname{Im}\tau_{IJ}(\phi) = g_{I\bar{J}}(\phi,\bar{\phi}) = \frac{\partial^2 K(\phi,\bar{\phi})}{\partial\phi^I\partial\bar{\phi}^{\bar{J}}} \qquad (15.3.7)$$

Since $\tau_{IJ}(\phi)$ is locally holomorphic, this relation implies that the Kähler potential K must be of the *special Kähler* type, so that $\tau_{IJ}(\phi)$ and K are given by,

$$\tau_{IJ}(\phi) = \frac{\partial^2\mathcal{F}(\phi)}{\partial\phi^I\partial\phi^J} \qquad\qquad \phi_{DI} = \frac{\partial\mathcal{F}(\phi)}{\partial\phi^I}$$

$$K(\phi,\bar{\phi}) = \frac{1}{2i}\left(\bar{\phi}^I\,\phi_{DI} - \phi^I\,\bar{\phi}_{DI}\right) \qquad (15.3.8)$$

where $\mathcal{F}(\phi)$ is a locally holomorphic function of ϕ^I referred to as the *prepotential*. Since the prepotential $\mathcal{F}$ depends only on the field ϕ and not on any derivatives of ϕ, its expression is entirely determined by evaluating $\mathcal{F}$ on the vacuum expectation values of the field ϕ, given in (15.3.2).

15.4 BPS states, monopoles, and dyons

In addition to the massless states described by the canonical fields of the $\mathcal{N}=2$ gauge multiplet, the theory also contains the massive gauge boson states as well as the 't Hooft–Polyakov magnetic monopoles produced by the Higgs mechanism $\mathfrak{g} \to \mathfrak{u}(1)^r$. These states are BPS states and carry nonzero electric and/or magnetic charges under the various $\mathfrak{u}(1)$ fields. In this section, we shall review the properties of these states.

The massive vector bosons, and their superpartners, that arise via the Higgs mechanism from the spontaneous symmetry breaking $\mathfrak{g} \to \mathfrak{u}(1)^r$ correspond to the root generators $\mathfrak{e}_\alpha$ in the decomposition[5] of the generators of $\mathfrak{g}$, while the massless gauge bosons of $\mathfrak{u}(1)^r$ correspond to the Cartan generators $\mathfrak{h}_I$ identified in (15.3.2). For each root $\boldsymbol{\alpha}$ there is a unique massive $\mathcal{N}=2$ vector boson multiplet in the spectrum. Counting the number of states for each spin, we see that these massive vector bosons must belong to BPS multiplets. As a result, the BPS formula gives their masses in terms of their central charges,

$$M_\alpha^W = |\boldsymbol{\alpha}\cdot\mathbf{a}| \qquad\qquad \mathbf{a} = (a_1,\cdots,a_r) \qquad (15.4.1)$$

---

[5] Our notation for Cartan generators and roots for a rank r Lie algebra $\mathfrak{g}$ in the Cartan basis is as follows. The Cartan generators will be denoted $\mathfrak{h}_I$ with $I = 1,\cdots,r$, as we have already used in (15.3.2) when parametrizing the vacua of the Coulomb branch. The root generators are denoted $\mathfrak{e}_\alpha$, where the corresponding root vector is an r-dimensional vector with components α_I. Their commutators may now be expressed as follows, $[\mathfrak{h}_I,\mathfrak{h}_J]=0$, $[\mathfrak{h}_I,\mathfrak{e}_\alpha]=\alpha_I\mathfrak{e}_\alpha$, and $[\mathfrak{e}_\alpha,\mathfrak{e}_{-\alpha}]=\sum_I\alpha_I\mathfrak{h}_I$.

a formula that is a direct consequence of the Higgs mechanism by a scalar ϕ in the adjoint representation of the gauge algebra $\mathfrak{g}$.

The massive magnetic monopole states, and their superpartners, also arise via the Higgs mechanism $\mathfrak{g} \to \mathfrak{u}(1)^r$ as 't Hooft–Polyakov magnetic monopoles. In the classical limit, they arise as static solutions of finite mass that saturate the Bogomolnyi bound,

$$M = \frac{2}{g^2} \left| \int d^3x \, \boldsymbol{\nabla} \cdot \mathrm{tr}\big(\phi \, \mathbf{B}^I\big) \right| \qquad (15.4.2)$$

where $\mathbf{B}^I$ is the magnetic field component of the field strength $F_{\mu\nu}^I$ and the integral gives the magnetic charge of the field $F_{\mu\nu}^I$. Precisely one magnetic monopole arises for each possible embedding of $\mathfrak{su}(2)$ into $\mathfrak{g}$, that is, for each root $\boldsymbol{\beta}$ of $\mathfrak{g}$. This construction will give us a BPS formula (at the semi-classical level) for the mass of the 't Hooft–Polyakov magnetic monopole in terms of the vacuum expectation value of the dual gauge scalar ϕ_{DI}. Applying this construction for each SU(2) subgroup of G, namely, for each root $\boldsymbol{\beta}$ of $\mathfrak{g}$, we find the semiclassical magnetic monopole BPS mass formula,

$$M_{\boldsymbol{\beta}}^M = |\boldsymbol{\beta} \cdot \mathbf{a}_D| \qquad\qquad \mathbf{a}_D = (a_{D1}, \cdots, a_{Dr}) \qquad (15.4.3)$$

A dyon may be specified by two roots: one root $\boldsymbol{\alpha}$ for its electric charge as in the construction of vector boson masses and another (co)root $\boldsymbol{\beta}$ for its magnetic charge as in the construction of the magnetic monopole masses. The full quantum BPS mass formula for such a dyon is,

$$M_{\boldsymbol{\alpha},\boldsymbol{\beta}}^D = |\boldsymbol{\alpha} \cdot \mathbf{a} + \boldsymbol{\beta} \cdot \mathbf{a}_D| \qquad (15.4.4)$$

The theory further contains an infinite number of states that may be neutral or carry arbitrary electric and magnetic charges, but that do not saturate the BPS bound and are thus non-BPS states. No simple formula for their mass in terms of the other quantum numbers is known to exist. Henceforth, we restrict attention to the BPS spectrum only.

15.5 SL(2, $\mathbb{Z}$)-duality of the $\mathcal{N} = 4$ theory

For the $\mathcal{N} = 4$ theory, the prepotential is given by,

$$\mathcal{F}(\phi) = \frac{1}{2}\tau \sum_I \phi^I \phi^I \qquad (15.5.1)$$

so that $a_{DI} = \tau a_I$ and $\tau_{IJ} = \tau \delta_{IJ}$ with τ independent of ϕ. The spectrum of vector boson, magnetic monopole, and dyon BPS states is then given by,

$$M_{\boldsymbol{\alpha},\boldsymbol{\beta}}^D(\tau) = |\boldsymbol{\alpha} \cdot \mathbf{a} + \tau\boldsymbol{\beta} \cdot \mathbf{a}| \qquad (15.5.2)$$

The roots $\boldsymbol{\alpha}$ and $\boldsymbol{\beta}$ may be decomposed in a basis of simple roots $\boldsymbol{\alpha}_I$,

$$\boldsymbol{\alpha} = \sum_I n_I \boldsymbol{\alpha}_I \qquad \boldsymbol{\beta} = \sum_I m_I \boldsymbol{\alpha}_I \qquad (15.5.3)$$

where the coefficients m_I and n_I are integers. The masses of the BPS states are then given by,

$$M_{m,n}(\tau) = |Z| \qquad Z = \sum_I (n_I + m_I \tau) \, \boldsymbol{\alpha}_I \cdot \mathbf{a} \qquad (15.5.4)$$

One of the original indications of the SL(2, ℤ) Montonen–Olive duality symmetry of the maximally supersymmetric $\mathcal{N} = 4$ Yang–Mills theory was that the spectrum of BPS states is invariant under SL(2, ℤ) transformations of the complexified coupling τ,

$$M_{m',n'}(\tau') = M_{m,n}(\tau) \qquad \tau \to \tau' = \frac{a\tau + b}{c\tau + d}$$

$$\mathbf{a} \to \mathbf{a}' = \frac{\mathbf{a}}{c\tau + d} \qquad (15.5.5)$$

provided the charges of the states are mapped as follows,

$$m_I = am'_I + cn'_I$$
$$n_I = bm'_I + dn'_I \qquad (15.5.6)$$

We note, however, that SL(2, ℤ) does not in general map $\mathcal{N} = 4$ Yang–Mills with gauge group G to itself. Instead, as we shall now explain, it transforms between theories with the same Lie algebra $\mathfrak{g}$ but different global structures.

15.5.1 The global structure of 𝒩 = 4 theories

The Lie algebra $\mathfrak{g}$ of an $\mathcal{N} = 4$ Yang–Mills theory does not uniquely specify the theory. Rather there remain different choices for the *global structure* of the theory, including (but not limited to) the particular choice of gauge group G corresponding to the algebra $\mathfrak{g}$.[6] The various Lie groups associated with a Lie algebra $\mathfrak{g}$ may be obtained by first identifying the unique simply-connected Lie group $\mathcal{G}$ whose Lie algebra is $\mathfrak{g}$ and then identifying the center $Z(\mathcal{G})$ of $\mathcal{G}$. All of the other gauge groups corresponding to the algebra $\mathfrak{g}$ are then of the form $G_H = \mathcal{G}/H$ for H a subgroup of $Z(\mathcal{G})$. Whenever H is nontrivial, that is, contains at least one element other than the identity, the

---

[6] Throughout this chapter, we shall restrict attention to gauge groups that are connected and compact, but not necessarily simply connected. An example in which the global structure of the gauge group alone is not enough to fully specify the global structure is provided by the case of SO(3) super Yang–Mills, which will be discussed below.

group G_H is nonsimply connected and a standard argument shows that its first homotopy group $\pi_1(G_H)$ is isomorphic to H.

The weight lattice Λ_w of a Lie algebra $\mathfrak{g}$ is the lattice spanned by the weights of all possible finite-dimensional representations of $\mathfrak{g}$ and coincides with the weight lattice of the simply connected Lie group $\mathcal{G}$ so that $\Lambda_w = \Lambda_w^{\mathcal{G}}$. The root lattice Λ_r is the weight lattice of the adjoint representation of $\mathfrak{g}$ and satisfies $\Lambda_r \subset \Lambda_w$. The weight lattice of a nonsimply connected subgroup $G_H \subset \mathcal{G}$ will be denoted $\Lambda_w^{G_H}$. Every representation of G_H is a representation of $\mathcal{G}$ so that $\Lambda_w^{G/H} \subset \Lambda_w^{\mathcal{G}}$, but the converse is not true, so that the inclusion is a strict one whenever H is nontrivial. The root lattice Λ_r of $\mathfrak{g}$ is a sublattice of $\Lambda_w^{G/H}$ for every $H \subset Z(\mathcal{G})$.

We shall now consider the $\mathcal{N} = 4$ Yang–Mills theories with gauge algebra $\mathfrak{g}$ and gauge group $G = G_H$ for the different choices of H. An $SL(2, \mathbb{Z})$ transformation maps the gauge algebra $\mathfrak{g}$ to itself and maps the gauge group G_H to a gauge group $G_{H'}$, where H' is not necessarily equal to H. This means that the allowed representations of G_H and $G_{H'}$ will differ from one another when $H' \neq H$ as is most easily seen by analyzing the spectrum of line operators in each theory. These line operators come in the following types.

- **Wilson lines:** To each center subgroup H and corresponding gauge group G_H is associated a set of *Wilson lines*, which correspond physically to the worldlines of heavy probe particles that transform under a representation $\mathcal{R}$ of G_H. The Wilson line may be expressed as the holonomy of the gauge field A in the representation $\mathcal{R}$ of G_H,

$$W_{\mathcal{R}}(\gamma) = \mathrm{Tr}_{\mathcal{R}} \left(P \exp\left\{ i \oint_{\gamma} A \right\} \right) \tag{15.5.7}$$

  with γ a 1-cycle in the four-dimensional space-time. Defined in this way, each Wilson line is manifestly invariant under H since A is invariant, but the selection of the representation $\mathcal{R}$ of G_H depends on H, as explained earlier. Note that since Wilson lines are worldlines of *probe* particles, their existence does not depend on the dynamical matter content of the theory: Once the gauge group G_H is fixed, Wilson lines will exist for every representation $\mathcal{R}$ of G_H. In terms of the weight lattice Λ_w of $\mathfrak{g}$, the Wilson lines correspond to points in $\Lambda_w^{G_H}/\mathcal{W}(\mathfrak{g})$, with $\Lambda_w^{G_H} \subset \Lambda_w$ the sublattice of weights of G_H and $\mathcal{W}(\mathfrak{g})$ the Weyl group of $\mathfrak{g}$.

- **'t Hooft lines:** Gauge theories also come equipped with *'t Hooft lines*, which are the *magnetic* counterparts of the Wilson lines. Unlike Wilson lines, they cannot be written as local functionals of the fields of the gauge theory, but they are labeled in an analogous way. In particular, let us denote the magnetic dual (also referred to as the Langlands or Goddard–Nuyts–Olive (GNO) dual) algebra of $\mathfrak{g}$ by $\mathfrak{g}^\vee$. The algebra $\mathfrak{g}^\vee$ is dual to $\mathfrak{g}$ in the sense that the weight lattice of $\mathfrak{g}^\vee$, which we denote by $\Lambda_w^\vee$, is the dual of the root lattice of $\mathfrak{g}$. For simply-laced $\mathfrak{g}$, namely, $\mathfrak{a}_r, \mathfrak{d}_r, \mathfrak{e}_6, \mathfrak{e}_7$, and $\mathfrak{e}_8$, we have $\mathfrak{g} = \mathfrak{g}^\vee$, whereas for nonsimply-laced algebras, one has

$$\mathfrak{b}_r^\vee = \mathfrak{c}_r \qquad \mathfrak{c}_r^\vee = \mathfrak{b}_r \qquad \mathfrak{g}_2^\vee = \mathfrak{g}_2 \qquad \mathfrak{f}_4^\vee = \mathfrak{f}_4 \qquad (15.5.8)$$

in the standard Cartan notation in which $\mathfrak{b}_r, \mathfrak{c}_r, \mathfrak{g}_2$, and $\mathfrak{f}_4$ are the Lie algebras of the Lie groups $\mathrm{SO}(2r + 1)$, $\mathrm{Sp}(2r)$, G_2, and F_4, respectively. At the level of groups, the correspondence is more subtle, with the results collected in Table 15.1. Note that,

$$\mathcal{W}(\mathfrak{g}^\vee) = \mathcal{W}(\mathfrak{g}) \qquad\qquad Z(\mathcal{G}^\vee) = Z(\mathcal{G}) \qquad (15.5.9)$$

The 't Hooft lines are then labeled by a particular subset of the points in $\Lambda_w^\vee / \mathcal{W}(\mathfrak{g})$, specified by the global structure of the magnetic gauge group $G^\vee$. The way to determine this subset will be explained below.

- **Dyonic lines:** Finally, we can consider mixed Wilson and 't Hooft lines, labeled by a subset of the electric and magnetic charges,

$$(q_e, q_m) \in \Lambda_w \times \Lambda_w^\vee \qquad (15.5.10)$$

subject to the identification,

$$(q_e, q_m) \sim (wq_e, wq_m) \qquad w \in \mathcal{W}(\mathfrak{g}) \qquad (15.5.11)$$

Note that this set can be larger than the set of pairs of representations of $\mathfrak{g}$ and $\mathfrak{g}^\vee$, which is labeled instead by $(\Lambda_w / \mathcal{W}(\mathfrak{g})) \times (\Lambda_w^\vee / \mathcal{W}(\mathfrak{g}))$.

We now explain how to determine the subset of allowed 't Hooft and dyonic lines. Having fixed a global form G_H for the gauge group, we know that the spectrum of Wilson lines is the set of all lines $(q_e, 0)$ with $q_e \in \Lambda_w^{G_H} / \mathcal{W}(\mathfrak{g})$. The spectrum of 't Hooft and dyonic lines is then given by the full spectrum of lines that are *mutually local* to the Wilson lines. By mutual locality, we mean that if one computes the correlation function of two lines supported on γ and γ', and then moves γ' in a loop around γ, the correlation function is unchanged. By contrast, correlators of mutually nonlocal operators would have to vanish by the same argument.

To make this more concrete note that, independent of the particular global group G_H, Wilson lines labeled by *roots* of $\mathfrak{g}$ must always be present, since roots are elements of $\Lambda_w^{G_H}$ for any choice of subgroup H. Likewise, even without solving the mutual locality constraint, it is clear that we must have at least the 't Hooft lines labeled by roots of $\mathfrak{g}^\vee$. Because the charge lattice

G	$G^\vee$
$SU(NM)/\mathbb{Z}_N$	$SU(NM)/\mathbb{Z}_M$
$SO(2N)$	$SO(2N)$
$\text{Spin}(2N)$ N odd	$\text{Spin}(2N)/\mathbb{Z}_4$
$\text{Spin}(2N)$ N even	$SO(2N)/\mathbb{Z}_2$
$SO(2N+1)$	$USp(2N)$
$\text{Spin}(2N+1)$	$USp(2N)/\mathbb{Z}_2$
E_6	$E_6/\mathbb{Z}_3$
E_7	$E_7/\mathbb{Z}_2$

Table 15.1   Langlands or GNO dual groups $G^\vee$ for various gauge groups G, with $(G^\vee)^\vee = G$. The gauge groups that do not appear in this table, namely, G_2, F_4, and E_8, have trivial center and hence only a single global form, meaning that $G^\vee = G$.

is closed under addition and inversion – that is, if (q_e, q_m) is present then so is $(-q_e, -q_m)$, and if furthermore (q'_e, q'_m) is present then so is $(q_e+q'_e, q_m+q'_m)$ – it is useful to organize lines into families labeled by the weight lattice modulo the root lattice. The weight lattice modulo the root lattice is nothing but the center, and recalling that $Z(\mathcal{G}) = Z(\mathcal{G}^\vee)$, we conclude that we may label families of lines by pairs,

$$(z_e, z_m) \in Z(\mathcal{G}) \times Z(\mathcal{G}) \tag{15.5.12}$$

If one element in such a family exists, so do all of the other elements, obtained by adding arbitrary root vectors.

For charges valued in the center, the mutual locality condition mentioned in the previous two paragraphs takes a simple form. For $Z(\mathcal{G}) = \mathbb{Z}_N$, which is the case for all simple Lie groups except for $SO(4N + 2)$,[7] then for lines (z_e, z_m) and (z'_e, z'_m), the mutual locality constraint reads [AST13],

$$z_e z'_m - z_m z'_e = 0 \mod N \tag{15.5.13}$$

We note that even upon fixing the global form of the gauge group G_H, and hence the lattice of allowed electric charges z_e, there can be multiple solutions to the mutual locality constraint, giving different lattices of dyonic lines. To fully specify the $\mathcal{N} = 4$ theory, one must specify all of this additional global data.

---

[7] In the case of $SO(4N + 2)$, the center is $\mathbb{Z}_2 \times \mathbb{Z}_2$. This case is more complicated and will not be discussed here; see [AST13] for details.

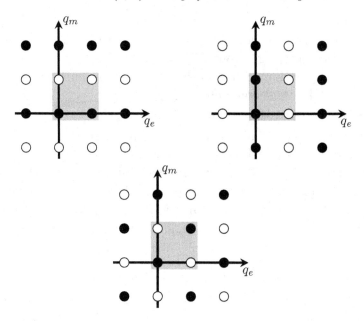

Figure 15.1 Lattice of line operators for the $SU(2)$, $SO(3)_+$, and $SO(3)_-$ theories. The black dots represent occupied lattice points. The gray square represents the sublattice of elements $(z_e, z_m) \in Z(SU(2)) \times Z(SU(2))$ valued in the center, which can be used to obtain the entire lattice by adding arbitrary root vectors.

15.5.2 Example: $\mathfrak{su}(2)$ SYM

Let us illustrate the above through the example of $\mathfrak{g} = \mathfrak{su}(2)$. In this case, $\mathcal{G} = SU(2)$ and $Z(\mathcal{G}) = \mathbb{Z}_2$. If we take $H \subset Z(\mathcal{G})$ to be trivial, then $G_H = SU(2)$ itself. Aside from the identity line $(z_e, z_m) = (0, 0)$, the spectrum of Wilson lines is generated by elements of the nontrivial center, which is just $(z_e, z_m) = (1, 0)$. The mutual locality condition requires that any other family of lines labeled by (z'_e, z'_m) satisfies $z'_m = 0$ mod 2, and hence that no other nontrivial representation of the center is allowed. The lattice of charges for $G_H = SU(2)$ is shown in the gray square in the upper left part of Figure 15.1 – the full lattice is obtained from this by adding elements of the root lattice.

We can instead choose $H = \mathbb{Z}_2$, in which case $G_H = SU(2)/\mathbb{Z}_2 = SO(3)$. The Wilson lines are now labeled by representations of $SO(3)$, which have even electric charges, and hence which, when reduced to the center, are all trivial $(z_e, z_m) = (0, 0)$. Because of this, demanding mutual locality with Wilson lines does not impose any constraints on the spectrum of 't Hooft and dyonic lines operators, and we may choose to have either $(z_e, z_m) = (0, 1)$

or $(1, 1)$ occupied (but not both, since the two are not mutually local). This gives the upper right and lower lattices in Figure 15.1. We note here that even upon fixing the gauge group $G_H = \mathrm{SO}(3)$, there are still two distinct choices for the spectrum of line operators, and one must specify the full lattice of lines before one completely specifies the theory. The two theories in this case are usually denoted by $\mathrm{SO}(3)_+$ and $\mathrm{SO}(3)_-$.

Let us point out here that the only ingredient entering in the above analysis was the *center* of $\mathcal{G}$. We may thus conclude that the case of $\mathfrak{g} = \mathfrak{e}_7$, for which the center is again $\mathbb{Z}_2$, is exactly identical, again with three distinct global variants. For groups with larger centers the analysis is more involved, but conceptually straightforward. On the other hand, when the center is trivial as for $\mathfrak{g} = \mathfrak{e}_8$, there is only a single variant of the theory.

15.5.3 $\mathrm{SL}(2, \mathbb{Z})$ *on line operators*

Having understood the spectrum of line operators, we may now return to the question of the action of $\mathrm{SL}(2, \mathbb{Z})$. The S and T transformations act in a straightforward way on the spectrum of line operators,

$$S : (q_e, q_m) \to (q_m, -q_e) \qquad T : (q_e, q_m) \to (q_e + q_m, q_m) \quad (15.5.14)$$

The first of these is the famous result that the S transformation implements electromagnetic duality or so-called *Montonen–Olive duality* on the theory, while the second is the effect by which magnetic monopoles become dyonic upon shift of the theta angle.

In the context of the example of $\mathfrak{g} = \mathfrak{su}(2)$, examining the action of S and T on the charge lattices in Figure 15.1 reveals that[8]

$$
\begin{aligned}
S: \quad & \mathrm{SU}(2) \to \mathrm{SO}(3)_+ & T: \quad & \mathrm{SU}(2) \to \mathrm{SU}(2) \\
& \mathrm{SO}(3)_+ \to \mathrm{SU}(2) & & \mathrm{SO}(3)_+ \to \mathrm{SO}(3)_- \\
& \mathrm{SO}(3)_- \to \mathrm{SO}(3)_- & & \mathrm{SO}(3)_- \to \mathrm{SO}(3)_+ \quad (15.5.15)
\end{aligned}
$$

Thus we see that no individual theory is actually self-dual under $\mathrm{SL}(2, \mathbb{Z})$. Instead, one can think of the set of three theories as transforming as a three-dimensional vector-valued modular function of $\mathrm{SL}(2, \mathbb{Z})$. Alternatively, each individual theory can be said to be invariant under an appropriate congruence subgroup. For example, for $G_H = SU(2)$, the theory is invariant under T and $ST^2 S$, which together generate the subgroup $\Gamma_0(2) \subset \mathrm{SL}(2, \mathbb{Z})$. Thus the $SU(2)$ theory is self-dual under $\Gamma_0(2)$.

---

[8] These maps hold up to local counter-terms built from the background gauge field for the $\mathbb{Z}_2$ center symmetry. Accounting for these counter-terms modifies, for example, the $\Gamma_0(2)$ appearing in the following paragraph to $\Gamma_0(4)$.

Let us close by mentioning that, in the cases in which there is only a single global variant of the theory, for example, for $\mathfrak{g} = \mathfrak{e}_8$ mentioned above, then the full $\mathrm{SL}(2,\mathbb{Z})$ really does act as a duality group for the theory.

15.6 The Seiberg–Witten solution

The Seiberg–Witten solution for an $\mathcal{N} = 2$ supersymmetric Yang–Mills theory in four space-time dimensions with gauge group G of rank r, is given in terms of the vacuum expectation values of the complex scalars a_I of the gauge multiplet and their magnetic duals a_{DI} in the Coulomb phase. When hypermultiplets are present in a representation $\mathcal{R} \oplus \bar{\mathcal{R}}$ of the gauge group G, we shall denote their G-invariant masses by m_f, where the index f runs over the various irreducible components $\mathcal{R}_f$ of $\mathcal{R}$. The prepotential $\mathcal{F}$ and the matrix τ_{IJ} for $I, J = 1, \cdots, r$ of effective couplings and mixings of the unbroken $U(1)^r$ gauge multiplets are related to these expectation values by,

$$a_{DI} = \frac{\partial \mathcal{F}}{\partial a_I} \qquad \tau_{IJ} = \frac{\partial a_{DI}}{\partial a_J} = \frac{\partial^2 \mathcal{F}}{\partial a_I\, \partial a_J} \qquad (15.6.1)$$

The original Seiberg–Witten solution was obtained for gauge group $\mathrm{SU}(2)$, but the results were soon thereafter generalized to other semi-simple compact gauge groups G, as detailed in the bibliographical notes.

The building blocks of the Seiberg–Witten solution are a curve $\mathcal{C}$ and a differential λ on $\mathcal{C}$. More precisely, $\mathcal{C} = \mathcal{C}(u, m)$ is a family of genus r Riemann surfaces parametrized by G-invariant local complex coordinates $u_1, \ldots, u_r$ and hypermultiplet masses m_f. A convenient definition of the u_I may be given in terms of the gauge-invariant traces $\mathrm{tr}(\phi^n)$ of the gauge scalar ϕ. The differential λ is meromorphic on $\mathcal{C}$ and its partial derivatives with respect to the parameters u_I are holomorphic Abelian differentials, possibly up to the addition of exact differentials of meromorphic functions on $\mathcal{C}$. In the presence of hypermultiplets, λ has poles whose residues are given by the masses m_f.

The Seiberg–Witten solution gives the vacuum expectation values of the gauge scalars a_I and their magnetic duals a_{DI} in terms of the periods of the differential λ on a canonical basis of homology cycles $\mathfrak{A}_I$ and $\mathfrak{B}_I$ of the curve $\mathcal{C}$, which satisfy canonical intersection pairings $\#(\mathfrak{A}_I, \mathfrak{B}_J) = \delta_{IJ}$, and $\#(\mathfrak{A}_I, \mathfrak{A}_J) = \#(\mathfrak{B}_I, \mathfrak{B}_J) = 0$ with $I, J = 1, \cdots, r = \mathrm{rank}(G)$,

$$2\pi i a_I = \oint_{\mathfrak{A}_I} \lambda \qquad 2\pi i a_{DI} = \oint_{\mathfrak{B}_I} \lambda \qquad (15.6.2)$$

By the very construction of this setup, the matrix τ_{IJ} has positive imaginary part as required on physical grounds by the fact that it provides the couplings for the U(1)^r gauge field strengths $F_{\mu\nu}^I$ of (15.3.6), which must be positive. To see this, we combine the following formulas,

$$2\pi i \frac{\partial a_I}{\partial u_K} = \oint_{\mathfrak{A}_I} \frac{\partial \lambda}{\partial u_K} \qquad 2\pi i \frac{\partial a_{DI}}{\partial u_K} = \oint_{\mathfrak{B}_I} \frac{\partial \lambda}{\partial u_K} \qquad (15.6.3)$$

to observe that τ_{IJ} is the period matrix of the curve $\mathcal{C}(u)$ given by,

$$\tau_{IJ} = \sum_K \frac{\partial a_{DI}}{\partial u_K} \frac{\partial u_K}{\partial a_J} \qquad (15.6.4)$$

whose imaginary part is positive by the Riemann bilinear relations; see Appendix B.1.5 for a detailed discussion.

To illustrate the construction outlined above, we present the Seiberg–Witten curves and differentials for the gauge group $G = \mathrm{SU}(N)$ in two examples. The first has $N_f < 2N$ hypermultiplets with masses m_f in the defining representation of G and is asymptotically free. The second has one hypermultiplet with mass m in the adjoint representation of G and is referred to as the $\mathcal{N} = 2^*$ theory. The latter is UV finite and reduces to the $\mathcal{N} = 4$ super Yang–Mills theory upon setting the mass m of the adjoint hypermultiplet to zero, and to the pure $\mathcal{N} = 2$ theory without hypermultiplets by sending the mass m to infinity while suitably scaling the gauge coupling.

15.6.1 The $N_f < 2N$ theory for gauge group $\mathrm{SU}(N)$

For the theory with $N_f < 2N$ hypermultiplets in the fundamental representation of $\mathrm{SU}(N)$ with masses m_f with $f = 1, \cdots, N_f$, the differential λ and the curve $\mathcal{C}$ are given by,

$$\lambda = \left(A' - \frac{1}{2}(A - y)\frac{B'}{B} \right) \frac{x dx}{y} \qquad y^2 = A(x)^2 - B(x) \qquad (15.6.5)$$

where $A(x)$ and $B(x)$ are polynomials in x given as follows,

$$A(x) = \prod_{i=1}^{N} (x - \bar{a}_i) \qquad B(x) = \Lambda^{2N - N_f} \prod_{f=1}^{N_f} (x - m_f) \qquad (15.6.6)$$

and Λ is the renormalization scale of the asymptotically free theory (not to be confused with lattices which are also denoted by the letter Λ). The gauge-invariant coordinates u_I are given in terms of symmetric polynomials in the variables $\bar{a}_i$ with $i = 1, \cdots, N$ which, since the gauge group is $\mathrm{SU}(N)$, must satisfy $\sum_i \bar{a}_i = 0$. The curve $\mathcal{C}$ is hyperelliptic of genus $N-1$. The differential

λ has simple poles with residue m_f at the points $(x, y) = (m_f, -A(m_f))$ on the second sheet of the hyperelliptic curve but is regular at the points $(x, y) = (m_f, A(m_f))$ on the first sheet of $\mathcal{C}$.

15.7 The $\mathcal{N} = 2^*$ theory for gauge group SU(N)

The $\mathcal{N} = 2^*$ theory possesses a rich modular structure, in part because it may be obtained as an ultraviolet finite mass deformation of the $\mathcal{N} = 4$ theory and has a marginal deformation controlled by the complex gauge coupling τ. Its Seiberg–Witten theory may be developed in terms of integrable mechanical systems, of the Hitchin or elliptic Calogero–Moser type. It is the latter approach that we shall follow here.

15.7.1 The Seiberg–Witten solution

To formulate the Seiberg–Witten solution for the $\mathcal{N} = 2^*$ theory for gauge group SU(N), we introduce a torus $\mathbb{C}/\Lambda$ for a lattice $\Lambda = \mathbb{Z}\omega_1 \oplus \mathbb{Z}\omega_2$, where the modulus $\tau = \omega_2/\omega_1$ is related to the gauge coupling g and the θ angle of the non-Abelian gauge theory by,

$$\tau = \frac{\theta}{2\pi} + \frac{4\pi i}{g^2} \tag{15.7.1}$$

Denoting the mass of the adjoint hypermultiplet by m, the differential λ and the curve $\mathcal{C}$ are given in a local complex coordinate z on $\mathbb{C}/\Lambda$ by,

$$\lambda = k\,dz \qquad\qquad R_N(k, z|\Lambda) = 0 \tag{15.7.2}$$

The function $R_N(k, z|\Lambda)$ is given by the characteristic polynomial for the Lax operator of the elliptic Calogero–Moser system for the root system $\mathfrak{a}_{N-1}$ of the Lie algebra $\mathfrak{su}(N)$,

$$R_N(k, z|\Lambda) = \det\left(kI - L(z|\Lambda)\right) \tag{15.7.3}$$

The components of the $N \times N$ matrix $L(z|\Lambda)$ and its conjugate Lax matrix $M(z|\Lambda)$ are given for $i, j = 1, \cdots, N$ by,

$$L_{ij}(z|\Lambda) = p_i \delta_{ij} - m(1 - \delta_{ij})\Phi(x_i - x_j, z|\Lambda) \tag{15.7.4}$$

$$M_{ij}(z|\Lambda) = m(1 - \delta_{ij})\Phi'(x_i - x_j, z|\Lambda) + m\delta_{ij}\sum_{k \neq i} \wp(x_i - x_k|\Lambda)$$

where $\wp$ is the Weierstrass function and Φ is the Lamé function,

$$\Phi(x, z|\Lambda) = \frac{\sigma(z - x|\Lambda)}{\sigma(z|\Lambda)\sigma(x|\Lambda)}\, e^{x\zeta(z|\Lambda)} \tag{15.7.5}$$

given in terms of the Weierstrass ζ- and σ-functions for the lattice Λ; see Section 2.4.3 for definitions of ζ and σ. The variables x_i and p_i are the positions and momenta for the Calogero–Moser system. The integrability condition $\dot{L} = [M, L]$ guarantees that $R_N(k, z)$ is a conserved quantity for all values of z, k, and Λ.

15.7.2 The Seiberg–Witten curve

The curve $R(k, z|\Lambda)$ admits an explicit presentation in terms of ϑ-functions,

$$R_N(k, z|\Lambda) = \vartheta_1\left(\frac{z}{\omega_1}\Big|\tau\right)^{-1} \vartheta_1\left(\frac{z}{\omega_1} - m\frac{\partial}{\partial\tilde{k}}\Big|\tau\right) P(\tilde{k}) \qquad (15.7.6)$$

where $\tilde{k} = k + m h_1(z|\Lambda)$. Here $P(\tilde{k})$ is a monic polynomial in $\tilde{k}$ of degree N with vanishing $\tilde{k}^{N-1}$ term, reflecting the fact that the gauge group is $\mathrm{SU}(N)$ as opposed to $\mathrm{U}(N)$,

$$P(\tilde{k}) = \tilde{k}^N + \sum_{n=0}^{N-2} u_n \tilde{k}^n \qquad (15.7.7)$$

The $N - 1$ remaining independent coefficients u_n are the moduli of the Coulomb branch. One may think of the roots of the polynomial as the vacuum expectation values of the gauge scalars a_i subject to $\sum_{i=1}^N a_i = 0$ when $m = 0$. Finally, the functions h_n are defined by,

$$h_n(z|\Lambda) = \vartheta_1\left(\frac{z}{\omega_1}\Big|\tau\right)^{-1} \frac{\partial^n}{\partial z^n} \vartheta_1\left(\frac{z}{\omega_1}\Big|\tau\right) \qquad (15.7.8)$$

The corresponding curves for $N = 2, 3, 4$ are given by,[9]

$$\begin{aligned}
R_2(k, z|\Lambda) &= k^2 + u_0 - m^2 \wp \\
R_3(k, z|\Lambda) &= k^3 + (u_1 - 3m^2\wp)k + u_0 + m^3\wp' \\
R_4(k, z|\Lambda) &= k^4 + (u_2 - 6m^2\wp)k^2 + (u_1 + 4m^3\wp')k \\
&\quad + u_0 - m^4\wp'' + 3m^4\wp^2
\end{aligned} \qquad (15.7.9)$$

The SW curve may thus be constructed from N copies, or sheets, of the underlying torus $\mathbb{C}/\Lambda$, namely, one for each solution k_i with $i = 1, \cdots, N$, glued together wherever two sheets $i \neq j$ intersect one another.

---

[9] The Weierstrass function $\wp(z|\tau)$ differs from $h_2(z|\Lambda) - h_1(z|\Lambda)^2$ by a z-independent term given by $\wp(z|\Lambda) = -\partial_z^2 \ln \vartheta_1(z/\omega_1|\tau) - \mathsf{E}_2(\tau)/(3\omega_1^2)$, as may be established using (4.1.7). In the final expressions given in (15.7.9), we have used the freedom to redefine the moduli parameters u_n by Λ-dependent but z-independent shifts, such as $u_0 \to u_0 - m^2 \mathsf{E}_2/(3\omega_1^2)$ for $N = 2$. These shifts are immaterial as the u_n provide an arbitrary coordinate system for the moduli of the Coulomb branch. The resulting shifted versions of (15.7.9) have the advantage of enjoying transparent modular transformation properties.

15.7.3 The vacuum expectation values

The vacuum expectation values a_I and a_{DI} of the gauge scalar ϕ and its magnetic dual ϕ_D are given by the general formulas (15.6.2) of the Seiberg–Witten solution. To obtain the corresponding formulas for the $\mathcal{N} = 2^*$ theory, we begin by spelling out a suitable choice of canonical homology basis.

A canonical choice for $\mathfrak{A}, \mathfrak{B}$ cycles on the torus $\mathbb{C}/\Lambda$ may be uplifted to a canonical homology basis of cycles $\mathfrak{A}_I, \mathfrak{B}_I$ for $I = 1, \cdots, N - 1$ on the sheet corresponding to the solution k_I. An element γ in the modular group SL$(2, \mathbb{Z})$ acts on the cycles $\mathfrak{A}, \mathfrak{B}$ by,

$$\begin{pmatrix} \mathfrak{B} \\ \mathfrak{A} \end{pmatrix} \to \begin{pmatrix} \mathfrak{B}' \\ \mathfrak{A}' \end{pmatrix} = \gamma \begin{pmatrix} \mathfrak{B} \\ \mathfrak{A} \end{pmatrix} \qquad \gamma = \begin{pmatrix} a & b \\ c & d \end{pmatrix} \qquad (15.7.10)$$

and this transformation may be lifted to a transformation on the homology cycles $\mathfrak{A}_I$ and $\mathfrak{B}_I$ on the curves $\mathcal{C}(u)$ by a subgroup SL$(2, \mathbb{Z})$ of the full modular group of the curve Sp$(2N - 2, \mathbb{Z})$. However, independent of this induced action of SL$(2, \mathbb{Z})$, we may act on the cycles $\mathfrak{A}_I, \mathfrak{B}_I$ of the curve $\mathcal{C}(u)$ by the full modular group Sp$(2N-2, \mathbb{Z})$. Thus, the action of SL$(2, \mathbb{Z})$ on $\mathfrak{A}, \mathfrak{B}$ and the action of Sp$(2N - 2, \mathbb{Z})$ on $\mathfrak{A}_I, \mathfrak{B}_I$ may be viewed as independent of one another, producing a larger modular structure governed by the group,

$$\text{SL}(2, \mathbb{Z}) \ltimes \text{Sp}(2N - 2, \mathbb{Z}) \qquad (15.7.11)$$

The corresponding construction gives rise to the notion of *bimodular forms* and may be extended to the superconformal $\mathcal{N} = 2$ theory with gauge group SU(N) and $2N$ hypermultiplets in the defining representation of SU(N).

In terms of these data, the expectation values a_I and their magnetic duals a_{DI} are given by,

$$2\pi i a_I = \oint_{\mathfrak{A}_I} k \, dz = \oint_{\mathfrak{A}} k_I dz$$

$$2\pi i a_{DI} = \oint_{\mathfrak{B}_I} k \, dz = \oint_{\mathfrak{B}} k_I dz \qquad (15.7.12)$$

As an aside, we mention a simple differential equation satisfied by the pre-potential $\mathcal{F}$ in terms of the modulus τ,

$$\left. \frac{\partial \mathcal{F}}{\partial \tau} \right|_{m,a_I} = \frac{1}{4\pi i} \sum_{I=1}^{N-1} \oint_{\mathfrak{A}_I} dz \, k^2 = \frac{1}{4\pi i} \oint_{\mathfrak{A}} dz \sum_{I=1}^{N-1} k_I^2 \qquad (15.7.13)$$

The partial derivative is considered at fixed m and a_I. Note that $\partial \mathcal{F}/\partial \tau$ receives no perturbative contributions, as those are purely one-loop order and independent of τ. The advantage of this formula is that we can now evaluate a_I and $\partial\mathcal{F}/\partial\tau$ entirely in terms of integrals over $\mathfrak{A}$-periods without having to carry out the rather complicated analysis of turning points and regularization required when evaluating the $\mathfrak{B}$-periods.

In the limit $m \to 0$, we recover the $\mathcal{N} = 4$ theory and we have $R(k, z|\tau) = P(k)$, so that the roots of $P(k)$ correspond to the vacuum expectation values of the gauge scalar of the $\mathcal{N} = 2$ gauge multiplet. Since we have $\oint_{\mathfrak{A}} dz = \omega_1$ by definition of these periods, we also have $2\pi i a_I = \omega_1 k_I$ in the $m \to 0$ limits, and thus we set $\omega_1 = 2\pi i$.

15.8 The $\mathcal{N} = 2^*$ theory for gauge group $\mathrm{SU}(2)$

It is instructive to analyze the action of the duality group $\mathrm{SL}(2, \mathbb{Z})$ for the case of gauge group $\mathrm{SU}(2)$ where many formulas and transformation laws can be made completely explicit. The curve $\mathcal{C}$ of (15.7.9) and differential λ may be parametrized as follows,

$$k^2 = \kappa^2 + m^2 \wp(z|\Lambda) \qquad\qquad \lambda = k\,dz \qquad\qquad (15.8.1)$$

where we have set $u_0 = -\kappa^2$ for later convenience. As $z \to 0$, we have $k \sim \pm m/z$ which produces the pole in the differential λ whose residue is the hypermultiplet mass, as expected on the basis of the general construction outlined earlier. To evaluate the periods a and a_D, we begin by evaluating the following elliptic integrals, using the relation $\wp(z|\Lambda) = -\zeta'(z|\Lambda)$ and the monodromy relations of $\zeta(z|\Lambda)$ given in (2.4.17),

$$A_1 = \frac{1}{2\pi i} \oint_{\mathfrak{A}} dz\, \wp(z|\Lambda) = -\frac{2\zeta(\frac{1}{2}|\tau)}{(2\pi i)^2} = \frac{1}{12} \mathsf{E}_2(\tau)$$

$$B_1 = \frac{1}{2\pi i} \oint_{\mathfrak{B}} dz\, \wp(z|\Lambda) = -\frac{2\zeta(\frac{\tau}{2}|\tau)}{(2\pi i)^2} = \frac{\tau}{12} \mathsf{E}_2(\tau) + \frac{1}{2\pi i} \qquad (15.8.2)$$

where the last equality was obtained by combining the formulas (2.6.22), (3.8.2), (3.8.7), and (4.1.6). This immediately allows us to compute the right side of the RG equation (15.7.13),

$$\frac{\partial\mathcal{F}}{\partial\tau}\bigg|_{m,a} = \frac{1}{4\pi i} \oint_{\mathfrak{A}} dz k^2 = \frac{1}{2}\kappa^2 + \frac{m^2}{24} \mathsf{E}_2 \qquad (15.8.3)$$

15.8.1 Expansion in powers of the mass m

To obtain the periods a and a_D, we expand the SW differential λ for SU(2) in powers of m,

$$\lambda = \kappa \sum_{n=0}^{\infty} \frac{\Gamma(n - \frac{1}{2})}{\Gamma(-\frac{1}{2})n!} \left(-\frac{m^2}{\kappa^2}\right)^n \wp(z|\Lambda)^n \, dz \qquad (15.8.4)$$

This expansion will be uniformly convergent for small enough m provided z is kept away from 0, which can always be achieved when we integrate over suitably chosen $\mathfrak{A}$ and $\mathfrak{B}$ cycles. To evaluate the periods a and a_D, we compute the integrals,

$$A_n = \frac{1}{2\pi i} \oint_{\mathfrak{A}} dz \, \wp(z|\Lambda)^n \qquad B_n = \frac{1}{2\pi i} \oint_{\mathfrak{B}} dz \, \wp(z|\Lambda)^n \qquad (15.8.5)$$

Clearly, we have $A_0 = 1$ and $B_0 = \tau$ while A_1 and B_1 were already evaluated in (15.8.2). For $n \geq 2$, we make use of the following recursion relation,

$$(8n - 4)\wp^n = (2n - 3)g_2 \wp^{n-2} + (2n - 4)g_3 \wp^{n-3} + 2(\wp' \wp^{n-2})' \qquad (15.8.6)$$

which is established using the defining equation $(\wp')^2 = 4\wp^3 - g_2 \wp - g_3$. Eliminating g_2 and g_3 in favor of E_4 and E_6 for the period normalization $\omega_1 = 2\pi i$,

$$g_2 = \frac{4\pi^4 \, \mathsf{E}_4}{3\omega_1^4} = \frac{\mathsf{E}_4}{12} \qquad g_3 = \frac{8\pi^6 \, \mathsf{E}_6}{27\omega_1^6} = -\frac{\mathsf{E}_6}{216} \qquad (15.8.7)$$

and integrating the relation (15.8.6), we obtain the following recursion relation, satisfied by both A_n and B_n with $n \geq 2$,

$$(8n - 4)A_n = \frac{2n - 3}{12} \mathsf{E}_4 A_{n-2} - \frac{n - 2}{108} \mathsf{E}_6 A_{n-3} \qquad (15.8.8)$$

The coefficients of this recursion relation are modular forms, but the initial conditions for A_1 and B_1 involve the quasi-modular form E_2. Linearity of (15.8.8) guarantees that the solutions A_n and B_n are linear in E_2 and may be cast in the following form,

$$A_n = K_n + \frac{\mathsf{E}_2}{12} L_{n-1} \qquad B_n = \tau A_n + \frac{L_{n-1}}{2\pi i} \qquad (15.8.9)$$

where K_n and L_n are modular forms of weight $2n$. The functions A_n and L_n satisfy (15.8.8) for A_n with initial conditions A_0 and A_1, while for L_n with initial conditions $B_0 - \tau A_0 = 0$ and $B_1 - \tau A_1 = (2\pi i)^{-1}$, respectively. Substituting these expressions into the expansion in powers of m of the period integrals, we obtain the a-period,

$$a = \kappa \sum_{n=0}^{\infty} \frac{\Gamma(n - \frac{1}{2})}{\Gamma(-\frac{1}{2})n!} \left(-\frac{m^2}{\kappa^2}\right)^n \left(K_n + \frac{\mathsf{E}_2}{12} L_{n-1}\right) \qquad (15.8.10)$$

and the following result for the combination $a_D - \tau a$ of the a_D and a periods,

$$a_D - \tau a = \frac{\kappa}{2\pi i} \sum_{n=0}^{\infty} \frac{\Gamma(n-\frac{1}{2})}{\Gamma(-\frac{1}{2})n!} \left(-\frac{m^2}{\kappa^2}\right)^n L_{n-1} \tag{15.8.11}$$

We shall establish below that the combination $a_D - \tau a$ of the periods transforms as a modular form in the modulus τ.

15.8.2 Low orders and perturbative contribution

The Eisenstein series $\mathsf{E}_{2n}(\tau)$ for $n \geq 1$ are normalized to 1 at the cusp $\tau \to i\infty$ and admit a series expansion at the cusp in powers of $q = e^{2\pi i\tau}$. The following coefficients vanish,

$$K_1 = L_{-1} = L_1 = 0 \tag{15.8.12}$$

and the nonvanishing low-order coefficients are given by $K_0 = L_0 = 1$ and,

$$K_2 = \frac{\mathsf{E}_4}{144} \quad K_3 = -\frac{\mathsf{E}_6}{2160} \quad K_4 = \frac{5\,\mathsf{E}_4^2}{48384} \quad K_5 = -\frac{\mathsf{E}_4 \mathsf{E}_6}{77760}$$

$$L_2 = \frac{\mathsf{E}_4}{80} \quad L_3 = -\frac{\mathsf{E}_6}{1512} \quad L_4 = \frac{7\,\mathsf{E}_4^2}{34560} \quad L_5 = -\frac{29\,\mathsf{E}_4 \mathsf{E}_6}{1330560} \tag{15.8.13}$$

To solve for the prepotential, we may either integrate the RG equation, or integrate $a_D = \partial \mathcal{F}/\partial a$. In either case, we need to express κ as a function of τ, m, a, which can be done order by order in m by inverting the series given in (15.8.10) for a as a function of τ, m, and κ,

$$\frac{\kappa}{a} = 1 - \frac{\mathsf{E}_2}{6} \frac{m^2}{(2a)^2} + \frac{\mathsf{E}_4 - 2\mathsf{E}_2^2}{72} \frac{m^4}{(2a)^4}$$

$$- \frac{20\mathsf{E}_2^3 - 11\mathsf{E}_2\mathsf{E}_4 - 4\mathsf{E}_6}{216} \frac{m^6}{(2a)^6} + \mathcal{O}(m^8) \tag{15.8.14}$$

The expression for $\mathcal{F}$ derived from the RG equation is given by,

$$\mathcal{F}(\tau, m, a) = \frac{1}{2}\tau a^2 + \mathcal{F}_0(m, a) - \frac{1}{2\pi i}\frac{\mathsf{E}_2}{24}\frac{m^4}{(2a)^2}$$

$$- \frac{1}{2\pi i}\left(\frac{\mathsf{E}_4}{720} + \frac{\mathsf{E}_2^2}{144}\right)\frac{m^6}{(2a)^4} + \mathcal{O}(m^8) \tag{15.8.15}$$

where $\mathcal{F}_0$ is independent of τ and undetermined by the RG equation. However, it may be determined from the expressions for a_D and a that we have already obtained, and we have,

$$\mathcal{F}_0(m, a) = cm^2 + \frac{m^2}{4\pi i}\ln\frac{a}{m} \tag{15.8.16}$$

In the weak coupling limit, where $q \to 0$ and $E_2, E_4, E_6 \to 1$, we evaluate the effective gauge coupling τ_{IR} as follows,

$$\tau_{\text{IR}} = \tau - \frac{1}{\pi i} \sum_{n=1}^{\infty} \frac{1}{n} \left(\frac{m^2}{4a^2} \right)^n + \mathcal{O}(q)$$

$$= \tau + \frac{1}{\pi i} \ln \left(1 - \frac{m^2}{4a^2} \right) + \mathcal{O}(q) \qquad (15.8.17)$$

Thus, the perturbative logarithm at the mass of the massive vector boson $m^2 = 4a^2$ is the result of resumming an infinite series of genuine modular form contributions. The logarithmic term produced by $\mathcal{F}_0$ is required to obtain the $n = 1$ term needed to complete the series for the logarithm.

15.8.3 Modular properties

Exceptionally, in this subsection only, we shall parametrize SL$(2, \mathbb{Z})$ transformations with capital letters in order to avoid a clash with the notation for the periods a and a_D. Under a modular transformation $M \in$ SL$(2, \mathbb{Z})$,

$$\tilde{\tau} = \frac{A\tau + B}{C\tau + D} \qquad \begin{pmatrix} \tilde{\omega}_2 \\ \tilde{\omega}_1 \end{pmatrix} = M \begin{pmatrix} \omega_2 \\ \omega_1 \end{pmatrix} \qquad M = \begin{pmatrix} A & B \\ C & D \end{pmatrix} \qquad (15.8.18)$$

we have the following transformations of Eisenstein series for $n \geq 1$,

$$\mathsf{E}_{2n}(\tilde{\tau}) = (C\tau + D)^{2n} \mathsf{E}_{2n}(\tau) + \frac{12}{2\pi i} C(C\tau + D) \delta_{n,1} \qquad (15.8.19)$$

and the Weierstrass function, in which we keep the period $\omega_1 = 2\pi i$ fixed,

$$\wp(\tilde{z}|\tilde{\tau}) = (C\tau + D)^2 \wp(z|\tau) \qquad \tilde{z} = \frac{z}{C\tau + D} \qquad (15.8.20)$$

Furthermore, K_n and L_n transform as modular forms of weight $(2n, 0)$. The transformation laws of the SW data may be deduced as follows. First of all, modular invariance of the SW differential implies the following transformation law for k^2,

$$\lambda = k\,dz = \tilde{k}\,d\tilde{z} \qquad \tilde{k}^2 = (C\tau + D)^2 k^2 \qquad (15.8.21)$$

Using the relation between z, k, and κ of (15.8.1), we readily obtain,

$$\tilde{\kappa} = (C\tau + D)\kappa \qquad (15.8.22)$$

Here, the sign ambiguity introduced by taking the square root of κ^2 amounts to implementing the transformation $-I \in$ SL$(2, \mathbb{Z})$ on κ. The transformation laws on a_D, a may be deduced from their explicit expressions in terms of κ

and τ. We begin by observing in (15.8.11) that the combination $a_D - \tau a$ transforms as a modular form of weight $(-1,0)$,

$$\tilde{a}_D - \tilde{\tau}\tilde{a} = (C\tau + D)^{-1}(a_D - \tau a) \tag{15.8.23}$$

Next, we use the explicit expression for a from (15.8.10) along with the transformation laws of κ, K_{2n}, and L_{2n-2} as modular forms of weight $(1,0)$, $(2n,0)$, and $(2n-2,0)$ respectively, as well as the transformation law of E_2, to obtain,

$$\tilde{a} = (C\tau + D)a + C(a_D - \tau a) = C\,a_D + D\,a \tag{15.8.24}$$

where the second term in the middle equality arises from the inhomogeneous transformation term of E_2. Assembling both transformation laws, we obtain,

$$\begin{pmatrix} \tilde{a}_D \\ \tilde{a} \end{pmatrix} = \begin{pmatrix} A & B \\ C & D \end{pmatrix} \begin{pmatrix} a_D \\ a \end{pmatrix} \tag{15.8.25}$$

which is precisely the transformation law induced by the action of the modular group on the cycles $\mathfrak{A}$ and $\mathfrak{B}$ for a modular invariant SW differential.

15.9 Linear quiver chains from a limit of $\mathcal{N}=2^*$

We now return to the $\mathcal{N}=2^*$ theory for the gauge group $SU(N)$, whose independent free parameters are the UV gauge coupling τ, the hypermultiplet mass m, and the vacuum expectation values a_I of the gauge scalars, or equivalently the roots of the polynomial $P(k)$. The masses of the gauge boson BPS states are $|a_I - a_J|$, while the masses of the BPS hypermultiplet states are $|a_I - a_J - m|$. Several interesting limits may be taken in this setup.

A first interesting limit is obtained by keeping the vacuum expectation values a_I fixed while sending the hypermultiplet mass $m \to \infty$ and letting $\tau \to i\infty$ at the rate dictated by the one-loop renormalization group β-function with renormalization scale Λ (not to be confused with the lattice Λ), $\Lambda^{2N} = (-)^N m^{2N} q$, where $q = e^{2\pi i \tau}$. This gives the pure $\mathcal{N}=2$ theory with gauge group $SU(N)$.

A second interesting limit is obtained by letting $m \to \infty$ and $\tau \to i\infty$, but now also sending some of the vacuum expectation values a_I to infinity. Upon carefully tuning the correlations between these limits, one can send some of the hypermultiplet masses and some of the gauge boson masses to ∞, while keeping the masses of others finite. This limit will break the gauge symmetry to $SU(N_1) \times \cdots \times SU(N_P)$ with $N = N_1 + \cdots + N_P$ and $N_p \geq 2$ for all $p = 1, \cdots, P$, while any $U(1)$ factors in the gauge group that arise may be shown to decouple. To obtain the above breaking pattern, we arrange the classical values for the expectation values in a linear chain. We

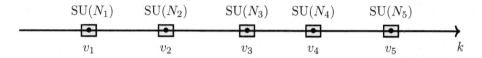

Figure 15.2 The distribution of vacuum expectation values for a breaking from $SU(N)$ to $SU(N_1) \times \cdots \times SU(N_5)$ with $N = N_1 + \cdots + N_5$, where the v_p are separated from their nearest neighbors by $\pm m$ plus a term Λ_p that is held fixed as $m, v_p \to \infty$. The gray rectangles schematically represent the ranges in which the $x_{i_p}^p$ can vary.

partition all the order parameters k_I for $I = 1, \cdots, N$ into P groups with $I = i_1 = 1, \cdots, N_1$ labeling the indices in the first group, $I = N_1 + i_2$ with $i_2 = 1, \cdots, N_2$ those in the second group; $I = N_1 + \cdots + N_{p-1} + i_p$ with $i_p = 1, \cdots, N_p$ those in group p, and so on until $I = N_1 + \cdots N_{P-1} + i_P$ with $i_P = 1, \cdots, N_P$. We now set,

$$k_{N_1 + \cdots + N_p + i_p} = v_p + x_{i_p}^p \qquad\qquad p = 1, \cdots, P \qquad (15.9.1)$$

The decomposition is not unique as it stands since v_p and $x_{i_p}^p$ may be shifted by an i_p-independent amount such that the corresponding k_I is unchanged. To make the decomposition unique, we impose the further conditions,

$$\sum_{p=1}^{P} N_p v_p = 0 \qquad\qquad \sum_{i_p=1}^{N_p} x_{i_p}^p = 0 \qquad (15.9.2)$$

We then order the values of v_p such that,

$$v_{p+1} - v_p = m + \Lambda_p \qquad (15.9.3)$$

and then send $v_p, m \to \infty$ along with $\tau \to i\infty$ while keeping all Λ_p and $x_{i_p}^p$ fixed. The result is a linear quiver chain, depicted in Figure 15.2. The gauge group is $SU(N_1) \times \cdots \times SU(N_P)$, and the theory contains hypermultiplets in bi-fundamental representations,

$$\bigoplus_{p=1}^{P-1} \left\{ (\mathbf{N}_p, \overline{\mathbf{N}}_{p+1}) \oplus (\overline{\mathbf{N}}_p, \mathbf{N}_{p+1}) \right\} \qquad (15.9.4)$$

Depending on the values of N_p, hypermultiplets in the fundamental of $SU(N)$ may also arise.

15.10 $\mathcal{N} = 2$ dualities

In the remainder of this chapter, we discuss a set of dualities between $\mathcal{N} = 2$ theories that are related to moduli spaces of Riemann surfaces. The theories in question will be *superconformal*, that is, theories with both supersymmetry and conformal symmetry. We begin by briefly reviewing these notions.

15.10.1 Conformal symmetry

The conformal algebra in four dimensions is given by $\mathfrak{so}(2,4) \approx \mathfrak{su}(2,2)$, generated by Lorentz transformations $L_{\mu\nu}$, translations P_μ, dilations D, and special conformal transformations K_μ. These generators satisfy $[P_\mu, P_\nu] = [K_\mu, K_\nu] = [D, L_{\mu\nu}] = 0$ as well as the following commutation relations,

$$[L_{\mu\nu}, L_{\rho\sigma}] = i(\eta_{\nu\rho}L_{\mu\sigma} + \eta_{\mu\sigma}L_{\nu\rho} - (\mu \leftrightarrow \nu)) \quad [K_\mu, P_\nu] = 2i(\eta_{\mu\nu}D - L_{\mu\nu})$$

$$[L_{\mu\nu}, P_\rho] = i(\eta_{\nu\rho}P_\mu - \eta_{\mu\rho}P_\nu) \qquad\qquad\qquad [D, P_\mu] = iP_\mu$$

$$[L_{\mu\nu}, K_\rho] = i(\eta_{\nu\rho}K_\mu - \eta_{\mu\rho}K_\nu) \qquad\qquad\qquad [D, K_\mu] = -iK_\mu \quad (15.10.1)$$

Dilations D generate a noncompact Abelian subalgebra $\mathfrak{so}(1,1) \subset \mathfrak{so}(2,4)$ that commutes with the Lorentz subalgebra $\mathfrak{so}(1,3)$ generated by $L_{\mu\nu}$. Therefore, we may choose the Cartan subalgebra of $\mathfrak{so}(2,4)$ to be generated by D and by the Cartan subalgebra of $\mathfrak{so}(1,3)$. The dilation generator can be used to assign dimensions to the other generators. The eigenvalue of D is customarily denoted by Δ (not to be confused with either the Laplacian or the discriminant cusp form) and referred to as the *conformal dimension*. The final two commutation relations are the statement that the momenta P_μ and special conformal generators K_μ have conformal dimensions 1 and -1, respectively. Furthermore, they imply that P_μ is *not* in the Cartan subalgebra, meaning that operators or states in a conformal field theory cannot be labeled by an energy/mass – indeed, such quantities would introduce a dimensionful scale to the theory and thereby violate conformal symmetry.

Conformal fields have already been introduced in Section 5.1.1, where they were labeled by $(h, \tilde{h}$, related to the conformal dimension Δ via $\Delta = h + \tilde{h}$. An important difference between the discussion here and in Section 5.1.1 is that in the latter case the discussion was specific to two space-time dimensions, for which the "global" conformal group discussed in this section is enhanced to (two copies of) an infinite-dimensional Virasoro symmetry, given in (5.1.3). In higher dimensions such an enhancement does not occur, though the notion of conformal primary introduced in Section 5.1.1 persists. In the current language, a state $|[j_1, j_2], \Delta\rangle$ labeled by its conformal dimension Δ and representation of the Lorentz group $\mathfrak{so}(1,3) \cong \mathfrak{su}(2) \oplus \mathfrak{su}(2)$ is a conformal primary if it satisfies,

$$K_\mu|[j_1, j_2], \Delta\rangle = 0 \qquad\qquad (15.10.2)$$

Starting from this highest weight state, a series of *conformal descendants* can be obtained via repeated application of P_μ.

15.10.2 Superconformal symmetry

When $\mathcal{N}$ supercharges transforming in the Weyl spinor representation of $\mathfrak{so}(1,3)$ are added to the conformal algebra $\mathfrak{so}(2,4)$, one obtains the superalgebra $\mathfrak{su}(2,2|\mathcal{N})$. In the current chapter, we will be mainly interested in the case of $\mathcal{N} = 2$, for which the maximal bosonic subalgebra of the superconformal algebra is,

$$\mathfrak{so}(2,4) \oplus \mathfrak{u}(2)_R \subset \mathfrak{su}(2,2|2) \tag{15.10.3}$$

The $\mathfrak{u}(2)_R$ factor is referred to as the *R-symmetry*, and its generators will be denoted by R_J^I. They satisfy the standard commutation relations,

$$[R_J^I, R_L^K] = \delta_J^K R_L^I - \delta_L^I R_J^K \tag{15.10.4}$$

As for the fermionic generators, in addition to the Poincaré supercharges Q_α^I, $\overline{Q}_{\dot\alpha I}$, we now have the superconformal partners $S_{\alpha I}$, $\overline{S}_{\dot\alpha}^I$. The anticommutation relations involving these generators take the form,

$$\{Q_\alpha^I, Q_\beta^J\} = \{S_{\alpha I}, S_{\beta J}\} = \{Q_\alpha^I, \overline{S}_{\dot\beta}^I\} = 0$$
$$\{Q_\alpha^I, \overline{Q}_{\dot\beta J}\} = 2\sigma_{\alpha\dot\beta}^\mu P_\mu \delta_J^I$$
$$\{S_{\alpha I}, \overline{S}_{\dot\beta}^J\} = 2\sigma_{\alpha\dot\beta}^\mu K_\mu \delta_I^J$$
$$\{Q_\alpha^I, S_{\beta J}\} = \epsilon_{\alpha\beta}(\delta_J^I D + R_J^I) + \tfrac{1}{2}\delta_J^I L_{\mu\nu} \sigma_{\alpha\beta}^{\mu\nu} \tag{15.10.5}$$

Note that, unlike in the nonconformal case, the superconformal algebra does not admit a deformation by central charges Z, since these would introduce a scale and thus would break conformal symmetry.

The states of an $\mathcal{N} = 2$ superconformal field theory can be labeled by their conformal dimension Δ and Lorentz quantum numbers $[j_1, j_2]$, as well as additional labels (R, r) for the $\mathfrak{u}(2)_R \cong \mathfrak{su}(2)_R \times \mathfrak{u}(1)_r$ R-symmetry. We will denote such states by $|[j_1, j_2], \Delta, R, r\rangle$. In analogy to a conformal primary, one defines a *superconformal primary* via

$$S_{\alpha I}|[j_1, j_2], \Delta, R, r\rangle = \overline{S}_{\dot\alpha}^I|[j_1, j_2], \Delta, R, r\rangle = 0 \tag{15.10.6}$$

These conditions, together with the anticommutation relation in (15.10.5), imply that all superconformal primaries satisfy $K_\mu|[j_1, j_2], \Delta, R, r\rangle = 0$ and hence are particular cases of conformal primaries in the definition of

(15.10.2). A superconformal multiplet is obtained by beginning with a super-conformal primary and applying arbitrary combinations of Q_α^I and $\overline{Q}_{\dot\alpha I}$. By the anticommutation relations in (15.10.5), this also includes all applications of the momentum operator P_μ and hence includes all conformal descendants.

15.10.3 Superconformal field theories

We may construct $\mathcal{N} = 2$ superconformal Lagrangians by combining the $\mathcal{N} = 2$ supermultiplets discussed in Section 15.1.2 in such a way as to cancel any potential nonconformality. Setting all masses for the hypermultiplets to zero, the only parameter left in an $\mathcal{N} = 2$ Lagrangian is the coupling τ, which is classically dimensionless. Thus any such Lagrangian would naively seem to have conformal invariance. While this is true at the classical level, quantum mechanically the coupling τ – or rather only g, since θ is topolog-ically protected – can receive radiative corrections making it dimensionful. The dependence of the quantum-corrected coupling on the energy scale μ is captured by the β-function,

$$\beta(g) = \mu \frac{d}{d\mu} \left(\frac{8\pi^2}{g^2} \right) \tag{15.10.7}$$

which, for a generic four-dimensional Yang–Mills theory with N_f Weyl fermions in a representation $\mathcal{R}_f$ and N_s scalars in a representation $\mathcal{R}_s$, is given by,

$$\beta(g) = \tfrac{11}{3} C(\mathfrak{g}) - \tfrac{2}{3} N_f C(\mathcal{R}_f) - \tfrac{1}{3} N_s C(\mathcal{R}_s) + \mathcal{O}(g) \tag{15.10.8}$$

Here $C(\mathcal{R})$ stands for the Dynkin index of the representation $\mathcal{R}$, defined via $\mathrm{Tr}_\mathcal{R}(T^a T^b) = C(\mathcal{R})\delta^{ab}$ with T^a the generators of the gauge algebra. By a slight abuse of notation, we denote the adjoint representation of $\mathfrak{g}$ by the same symbol $\mathfrak{g}$. In particular, $C(\mathfrak{g}) = h^\vee$ is the dual Coxeter number.

For an $\mathcal{N} = 1$ theory, the vector multiplet, which contains one vector and one Weyl fermion in the adjoint representation, contributes,

$$\tfrac{11}{3} C(\mathfrak{g}) - \tfrac{2}{3} C(\mathfrak{g}) = 3\, h^\vee \tag{15.10.9}$$

to the β-function. The contribution for a chiral multiplet, containing a Weyl fermion and a scalar both in the representation $\mathcal{R}$, is given by,

$$-\tfrac{2}{3} C(\mathcal{R}) - \tfrac{1}{3} C(\mathcal{R}) = -C(\mathcal{R}) \tag{15.10.10}$$

For example, an $\mathcal{N} = 1$ super Yang–Mills theory with gauge group $SU(N)$ (for which $h^\vee = N$) with N_f chiral multiplets in the representation $\mathcal{R}$ has the following β-function,

$$\beta(g) = 3N - N_f C(\mathcal{R}) + \mathcal{O}(g) \tag{15.10.11}$$

For conformality, we see that we must require $N_f C(\mathcal{R}) = 3N$.

We now move on to the main case of interest to us, namely, $\mathcal{N} = 2$ theories. The contribution to the β-function of an $\mathcal{N} = 2$ vector multiplet is,

$$\tfrac{11}{3} C(\mathfrak{g}) - 2 \times \tfrac{2}{3} C(\mathfrak{g}) - \tfrac{1}{3} C(\mathfrak{g}) = 2 \, h^\vee \qquad (15.10.12)$$

while the contribution from a hypermultiplet in the representation $\mathcal{R}$ is,

$$-2 \times \tfrac{2}{3} C(\mathcal{R}) - 2 \times \tfrac{1}{3} C(\mathcal{R}) = -2 \, C(\mathcal{R}) \qquad (15.10.13)$$

We thus see that for an $\mathcal{N} = 2$ gauge theory with gauge group $\mathrm{SU}(N_c)$ and N_f hypermultiplets in the representation $\mathcal{R}$ of $\mathrm{SU}(N_c)$ we have,

$$\beta(g) = 2N_c - 2N_f C(\mathcal{R}) + \mathcal{O}(g) \qquad (15.10.14)$$

Taking in particular the case in which all hypermultiplets are in the fundamental representation $\mathcal{R}_1$ for which $C(\mathcal{R}_1) = \tfrac{1}{2}$, we see that conformality can be achieved only if $N_f = 2N$.

The theories that we will study in the rest of this section will be built on theories of this type, namely, $\mathcal{N} = 2$ theories with $\mathrm{SU}(N)$ gauge group and N_f fundamental hypermultiplets. The data of such a theory can be captured in a so-called *quiver diagram* – a diagram whose internal nodes are gauge groups, whose external nodes are fundamental hypermultiplets, and whose internal edges are bi-fundamental hypermultiplets, with the structure of the quiver chosen to be compatible with the vanishing of all beta functions.[10] For example, $(N) - [2N]$ represents $\mathrm{SU}(N)$ gauge theory with $2N$ fundamentals and has a manifestly vanishing beta function. Likewise the quiver,

$$[N] - (N) - (N) - [N] \qquad (15.10.15)$$

represents $\mathrm{SU}(N) \times \mathrm{SU}(N)$ gauge theory with one bi-fundamental hypermultiplet corresponding to the internal edge, and with N additional fundamentals for each copy of $\mathrm{SU}(N)$. Since the internal edge can be interpreted as giving N fundamental hypers to the left gauge node and N anti-fundamental hypers to the right gauge node, the beta functions for both gauge couplings indeed vanish.

We will focus on the case of $N = 2$, for which something rather special happens. In particular, note that in this case each bi-fundamental has two $\mathrm{SU}(2)$ gauges indices and one $\mathrm{SU}(2)$ flavor index, that is, we can write them as

---

[10] One could ask why we allow for only bi-fundamental edges in the graph. It turns out that for more complicated edges, it is impossible to satisfy both anomaly cancellation and vanishing of the beta functions. The only exception is the case of tri-fundamentals in theories with $\mathrm{SU}(2)$ gauge nodes, which we will return to below.

$\hat{Q}_{ijs}$ with indices as defined above. We can imagine weakly gauging the flavor symmetry, or conversely ungauging one of the gauge symmetries, which
makes it clear that all three indices should be treated democratically. In
other words, it is useful to think of the elementary building blocks as *tri-
fundamental* matter, which we will represent pictorially by,

$$\hat{Q}_{ijs} \quad \Rightarrow$$

With this in mind, it is possible to replace any SU(2) quiver diagram with
a trivalent network, for example

$$[2] - (2) - [2] \quad \Leftrightarrow \qquad\qquad (2) = (2) \quad \Leftrightarrow$$

$$[2] - (2) - (2) - [2] \quad \Leftrightarrow$$

Conversely, any trivalent graphs can be given an interpretation as a superconformal field theory with SU(2) nodes.

15.10.4 SU(2) *quivers and Riemann surfaces*

Consider a trivalent graph Γ with g faces and n external legs. We will denote
the theories obtained from such graphs by $\mathcal{T}_{g,n}$. There is one gauge SU(2)
per internal edge, and it can be shown that the number of such internal
edges is $3g - 3 + n$. To each of these internal nodes is associated a gauge
coupling τ. The quantity $3g - 3 + n$ appearing here may seem familiar –
indeed, if $n = 0$ then this precisely the dimension of the moduli space of
genus g Riemann surfaces, as given in (C.7.9). For nonzero n, this gives the
dimension of the moduli space of genus g Riemann surfaces *with n marked
points*.

The mapping between the moduli space of 4d $\mathcal{N} = 2$ SU(2) quiver theories
and that of punctured Riemann surfaces can be made more precise as follows.
We identify each trivalent vertex with a three-punctured sphere,

Gauging involves gluing two such trivalent vertices together, which in other words corresponds to plumbing together two three-punctured spheres.

Indeed, let us consider gluing two trivalent vertices together to obtain the SU(2) $N_f = 4$ theory. In the corresponding Riemann surface picture, we would like to plumb together two spheres along their punctures. Say that the two spheres have complex coordinates w and z, with the three punctures being located at $w, z = 0, 1, \infty$. We will glue a neighborhood of the point at $w = \infty$ to a neighborhood of the point at $z = 0$,

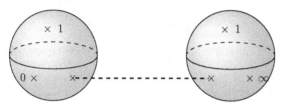

To glue them, we begin by changing to the local coordinate $\tilde{w} = \frac{1}{w}$ which describes the region around the $w = \infty$ puncture, and then we identify

$$z\tilde{w} = q \qquad (15.10.16)$$

with q some complex parameter. This parameter is identified with the coupling in the field theory via $q = e^{i\pi\tau}$.

From this picture, it is clear that the conformal manifold of the SU(2) $N_f = 4$ theory – that is to say the space in which τ can take values – can be identified with the moduli space of a four-punctured sphere. The latter is none other than the usual fundamental domain, shown in Figure 3.2. The cusp at $\tau = i\infty$, that is, $q \to 0$, corresponds to the separating limit in which the four-punctured sphere pinches off into two three-punctured spheres.

It is also useful to consider the moduli space of a four-punctured sphere when the four punctures are taken to be *distinguishable*. In that case, the moduli space takes the form shown in Figure 15.3. There are now three cusps, at $\tau = 0, 1, i\infty$, which correspond to configurations in which different pairs of marked points collide. It is interesting to consider the behavior of the field theory at each of these three cusps.

Let us begin with the weakly coupled cusp at $\tau = i\infty$. At this point the theory is well described by the usual SU(2) $N_f = 4$ theory discussed earlier. This theory has an SO(8) flavor symmetry, under which the hypermultiplets transform in the $\mathbf{8}_v$. We consider the following chain of subgroups of SO(8),

$$\text{SO}(8) \supset \text{SO}(4) \times \text{SO}(4) \supset \text{SU}(2)_a \times \text{SU}(2)_b \times \text{SU}(2)_c \times \text{SU}(2)_d$$

which is manifest in the trivalent diagram. In particular, each of the four

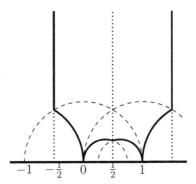

Figure 15.3 The fundamental domain for a sphere with four distinguishable marked points.

external legs carries one copy of SU(2). Let us assign them as follows,

$$\tau \to i\infty : \qquad \begin{array}{cc} b & c \\ & \diagdown\!\!\!\diagup \\ a & d \end{array}$$

In the corresponding four-punctured sphere, the points labeled by a and b are close to one another and separated by a long thin tube from the points c and d. We now move toward the point in moduli space with $\tau = 0$. In this case, the four-punctured sphere changes to a configuration in which the points a and c approach one another and are separated by a long tube from b and d. Likewise as we approach the cusp at $\tau = 1$, the points a and d approach one another and are separated by a long tube from c and b. In terms of trivalent diagrams, this is

$$\tau \to 0 : \qquad \begin{array}{cc} c & b \\ & \diagdown\!\!\!\diagup \\ a & d \end{array} \qquad\qquad \tau \to 1 : \qquad \begin{array}{cc} d & c \\ & \diagdown\!\!\!\diagup \\ a & b \end{array}$$

We see that in all cases, the quiver is of the same general form as before, suggesting that the strong-coupling regime of SU(2) $N_f = 4$ is again described by SU(2) $N_f = 4$. However, distinguishing between the marked points allows us to make an important observation about these strong-coupling dualities. In particular, let us recall that in the original weakly coupled theory, the hypermultiplets transformed in the $\mathbf{8}_v$ of SO(8). Decomposing this representation into those of the SU(2) subgroups, we have

$$\mathbf{8}_v = \mathbf{4} \oplus \mathbf{4} = (\mathbf{2}_a \otimes \mathbf{2}_b) \oplus (\mathbf{2}_c \otimes \mathbf{2}_d) \qquad (15.10.17)$$

On the other hand, one has the following decompositions,

$$\mathbf{8}_s = (\mathbf{2}_a \otimes \mathbf{2}_c) \oplus (\mathbf{2}_b \otimes \mathbf{2}_d)$$
$$\mathbf{8}_c = (\mathbf{2}_a \otimes \mathbf{2}_d) \oplus (\mathbf{2}_b \otimes \mathbf{2}_c) \tag{15.10.18}$$

with $\mathbf{8}_s$ and $\mathbf{8}_c$ the spinor and conjugate spinor representations of SO(8). Thus swapping SU(2)$_b$ with SU(2)$_c$ has the effect of switching from the vector to the spinor representation of SO(8), and likewise for swapping SU(2)$_b$ with SU(2)$_d$. We thus see that upon approaching the strongly coupled cusps in moduli space, we reobtain SU(2) $N_f = 4$, but with the hypermultiplets transforming in spinor representations of the flavor symmetry group.

Similar statements hold for the $[2] - (2) - (2) - [2]$ quiver whose trivalent diagram was given earlier. Again, approaching any cusp gives rise to a quiver in the same form, but with a rearrangement of the SU(2) subgroups of the total flavor symmetry. In general, though, the story is far richer. Indeed, it is often possible that the theory emerging at a cusp has a completely distinct quiver than that of the original theory. This, for example, happens in the case of the three node quiver $[2] - (2) - (2) - (2) - [2]$. In particular, if we consider the strong-coupling limit of the middle node, while keeping the outer nodes weakly coupled, the quiver transforms as,

which gives a quiver of an entirely new type.

We close this section by noting that a physical origin of the correspondence between 4d $\mathcal{N} = 2$ theories and Riemann surfaces can be obtained via M-theory. In particular, if we wrap an M5-brane on the relevant Riemann surfaces, with the insertion of appropriate defect operators at the marked points, then reducing the worldvolume theory from six- to four-dimension gives precisely the four-dimensional $\mathcal{N} = 2$ theories in question. Due to their six-dimensional origins, the four-dimensional $\mathcal{N} = 2$ theories constructed in this way are referred to as *Class $\mathcal{S}$* theories.

• Bibliographical notes

Comprehensive expositions of supersymmetric quantum field theories and supersymmetric Yang Mills theories may be found in the books by Wess and Bagger [WB92], Weinberg [Wei13], Freedman and Van Proeyen [FVP12], and Shifman [Shi22], and in the lecture notes [Wit99].

Goddard–Nuyts–Olive duality was introduced in [GNO77] and realized in $\mathcal{N} = 4$ Yang–Mills theory as Montonen–Olive duality in [MO77, Osb79]. The effect of topological charges on the central charge of the supersymmetry algebra was obtained in [WO78], while the effect by which magnetic monopoles in the presence of a θ-angle become dyons was derived in [Wit79]. Subsequent tests of S-duality in $\mathcal{N} = 4$ were carried out in [VW94]. The relation between electric–magnetic duality and the geometric Langlands program was established in [KW07]. Integrated correlators in the superconformal phase of $\mathcal{N} = 4$ at fixed gauge coupling provide explicit realizations of S-duality [CGP$^+$21, DGW21a, DGW21b, DGWX23]. For a SAGEX overview of modular invariance in $\mathcal{N} = 4$ Yang–Mills theory and Type IIB superstring theory, see [DGW22].

Seiberg–Witten theory was originally formulated for gauge group SU(2) in [SW94a] and [SW94b]. Generalizations to other gauge groups with various hypermultiplet contents were soon thereafter produced and are reviewed in the lecture notes [IS96, AGH97, DP99, Mar20] and in the book by Tachikawa [Tac13], where extensive bibliographical references may also be found.

Construction of the prepotential using localization techniques was initiated for the instanton part in [Nek03] and generalized to include also the perturbative contributions in [Pes12]. An overview of localization techniques is provided by the collection of essays in [P$^+$17].

The relation between integrable systems and the Seiberg–Witten solution for the $\mathcal{N} = 2^*$ theory was put forward in [DW96], [Mar96], [GKM$^+$95], and [GMMM98]. The relation between the Hitchin system, D-branes, and ALE instantons was expounded upon in [DM96]. The implementation using the Calogero–Moser system for SU(N) gauge group was carried out in [DP98b] and for other gauge groups in [DP98a] and [DP98c] where the behavior under SL$(2, \mathbb{Z})$ transformations was also analyzed. The derivations in Section 15.8 closely follow [DP98b] and [MMZ20] and reproduce the hypermultiplet mass expansion of [MNW98]. The modular properties of the Seiberg–Witten solution for gauge group SU(2) and $N_f \leq 4$ fundamentals were analyzed in detail in [AFM22a] and [AFM22b] and for SU(3) in [AFM21]. The role of bimodular forms was discussed in those references, in [MM21], and in the mathematical literature in [SZ06]. The Seiberg–Witten curve for linear quivers and gauge group SU(N) was constructed from M-theory in [Wit97]. Relations with the geometric Langlands program are exhibited in [GW06].

The dualities of $\mathcal{N} = 2$ superconformal Yang–Mills theories were developed in [Gai12] and are reviewed in [AATM$^+$22]. The global structure and spectrum of line operators in $\mathcal{N} = 4$ Yang–Mills is clearly presented in

[AST13]. Useful lecture notes on the dynamics of four-dimensional super-symmetric gauge theories may be found in [RSSZ22]. Derivations from compactifications on Calabi–Yau manifolds are studied in [GVW00].

Besides the beautiful mathematics that arose from Seiberg–Witten theory, various exciting physical results also emerged, including the discovery of a number of strongly coupled, non-Lagrangian theories known as *Argyres-Douglas theories* [AD95], as well as the later-discovered *Minahan-Nemeschansky theories* [MN96, MN97]. These bear close relation to complex-multiplication and to Kodaira's theory of the degeneration of elliptic fibers.

Finally, in the last decade, an interesting connection between four-dimensional $\mathcal{N} = 2$ theories and 2d vertex operator algebras (VOAs) has been uncovered [BLL+15], which has given even more direct connections between 4d $\mathcal{N} = 2$ theories and modularity. Indeed, by studying the modular differential equations satisfied by the corresponding VOAs [BR18, KMRW22], it has been possible to provide a partial classification for low-rank 4d $\mathcal{N} = 2$ theories. The full power of these techniques remains to be explored.

Exercises

15.1 For the case of $U(1)$ gauge group and trivial Kähler and superpotentials, evaluate the integrals over superspace in (15.2.3) to obtain a Lagrangian in terms of component and auxiliary fields.

15.2 By coupling free Maxwell theory to a field B with new gauge redundancy $\mathcal{G} : A \to A + \Lambda$, $B \to B + d\Lambda$ and integrating out the original Maxwell field, derive, at the level of the path integral, the electro–magnetic duality relating the theories at couplings g^2 and $1/g^2$.

15.3 Work out the global variants for $\mathfrak{g} = \mathfrak{su}(3)$ $\mathcal{N} = 4$ Yang–Mills theory and show how they map into one another under the action of $\mathrm{SL}(2, \mathbb{Z})$.

15.4 Starting from the spectrum of line operators, identify a and a_D for the $\mathrm{SO}(3)_\pm$ theories in terms of those of the $\mathrm{SU}(2)$ theory. Use this to identify the low-energy couplings τ, as well as the transformations of the $\mathfrak{A}$ and $\mathfrak{B}$ cycles of the Seiberg–Witten curve.

15.5 Consider a rank three self-dual antisymmetric tensor field H on a six-dimensional manifold of the form $\mathbb{R}^4 \times \mathbb{T}^2$, where $\mathbb{R}^4$ has Minkowski signature, which obeys the linear field equation $dH = 0$. Show that toroidal compactification on $\mathbb{T}^2$, where only the zero modes of the fields on $\mathbb{T}^2$ are retained, is equivalent to the free Maxwell equations on $\mathbb{R}^4$. Derive the implications of the $\mathrm{SL}(2, \mathbb{Z})$ modular group symmetry of $\mathbb{T}^2$ on the electromagnetic fields in $\mathbb{R}^4$.

16

Basic Galois theory

We close with a brief introduction to Galois theory, largely following the presentations of [Esc01] and [Hun12], illustrating the application of these mathematical ideas in physics through examples from conformal field theory.

16.1 Fields

In mathematics, a *field*[1] is a set F equipped with two binary operations: $F \times F \to F$ on elements $a, b \in F$, referred to as addition $a + b$ and multiplication $ab = a \cdot b$. Under addition F forms a group with unit element 0, while under multiplication $F \setminus \{0\}$ forms a group with unit 1. The two operations are commutative and are intertwined by the property of distributivity which requires $a \cdot (b + c) = a \cdot b + a \cdot c$ for all $a, b, c \in F$.

A field F is said to have *characteristic* 0 if there exists no integer n such that the n-fold sum of the unit 1 vanishes. If there does exist a positive integer n such that the n-fold sum of the unit 1 vanishes, then the smallest such integer may be shown to be a prime number p and the characteristic of F is defined to be p. If F has characteristic p, then we have $p \cdot a = 0$ for all $a \in F$. The simplest finite fields of characteristic p prime are $\mathbb{F}_p = \mathbb{Z}/(p\mathbb{Z})$.

Well-known examples of fields of characteristic 0 include the rational numbers $\mathbb{Q}$, the real numbers $\mathbb{R}$, the complex numbers $\mathbb{C}$, and the algebraic numbers $\bar{\mathbb{Q}}$ (namely, complex numbers that satisfy a polynomial equation with rational coefficients). Rational functions of a single variable x form fields of characteristic 0, namely, $\mathbb{Q}(x)$, $\mathbb{R}(x)$, $\mathbb{C}(x)$, and $\bar{\mathbb{Q}}(x)$ depending on whether the coefficients in the rational function are in $\mathbb{Q}$, $\mathbb{R}$, $\mathbb{C}$, of $\bar{\mathbb{Q}}$, respectively.

A subset $F \subset K$ of a field K is referred to as a *subfield* if F is a field with respect to the binary operations of the field K. In the above examples, we

---

[1] Not to be confused with the fields of physics, such as the electromagnetic field.

have the subfield inclusions $\mathbb{Q} \subset \mathbb{R} \subset \mathbb{C}$, and $\mathbb{Q}(x) \subset \mathbb{R}(x) \subset \mathbb{C}(x)$, as well as $\mathbb{Q} \subset \mathbb{Q}(x)$, $\mathbb{R} \subset \mathbb{R}(x)$, and $\mathbb{C} \subset \mathbb{C}(x)$. The intersection $F_1 \cap F_2$ of two subfields $F_1, F_2 \subset K$ is a subfield of K.

16.2 Field extensions

A field K is said to be an *extension* of a field F if F is a subfield of K, namely, $F \subset K$. The *degree* of the extension, denoted $[K : F]$, is the dimension of K viewed as a vector space over the field F. If F is a subfield of a field K which itself is a subfield of a field L, then the following relation holds between the degrees of extension,

$$F \subset K \subset L \qquad\qquad [L : F] = [L : K][K : F] \qquad (16.2.1)$$

Equivalently, these relations hold when L is an extension of K which in turn is an extension of F. For example, $\mathbb{C}$ is a field extension of $\mathbb{R}$ by $i = \sqrt{-1}$, and the degree of the extension is $[\mathbb{C} : \mathbb{R}] = 2$ since $\{1, i\}$ is a basis of $\mathbb{C}$ over $\mathbb{R}$. On the other hand, $\mathbb{R}$ is a field extension of $\mathbb{Q}$ of infinite degree.

We denote the ring of polynomials in a variable x with coefficients in the field F by $F[x]$. A polynomial $f(x) \in F[x]$ of degree $n \geq 0$ takes the form,

$$f(x) = a_n x^n + a_{n-1} x^{n-1} + \cdots a_1 x + a_0 \qquad (16.2.2)$$

with $a_0, a_1, \cdots, a_{n-1}, a_n \in F$ and $a_n \neq 0$. The polynomial $f(x)$ is *monic* provided $a_n = 1$. A polynomial $f(x) \in F[x]$ is *reducible* over F if $f(x)$ may be written as the product $f(x) = f_1(x) f_2(x)$ of polynomials $f_1(x), f_2(x) \in F[x]$ of nonvanishing degrees, $\deg f_1, \deg f_2 \geq 1$, and $f(x)$ is *irreducible* over F otherwise. Clearly, every polynomial of degree 1 is irreducible.

An element $y \in K$ is said to be *algebraic* over F if there exists a polynomial $p(x) \in F[x]$ of degree $\deg p \geq 1$ such that $p(y) = 0$. Given an extension K of F containing an element y that is algebraic over F, there exists a unique monic irreducible polynomial $f(x) \in F[x]$ such that $f(y) = 0$. The polynomial $f(x)$ is known as the *minimal polynomial of y over F*. The uniqueness is a nontrivial result – it implies that once we have found any monic irreducible polynomial in $F[x]$ with y as a root, it must be the minimal polynomial.

As an example, consider $K = \mathbb{Q}(\sqrt{3})$ and $F = \mathbb{Q}$. The polynomial $f(x) = x^2 - 3$ is monic and irreducible in $\mathbb{Q}[x]$ and has $\sqrt{3}$ as a root. It is thus the minimal polynomial of $\sqrt{3}$ over $\mathbb{Q}$. This simple example illustrates an important fact: The minimal polynomial is dependent on the choice of base field F. Indeed, if we had chosen $F = \mathbb{R}$, then $f(x)$ would have been reducible, splitting to $f(x) = (x - \sqrt{3})(x + \sqrt{3})$. Thus the minimal polynomial of $\sqrt{3}$ over $\mathbb{R}$ is instead $\tilde{f}(x) = x - \sqrt{3}$.

16.2.1 Simple extensions

Let K be an extension of F and y be an element in K. We denote by $F(y)$ the intersection of all subfields of K which contain both F and y. This intersection is a field, since the intersection of subfields of K is itself a field. Since $F(y)$ is contained in all subfields containing F and y, it is by definition the smallest subfield containing F and y. We refer to such a minimal extension $F(y)$ as a *simple extension* of F by y. In fact, *any* finite extension of a field F of characteristic 0 is a simple extension. For algebraic y, namely y satisfying $f(y) = 0$ for the minimal polynomial f of degree n with coefficients in F, one has the following result.

Theorem 16.2.1. *If y is algebraic over F, and K is a field extension of F by the element y, satisfying $f(y) = 0$ for the minimal polynomial $f(x) \in F[x]$ of degree n, then the simple extension $F(y)$ does not depend on K, and satisfies,*

$$F(y) \cong F[x]/f(x) \qquad\qquad [F(y) : F] = n \qquad\qquad (16.2.3)$$

In other words, the field extension $F(y)$ is completely determined by $F[x]$ and the minimal polynomial f of y.

An analogous theorem holds for finitely generated field extensions of F by elements $y_1, \cdots, y_n$, denoted $F(y_1, y_2, \ldots, y_n)$, as long as K is an *algebraic extension* of F, that is, if every element in K is algebraic over F.

An immediate consequence of Theorem 16.2.1 is that if y and w have the same minimal polynomial $f(x)$ in $F[x]$, then $F(y)$ is isomorphic to $F(w)$.

16.2.2 Splitting fields

So far we have considered extension fields of F containing a root y of a polynomial $f(x) \in F[x]$. We now consider extension fields that contain *all* the roots $y_1, \cdots, y_n$ of $f(x)$. In particular, let K be an extension field of F and $f(x)$ be a nonconstant polynomial of degree n in $F[x]$. If $f(x)$ factors in $K[x]$ as $f(x) = \alpha(x - y_1)(x - y_2) \ldots (x - y_n)$, then we say that $f(x)$ *splits* in K. If K is the smallest extension containing all the roots of $f(x)$, then we refer to K as the *splitting field* for $f(x)$.

By Theorem 16.2.1, any two splitting fields of a polynomial in $F[x]$ are isomorphic. Furthermore, every polynomial $f(x) \in F[x]$ admits a splitting field over F.

Theorem 16.2.2. *Let F be a field and $f(x)$ be a nonconstant polynomial of degree n in $F(x)$. Then there exists a splitting field K of $f(x)$ over F such that $[K : F] \leq n!$.*

Of course, we already know that, if K is a simple extension of F and the polynomial $f(x)$ has n distinct roots, then $[K : F] = n$ by Theorem 16.2.1.

We have just noted that every polynomial admits a splitting field; it is then natural to ask if there exists a field extension of F over which *every* polynomial splits. If such a universal splitting field exists, the field F is said to be *algebraically closed*. A well-known example is $F = \mathbb{C}$. Another is the field $\mathbb{A}$ of all algebraic numbers over $\mathbb{Q}$.

16.2.3 Normal extensions

We close this section by introducing the concept of a *normal extension*. An algebraic extension K of F is said to be normal provided that whenever an irreducible polynomial has one root in K, it has all roots in K (i.e., it splits over K). For example, $\mathbb{Q}(\sqrt{3})$ is a normal extension of $\mathbb{Q}$ since both of the roots $\pm\sqrt{3}$ of the minimal polynomial $x^2 - 3$ are in $\mathbb{Q}(\sqrt{3})$. On the other hand, the extension $\mathbb{Q}(\sqrt[3]{2})$ is not a normal extension of $\mathbb{Q}$ since two roots of the minimal polynomial $x^3 - 2$ are complex and not contained in $\mathbb{Q}(\sqrt[3]{2})$.

16.3 Field automorphisms and the Galois group

An automorphism of a field K is a bijection $\sigma : K \to K$ such that for all $a, b \in K$ we have,

$$\sigma(a + b) = \sigma(a) + \sigma(b)$$
$$\sigma(ab) = \sigma(a)\,\sigma(b) \tag{16.3.1}$$

Two automorphisms $\sigma, \tilde{\sigma}$ are distinct if $\sigma(a) \neq \tilde{\sigma}(a)$ for at least one element $a \in K$. Distinct automorphisms are linearly independent – that is, if $\sigma_1, \ldots, \sigma_n$ are distinct automorphisms of K, then a linear relation valid for all $y \in K$,

$$a_1\sigma_1(y) + \cdots + a_n\sigma_n(y) = 0 \tag{16.3.2}$$

implies that $a_1 = \cdots = a_n = 0$.

Let K be an extension field of F. An *F-automorphism of K* is an automorphism of the field K, namely, $\sigma : K \to K$, that fixes each element of F. The set of all F-automorphisms will be denoted $\mathrm{Gal}(K, F)$ and is called the *Galois group of K over F*. The use of the word "group" is justified since $\mathrm{Gal}(K, F)$ can be endowed with a group structure via function composition. Indeed, if $\sigma_1, \sigma_2 \in \mathrm{Gal}(K, F)$, then clearly $\sigma_1 \circ \sigma_2$ is also an automorphism of K leavings F invariant. We begin with the following result.

Theorem 16.3.1. *Consider a polynomial $f(x) \in F[x]$. If $y \in K$ is a root of $f(x)$ and $\sigma \in \mathrm{Gal}(K, F)$, then $\sigma(y)$ is also a root of $f(x)$.*

To prove this theorem, let $f(x)$ be an arbitrary polynomial of arbitrary degree n,

$$f(x) = c_n x^n + \cdots + c_1 x + c_0 \tag{16.3.3}$$

Since $f(y) = 0$, and σ is a homomorphism with $\sigma(c_i) = c_i$ for each $c_i \in F$, we have,

$$
\begin{aligned}
0 = \sigma(0) &= \sigma\left(c_n y^n + \ldots c_1 y + c_0\right) \\
&= \sigma(c_n)\sigma(y^n) + \cdots + \sigma(c_1)\sigma(y) + \sigma(c_0) \\
&= c_n \sigma(y)^n + \cdots + c_1 \sigma(y) + c_0
\end{aligned}
\tag{16.3.4}
$$

and hence $\sigma(y)$ is a root of $f(x)$.

Theorem 16.3.1 shows that every image of a root y of $f(x)$ under an automorphism of the Galois group is also a root of $f(x)$. Conversely, we could ask if every root of $f(x)$ can be obtained as the image of some root y under an automorphism $\sigma \in \mathrm{Gal}(K, F)$. In some instances, this is indeed the case, as summarized by the following theorem.

Theorem 16.3.2. *If K is a splitting field for a polynomial in $F[x]$ and $y_1, y_2 \in K$ are two roots, then there exists a $\sigma \in \mathrm{Gal}(K, F)$ such that $\sigma(y_1) = y_2$ if and only if y_1 and y_2 have the same minimal polynomial over F.*

The forward direction follows immediately from Theorem 16.3.1. For the converse, note from Theorem 16.2.1 that if y_1 and y_2 have the same minimal polynomial, then there is an isomorphism $\sigma : F(y_1) \cong F(y_2)$. This isomorphism can be chosen such that $\sigma(y_1) = y_2$, while fixing F element-wise. Since K is a splitting field of some polynomial in $F[x]$, it is also a splitting field for the same polynomial over $F(y_1)$ and $F(y_2)$, and thus σ extends to an F-automorphism of K, that is, $\sigma \in \mathrm{Gal}(K, F)$.

An important fact of which we make repeated use is the following,

Theorem 16.3.3. *If $K = F(y_1, \ldots, y_n)$ is an algebraic extension of F, then the automorphisms in $\mathrm{Gal}(K, F)$ are determined completely by their action on $y_1, \ldots, y_n$. That is, if $\sigma_1, \sigma_2 \in \mathrm{Gal}(K, F)$ satisfy $\sigma_1(y_i) = \sigma_2(y_i)$ for all $i = 1, \ldots, n$, then $\sigma_1 = \sigma_2$.*

Together with Theorem 16.3.1, this result can be used to at least partially construct concrete Galois groups. For example, consider the Galois group of $\mathbb{Q}(\sqrt{3}, \sqrt{5})$ over $\mathbb{Q}$. By Theorem 16.3.3, the Galois automorphisms are determined completely by their action on $\sqrt{3}$ and $\sqrt{5}$. The minimal polynomial of $\sqrt{3}$ is $x^2 - 3$, and hence by Theorem 16.3.2, we must have $\sqrt{3} \to \pm\sqrt{3}$. Similar statements hold for $\sqrt{5}$. Hence the Galois group $\mathrm{Gal}(\mathbb{Q}(\sqrt{3}, \sqrt{5}), \mathbb{Q})$ has at most four elements,

$$\sqrt{3} \overset{1}{\to} \sqrt{3} \qquad \sqrt{3} \overset{\alpha}{\to} -\sqrt{3} \qquad \sqrt{3} \overset{\beta}{\to} \sqrt{3} \qquad \sqrt{3} \overset{\gamma}{\to} -\sqrt{3}$$
$$\sqrt{5} \to \sqrt{5} \qquad \sqrt{5} \to \sqrt{5} \qquad \sqrt{5} \to -\sqrt{5} \qquad \sqrt{5} \to -\sqrt{5} \quad (16.3.5)$$

In fact, it turns out that in this case there are exactly four elements, and $\mathrm{Gal}(\mathbb{Q}(\sqrt{3}, \sqrt{5}), \mathbb{Q}) \cong \mathbb{Z}_2 \times \mathbb{Z}_2$, though we will not prove this here.

In the above example, every automorphism of the Galois group permutes the four roots $\pm\sqrt{3}, \pm\sqrt{5}$. This is a particular case of a more general result,

Theorem 16.3.4. *If K is the splitting field of a separable polynomial $f(x)$ of degree n in $F[x]$ (a separable polynomial of degree n has exactly n distinct roots in K), then $\mathrm{Gal}(K, F)$ is isomorphic to a subgroup of S_n.*

That is, every element of $\mathrm{Gal}(K, F)$ produces a permutation of the roots of $f(x)$. Of course, the converse is not necessarily true. For example, in the case of $\mathbb{Q}(\sqrt{3}, \sqrt{5})$, the permutation mapping $\pm\sqrt{3} \to \pm\sqrt{5}, \pm\sqrt{5} \to \pm\sqrt{3}$ does not correspond to an F-automorphism of K.

16.3.1 Intermediate fields

Let K be an extension field of F. A field E such that $F \subseteq E \subseteq K$ is called an *intermediate field* of the extension. It can be shown that the Galois group $\mathrm{Gal}(E, F)$ associated to the intermediate field is a subgroup of $\mathrm{Gal}(K, F)$. Hence to every intermediate field, we can associate a subgroup of the Galois group $\mathrm{Gal}(K, F)$. Conversely, if H is a subgroup of the Galois group, one can associate to it an intermediate field E_H such that,

$$E_H = \{k \in K \mid \sigma(k) = k \quad \text{for all} \ \sigma \in H\} \qquad (16.3.6)$$

known as the *fixed field of H*. That E_H is indeed a field is easy to check, since for example the equalities $\sigma(0) = 0$ and $\sigma(1) = 1$ imply that 0 and 1 are contained in E_H, and since,

$$\sigma(c + d) = \sigma(c) + \sigma(d) = c + d$$
$$\sigma(cd) = \sigma(c)\sigma(d) = cd \tag{16.3.7}$$

imply that E_H is closed under addition and multiplication. The presence of inverses follows simply as well. A somewhat stronger result is the following,

Theorem 16.3.5. *Let K be a finite-dimensional extension field of F. If H is a subgroup of $\mathrm{Gal}(K, F)$ and E_H is the fixed field of H, then $H = \mathrm{Gal}(K, E_H)$ and $|H| = [K : E]$.*

We may now return to the example of $\mathrm{Gal}(\mathbb{Q}(\sqrt{3}, \sqrt{5}), \mathbb{Q})$. In this case, $\mathbb{Q}(\sqrt{3})$ is an intermediate field of the extension $\mathbb{Q}(\sqrt{3}, \sqrt{5})$ of $\mathbb{Q}$. Recall that $\mathrm{Gal}(\mathbb{Q}(\sqrt{3}, \sqrt{5}), \mathbb{Q}) = \{1, \alpha, \beta, \gamma\}$ as defined in (16.3.5). On the other hand, $\mathrm{Gal}(\mathbb{Q}(\sqrt{3}, \sqrt{5}), \mathbb{Q}(\sqrt{3}))$ should consist of only the transformations leaving the base field $\mathbb{Q}(\sqrt{3})$ invariant. Hence we have that

$$\mathrm{Gal}(\mathbb{Q}(\sqrt{3}, \sqrt{5}), \mathbb{Q}(\sqrt{3})) = \{1, \beta\} \tag{16.3.8}$$

which is clearly a subgroup of $\mathrm{Gal}(\mathbb{Q}(\sqrt{3}, \sqrt{5}), \mathbb{Q})$. Conversely, the fixed field of $H = \{1, \beta\}$ is precisely $E_H = \mathbb{Q}(\sqrt{3})$.

16.4 The fundamental theorem of Galois theory

Consider a finite-dimensional extension field K of F. Let S be the set of all intermediate fields, and T be the set of all subgroups of the Galois group $\mathrm{Gal}(K, F)$. We introduce a function $\varphi : S \to T$ such that for each intermediate field E,

$$\varphi(E) = \mathrm{Gal}(K, E) \tag{16.4.1}$$

The function φ is referred to as a *Galois correspondence*. In this correspondence, K is mapped to the identity subgroup of $\mathrm{Gal}(K, F)$, while F is mapped to the entire group $\mathrm{Gal}(K, F)$. By Theorem 16.3.5, the Galois correspondence is surjective.

While surjectivity of the Galois correspondence is guaranteed, injectivity is not. For example, consider the extension field $\mathbb{Q}(\sqrt[3]{2})$ over $\mathbb{Q}$. Every automorphism in the Galois group maps $\sqrt[3]{2}$ to a root of the minimal polynomial $x^3 - 2$ by Theorem 16.3.1. Since $\sqrt[3]{2}$ is the only real root, and since all elements of $\mathbb{Q}(\sqrt[3]{2})$ are real, every automorphism must map $\sqrt[3]{2}$ to itself.

Thus by Theorem 16.3.3, we conclude that $\mathrm{Gal}(\mathbb{Q}(\sqrt[3]{2}), \mathbb{Q})$ is trivial. This means that under the Galois correspondence, both $\mathbb{Q}(\sqrt[3]{2})$ and $\mathbb{Q}$ must be associated to the same group, that is, the trivial group $\{1\}$. Therefore the Galois correspondence is not injective.

In the cases in which the Galois correspondence *is* injective, the extension K of F is called a *Galois extension*. Given a Galois extension K, if E_1 and E_2 are intermediate fields such that $\mathrm{Gal}(K, E_1) = \mathrm{Gal}(K, E_2)$, then we can conclude that $E_1 = E_2$. Note that in characteristic 0, a Galois extension is simply a splitting field.

As an example, consider the extension $\mathbb{Q}(\sqrt{3}, \sqrt{5})$ over $\mathbb{Q}$. This extension is Galois because it is the splitting field of $f(x) = (x^2 - 3)(x^2 - 5)$. In this case, the Galois correspondence φ must be bijective, and indeed we have

$$\varphi : \mathbb{Q}(\sqrt{3}, \sqrt{5}) \mapsto \mathrm{Gal}(\mathbb{Q}(\sqrt{3}, \sqrt{5}), \mathbb{Q}(\sqrt{3}, \sqrt{5})) = 1$$
$$\mathbb{Q}(\sqrt{3}) \mapsto \mathrm{Gal}(\mathbb{Q}(\sqrt{3}, \sqrt{5}), \mathbb{Q}(\sqrt{3})) = \{1, \beta\}$$
$$\mathbb{Q}(\sqrt{5}) \mapsto \mathrm{Gal}(\mathbb{Q}(\sqrt{3}, \sqrt{5}), \mathbb{Q}(\sqrt{5})) = \{1, \alpha\}$$
$$\mathbb{Q}(\sqrt{15}) \mapsto \mathrm{Gal}(\mathbb{Q}(\sqrt{3}, \sqrt{5}), \mathbb{Q}(\sqrt{5})) = \{1, \gamma\}$$
$$\mathbb{Q} \mapsto \mathrm{Gal}(\mathbb{Q}(\sqrt{3}, \sqrt{5}), \mathbb{Q}) = \{1, \alpha, \beta, \gamma\} \qquad (16.4.2)$$

In this example, we observe that all intermediate fields appearing are themselves Galois extensions of $\mathbb{Q}$. For example, $\mathbb{Q}(\sqrt{3})$ is a Galois extension since it is the splitting field of $x^2 - 3$. Furthermore, the corresponding subgroups of the Galois group are all normal. The general version of this statement is given by the Fundamental Theorem of Galois Theory,

Theorem 16.4.1. *Let K be a Galois extension of F. Then*

1. *The map $\varphi : S \to T$ between the set S of intermediate fields and the set T of subgroups of the Galois group $\mathrm{Gal}(K, F)$ defined by $\varphi(E) = \mathrm{Gal}(K, E)$ is a bijection. One has*

$$[K : E] = |\mathrm{Gal}(K, E)|$$
$$[E : F] = [\mathrm{Gal}(K, F) : \mathrm{Gal}(K, E)]$$
$$[K : F] = |\mathrm{Gal}(K, F)| \qquad (16.4.3)$$

2. *An intermediate field E is a normal extension of F if and only if the corresponding group $\mathrm{Gal}(K, E)$ is a normal subgroup of $\mathrm{Gal}(K, F)$, in which case $\mathrm{Gal}(E, F) \cong \mathrm{Gal}(K, F)/\mathrm{Gal}(K, E)$.*

16.5 Solvability by radicals

The most famous application of Galois theory, and the reason for its initial development, is to the study of roots of polynomial equations. In particular, Galois theory provides a criterion for a polynomial to be solvable by radicals, as we now discuss. Let us first define the notion of solvability by radicals. A field K is said to be a *radical extension* of a field F if there exists a chain of fields,

$$F = F_0 \subseteq F_1 \subseteq F_1 \subseteq F_2 \subseteq \cdots \subseteq F_r = K \qquad (16.5.1)$$

such that for each $i = 1, 2, \ldots, r$ we have $F_i = F_{i-1}(y_i)$ and some power of y_i is in F_{i-1}. Now let $f(x) \in F[x]$. The equation $f(x) = 0$ is said to be *solvable by radicals* if there exists a radical extension of F that contains a splitting field of $f(x)$.

On the other hand, a group G is said to be solvable if there exists a chain of subgroups,

$$G = G_0 \supseteq G_1 \supseteq G_2 \supseteq \cdots \supseteq G_{n-1} \supseteq G_n = \{1\} \qquad (16.5.2)$$

such that G_i is a normal subgroup of the preceding group G_{i-1} and the quotient group G_{i-1}/G_i is abelian. For example, every Abelian group G is solvable since $G \supseteq \{1\}$ is a chain of the required form. As a more nontrivial example, consider the cyclic subgroup $\langle (123) \rangle$ of order 3 in S_3. The chain,

$$S_3 \supseteq \langle (123) \rangle \supseteq \langle (1) \rangle \qquad (16.5.3)$$

shows that S_3 is solvable. On the other hand, it can be shown that for $n \geq 5$ the permutation group S_n is *not* solvable.

We now state the famous Galois Criterion for solvability by radicals.

Theorem 16.5.1. *Let F be a field of characteristic 0 and $f(x) \in F[x]$. Then $f(x) = 0$ is solvable by radicals if and only if the Galois group of $f(x)$ is solvable.*

As a simple example, consider the polynomial $f(x) = 2x^5 - 10x + 5 \in \mathbb{Q}[x]$. By Theorem 16.3.4, the Galois group must be a subgroup of S_5. We will now show that it is in fact *equal* to S_5. To do so, note that the function $f(x)$ has a maximum at $x = -1$, a minimum at $x = 1$, and an inflection point at $x = 0$. This means that $f(x)$ crosses the x-axis exactly three times, and hence that $f(x)$ has exactly three real roots. If K is a splitting field of $f(x)$ over $\mathbb{Q}$, then $|\mathrm{Gal}(K, \mathbb{Q})| = [K : \mathbb{Q}]$ by Theorem 16.4.1. Taking any root r

of $f(x)$, we have

$$[K:\mathbb{Q}] = [K:\mathbb{Q}(r)][\mathbb{Q}(r):\mathbb{Q}] = 5[K:\mathbb{Q}(r)] \qquad (16.5.4)$$

by Theorem 16.2.1. From this it follows that 5 divides $|\mathrm{Gal}(K,\mathbb{Q})|$, and thus that $\mathrm{Gal}(K,\mathbb{Q})$ contains an element of order 5. In other words, $\mathrm{Gal}(K,\mathbb{Q})$ contains a 5-cycle of S_5. In addition, $\mathrm{Gal}(K,\mathbb{Q})$ also contains complex conjugation, which fixes the three real roots of $f(x)$ but interchanges the two nonreal roots. Thus $\mathrm{Gal}(K,\mathbb{Q})$ contains a transposition. But in fact, the only subgroup of S_5 containing both a 5-cycle and a transposition is S_5 itself. We thus conclude that $\mathrm{Gal}(K,\mathbb{Q}) = S_5$. Then using the fact that S_n is not solvable for $n \geq 5$, we conclude from Theorem 16.5.1 that $f(x)$ does not admit a solution in radicals. This implies the famous result that there is no universal formula (involving only field operations and roots) for the solutions of fifth-degree polynomial equations.

16.6 Cyclotomic fields and Abelian extensions

The Nth roots of unity ζ satisfy $\zeta^N = 1$ and form a multiplicative group that is isomorphic to the additive group $\mathbb{Z}_N$. Any such root is of the form $e^{2\pi i k/N}$ for some $k \in \{0, 1, \cdots, N-1\}$. An Nth root of unity is a *primitive root* if it generates the entire group of Nth roots. The primitive roots are given by $e^{2\pi i k/N}$ with $\gcd(k, N) = 1$. Clearly, when N is prime, every root different from 1 is a primitive root. The extension $\mathbb{Q}(\zeta)$ of the rationals by any primitive Nth root of unity ζ is referred to as a cyclotomic field. Clearly, the degree of extension is $[\mathbb{Q}(\zeta):\mathbb{Q}] = N$. The Galois group $\mathrm{Gal}(\mathbb{Q}(\zeta),\mathbb{Q})$ is Abelian, as the automorphisms are just the representations of $\mathbb{Z}_N$.

Denoting the fundamental Nth root by $\zeta = e^{2\pi i/N}$, we have the following factorization of the polynomial $x^N - 1$,

$$x^N - 1 = \prod_{n=0}^{N-1} (x - \zeta^n) = (x-1) \prod_{n=1}^{N-1} (x - \zeta^n) \qquad (16.6.1)$$

The Nth cyclotomic polynomial $\Phi_N(x)$, defined for $\zeta^N = 1$ by,

$$\Phi_N(x) = \prod_{\zeta \text{ primitive}} (x - \zeta) \qquad (16.6.2)$$

is irreducible over $\mathbb{Q}$ and has its degree given by the Euler totient function $\deg(\Phi_N) = |\mathbb{Z}_N^*| = \phi(N)$. Furthermore, $\Phi_N(x)$ is the minimal polynomial for the primitive Nth roots of unity, and provides the following factorization,

$$x^N - 1 = \prod_{d|N} \Phi_d(x) \tag{16.6.3}$$

The degrees on both sides match thanks to the identity $\sum_{d|N} \phi(d) = N$. For the special case where $N = p$ is prime, we have $\Phi_p(x) = \sum_{n=0}^{p-1} x^n$.

The universal importance of cyclotomic fields may best be perceived from the Kronecker–Weber theorem.

Theorem 16.6.1. *Let F be a field extension of $\mathbb{Q}$ whose Galois group $\mathrm{Gal}(F, \mathbb{Q})$ is Abelian. Then there exists a cyclotomic extension field $\mathbb{Q}(\zeta)$ of $\mathbb{Q}$ such that $F \subset \mathbb{Q}(\zeta)$. In other words, Galois extensions of $\mathbb{Q}$ with Abelian Galois group are precisely the subfields of the cyclotomic fields.*

For a proof we refer to [Mar77]; see also [Cox11a] and [Esc01] for partial proofs, examples, and applications.

16.7 Galois theory in rational conformal field theory

We close with an application of Galois theory to physics. As discussed in Sections 8.3.2 and 10.10, the torus partition function of a rational conformal field theory (fields now in the sense of physics!) can be expressed in terms of a finite number of characters $\chi(\tau) = (\chi_1(\tau), \ldots, \chi_d(\tau))^t$, transforming in a vector representation of the modular group. The action of $S : \tau \to -1/\tau$ in particular is implemented by a matrix known as the modular matrix $\boldsymbol{S}$,

$$\chi(-1/\tau) = \boldsymbol{S}\,\chi(\tau) \tag{16.7.1}$$

A crucial constraint on $\boldsymbol{S}$ is that it satisfy the Verlinde formula,

$$N_{pq}^r = \sum_{s \in \mathcal{I}} \frac{\boldsymbol{S}_{p,s}\boldsymbol{S}_{q,s}\boldsymbol{S}_{r,s}^\dagger}{\boldsymbol{S}_{0,s}} \tag{16.7.2}$$

where $p, q, s \in \mathcal{I}$ labels the finite set of primary fields ϕ_s of the theory and N_{pq}^r are fusion coefficients of primaries, namely, they satisfy,

$$\phi_p \times \phi_q = \sum_r N_{pq}^r \phi_r \tag{16.7.3}$$

In particular, the operator product coefficients N_{pq}^r are integers. Treating N_p with entries $(N_p)_q^r = N_{pq}^r$ as matrices, we may write their characteristic polynomials as follows,

$$\det(\lambda_p \mathbb{1} - N_p) = 0 \tag{16.7.4}$$

The polynomial in each λ_p is a monic with integer coefficients. From the Verlinde formula, it follows that the roots are given by,

$$\lambda_p^{(q)} = \frac{S_{p,q}}{S_{0,q}} \tag{16.7.5}$$

Whereas the entries of the matrix N_p are integers, the ratios of S-matrix elements $\lambda_p^{(q)}$ are generically not integers, and in fact not even rational. Consider the extension field $\mathbb{Q}(S_{p,q})$ over $\mathbb{Q}$, where $\mathbb{Q}(S_{p,q})$ denotes extension by the matrix elements of the modular S-matrix. By Theorem 16.3.1, the elements $\sigma \in \mathrm{Gal}(\mathbb{Q}(S_{p,q}), \mathbb{Q})$ act on the $\lambda_p^{(q)}$ by permuting them among themselves. In particular, we have,

$$\sigma\left(\frac{S_{p,q}}{S_{0,q}}\right) = \frac{\sigma(S_{p,q})}{\sigma(S_{0,q})} = \frac{S_{p,\sigma(q)}}{S_{0,\sigma(q)}} \tag{16.7.6}$$

Note that here we have associated the permutation of the roots $\lambda_p^{(q)}$ to a permutation of the underlying primary fields. Indeed, it is possible to prove that there exists a set of signs $\epsilon_\sigma(q) \in \{\pm 1\}$ such that,

$$\sigma(S_{p,q}) = \epsilon_\sigma(q) S_{p,\sigma(q)} \tag{16.7.7}$$

that is, Galois conjugation acts on the S-matrix by signed permutation.

This so-called *Galois symmetry* of the rational conformal field theory leads to the following physical constraints. Recall that the partition function of a rational conformal field theory is obtained by gluing together the chiral and antichiral characters via,

$$Z(\tau, \bar{\tau}) = \sum_{\in \mathcal{I}} \sum_{\bar{p} \in \bar{\mathcal{I}}} \mathcal{N}_{p,\bar{p}} \, \chi_p(\tau) \, \overline{\chi}_{\bar{p}}(\bar{\tau}) \tag{16.7.8}$$

for some Hermitian matrix $\mathcal{N}$ with entries $\mathcal{N}_{p,\bar{p}}$. By modular invariance the gluing matrix $\mathcal{N}_{p,\bar{p}}$ is required to satisfy $(\mathcal{N}\overline{S})_{p,\bar{q}} = (S\mathcal{N})_{p,\bar{q}}$. We now apply a Galois automorphism σ to this equation, giving,

$$\sum_{\bar{m} \in \bar{\mathcal{I}}} \mathcal{N}_{p,\bar{m}} \, \bar{\epsilon}_\sigma(\bar{m}) \, \overline{S}_{\sigma(\bar{m}),\bar{q}} = \epsilon_\sigma(p) \sum_m S_{\sigma(p),m} \, \mathcal{N}_{m,\bar{q}}$$

$$= \epsilon_\sigma(p) \sum_{\bar{m}} \mathcal{N}_{\sigma(p),\bar{m}} \, \overline{S}_{\bar{m},\bar{q}}$$

$$= \epsilon_\sigma(p) \sum_{\bar{m}} \mathcal{N}_{\sigma(p),\sigma(\bar{m})} \, \overline{S}_{\sigma(\bar{m}),\bar{q}} \tag{16.7.9}$$

Using invertibility of S, we then obtain,

$$\mathcal{N}_{\sigma(p),\sigma(\bar{q})} = \epsilon_\sigma(p)\bar{\epsilon}_\sigma(\bar{q})\mathcal{N}_{p,\bar{q}} \tag{16.7.10}$$

Because $\mathcal{N} \geq 0$, we conclude that $\mathcal{N}_{p,\bar{q}}$ can only be nonzero when $\epsilon_\sigma(p) = \bar{\epsilon}_\sigma(\bar{q})$ for all σ. This turns out to be a very useful result, leading to powerful selection rules for possible modular invariant combinations of characters. In the literature on conformal field theory, it is referred to as the "parity rule."

• Bibliographical notes

An all-time classic introduction to Galois theory may be found in the book by Artin [AM98], while a general treatment of modern algebra pertinent to Galois theory is given in the book by van der Waerden [VdW91]. The books by Escofier [Esc01] and Cox [Cox11a] provide clear and systematic introductions with many examples, problems sets, and historical notes. We also refer to the recent books by Hungerford [Hun12] and Marcus [Mar77].

In the context of rational conformal field theory, Galois symmetry first appeared in [DBG91], with numerous follow-up works including [CG94], [CG99], and [Ban03]. The connection to Hecke operators was discussed in [HW18, HHW20]. Galois symmetry also plays a role in the study of three-dimensional topological quantum field theories, as discussed in [BR20], [BR22], and [KKO+22].

Exercises

16.1 Consider an $n \times n$ matrix M with entries in $\mathbb{Q}$ and $\det M \neq 0$. While a linear equation $MX = A$, with $A \in \mathbb{Q}^n$, is solved for $X \in \mathbb{Q}^n$ by the methods of linear algebra, obtaining the eigenvalues and eigenvectors of M requires, in general, a field extension of $\mathbb{Q}$. Show that the field extension $F = \mathbb{Q}(\lambda_1, \cdots, \lambda_n)$ of $\mathbb{Q}$ by the eigenvalues of M is Galois, and that the entries of the eigenvectors of M take values in F.

16.2 In Exercise 16.1, consider the matrix of frequencies of a chain of coupled harmonic oscillators $M_{ij} = 2\delta_{i,j} - \delta_{i,j+1} - \delta_{i,j-1}$, where $\delta_{i,j}$ is the Kronecker δ-symbol with period N, so that $\delta_{i,j} = \delta_{i+N,j} = \delta_{i,j+N}$. Characterize the field extension that contains all the eigenvalues of M.

16.3 Consider the modular functions $h_N(v|\tau)$ of $\Gamma(N)$ introduced in (7.3.1) for the case $N = 2$. Show that the functions for $v_1 = (\frac{1}{2}, 0)$, $v_2 = (0, \frac{1}{2})$, and $v_3 = (\frac{1}{2}, \frac{1}{2})$ obey the same cubic equation in $\mathbb{C}(j)$ with rational coefficients, that is, actually in $\mathbb{Q}(j)$. Compute the Galois group of the extension of $\mathbb{Q}(j)$ by $h_N(v_i|\tau)$ for $i = 1, 2, 3$.

16.4 Generalizing Exercise 16.3 to the case of arbitrary N, show that $h_N(v|\tau)$ for $Nv \in \mathbb{Z}^N$ and $v \not\equiv 0$ are algebraic in $\mathbb{Q}(j)$.

16.5 Determine the Galois group of the extension of $\mathbb{Q}(j)$ by the $h_N(v|\tau)$ for which $v \not\equiv 0$ and $Nv \in \mathbb{Z}^N$ as given in Exercise 16.3.

Part III
Appendix

Four relatively short Appendixes collect basic mathematical material that, while central to various parts of the core material, is not concerned directly with either modular forms or string theory. A fifth, Appendix E, provides solutions to the exercises posed in Chapters 2 through 16.

Appendix A is primarily included for the benefit of the reader with a physics training and is presumably familiar to readers with a mathematics training. It includes a basic review of arithmetic modulo N, the Chinese remainder theorem, solving polynomial equations in modular arithmetic, quadratic residues, Gauss sums, the quadratic reciprocity theorem, Dirichlet characters, and L-functions. This material is especially relevant to the discussions of congruence subgroups in Chapters 6, 7, 10, and 11.

Appendix B provides a review of some of the basic definitions, theorems, and general results on compact Riemann surfaces needed in the main text, including the topological, metrical, and complex structural aspects of Riemann surfaces and their construction in terms of Fuchsian groups. This material is directly relevant to conformal field theory in Section 5; the structure of the modular curves in Section 6; string amplitudes in Chapter 12; Seiberg–Witten theory in Chapter 15; and to the subsequent Appendixes C and D.

Appendix C provides a review of holomorphic line bundles on compact Riemann surfaces, the Riemann–Roch theorem and vanishing theorems, with applications to evaluating the dimension of moduli space and describing the spaces of tensor and spinor fields on Riemann surfaces. This material is directly relevant to the physical applications discussed in Chapter 5, to string amplitudes in Chapter 12, and to the material in Appendix D.

Appendix D provides an overview of modular geometry, the Siegel upper half-space, Riemann ϑ-functions, the Riemann relations on ϑ-functions, the Abel map, the Jacobian variety, and their relation to higher genus Riemann surfaces via the Riemann vanishing theorem. Holomorphic and meromorphic differentials of arbitrary weight on Riemann surfaces of arbitrary genus are constructed using Riemann ϑ-functions and the prime form. As a physics application, the correlation functions for the bc system, which were already obtained for genus one in Chapter 5, are evaluated explicitly using the properties of meromorphic differentials and the operator product expansion.

Appendix A

Some arithmetic

In this appendix we collect some basic results in number theory, including the Chinese remainder theorem, its application to solving polynomial equations, the Legendre, Jacobi, and Kronecker symbols, quadratic reciprocity, its application to solving quadratic equations mod N, and a brief introduction to Dirichlet characters and L-functions. Throughout, the set of positive integers is denoted $\mathbb{N} = \{1, 2, 3, \cdots\}$ and does not include 0.

A.1 Arithmetic mod N

We denote by $\mathbb{Z}_N = \mathbb{Z}/N\mathbb{Z} = \{0, 1, 2, \cdots, N-1\}$ the additive group of integers modulo N. Addition of two elements $a, b \in \mathbb{Z}_N$ is denoted by $a + b \pmod{N} \in \mathbb{Z}_N$, the identity is 0, and the inverse of $a \in \mathbb{Z}_N$ is $-a \pmod{N}$. The group $\mathbb{Z}_N$ is isomorphic to the multiplicative group of N-th roots of unity $\{1, \varepsilon, \varepsilon^2, \cdots, \varepsilon^{N-1}\}$, where $\varepsilon = e^{2\pi i/N}$, under multiplication in $\mathbb{C}$.

For $p \in \mathbb{N}$ a prime number, the set $\mathbb{Z}_p^* = \{1, 2, \cdots, p-1\}$ forms a group under multiplication of integers $\pmod{p}$, and $\mathbb{F}_p = \{0, 1, 2, \cdots, p-1\}$ forms a *finite field* under addition in $\mathbb{Z}_p$ and multiplication in $\mathbb{Z}_p^*$.

For arbitrary $N \in \mathbb{N}$, we denote by $\mathbb{Z}_N^*$ the maximal multiplicative subgroup of $\mathbb{Z}_N$. When N is prime, the definition coincides with the one provided above. When N is not prime, however, there are elements in $\mathbb{Z}_N$ other than 0 which are not invertible, and have to be removed to produce a multiplicative group. Consider an element $a \in \mathbb{Z}_N$ which is neither zero nor relatively prime with N, namely, such that $\gcd(a, N) > 1$, then it follows that the element $b = N/\gcd(a, N) \in \mathbb{Z}_N$ satisfies $ab = N$ so that $ab \equiv 0 \pmod{N}$ and therefore the element a is not invertible in $\mathbb{Z}_N$. Every invertible element $a \in \mathbb{Z}_N$ must be relatively prime to N, so that the multiplicative subgroup is given by,

$$\mathbb{Z}_N^* = \{a \in \mathbb{Z}_N \setminus \{0\}, \ \gcd(a, N) = 1\} \qquad |\mathbb{Z}_N^*| = \phi(N) \quad \text{(A.1.1)}$$

Here, $\phi(N)$ is Euler's totient function counting the number of integers less than N that are relatively prime to N, and $|S|$ denotes the cardinality of S. For example, we may represent the group $\mathbb{Z}_{15}^*$ by listing its elements in pairs of opposites modulo 15 as follows, $\mathbb{Z}_{15}^* = \{\pm 1, \pm 2, \pm 4, \pm 7\}$. The elements ± 1 and ± 4 square to 1 (mod 15), while $7 \times (-2) \equiv (-7) \times 2 \equiv 1 \pmod{15}$.

A.2 Chinese remainder theorem

Theorem A.2.1. *Let $n_1, \cdots, n_r$ be positive integers which are relatively prime in pairs so that they satisfy $\gcd(n_i, n_j) = 1$ for all $1 \le i \ne j \le r$ and let $b_1, \cdots, b_r$ be arbitrary integers. Then the system of congruences,*

$$x \equiv b_i \pmod{n_i} \qquad \text{for all } \ i = 1, \cdots, r \qquad \text{(A.2.1)}$$

has a unique solution modulo the product $n_1 \cdots n_r = N$. An equivalent statement is that the map,

$$x \pmod{N} \ \mapsto \ \big(x \pmod{n_1}, \cdots, x \pmod{n_r}\big) \qquad \text{(A.2.2)}$$

is a ring isomorphism,

$$\mathbb{Z}_N \cong \mathbb{Z}_{n_1} \times \cdots \times \mathbb{Z}_{n_r} \qquad \text{(A.2.3)}$$

To prove the theorem, we begin by proving that the solution is unique, so that the map is injective. If we had two solutions x, y satisfying the same congruence conditions, then $x - y \equiv 0 \pmod{n_i}$ for all i. Since the n_i are pairwise coprime their product N also divides $x - y$ so that $x - y \equiv 0 \pmod{N}$. Since the cardinality of the left and right sides are equal to one another, thanks to the multiplicative property of the Euler totient function, the map must be surjective as well, and is thereby an isomorphism.

A.3 Solving polynomial equations

Theorem A.3.1. *Let $f(x)$ be a polynomial in x with integer coefficients $f \in \mathbb{Z}[x]$, let $n_1, \cdots, n_r$ be positive integers which are relatively prime in pairs $\gcd(n_i, n_j) = 1$ for all $1 \le i \ne j \le r$, and denote the product $n_1 \cdots n_r = N$. Then the congruence,*

$$f(x) \equiv 0 \pmod{N} \qquad \text{(A.3.1)}$$

has a solution if and only if each one of the congruences,

$$f(x) \equiv 0 \;(\text{mod } n_i) \qquad i = 1, \cdots, r \qquad (\text{A.3.2})$$

has a solution. The number of solutions $\nu(N)$ *to (A.3.1) is given in terms of the number of solutions* $\nu(n_i)$ *for each of the congruences by,*

$$\nu(N) = \nu(n_1) \cdots \nu(n_r) \qquad (\text{A.3.3})$$

To prove the theorem, we prove both implications. Clearly if x satisfies (A.3.1) then it satisfies (A.3.2). Conversely, let x_i (mod n_i) be a solution to each congruence (A.3.2), which exists by the assumption of the theorem, then there exists x (mod N) which solves,

$$x \equiv x_i \;(\text{mod } n_i) \qquad (\text{A.3.4})$$

by the Chinese remainder theorem.

Applying Theorem A.3.1 to polynomial equations modulo an integer N whose prime factorization is given by $N = \prod_i p_i^{\alpha_i}$ for p_i all distinct primes and $\alpha_i \geq 1$, we have the equivalence,

$$f(x) \equiv 0 \;(\text{mod } N) \qquad \Leftrightarrow \qquad \left\{ f(x) \equiv 0 \;(\text{mod } p_i^{\alpha_i}) \text{ for all } i \right\} \; (\text{A.3.5})$$

The individual congruences for each prime power are obtained by induction on the exponent α as follows.

Theorem A.3.2. *Assume* p *prime,* $\alpha \geq 2$, *and* $f(x)$ *a polynomial in* x *of degree* M, *and let* r *be a solution of the congruence,*

$$f(x) \equiv 0 \;(\text{mod } p^{\alpha-1}) \qquad (\text{A.3.6})$$

with $0 \leq r < p^{\alpha-1}$. *The existence of solutions proceeds as follows.*

(a) If $f'(r) \not\equiv 0$ (mod p)

- *then* r *lifts in a unique way from* (mod $p^{\alpha-1}$) *to* (mod p^α).

(b) If $f'(r) \equiv 0$ (mod p)

- $f(r) \equiv 0$ (mod p^α): *then* r *lifts from* (mod $p^{\alpha-1}$) *to* (mod p^α) *in* p *different ways.*
- $f(r) \not\equiv 0$ (mod p^α): *then* r *cannot be lifted from* (mod $p^{\alpha-1}$) *to* (mod p^α).

To prove the theorem, we consider all possible lifts of r from (mod $p^{\alpha-1}$) to (mod p^α), which are given by $r + qp^{\alpha-1}$ (mod p^α) for some integer $0 \le q < p$. Taylor expanding the polynomial f of degree M,

$$f(r + qp^{\alpha-1}) = f(r) + f'(r)qp^{\alpha-1} + \sum_{n=2}^{M} \frac{1}{n!} f^{(n)}(r)q^n p^{n(\alpha-1)} \quad (A.3.7)$$

and reducing this equation (mod p^α) we obtain,

$$f(r + qp^{\alpha-1}) \equiv f(r) + f'(r)qp^{\alpha-1} \ (\text{mod } p^\alpha) \qquad (A.3.8)$$

Since $f(r) \equiv 0$ (mod $p^{\alpha-1}$) by the assumption (A.3.6), we must have $f(r) = kp^{\alpha-1}$ for some integer k. Therefore the number of solutions to $f(x) \equiv 0$ (mod p^α) equals the number of solutions to the congruence $k + qf'(r) \equiv 0$ (mod p). (a) If $f'(r) \not\equiv 0$ (mod p), then there is a unique solution q for every k. (b) If $f'(r) \equiv 0$ (mod p) and $f(r) \equiv 0$ (mod p^α), then $k \equiv 0$ (mod p) and there are p solutions q, while if $f(r) \not\equiv 0$ (mod p^α), then no solutions q can exist, which completes the proof of the theorem.

A.4 Quadratic residues and quadratic residue symbols

If $m, N \in \mathbb{N}$ and $n \in \mathbb{Z}$ with $\gcd(n, N) = 1$, then n is an mth power residue mod N if $x^m \equiv n$ (mod N) admits a solution, and a nonresidue if it admits no solutions. In the special case of quadratic residues, where $m = 2$ and $N = p$ is an odd prime, an integer $n \not\equiv 0$ (mod p) is a quadratic residue provided the equation,

$$x^2 \equiv n \ (\text{mod } p) \qquad (A.4.1)$$

has a solution, and a quadratic nonresidue if the equation has no solutions. If $n \not\equiv 0$ (mod p) is a quadratic residue and x is a solution to (A.4.1), then $-x$ is also a solution to the same equation. When $n \equiv 0$ (mod p), then equation (A.4.1) has one solution, namely, $x \equiv 0$ (mod p).

The *Legendre symbol* $(n|p)$ for p prime is defined as follows,[1]

$$(n|p) = \begin{cases} +1 & \text{if } n \text{ is a quadratic residue mod } p \\ -1 & \text{if } n \text{ is a quadratic non-residue mod } p \\ 0 & \text{if } n \equiv 0 \ (\text{mod } p) \end{cases} \qquad (A.4.2)$$

The number of solutions x to equation (A.4.1) is $1 + (n|p)$ in all cases.

Proposition A.4.1. *Further properties of quadratic residues and the Legendre symbol for an arbitrary odd prime p are as follows.*

---

[1] Frequently used alternative notations for the Legendre symbol are $(\frac{n}{p})$ and (n/p).

1. *There are $(p-1)/2$ residues and $(p-1)/2$ non-residues in $\mathbb{Z}_p^*$.*
2. *For any $n \in \mathbb{Z}$, we have $(n|p) = n^{(p-1)/2} \pmod{p}$.*
3. *The first supplement is $(-1|p) = (-)^{(p-1)/2}$.*
4. *The second supplement is $(2|p) = (-)^{(p^2-1)/8}$.*
5. *For arbitrary $m, n \in \mathbb{Z}$ we have $(mn|p) = (m|p)(n|p)$.*

The proof proceeds as follows. Item 1 follows from the fact that the map $x \mapsto x^2$ is two-to-one in $\mathbb{Z}_p^*$. For Item 2, we first note that Fermat's little theorem, namely, that $n^{p-1} \equiv 1 \pmod{p}$ for any n not divisible by p, implies $n^{(p-1)/2} \equiv \pm 1 \pmod{p}$. For $(n|p) = 1$, there exists an x so that $x^2 \equiv n \pmod{p}$ and hence $n^{(p-1)/2} = x^{p-1} \equiv 1 = (n|p) \pmod{p}$. For $(n|p) = -1$, the degree $(p-1)/2$ polynomial congruence equation $x^{(p-1)/2} - 1 \equiv 0 \pmod{p}$ has $(p-1)/2$ solutions. The quadratic residues already provide $(p-1)/2$ solutions, so the nonresidues cannot be solutions. As a result, we have $x^{(p-1)/2} \not\equiv 1 \pmod{p}$ and hence $x^{(p-1)/2} \equiv -1 = (n|p) \pmod{p}$. Finally, Items 3, 4, and 5 follow from Item 2. Item 3 may also be derived from Fermat's theorem that a prime p is the sum of the squares of two integers if and only if $p \equiv 1 \pmod{4}$.

The *Jacobi symbol* $(n|N)$ generalizes the Legendre symbol to an arbitrary positive odd integer $N = \prod_i p_i^{\alpha_i}$, where p_i are distinct primes and $\alpha_i \geq 1$ and any $n \in \mathbb{Z}$ by,

$$(n|N) = \prod_i (n|p_i)^{\alpha_i} \tag{A.4.3}$$

where in each factor $(n|p_i)$ is the Legendre symbol. The Jacobi symbol is a completely multiplicative function in both of its arguments, assuming that M and N are both positive odd integers,

$$(mn|N) = (m|N)(n|N)$$
$$(m|MN) = (m|M)(m|N) \tag{A.4.4}$$

and satisfies the following relations,

$$(mM|M) = 0 \qquad\qquad (-1|M) = (-1)^{(M-1)/2}$$
$$(2|M) = (-1)^{(M^2-1)/8} \tag{A.4.5}$$

The *Kronecker symbol* further generalizes the Jacobi symbol by lifting the restrictions that N be odd and positive and is defined as follows. If $N = u \prod_i p_i^{\alpha_i}$, where $u = \pm 1$ and the product runs over all distinct primes including 2, then the Kronecker symbol is defined by,

$$(n|N) = (n|u) \prod_i (n|p_i)^{\alpha_i} \tag{A.4.6}$$

where $(n|p_i)$ stands for the Legendre symbol when p_i is an odd prime, while the remaining data are supplied by,

$$(n|-1) = \begin{cases} 1 & \text{if } n \geq 0 \\ -1 & \text{if } n < 0 \end{cases} \qquad (A.4.7)$$

and

$$(n|2) = \begin{cases} 0 & \text{if } n \equiv 0 \pmod 2 \\ +1 & \text{if } n \equiv \pm 1 \pmod 8 \\ -1 & \text{if } n \equiv \pm 3 \pmod 8 \end{cases} \qquad (A.4.8)$$

The Legendre, Jacobi, and Kronecker symbols are referred to as *quadratic residue symbols*.

A.4.1 Quadratic residues $x^2 \equiv -1$ (mod N)

We now analyze the particular example of quadratic residues of -1 modulo N. This example, as well as that in the following section, will be used extensively when we discuss congruence subgroups.

To obtain the number of solutions of the congruence $x^2 \equiv -1 \pmod N$ for an arbitrary integer N with prime factorization $N = \prod_i p_i^{\alpha_i}$, we consider the polynomial $f(x) = x^2 + 1$ with $f'(x) = 2x$ and use Theorem A.3.2.

- If p is an odd prime, and $0 \leq r < p$ solves $r^2 + 1 \equiv 0 \pmod p$ then $r \not\equiv 0 \pmod p$. As a result, $f'(r) = 2r \not\equiv 0 \pmod p$ so that there is a unique lift of r to $\pmod{p^2}$, and to any $\pmod{p^\alpha}$.

- If $p = 2$ then $r \equiv 1 \pmod 2$ so there is one solution $\pmod 2$. Now using Theorem A.3.2, $f'(r) \equiv 0 \pmod 2$, and the fact that every lift $f(1 + 2q) \equiv 1 \pmod 4$ implies that there are no solutions $\pmod 4$, and hence no solutions $\pmod{2^\alpha}$ for all $\alpha \geq 2$.

Hence the number of solutions to $x^2 \equiv -1 \pmod N$ is given by,

$$\left| \{ x \in \mathbb{Z}_N \mid x^2 \equiv -1 \} \right| = \begin{cases} 0 & 4|N \\ \prod_i \left(1 + (-1|p_i) \right) & 4 \nmid N \end{cases} \qquad (A.4.9)$$

where $|S|$ stands for the cardinality of the set S. The Legendre symbol $(-1|p)$ evaluates to $(-1)^{\frac{p-1}{2}}$ for any odd prime p. For $p = 2$, the number of solutions is 1, so we set $(-1|2) = 0$.

A.4.2 Quadratic residues $x^2 + x \equiv -1 \pmod{N}$

To obtain the number of solutions of the congruence $x^2 + x \equiv -1 \pmod{N}$ for an arbitrary integer N with prime factorization $N = \prod_i p_i^{\alpha_i}$, we solve the following equivalent problem by setting $y = 2x + 1$,

$$y = 2x + 1 \qquad\qquad y^2 + 3 \equiv 0 \pmod{4N} \qquad\qquad \text{(A.4.10)}$$

Applying the Chinese remainder theorem A.2.1, the above equation is equivalent to the following set of equations,

$$y^2 + 3 \equiv 0 \pmod{2^{\alpha_2 + 2}}$$
$$y^2 + 3 \equiv 0 \pmod{3^{\alpha_3}}$$
$$y^2 + 3 \equiv 0 \pmod{p_i^{\alpha_i}} \qquad\qquad p_i \geq 5 \qquad\qquad \text{(A.4.11)}$$

Next, we apply Theorem A.3.2 for the function $f(y) = y^2 + 3$ with $f'(y) = 2y$ to each equation in (A.4.11).

- Any solution r that may exist when $\alpha_2 = 0$ satisfies $f'(r) \equiv 0 \pmod 2$ and cannot be lifted to $\alpha_2 > 0$ since r is odd. As a result there are no solutions for even N.
- For $\alpha_3 = 1$, the equation $y^2 + 3 \equiv 0 \pmod 3$ has one solution, namely, $y \equiv 0 \pmod 3$. By Theorem A.3.2, this solution cannot be extended to $\alpha_3 = 2$ since $f'(y) \equiv 0 \pmod 3$ and $f(y) = 3 \not\equiv 0 \pmod 9$. Indeed, the quadratic residues $y^2 + 3 \pmod 9$ are $3, 4, 7, 1$, which does not include 0, so that the congruence $y^2 + 3 \equiv 0 \pmod 9$ has no solutions when $9 | N$.
- For an odd prime $p_i \geq 5$, the congruence $y^2 + 3 \equiv 0 \pmod{p_i}$ has $1 + (-3|p_i)$ solutions. When a solution r exists it satisfies $0 < r < p_i$ (which would not be the case for $p_i = 3$) so that $f'(r) = 2r \not\equiv 0 \pmod{p_i}$. By Theorem A.3.2, the solution r then extends uniquely to a solution mod $p_i^{\alpha_i}$ for arbitrary α_i. Thus, for all primes $p_i \geq 5$, the number of solutions to the third equation in (A.4.11) is $1 + (-3|p_i)$.

When $9 \nmid N$, the formula $1 + (-3|p_i)$ for the number of solutions for prime p_i is actually valid for $p_i = 3$ since this case has one solution and $(-3|3) = 0$. It is also valid for $p_i = 2$ since we have $(-3|2) = -1$ by the last line in (A.4.8), so that there are no solutions whenever N is even, as was established already earlier. Putting all together we have,

$$\left| \{ x \in \mathbb{Z}_N \mid x^2 + x \equiv -1 \} \right| = \begin{cases} 0 & 9 \,|\, N \\ \prod_i \left(1 + (-3|p_i) \right) & 9 \nmid N \end{cases} \qquad \text{(A.4.12)}$$

Note that, by quadratic reciprocity (to be discussed in Section A.6), if $p_i \neq 3$ then $(-3|p_i) = (-1|p_i)(3|p_i) = (p_i|3)$ which evaluates to ± 1 when $p_i \equiv$

$\pm 1 \pmod 3$. For $p_i \equiv -1 \pmod 3$, there are no solutions $y \in \mathbb{Z}$ and hence no solution of the form $y \equiv 2x + 1 \pmod{p_i}$. For $p_i \equiv 1 \pmod 3$, there are two solutions $y \pmod{p_i}$. For any given y, there is then a unique solution x to $y \equiv 2x + 1 \pmod{p_i}$.

A.5 Gauss sums

For an odd prime p, and an arbitrary $a \in \mathbb{Z}$, the Gauss sum $\mathfrak{g}_a$ is defined by,

$$\mathfrak{g}_a = \sum_{k=0}^{p-1} (k|p)\, \varepsilon^{ka} \qquad\qquad \varepsilon = e^{2\pi i/p} \qquad\qquad (A.5.1)$$

Gauss sums are directly related to the Legendre symbol,

$$\mathfrak{g}_a = (a|p)\mathfrak{g} \qquad\qquad \mathfrak{g} = \mathfrak{g}_1 \qquad\qquad (A.5.2)$$

Indeed, for a divisible by p we have $\mathfrak{g}_a = 0$ in view of the fact that the number of residues in $\mathbb{Z}_p^*$ equals the number of nonresidues. Hence the relation holds trivially when $p|a$ as both sides vanish. For $p \nmid a$, we use the fact that $(a|p)^2 = 1$, the definition of $\mathfrak{g}_a$, and the complete multiplication property of the Legendre symbol,

$$\mathfrak{g}_a = (a|p) \sum_{k=0}^{p-1}(a|p)(k|p)\,\varepsilon^{ka} = (a|p)\sum_{k=0}^{p-1}(ka|p)\,\varepsilon^{ka} = (a|p)\mathfrak{g} \quad (A.5.3)$$

The last equality follows, for $p \nmid a$, from the fact that the sum over k is identical to the sum over ka. Furthermore, the relation,

$$\mathfrak{g}^2 = (-)^{(p-1)/2} p \qquad\qquad (A.5.4)$$

follows from $\mathfrak{g}_a \mathfrak{g}_{-a} = (-1|p)\mathfrak{g}^2$ and summing over a in two different ways,

$$\sum_{a=0}^{p-1} \mathfrak{g}_a \mathfrak{g}_{-a} = (p-1)(-1|p)\mathfrak{g}^2 = \sum_{a,k,\ell=0}^{p-1} (k\ell|p)\varepsilon^{(k-\ell)a}$$

$$= p \sum_{k,\ell=0}^{p-1} (k\ell|p)\delta_{k,\ell} = p(p-1) \qquad\qquad (A.5.5)$$

where $\delta_{k,\ell}$ is the Kronecker δ. The result (A.5.4) follows upon using the first supplement of Proposition A.4.1, namely, $(-1|p) = (-)^{(p-1)/2}$.

A.6 Quadratic reciprocity

Quadratic reciprocity states that for p, q distinct odd primes, we have,

$$(p|q)(q|p) = (-)^{(p-1)(q-1)/4} \tag{A.6.1}$$

Many proofs exist of the law of quadratic reciprocity, of which Gauss produced at least 8. Here we shall use Gauss's sums as defined in the preceding section. For any odd prime $q \neq p$, we begin by using (A.5.4) to write,

$$\mathfrak{g}^{q-1} = (\mathfrak{g}^2)^{(q-1)/2} = (\mathfrak{g}^2|q)$$
$$= (-1|q)^{(p-1)/2}(p|q) = (-)^{(p-1)(q-1)/2}(p|q) \tag{A.6.2}$$

where we used the relation $(n|q) \equiv n^{(q-1)/2} \pmod{q}$ of property 2 of Proposition A.4.1 for $n = \mathfrak{g}^2$ in the second equality. Next, we use the fact that the prime power of a multiple sum evaluates as follows,

$$(a_1 + \cdots + a_n)^q = a_1^q + \cdots a_n^q \pmod{q} \tag{A.6.3}$$

This relation is valid for $a_1, \cdots, a_n \in \mathbb{Z}$, but we generalize it here to the ring of algebraic integers $\mathbb{Z}_p[\varepsilon]$. As a result we have,

$$\mathfrak{g}^q = \left(\sum_{k=0}^{p-1}(k|p)\varepsilon^k\right)^q = \sum_{k=0}^{p-1}(k|p)^q\varepsilon^{kq} = \sum_{k=0}^{p-1}(k|p)\varepsilon^{kq} = \mathfrak{g}_q \pmod{q} \tag{A.6.4}$$

By (A.5.2), we have $\mathfrak{g}_q = (q|p)\mathfrak{g}$, and upon combining with (A.6.2), we get,

$$\mathfrak{g}^{q-1} = (-)^{(p-1)(q-1)|2}(p|q) = (q|p) \tag{A.6.5}$$

The last equality implies the law of quadratic reciprocity using $(p|q)^2 = 1$.

A.7 Characters

Consider a finite Abelian group G. A function $\chi : G \to \mathbb{C}^*$ is said to be a *character* if it is a group homomorphism, that is, $\chi(gh) = \chi(g)\chi(h)$ for all $g, h \in G$. Characters themselves form a group $\hat{G} = \text{Hom}(G, \mathbb{C}^*)$, known as the Poincaré dual of G. We will denote the identity element of this group by χ_0, such that $\chi_0(g) = 1$ for all g. The inverse elements are obtained by complex conjugation $\chi^{-1}(g) = \overline{\chi(g)}$.

As an example, consider the case of a cyclic group $G = \mathbb{Z}_n$. If we denote the order-n generator by v, then it is clear that $\chi(v)^n = \chi(v^n) = 1$ and hence we can take $\chi(v)$ to be an nth root of unity. Conversely, every nth root of unity defines a character of $\mathbb{Z}_n$. Thus we conclude that $\hat{G}$ is in fact isomorphic to the original group, $\hat{G} \cong \mathbb{Z}_n$.

Characters satisfy the following relations,

$$\sum_{g \in G} \chi(g) = \begin{cases} |G| & \chi = \chi_0 \\ 0 & \text{otherwise} \end{cases}$$

$$\sum_{\chi \in \hat{G}} \chi(g) = \begin{cases} |\hat{G}| & g = 1 \\ 0 & \text{otherwise} \end{cases} \tag{A.7.1}$$

The first line of each of these is obvious. To prove the second line of the first relation, begin by choosing $h \in G$ such that $\chi(h) \neq 1$. Then we have,

$$\chi(h) \sum_{g \in G} \chi(g) = \sum_{g \in G} \chi(hg) = \sum_{g \in G} \chi(g) \tag{A.7.2}$$

where we have used the fact that χ is a homomorphism, as well as closure of the group G. Since $\chi(h) \neq 1$, we conclude $\sum_{g \in G} \chi(g) = 0$ as per the claim. The second relation is proven exactly analogously – namely, we multiply by $\chi'(g) \neq 1$ and used the fact that $\hat{G}$ is a group.

From these results we obtain the following Schur orthogonality relations,

$$\sum_{g \in G} \chi(g)\overline{\chi'(g)} = \begin{cases} |G| & \chi = \chi' \\ 0 & \text{otherwise} \end{cases}$$

$$\sum_{\chi \in \hat{G}} \chi(g)\overline{\chi(h)} = \begin{cases} |\hat{G}| & g = h \\ 0 & \text{otherwise} \end{cases} \tag{A.7.3}$$

Indeed, for the former note that if $\chi, \chi' \in \hat{G}$ then also $\chi(\chi')^{-1} \in \hat{G}$. We may thus use the relation (A.7.1) with χ replaced by $\chi(\chi')^{-1}$. Noting that,

$$\sum_{g \in G} (\chi(\chi')^{-1})(g) = \sum_{g \in G} \chi(g)\overline{\chi'(g)} \tag{A.7.4}$$

we obtain the desired result. Likewise for the latter, we use the fact that,

$$\sum_{\chi \in \hat{G}} \chi(gh^{-1}) = \sum_{\chi \in \hat{G}} \chi(g)\overline{\chi(h)} \tag{A.7.5}$$

This completes the proof of both relations in (A.7.1).

A.8 Dirichlet characters

A *Dirichlet character* with modulus $a \in \mathbb{N}$ is a function $\chi_a : \mathbb{Z} \to \mathbb{C}$ satisfying the following properties:

1. Periodicity: $\chi_a(n) = \chi_a(n + a)$ for all $n \in \mathbb{Z}$.

2. $\chi_a(n) \neq 0$ if and only if $\gcd(n, a) = 1$.
3. Multiplicativity: $\chi_a(nm) = \chi_a(n)\chi_a(m)$ for all $n, m \in \mathbb{Z}$.

Alternatively, Dirichlet characters may be thought of as characters for the multiplicative group $G = \mathbb{Z}_a^*$ of invertible elements of $\mathbb{Z}_a$. This is an Abelian group with $|G| = \phi(a)$ elements, where ϕ is Euler's totient function. In this case, our orthogonality relations read as follows,

$$\sum_{g=0}^{a-1} \chi_a(g)\overline{\chi_a'(g)} = \begin{cases} \phi(a) & \chi_a = \chi_a' \\ 0 & \text{otherwise} \end{cases}$$

$$\sum_{\chi_a \in \hat{G}} \chi_a(g)\overline{\chi_a(h)} = \begin{cases} \phi(a) & g \equiv h \pmod a \\ 0 & \text{otherwise} \end{cases} \tag{A.8.1}$$

The *principal character* $\chi_a^\pi(n)$ modulo a is defined by,

$$\chi_a^\pi(n) = \begin{cases} 1 & \gcd(n, a) = 1 \\ 0 & \gcd(n, a) > 1 \end{cases} \tag{A.8.2}$$

When n is restricted to $\mathbb{Z}_a^*$ the principal character reduces to $\chi_a^\pi(n) = 1$, whence its alternative designation as the *identity character*.

A positive integer $d|a$ is an *induced modulus* of a Dirichlet character χ_a with modulus a if $\chi_a(n) = 1$ for every n that satisfies $\gcd(n, a) = 1$ and $n \equiv 1 \pmod d$. A Dirichlet character χ_a with modulus a is a *primitive character* if it has no induced modulus $d < a$. The smallest induced modulus of χ_a is referred to as the *conductor* of χ_a. Thus the conductor of a primitive character χ_a is its modulus a. An arbitrary character χ_a with conductor d may be decomposed as follows,

$$\chi_a(n) = \psi_d(n) \, \chi_a^\pi(n) \tag{A.8.3}$$

where $\psi_d(n)$ is a primitive character with modulus d, and χ_a^π is the principal character with modulus a.

A.9 Dirichlet *L*-functions

We now introduce the Dirichlet L-function for a Dirichlet character $\chi_a(n)$,

$$L(s, \chi_a) = \sum_{n=1}^{\infty} \frac{\chi_a(n)}{n^s} \tag{A.9.1}$$

The Dirichlet L-function $L(s, \chi_a)$ converges and is analytic for $\text{Re}\, s > 1$

[IK21, Ser73]. In this region, we have the following Euler product formula,

$$L(s, \chi_a) = \prod_p \left(1 - \frac{\chi_a(p)}{p^s}\right)^{-1} \tag{A.9.2}$$

the product being over all prime numbers p. To prove this, we note that for each prime p, we have,

$$\left(1 - \frac{\chi_a(p)}{p^s}\right)^{-1} = \sum_{n=0}^{\infty} \chi_a(p)^n p^{-ns} = \sum_{n=0}^{\infty} \chi_a(p^n) p^{-ns} \tag{A.9.3}$$

Hence for fixed prime q, we have,

$$\prod_{p \leq q} \left(1 - \frac{\chi_a(p)}{p^s}\right)^{-1} = \sum_{n \in S_q} \frac{\chi_a(n)}{n^s} \tag{A.9.4}$$

with S_q the set of all natural numbers whose prime factors are less than or equal to q. Then for any $N \in \mathbb{N}$, we have,

$$\sum_{n=1}^{N} \frac{\chi_a(n)}{n^s} = \prod_{p \leq q} \left(1 - \frac{\chi_a(p)}{p^s}\right)^{-1} - \sum_{\substack{n \in S_q \\ n \geq N}} \frac{\chi_a(n)}{n^s} \tag{A.9.5}$$

with q the largest prime less than or equal to N. Taking the limit $N \to \infty$ for $\mathrm{Re}\,(s) > 1$ gives (A.9.2).

Using this product formula, it is possible to show that $L(s, \chi_a^\pi)$ extends to a meromorphic function on $\mathrm{Re}\,s > 0$, with the only pole at $s = 1$. We do not reproduce the proof here; it can be found in, for example, [IK21, Ser73].

A.9.1 Dirichlet's theorem on arithmetic progressions

We now use Dirichlet L-functions to prove Dirichlet's Theorem on arithmetic progressions.

Theorem 1. *Let m and a be relatively prime natural numbers. Then there are infinitely many prime numbers p such that $p \equiv m \pmod{a}$.*

In order to prove this, we aim to prove the following stronger result,

$$\sum_{p \equiv m\,(a)} \frac{1}{p} \quad \text{is divergent} \tag{A.9.6}$$

which implies the theorem. To do so, begin by considering the logarithm of the Dirichlet L-function, which by the product formula takes the form,

$$\ln L(s, \chi_a) = -\sum_p \ln\left(1 - \frac{\chi_a(p)}{p^s}\right) \tag{A.9.7}$$

valid for $\mathrm{Re}\, s > 1$. Taylor expanding the logarithm gives

$$\ln L(s, \chi_a) = \sum_p \frac{\chi_a(p)}{p^s} + \sum_p \sum_{n=2}^{\infty} \frac{\chi_a(p)^n}{n p^{ns}} \tag{A.9.8}$$

where, for reasons to become clear in a moment, we have split the $n = 1$ term from the $n \geq 2$ terms. Using the orthogonality of Dirichlet characters in (A.8.1), we find that,

$$\sum_{p \equiv m\,(a)} \frac{1}{p^s} = \frac{1}{\phi(a)} \sum_{\chi_a \in \hat{G}} \overline{\chi_a(m)} \sum_p \frac{\chi_a(p)}{p^s} \tag{A.9.9}$$

$$= \frac{1}{\phi(a)} \left[\sum_{\chi_a \in \hat{G}} \overline{\chi_a(m)} \ln L_a(s, \chi) - \sum_{\chi_a \in \hat{G}} \overline{\chi_a(m)} \sum_p \sum_{n=2}^{\infty} \frac{\chi_a(p)^n}{n p^{ns}} \right]$$

where $(a, m) = 1$ lest $\chi_a(m) = 0$. To prove the theorem, we must now show that the right-hand side is divergent as $s \to 1$. To do so, we begin by showing that $\sum_p \sum_{n=2}^{\infty} \frac{\chi_a(p)^n}{n p^{ns}}$ is bounded as $s \to 1$, and hence that the second term in the difference does not diverge. To see this, note that since $|\chi_a(p)| = 0$ or 1, we have $|\chi_a(p) p^{-s}| \leq |p^{-s}| \leq \frac{1}{2}$. Then we have,

$$\left| \sum_{n=2}^{\infty} \frac{\chi_a(p)^n}{n p^{ns}} \right| \leq \left| \frac{\chi_a(p)}{p^s} \right|^2 \sum_{n=2}^{\infty} \frac{1}{n} \left| \frac{\chi_a(p)}{p^s} \right|^{n-2}$$

$$\leq \left| \frac{\chi_a(p)}{p^s} \right|^2 \sum_{n=2}^{\infty} \frac{1}{2} \frac{1}{2^{n-2}} = \left| \frac{\chi_a(p)}{p^s} \right|^2 \leq \frac{1}{p^2} \tag{A.9.10}$$

Hence we conclude that,

$$\left| \sum_p \sum_{n=2}^{\infty} \frac{\chi_a(p)^n}{n p^{ns}} \right| \leq \sum_p \frac{1}{p^2} < \sum_{n=1}^{\infty} \frac{1}{n^2} \tag{A.9.11}$$

the latter of which is convergent.

Having shown that the second term on the right side of (A.9.9) is not divergent as $s \to 1$, we are now tasked with showing that the first term *is* divergent. At the end of Appendix A.9, we quoted the result that $L(s, \chi_a^\pi) \to$

∞ as $s \to 1$, and hence $\ln L(s, \chi_a^\pi)$ diverges in this limit as well. To complete the proof it suffices to show that $L(1, \chi_a) \neq 0$ for all $\chi_a \neq \chi_a^\pi$.

To show this, first consider a character χ_a such that χ_a^2 is not the principal character. Then we can consider the function,

$$\lambda(s) = L(s, \chi_a^\pi)^3 L(s, \chi_a)^4 L(s, \chi_a^2) \qquad (A.9.12)$$

By (A.9.8), we have,

$$|\lambda(s)| = \left| \exp\left[\sum_p \sum_{n=1}^\infty \frac{3 + 4\chi_a(p^n) + \chi_a^2(p^n)}{np^{ns}} \right] \right| \qquad (A.9.13)$$

Letting $\theta_{n,p} = \mathrm{Arg}\,(\chi_a(p^n))$, we then obtain,

$$|\lambda(s)| = \exp\left| \sum_p \sum_{n=1}^\infty \frac{3 + 4\cos\theta_{n,p} + \cos 2\theta_{n,p}}{np^{n\,\mathrm{Re}(s)}} \right| \qquad (A.9.14)$$

But for any real θ, we have,

$$3 + 4\cos\theta + \cos 2\theta = 2(1 + \cos\theta)^2 \geq 0 \qquad (A.9.15)$$

Thus we conclude that,

$$|\lambda(s)| \geq e^0 = 1 \qquad (A.9.16)$$

This gives the desired result. Indeed, if we had $L(1, \chi_a) = 0$, then since $L(s, \chi_a^\pi)$ has a simple pole at $s = 1$ but $L(s, \chi_a^2)$ does *not* (because $\chi_a^2 \neq \chi_a^\pi$ by assumption) the degree ≥ 4 zero wins against the degree 3 pole, and we would have $\lambda(1) = 0$, in contradiction to the inequality just obtained. Thus we must have $L(1, \chi_a) \neq 0$.

For the case of $\chi_a^2 = \chi_a^\pi$, a similar argument can be run using $\lambda(s) = L(s, \chi_a)L(s, \chi_a^\pi)$. In this case, one aims to show that $\ln \lambda(s) \geq \ln \zeta(2s)$. In total, one concludes that $L(1, \chi_a) \neq 0$ for all $\chi_a \neq \chi_a^\pi$, completing the proof of the theorem.

A.9.2 Perfect squares and quadratic residues

We close by using the various tools of this section, including Dirichlet's theorem of arithmetic progressions, to prove the following statement:

Theorem 2. *A number $N \in \mathbb{N}$ is a perfect square if and only if it is a quadratic residue for all primes p coprime to N.*

One direction of this statement is obvious: If N can be written as a perfect square, then it is a perfect square modulo any prime. We now focus on the other direction. Assume that N is a quadratic residue modulo all coprime p. Say that N is *not* a perfect square. We may then write N in the form $N = j^2 p_1 \ldots p_r$ for distinct primes $p_1, \ldots p_r$. Since any odd power of 2 is a quadratic nonresidue mod 3, we may assume that there is at least one odd factor. Take p_1 to be odd and consider nonzero integers m and n_i such that $(m|p_1) = -1$ and $(n_i|p_i) = 1$ for $i = 2, \ldots, r$. By the Chinese remainder theorem, there exists a solution to the combined set of congruences,

$$x \equiv 1 \pmod 4$$
$$x \equiv m \pmod{p_1}$$
$$x \equiv n_i \pmod{p_i} \qquad p_i \neq 2 \qquad \text{(A.9.17)}$$

If one of the prime factors $p_i = 2$, then we replace the final congruence with $x \equiv 1 \pmod 8$. Given a particular solution x_0 to these congruences, the general solution takes the form $x = x_0 + 4k p_1 \ldots p_r$ for integer k. By Dirichlet's theorem of arithmetic progressions, this takes an infinite number of prime values, and we choose one, say $x = p$, such that $p \nmid j$. By definition,

$$(p|p_1) = (m|p_1) = -1 \qquad (p|p_i) = (n_i|p_i) = 1 \qquad i = 2, \ldots, r \quad \text{(A.9.18)}$$

Then using quadratic reciprocity, together with the first congruence of (A.9.17), we conclude that,

$$(p_1|p) = -1 \qquad \text{and} \qquad (p_i|p) = 1 \qquad i = 2, \ldots, r \quad \text{(A.9.19)}$$

Thus $(N|p) = (p_1|p) \ldots (p_r|p) = -1$, and we have identified a prime p coprime to N for which N is not a quadratic residue, contrary to hypothesis. We conclude that N must be a perfect square.

• Bibliographical notes

An excellent overview of some classic problems in arithmetic is given in the book by Cox [Cox11b], where a historical perspective is also provided. An introductory text into analytic number theory, including Dirichlet characters, may be found in the book by Apostol [Apo76a]. Two other classics are the books by Serre [Ser73] and Hecke [Hec81]. Useful references for results in analytic number theory, including L-functions, may be found in [Car92] and in the books by Iwaniec [IK21], Rademacher [Rad70], and Siegel [Sie61].

Appendix B

Riemann surfaces

A Riemann surface is a connected complex manifold of two real dimensions or equivalently a connected complex manifold of one complex dimension. In this appendix, we shall review the topology of Riemann surfaces, their homotopy groups, homology groups, uniformization, construction in terms of Fuchsian groups, as well as their emergence from two-dimensional orientable Riemannian manifolds.

B.1 Topology

The topology of a Riemann surface Σ is the topology of the underlying orientable two-dimensional manifold Σ. Throughout, we shall restrict to Riemann surfaces whose boundary $\partial\Sigma$ is the union of a finite number b of one-dimensional components, and a finite number n of points referred to as punctures of Σ. Given a triangulation of Σ with F_Σ faces, E_Σ edges, and V_Σ vertices, the Euler number $\chi(\Sigma)$ is defined by,

$$\chi(\Sigma) = F_\Sigma - E_\Sigma + V_\Sigma \qquad (\text{B.1.1})$$

The Euler number is a topological invariant in the sense that its value is independent of the triangulation chosen for a given Σ. In suitably chosen triangulations, the operation of adding a puncture to Σ (namely, removing a point from Σ) leaves F_Σ and E_Σ unchanged but diminishes V_Σ by one. Similarly, adding a one-dimensional boundary component leaves E_Σ and V_Σ unchanged but diminishes F_Σ by one. The remaining contribution to $\chi(\Sigma)$ derives from the genus g, which is the number of handles on Σ,

$$\chi(\Sigma) = 2 - 2g - n - b \qquad (\text{B.1.2})$$

A Riemann surface with $b \neq 0$ may be studied by considering its double, which is obtained by gluing Σ and its mirror image Σ' into a Riemann surface $\hat{\Sigma}$ without one-dimensional boundary components. The surface $\hat{\Sigma}$ is endowed with an involution ι which swaps Σ and Σ'. Henceforth, we shall restrict attention to surfaces with $b = 0$.

Compact Riemann surfaces play a central role in both mathematics and physics. It will often be useful to consider the generalization to a *compact Riemann surface with n punctures* defined to be a compact Riemann surface with a finite number n of points removed, namely, with n punctures. Their study may be initiated by first studying the underlying compact surface and then removing n points.

B.1.1 The Riemann–Hurwitz formula

Holomorphic maps between Riemann surfaces allow us to relate different Riemann surfaces. We begin by considering a nonconstant holomorphic map $f : X \to Y$ between two compact Riemann surfaces X and Y. The degree of f is the multiplicity of the function f^{-1} given by the cardinality $d = |f^{-1}(y)|$ for all but finitely many points $y \in Y$. A formula valid for all points in Y is obtained in terms of the *degree of ramification* e_x of f at a point $x \in X$ which, in local coordinates, is given by $f(z) - f(x) \sim (z - x)^{e_x}$. The degree of f may then be defined for all $y \in Y$ by the following formula,

$$d = \sum_{x \in f^{-1}(y)} e_x \qquad (B.1.3)$$

and d is independent of y. The Riemann–Hurwitz formula relates the genera $g(X)$ and $g(Y)$ of the surfaces X and Y to the degree of f and the deviations from 1 of the degree of ramification,

$$2g(X) - 2 = d\big(2g(Y) - 2\big) + \sum_{x \in X} (e_x - 1) \qquad (B.1.4)$$

A hyperelliptic Riemann surface X, viewed as a double cover of the Riemann sphere Y, provides an illuminating example of the Riemann–Hurwitz formula for $d = 2$. In this case, we have $2n$ branch points at which $e_x = 2$ readily gives $g(X) = n - 1$. The formula (B.1.4) may be proven by using the expression for the Euler number $\chi(X)$ in (B.1.1) of X in terms of a triangulation by F_X faces, E_X edges, and V_X vertices (including all the ramification points), and similarly for $\chi(Y)$. Given that the degree of the map is d, we have $F_X = dF_Y$ and $E_X = dE_Y$ but $V_X = dV_Y - \sum_{x \in X}(e_x - 1)$ since a vertex with ramification degree e_x counts for only one vertex of X. Expressing the result in terms of the Euler numbers, we have $\chi(X) = d\chi(Y) - \sum_{x \in X}(e_x - 1)$. For a compact Riemann surface without boundary or punctures, the Euler numbers are related to the genus by $\chi(X) = 2 - 2g(X)$ and $\chi(Y) = 2 - 2g(Y)$ from which (B.1.4) follows.[1]

B.1.2 Homotopy

The homotopy groups $\pi_0(\Sigma)$ and $\pi_2(\Sigma)$ for a compact Riemann surface are trivial, and the interesting homotopy resides entirely in the first homotopy group $\pi_1(\Sigma)$, or *fundamental group*, which we now define.

A closed curve $\mathfrak{C} : [0, 1] \to \Sigma$ is a continuous map, parametrized by $\mathfrak{C}(s)$ with $s \in [0, 1]$, such that $\mathfrak{C}(1) = \mathfrak{C}(0)$. Two closed curves $\mathfrak{C}_0, \mathfrak{C}_1$ are said to be homotopic (to one another) iff there exists a family of curves $\mathfrak{F}(s, t)$ parametrized by $t \in [0, 1]$, such that $\mathfrak{F}(s, t) \subset \Sigma$, is continuous in s, t for all $s, t \in [0, 1]$ and interpolates as follows, $\mathfrak{F}(s, 0) = \mathcal{C}_0(s)$ and $\mathfrak{F}(s, 1) = \mathfrak{C}_1(s)$.

The homotopy relation among closed curves is clearly an equivalence relation, whose classes are the first homotopy classes. To make the set of classes into a group under composition of maps, we choose a base-point $P \in \Sigma$ and restrict to curves in each class that pass through P. Thus, we may choose the parametrization of each closed curve $\mathfrak{C}$ that passes through P so that $\mathfrak{C}(0) = \mathfrak{C}(1) = P$. The composition $\mathfrak{C}_1 \circ \mathfrak{C}_0$ of two curves $\mathfrak{C}_0, \mathfrak{C}_1$ is given by,

$$\mathfrak{C}_1 \circ \mathfrak{C}_0(s) = \begin{cases} \mathfrak{C}_0(2s) & \text{for} \quad s \in [0, \tfrac{1}{2}] \\ \mathfrak{C}_1(2s - 1) & \text{for} \quad s \in [\tfrac{1}{2}, 1] \end{cases} \tag{B.1.5}$$

The composition of closed curves induces a composition of first homotopy classes of curves, where the identity element e is the class of contractile curves, and the inverse is the class of curves with opposite orientation. The resulting group of classes is independent of the base-point[2] P and is referred to as the first homotopy group, or fundamental group $\pi_1(\Sigma)$.

Proposition B.1.1. *The Riemann sphere $\hat{\mathbb{C}}$, the complex plane $\mathbb{C}$, the Poincaré upper half plane $\mathcal{H}$, and the unit disc $\mathcal{D}$ are simply connected Riemann surfaces, as is any compact Riemann surface Σ_0 of genus 0, so that their first homotopy groups are given by,*

$$\pi_1(\hat{\mathbb{C}}) = \pi_1(\mathbb{C}) = \pi_1(\mathcal{H}) = \pi_1(\mathcal{D}) = \pi_1(\Sigma_0) = \{e\} \tag{B.1.6}$$

For a compact surface Σ of genus 1, namely a torus, we have $\pi_1(\Sigma) = \mathbb{Z}^2$.

---

[1] For the special case of an un-ramified covering the formula $\chi(X) = d\chi(Y)$ for $\chi < 0$ readily follows from the fact that the Euler number is proportional to the area for a metric of constant negative curvature.

[2] Independence of P requires Σ to be path-connected as we shall assume throughout.

For surfaces of genus $g \geq 2$, the fundamental group is non-Abelian, and will be discussed concretely in Appendix B.3.2 in terms of $SL(2, \mathbb{R})$ matrices.

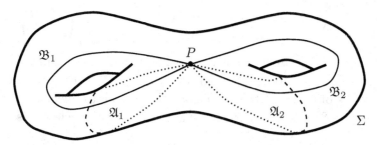

Figure B.1 A choice of homotopy generators $\mathfrak{A}_I$ and $\mathfrak{B}_I$, drawn in dotted and full thin lines, respectively, with common base point P is illustrated for a compact genus-two Riemann surface Σ.

B.1.3 Homology

The top and bottom homology groups of a Riemann surface are trivial. All the interesting homology resides in the first homology group which, for a compact Riemann surface of genus g with n punctures, is given by $H_1(\Sigma, \mathbb{Z}) = \mathbb{Z}^{2g+n}$. The primitive homology cycle associated with each puncture is homologous to a small circle centered at the puncture. Having chosen an orientation for the Riemann surface Σ, we introduce a binary intersection pairing $\mathfrak{J}(\mathfrak{C}_1, \mathfrak{C}_2)$ on oriented 1-cycles $\mathfrak{C}_1$ and $\mathfrak{C}_2$, which counts the number of intersections of the 1-cycles $\mathfrak{C}_1$ with $\mathfrak{C}_2$ weighed by ± 1 factors for the relative orientation of the cycles at their intersection points. The pairing $\mathfrak{J}(\mathfrak{C}_1, \mathfrak{C}_2)$ is odd under swapping $\mathfrak{C}_1, \mathfrak{C}_2$.

Specializing to the case of a compact Riemann surface Σ without punctures, the intersection form $\mathfrak{J}$ is nondegenerate, and we may choose a basis for $H_1(\Sigma, \mathbb{Z}) = \mathbb{Z}^{2g}$ consisting of homology cycles $\mathfrak{A}_I, \mathfrak{B}_J$ for $I, J = 1, \cdots, g$ such that,

$$\mathfrak{J}(\mathfrak{A}_I, \mathfrak{A}_J) = \mathfrak{J}(\mathfrak{B}_I, \mathfrak{B}_J) = 0$$
$$\mathfrak{J}(\mathfrak{A}_I, \mathfrak{B}_J) = -\mathfrak{J}(\mathfrak{B}_I, \mathfrak{A}_J) = \delta_{IJ} \qquad (B.1.7)$$

Two canonical bases $(\mathfrak{A}, \mathfrak{B})$ and $(\tilde{\mathfrak{A}}, \tilde{\mathfrak{B}})$ are related by a linear transformation that may be represented by a matrix M with integer entries,

$$\begin{pmatrix} \tilde{\mathfrak{B}} \\ \tilde{\mathfrak{A}} \end{pmatrix} = M \begin{pmatrix} \mathfrak{B} \\ \mathfrak{A} \end{pmatrix} \qquad (B.1.8)$$

Here, $\mathfrak{A}$ and $\mathfrak{B}$ are column matrices with entries $\mathfrak{A}_I$ and $\mathfrak{B}_I$, respectively.

The matrix M is an element of the rank g modular group $\mathrm{Sp}(2g, \mathbb{Z})$, which preserves the intersection matrix $\mathfrak{J}$,

$$M^t \mathfrak{J} M = \mathfrak{J} \qquad \mathfrak{J} = \begin{pmatrix} 0 & -I_g \\ I_g & 0 \end{pmatrix} \qquad M = \begin{pmatrix} A & B \\ C & D \end{pmatrix} \quad \text{(B.1.9)}$$

where A, B, C, and D are $g \times g$ matrices with integer entries, and I_g is the $g \times g$ identity matrix. An important subgroup of $\mathrm{Sp}(2g, \mathbb{Z})$ is the group $\mathrm{GL}(g, \mathbb{Z})$ which consists of those M that transform $\mathfrak{A}$-cycles into linear combinations of $\mathfrak{A}$-cycles and $\mathfrak{B}$-cycles into linear combinations of $\mathfrak{B}$-cycles. It is obtained by setting $B = C = 0$ and $D^{-1} = A^t$.

B.1.4 Cohomology

By the de Rham theorem, the first de Rham cohomology group is given by $H^1(\Sigma, \mathbb{R}) = \mathbb{R}^{2g}$ and is generated by $2g$ real-valued harmonic 1-forms on Σ. Since Σ is a complex manifold, harmonic 1-forms may be decomposed into holomorphic and anti-holomorphic 1-forms, and the cohomology may be decomposed into the Dolbeault cohomology groups,

$$H^1(\Sigma, \mathbb{R}) = H^{(1,0)}(\Sigma) \oplus H^{(0,1)}(\Sigma) \qquad \text{(B.1.10)}$$

generated by holomorphic $(1, 0)$ forms and anti-holomorphic $(0, 1)$ forms respectively. A canonical basis of $H^{(1,0)}(\Sigma, \mathbb{Z}) = \mathbb{Z}^g$ consists of holomorphic $(1, 0)$-forms ω_I with $I = 1, \cdots, g$ whose periods are normalized on the cycles $\mathfrak{A}_I$ of the canonical homology basis $(\mathfrak{A}, \mathfrak{B})$, as follows,

$$\oint_{\mathfrak{A}_I} \omega_J = \delta_{IJ} \qquad \text{(B.1.11)}$$

The periods on the remaining 1-cycles are then completely determined and give the matrix elements of the period matrix Ω of Σ,

$$\oint_{\mathfrak{B}_I} \omega_J = \Omega_{IJ} \qquad \text{(B.1.12)}$$

Under modular transformations $M \in \mathrm{Sp}(2g, \mathbb{Z})$, whose parametrization in terms of $g \times g$ matrices A, B, C, and D is given in (B.1.9), the matrix of holomorphic Abelian differentials ω, the period matrix Ω, and its imaginary part $Y = \mathrm{Im}(\Omega)$, transform as follows,

$$\tilde{\omega} = \omega(C\Omega + D)^{-1}$$
$$\tilde{\Omega} = (A\Omega + B)(C\Omega + D)^{-1}$$
$$\tilde{Y} = (\Omega C^t + D^t)^{-1} Y (C\Omega^* + D) \qquad \text{(B.1.13)}$$

Clearly, the transformation formula for Ω reduces to the standard Möbius transformation formula for the special case of $g = 1$.

B.1.5 Riemann bilinear relations

A useful formula relates the integral over Σ of a wedge product of closed forms ϖ_1 and ϖ_2 to their line integrals over canonical homology cycles,

$$\int_\Sigma \varpi_1 \wedge \varpi_2 = \sum_{K=1}^{g} \left(\oint_{\mathfrak{A}_K} \varpi_1 \oint_{\mathfrak{B}_K} \varpi_2 - \oint_{\mathfrak{B}_K} \varpi_1 \oint_{\mathfrak{A}_K} \varpi_2 \right) \quad \text{(B.1.14)}$$

The formula may be established by cutting the surface Σ along homology cycles $\mathfrak{A}_K$ and $\mathfrak{B}_K$ for $K = 1, \cdots, g$, chosen so as to have a point in common, into a simply connected domain Σ_{cut} in the plane, as illustrated in Figure B.2. Expressing one of the closed differentials as an exact differential of a local function in Σ_{cut}, for example, $\varpi_2 = df_2$, the formula may be derived using Green's theorem for an arbitrary 1-form ϖ,

$$\int_{\Sigma_{\text{cut}}} d\varpi = \oint_{\partial\Sigma_{\text{cut}}} \varpi \quad \text{(B.1.15)}$$

where the boundary consists of the union of cycles,

$$\partial\Sigma_{\text{cut}} = \bigcup_I \left(\mathfrak{A}_I \cup \mathfrak{B}_I \cup \mathfrak{A}_I^{-1} \cup \mathfrak{B}_I^{-1} \right) \quad \text{(B.1.16)}$$

as illustrated in Figure B.2 for genus $g = 2$.

The period matrix Ω is symmetric by the Riemann bilinear relations obtained by setting $\varpi_1 = \omega_I$ and $\varpi_2 = \omega_J$ in (B.1.14). Furthermore, the imaginary part of the period matrix,

$$Y = \text{Im } \Omega \quad \text{(B.1.17)}$$

is positive definite by the Riemann relations obtained by setting $\varpi_1 = \omega_I$ and $\varpi_2 = \bar{\omega}_J$ in (B.1.14), and using the fact that $\frac{i}{2} \int_\Sigma \varpi \wedge \varpi^*$ is positive definite for holomorphic 1-forms ϖ.

B.2 Metrics and complex structures

A different point of view on Riemann surfaces, with or without punctures, is obtained by considering connected two-dimensional orientable manifolds endowed with a Riemannian metric. This point of view is important in physics and is widely regarded as the starting point for string perturbation theory in the Polyakov formulation. Consider a two-dimensional oriented manifold Σ

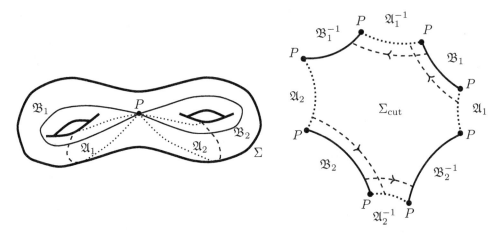

Figure B.2 Representation of the compact genus-two Riemann surface Σ, copied from Fig. B.1, in terms of a simply connected domain $\Sigma_{\text{cut}} \subset \mathbb{C}$ obtained by cutting Σ along the cycles $\mathfrak{A}_1, \mathfrak{A}_2$ and $\mathfrak{B}_1, \mathfrak{B}_2$ with common base point P, and inverse cycles pairwise identified under the dashed arrows.

endowed with a Riemannian metric $\mathfrak{g}$. In a system of local real coordinates $\xi^m = (\xi^1, \xi^2)$, the metric takes the form,[3]

$$\mathfrak{g} = \mathfrak{g}_{mn}(\xi)d\xi^m d\xi^n \qquad (\text{B.2.1})$$

We review the following notions:

- A *diffeomorphism* is a smooth map $\xi \to \xi'(\xi)$ under which $\mathfrak{g}$ is invariant, and the components $\mathfrak{g}_{mn}$ transform as follows,

$$\mathfrak{g}'_{mn}(\xi')\frac{\partial \xi'^m}{\partial \xi^p}\frac{\partial \xi'^n}{\partial \xi^q} = \mathfrak{g}_{pq}(\xi) \qquad (\text{B.2.2})$$

  Diffeomorphisms form an infinite-dimensional group $\text{Diff}(\Sigma)$.

- A *Weyl transformation* rescales the metric $\mathfrak{g}$ by an arbitrary everywhere positive function λ on Σ, namely, $\mathfrak{g}_{mn}(\xi) \to \mathfrak{g}'_{mn}(\xi) = \lambda(\xi)\mathfrak{g}_{mn}(\xi)$. Weyl transformations form an infinite-dimensional group $\text{Weyl}(\Sigma)$.

- Generally, a *conformal transformation* preserves all angles. All Weyl transformations are conformal transformations in this sense. A *conformal coordinate transformation*, or *conformal diffeomorphism* is a diffeomorphism which is conformal, which requires that $\mathfrak{g}'_{mn}(\xi') = \mu(\xi)\mathfrak{g}_{mn}(\xi)$ for some everywhere positive real function μ.

---

[3] Throughout, the Einstein convention is used to sum over a pair of repeated upper and lower indices.

- An *isometry* of $\mathfrak{g}$ preserves all distances. A diffeomorphism is an isometry of $\mathfrak{g}$ provided $\mathfrak{g}'_{mn}(\xi') = \mathfrak{g}_{mn}(\xi)$. Every conformal diffeomorphism is an isometry, but there exist conformal transformations, such as scaling by a constant, that are not isometries.

We have the following theorem.

Theorem B.2.1. *An orientable connected two-dimensional surface Σ endowed with a Riemannian metric is automatically a Riemann surface.*

To establish this result, we show that the combination of orientability and the presence of a Riemannian metric induces a unique complex structure. The orientation may be specified by choosing an oriented volume form,

$$d\mu_{\mathfrak{g}} = \sqrt{\det \mathfrak{g}}\, \varepsilon_{mn}\, d\xi^m \wedge d\xi^n \qquad \varepsilon_{mn} = -\varepsilon_{nm}, \ \varepsilon_{12} = +1 \quad (\text{B.2.3})$$

A reversal of orientation corresponds to reversing the sign of ε. We denote by $T_p(\Sigma)$ the tangent space to Σ at the point $p \in \Sigma$, and by $T_p^*(\Sigma)$ the cotangent space. A convenient basis is given by the partial derivatives $\partial_m = \partial/\partial\xi^m$ for $T_p(\Sigma)$ and by the differentials $d\xi^m$ for $T_p^*(\Sigma)$. The tangent space $T_p(\Sigma)$ may be viewed as the space of vectors $v^m \partial_m$ at p, while the cotangent space may be viewed as the space of forms $\omega_m d\xi^m$ at p. An *almost complex structure* $\mathcal{J}$ is a map from $T_p(\Sigma)$ to itself, or equivalently from $T_p^*(\Sigma)$ to itself, whose square is minus the identity map,

$$\begin{aligned}
\mathcal{J} : T_p(\Sigma) \to T_p(\Sigma) \qquad & v^m \to \mathcal{J}^m{}_n v^n \qquad & \mathcal{J}^2 = -I \\
\mathcal{J} : T_p^*(\Sigma) \to T_p^*(\Sigma) \qquad & \omega_m \to \mathcal{J}_m{}^n \omega_n \qquad & (\text{B.2.4})
\end{aligned}$$

Having a metric and an orientation, it is straightforward to exhibit this map,

$$\begin{aligned}
\mathcal{J}^m{}_n &= \sqrt{\det \mathfrak{g}}\, \varepsilon_{rn}\, \mathfrak{g}^{rm} \\
\mathcal{J}_m{}^n &= \sqrt{\det \mathfrak{g}}\, \varepsilon_{mr}\, \mathfrak{g}^{rn} \qquad\qquad (\text{B.2.5})
\end{aligned}$$

The metric is a covariantly constant tensor with respect to the affine connection and its associated covariant derivative ∇_s, so that we have $\nabla_s \mathfrak{g}_{mn} = 0$. As a result, $\mathcal{J}$ obeys $\nabla_s \mathcal{J}_m{}^n = 0$ and is therefore also a covariantly constant tensor acting on $T_p(\Sigma)$ and $T_p^*(\Sigma)$. Furthermore, $\mathcal{J}$ is invariant under arbitrary Weyl rescaling and therefore depends only on the *conformal class* of $\mathfrak{g}_{mn}$. Finally, $\mathcal{J}$ is integrable in the sense that the system of equations,

$$\mathcal{J}_m{}^n \partial_n f = i\, \partial_m f \qquad\qquad (\text{B.2.6})$$

is integrable and thereby defines locally holomorphic functions f on Σ which endow Σ with a *complex structure*.[4] We may use the solutions to (B.2.6) to

---

[4] In dimensions higher than 2, the system (B.2.6) has nontrivial integrability conditions, but in two dimensions, there are none since the operator $\mathcal{J} - iI$ has a one-dimensional null space.

define a system of local complex coordinates z and $\bar{z}$ in which the metric $\mathfrak{g}$ is locally conformally flat,

$$\mathfrak{g} = \mathfrak{g}_{z\bar{z}}|dz|^2 \tag{B.2.7}$$

for a real positive function $\mathfrak{g}_{z\bar{z}}$. The conformally flat form of the metric is invariant under locally holomorphic diffeomorphisms $z \to z'(z)$. The complex manifold Σ may be constructed by covering Σ with coordinate charts $\mathcal{U}_\alpha$ in each of which we have local complex coordinates $(z_\alpha, \bar{z}_\alpha)$. In the intersection $\mathcal{U}_\alpha \cap \mathcal{U}_\beta$ of the two coordinate charts, the local coordinates are related by transition functions $\varphi_{\alpha\beta}(z)$ which are holomorphic and nowhere vanishing via the relations $z_\alpha(z) = \varphi_{\alpha\beta}(z)z_\beta(z)$. Thus Σ is a Riemann surface.

B.2.1 The Euler number in terms of a metric

The Riemann tensor on a connected two-dimensional Riemannian manifold has only one independent component, which may be expressed in terms of the Gaussian curvature $R_\mathfrak{g}$. The Euler number of a compact Riemann surface Σ with metric $\mathfrak{g}$ may be expressed in terms of $R_\mathfrak{g}$ by the classic formula,

$$\chi(\Sigma) = \frac{1}{2\pi} \int_\Sigma d\mu_\mathfrak{g} \, R_\mathfrak{g} \tag{B.2.8}$$

The normalization of the Gaussian curvature $R_\mathfrak{g}$ is such that the round sphere S^2 of radius 1 has area 4π and $R_\mathfrak{g} = 1$, so that (B.2.8) corresponds to genus zero. The Euler number is a topological invariant since $d\mu_\mathfrak{g} \, R_\mathfrak{g}$ is locally an exact $(1,1)$-form, so that the integral in (B.2.8) is independent of the metric $\mathfrak{g}$, and only depends on the topology of Σ.

B.3 Uniformization

In this subsection, we present the uniformization theorem for simply connected Riemann surfaces and then use its results to produce concrete constructions for arbitrary compact Riemann surfaces with punctures.

Theorem B.3.1 (The uniformization theorem). *A simply connected Riemann surface is conformally isomorphic to either one of the following Riemann surfaces:*

1. *The Riemann sphere $\hat{\mathbb{C}} = \mathbb{C} \cup \{\infty\}$ (also denoted S^2, or $\mathbb{CP}^1$);*
2. *The complex plane $\mathbb{C}$;*
3. *The complex upper half plane $\mathcal{H} = \{z \in \mathbb{C}, \, \mathrm{Im}\,(z) > 0\}$ which is isomorphic to the unit disc $\mathcal{D} = \{z \in \mathbb{C}, \, |z| < 1\}$.*

For an arbitrary Riemannian metric $\mathfrak{g}$ on Σ in each case, there exists a Weyl transformation to a metric $\hat{\mathfrak{g}}$ such that $R_{\hat{\mathfrak{g}}} = 1$ for $\hat{\mathbb{C}}$; $R_{\hat{\mathfrak{g}}} = 0$ for $\mathbb{C}$; and $R_{\hat{\mathfrak{g}}} = -1$ for $\mathcal{H}$ and $\mathcal{D}$. The corresponding metrics are given as follows,

$$\hat{\mathfrak{g}}_{S^2} = \frac{4|dz|^2}{(1+|z|^2)^2} \qquad\qquad \hat{\mathfrak{g}}_{\mathbb{C}} = |dz|^2$$

$$\hat{\mathfrak{g}}_{\mathcal{H}} = \frac{|dz|^2}{(\mathrm{Im}\, z)^2} \qquad\qquad \hat{\mathfrak{g}}_{\mathcal{D}} = \frac{4|dz|^2}{(1-|z|^2)^2} \qquad (\mathrm{B.3.1})$$

as expressed in their respective global complex coordinates.

Each one of these spaces is maximally symmetric and may be expressed as the quotient of its isometry group by the isotropy group of a point,

$$\hat{\mathbb{C}} = \mathrm{SU}(2)/\mathrm{U}(1) \qquad\qquad \mathbb{C} = \mathrm{ISO}(2)/\mathrm{SO}(2)$$

$$\mathcal{H} = \mathrm{SL}(2,\mathbb{R})/\mathrm{SO}(2) \qquad\qquad \mathcal{D} = \mathrm{SU}(1,1)/\mathrm{U}(1) \qquad (\mathrm{B.3.2})$$

The isometry group $\mathrm{PSL}(2,\mathbb{R}) = \mathrm{SL}(2,\mathbb{R})/\mathbb{Z}_2$ of $\mathcal{H}$ and the isometry group $\mathrm{PSU}(1,1) = \mathrm{SU}(1,1)/\mathbb{Z}_2$ of $\mathcal{D}$ are isomorphic to one another. The action on the coordinate z in either case is given by,

$$z \to \frac{az+b}{cz+d} \qquad (\mathrm{B.3.3})$$

where $a,b,c,d \in \mathbb{R}$ with $ad - bc = 1$ for $\mathcal{H}$ and $c = \bar{b}, d = \bar{a}$ and $|a|^2 - |b|^2 = 1$ for $\mathcal{D}$. In both cases, the elements I and $-I$ leave z invariant and are identified under the center $\mathbb{Z}_2$.

The geodesics of $\hat{\mathbb{C}}$ are the great circles; those of $\mathbb{C}$ are straight lines; and those of $\mathcal{H}$ are half-circles centered on the real axis (see Exercise 3.1 for a derivation of this important result). The geodesics of $\mathcal{D}$ are the circle segments in $\mathcal{D}$, centered at any $\alpha \in \mathbb{C}$ with $|\alpha| > 1$, satisfying $|z - \alpha|^2 = |\alpha|^2 - 1$. They intersect the unit circle $|z| = 1$ orthogonally. The system of geodesics of $\mathcal{H}$ and $\mathcal{D}$ makes these spaces into standard models for two-dimensional hyperbolic (or no-Euclidean) geometry.

The uniformization theorem provides us with a powerful tool to classify and construct Riemann surfaces even when they are not simply connected. The key to the construction is the first homotopy group $\pi_1(\Sigma)$, which is trivial for simply connected Σ but nontrivial when Σ is not simply connected.

B.3.1 Genus one

Any compact Riemann surface Σ of genus one has $\pi_1(\Sigma) = \mathbb{Z}^2$ whose generators may be chosen to be the $\mathfrak{A}$ and $\mathfrak{B}$ cycles of the first homology group.

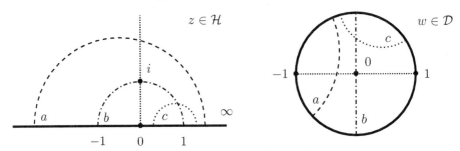

Figure B.3 Geodesics in the upper half plane $\mathcal{H}$ are semicircles centered on $\mathbb{R}$ or vertical lines, as depicted in the left panel, while their images under the conformal map $w = (1+iz)/(1-iz)$ from $\mathcal{H}$ to the unit disc $\mathcal{D}$ are arcs of circles intersecting ∂D at right angles, as depicted in the right panel.

The universal simply connected covering of a compact genus-one surface Σ is the complex plane $\mathbb{C}$. The fundamental group is Abelian and acts on $\mathbb{C}$ by translations via a representation $\rho : \mathbb{Z}^2 \mapsto \mathbb{C}$ which defines a lattice Λ. To specify the representation, and the lattice Λ, it suffices to assign the periods $\rho(1,0) = \omega_1$ and $\rho(0,1) = \omega_2$ so that the lattice is given by $\Lambda = \mathbb{Z}\omega_1 \oplus \mathbb{Z}\omega_2$. The lattice is two-dimensional (over $\mathbb{R}$) provided $\omega_2/\omega_1 = \tau \in \mathcal{H}$. The action of the fundamental group, via the representation ρ on $\mathbb{C}$, is transitive and the quotient $\mathbb{C}/\Lambda$ is the original genus-one compact Riemann surface Σ. Since translations in $\mathbb{C}$ are isometries of the flat Euclidean metric $|dz|^2$ on $\mathbb{C}$, the genus-one surface $\mathbb{C}/\Lambda$ inherits the flat metric $|dz|^2$ subject to the identifications $z \approx z + \omega_1$ and $z \approx z + \omega_2$.

B.3.2 Higher genus $g \geq 2$

Theorem B.3.2. *Every compact Riemann surface Σ of genus $g \geq 2$ admits a Riemannian metric $\hat{\mathfrak{g}}$ of constant negative curvature $R_{\hat{\mathfrak{g}}} = -1$.*

The curvature $R_{\mathfrak{g}}$ of an arbitrary metric $\mathfrak{g}$ on Σ is related to the curvature $R_{\hat{\mathfrak{g}}}$ of a metric $\hat{\mathfrak{g}}$ by a Weyl transformation on the metric $\mathfrak{g} = e^{2\sigma}\hat{\mathfrak{g}}$ via the Liouville equation,

$$R_{\mathfrak{g}} = e^{-2\sigma} R_{\hat{\mathfrak{g}}} - \Delta_{\mathfrak{g}}\sigma \tag{B.3.4}$$

where $\Delta_{\mathfrak{g}}$ is the Laplace–Beltrami operator on scalar functions for the metric $\mathfrak{g}$, given by,

$$\Delta_{\mathfrak{g}}\sigma = \frac{1}{\sqrt{\det \mathfrak{g}}} \partial_m(\sqrt{\det \mathfrak{g}}\, \mathfrak{g}^{mn} \partial_n \sigma) \tag{B.3.5}$$

To prove the theorem one shows that, for an arbitrary metric $\mathfrak{g}$, the Liouville equation has a solution for $R_{\hat{g}} = -1$. The Liouville equation first arose precisely in this context. Before presenting the construction of higher genus surfaces we introduce the concept of a Fuchsian group.

B.4 Fuchsian groups

A systematic approach to the classification and explicit construction of Riemann surfaces is through the use of Fuchsian groups. A Fuchsian group is defined to be isomorphic to a discrete subgroup of $\mathrm{SL}(2, \mathbb{R})$. Fuchsian groups are special cases of Kleinian groups that are discrete subgroups of $\mathrm{SL}(2, \mathbb{C})$. The elements $\gamma \neq \pm I$ of a Fuchsian group $\Gamma \subset \mathrm{SL}(2, \mathbb{R})$ belong to one of the following three types,

$$
\begin{array}{ll}
\gamma \ elliptic\ element & |\mathrm{tr}(\gamma)| < 2 \\
\gamma \ parabolic\ element & \mathrm{tr}(\gamma) = \pm 2 \\
\gamma \ hyperbolic\ element & |\mathrm{tr}(\gamma)| > 2
\end{array}
\tag{B.4.1}
$$

Since $\det(\gamma) = 1$ and $\gamma \neq \pm I$, the matrix γ must be conjugate, under $\mathrm{SL}(2, \mathbb{R})$, to one of the following matrices,

$$
\begin{pmatrix} \lambda & 0 \\ 0 & \lambda^{-1} \end{pmatrix}
\qquad\qquad
\begin{pmatrix} \pm 1 & 1 \\ 0 & \pm 1 \end{pmatrix}
\tag{B.4.2}
$$

The left matrix is for elliptic type with $|\lambda| = 1$ and hyperbolic type with $\lambda \in \mathbb{R}$, both with $\lambda \neq \pm 1$. The right matrix is for parabolic type.

B.4.1 Action of Γ on $\mathcal{H}$ and $\mathbb{R} \cup \{\infty\}$

Elements $\gamma \in \mathrm{SL}(2, \mathbb{R})$ act on $\mathcal{H}$ by Möbius transformation,

$$
\gamma z = \gamma(z) = \frac{az + b}{cz + d}
\qquad\qquad
\gamma = \begin{pmatrix} a & b \\ c & d \end{pmatrix}
\tag{B.4.3}
$$

Since the element $\gamma = -I$ leaves every point of $\mathcal{H}$ invariant, the action is effectively by the normal subgroup $\mathrm{PSL}(2, \mathbb{R}) = \mathrm{SL}(2, \mathbb{R})/\{\pm I\}$. The action of $\mathrm{PSL}(2, \mathbb{R})$ on $\mathcal{H}$ is transitive since an arbitrary point $z = x + iy \in \mathcal{H}$ is the image of the point $i \in \mathcal{H}$, namely, $z = \gamma_z(i)$, for a group element $\gamma_z \in \mathrm{PSL}(2, \mathbb{Z})$ whose entries satisfy $c = 0, d = a^{-1}$ and $x = ab, y = a^2$. The isotropy group of the point i is the $\mathrm{SO}(2)_i$ subgroup of $\mathrm{SL}(2, \mathbb{R})$ obtained by setting $c = -b$ and $d = a$. The isotropy subgroup $\mathrm{SO}(2)_z$ of an arbitrary point z is the conjugate of $\mathrm{SO}(2)_i$ by the element γ_z given by $\mathrm{SO}(2)_z = \gamma_z \mathrm{SO}(2)_i \gamma_z^{-1}$. As a result, $\mathcal{H}$ may be represented by the coset space,

$$
\mathcal{H} = \mathrm{SL}(2, \mathbb{R})/\mathrm{SO}(2)
\tag{B.4.4}
$$

The transitive action of $\mathrm{SL}(2,\mathbb{R})$ on $\mathcal{H}$ extends to a transitive action on $\mathbb{R} \cup \{\infty\}$ so that every point $x \in \mathbb{R} \cup \{\infty\}$ may be expressed as $x = \gamma_x(\infty)$ for some $\gamma_x \in \mathrm{SL}(2,\mathbb{R})$.

An important role will be played by the fixed points of various elements $\gamma \in \mathrm{SL}(2,\mathbb{R})$. The classification of the elements of $\mathrm{SL}(2,\mathbb{R})$ into elliptic, parabolic, and hyperbolic given in (B.4.1) is equivalent to the following classification in terms of their fixed points,

$$
\begin{array}{ll}
\textit{elliptic element} & \text{one fixed point } z \in \mathcal{H}, \text{ the other at } \bar{z} \notin \mathcal{H} \\
\textit{parabolic element} & \text{one fixed point, which is in } \mathbb{R} \cup \{\infty\} \\
\textit{hyperbolic element} & \text{two fixed points, which are in } \mathbb{R} \cup \{\infty\} \quad \text{(B.4.5)}
\end{array}
$$

In particular, the classifications of (B.4.1) and (B.4.5) apply to the elements of an arbitrary discrete subgroup $\Gamma \subset \mathrm{SL}(2,\mathbb{R})$. One defines,

$$
\begin{array}{ll}
\textit{elliptic point of } \Gamma & \text{is a point } z \in \mathcal{H} \text{ such that there exists} \\
& \text{an elliptic } \gamma \in \Gamma \text{ with } \gamma(z) = z \\
\textit{cusp of } \Gamma & \text{is a point } x \in \mathbb{R} \cup \{\infty\} \text{ such that there exists} \\
& \text{a parabolic } \gamma \in \Gamma \text{ with } \gamma(x) = x \quad \text{(B.4.6)}
\end{array}
$$

If z is an elliptic point of Γ, then $\gamma(z)$ is an elliptic point of Γ for all $\gamma \in \Gamma$. If x is a cusp of Γ, then $\gamma(x)$ is a cusp of Γ for every $\gamma \in \Gamma$.

Proposition B.4.1. *(a) If $z \in \mathcal{H}$ is an elliptic point of Γ then the set $\Gamma_z = \{\gamma \in \Gamma \,|\, \gamma(z) = z\}$ is a finite cyclic group. (b) The set of elements of finite order in Γ consist of the elliptic elements of Γ together with $\pm I$.*

To prove (a) we note that Γ_z is a subgroup of Γ. Representing $z = \gamma_z(i)$, we see that Γ_z is the set of $\gamma \in \Gamma$ for which $\gamma_z^{-1}\gamma\gamma_z(i) = i$. The isotropy group of the point z was earlier identified as $\mathrm{SO}(2)_z$ so that $\Gamma_z \subset \mathrm{SO}(2)_z$. Since Γ_z is discrete and a subgroup of the compact group $\mathrm{SO}(2)_z$, it must be finite. Since all finite subgroups of $\mathrm{SO}(2)$ are cyclic, Γ_z must also be cyclic. To prove (b), we use the fact that an element $\gamma \in \Gamma$ is conjugate under $\mathrm{SL}(2,\mathbb{C})$ to a diagonal matrix with diagonal entries $\lambda, \lambda^{-1} \in \mathbb{C}$ as in (B.4.2). Finite order implies that there exists a positive integer n such that $\lambda^n = 1$. When $n = 1, 2$ we have $\gamma = \pm I$ while for $n > 2$, γ must be elliptic. Conversely, every elliptic element is of finite order by (a).

B.5 Construction of Riemann surfaces via Fuchsian groups

The fundamental group $\pi_1(\Sigma)$ for an arbitrary compact Riemann surface Σ of genus $g \geq 2$ is discrete and non-Abelian. The universal (i.e., simply connected) covering space $\hat{\Sigma}$ of Σ may be realized as a fiber bundle over Σ with structure group isomorphic to $\pi_1(\Sigma)$. Since Σ admits a metric $\hat{g}$ of constant negative curvature $R_{\hat{g}} = -1$ by Theorem B.3.2, the universal covering surface admits a metric of constant negative curvature as well. Therefore, by the uniformization theorem, we conclude that $\hat{\Sigma} = \mathcal{H}$ or the conformally isomorphic $\hat{\Sigma} = \mathcal{D}$. Conversely, the fundamental group $\pi_1(\Sigma)$ must act on $\hat{\Sigma}$ by an isometry of the constant curvature metric. Since we have $\hat{\Sigma} = \mathcal{H}$, whose isometry group is $\mathrm{PSL}(2, \mathbb{R})$, or $\hat{\Sigma} = \mathcal{D}$ whose isometry group is $\mathrm{PSU}(1,1)$, the actions of $\pi_1(\Sigma)$ on $\mathcal{H}$ or on $\mathcal{D}$ must be by representation $\rho_{\mathcal{H}}$ or $\rho_{\mathcal{D}}$,

$$\rho_{\mathcal{H}} : \pi_1(\Sigma) \mapsto \mathrm{PSL}(2, \mathbb{R})$$
$$\rho_{\mathcal{D}} : \pi_1(\Sigma) \mapsto \mathrm{PSU}(1,1) \tag{B.5.1}$$

The representation $\rho_{\mathcal{H}}$ assigns element of $\mathrm{PSL}(2, \mathbb{R})$ and $\rho_{\mathcal{D}}$ assigns an element of $\mathrm{PSU}(1,1)$ to each equivalence class of closed curves on Σ with a specified, though arbitrary, base point P. To realize these assignments concretely, we choose a base point $P \in \Sigma$, as well as a set of homology generators $\mathfrak{A}_I$ and $\mathfrak{B}_J$ for $I, J = 1, \cdots, g$ passing through P and intersecting one another no-where else, as shown in Figure B.1. The image of $\pi_1(\Sigma)$ under either representation $\rho_{\mathcal{H}}$ or $\rho_{\mathcal{D}}$ forms a discrete subgroup of $\mathrm{SL}(2, \mathbb{R})$ or $\mathrm{SU}(1,1)$ which is a Fuchsian group Γ. When Σ is compact, all elements of Γ other than $\pm I$ are hyperbolic.

To be concrete, we shall use $\hat{\Sigma} = \mathcal{D}$ and $\rho = \rho_{\mathcal{D}}$ and cut the surface Σ along each one of the closed curves $\mathfrak{A}_I$ and $\mathfrak{B}_J$ to unfold Σ into a polygonal simply connected domain contained in $\mathcal{D}$. This domain will be denoted $\Sigma_{\mathcal{D}}$ and is referred to as the *fundamental domain* for the Fuchsian group Γ. It is shown schematically for genus 2 in Figure B.4. To reconstruct the surface Σ from the fundamental domain $\Sigma_{\mathcal{D}}$ represented in Figure B.4, we pairwise identify the curves corresponding to $\mathfrak{A}_I$ and $\mathfrak{A}_I^{-1}$ by the transformation $\rho(\mathfrak{B}_I)$, thereby closing the curves $\mathfrak{B}_I$. Similarly, we identify the curves $\mathfrak{B}_I$ and $\mathfrak{B}_I^{-1}$ by the transformation $\rho(\mathfrak{A}_I^{-1})$. Clearly, the composition of all commutators must give the identity, so we must have,

$$I = \rho(\mathfrak{C}_1)\rho(\mathfrak{C}_2) \cdots \rho(\mathfrak{C}_g)$$
$$\rho(\mathfrak{C}_I) = \rho(\mathfrak{A}_I)\rho(\mathfrak{B}_I)\rho(\mathfrak{A}_I)^{-1}\rho(\mathfrak{B}_I)^{-1} \tag{B.5.2}$$

Each homotopy class of $\pi_1(\Sigma)$ contains a unique geodesic, or closed curve of minimal length in the hyperbolic metric. Choosing the closed curves $\mathfrak{A}_I$ and $\mathfrak{B}_J$ to be geodesics, we obtain a representation of Σ in $\mathcal{D}$ by a domain $\Sigma_{\mathcal{D}}$ whose boundary is the union of circular arcs each of which is a geodesic of $\mathcal{D}$, as represented in Figure B.4. The hyperbolic length of the closed curve

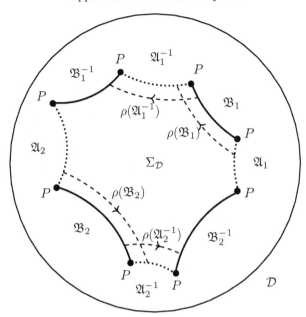

Figure B.4 A fundamental domain $\Sigma_\mathcal{D}$ for a compact genus-two Riemann surface Σ in the hyperbolic disc $\mathcal{D}$. The surface Σ is recovered by identifying the curves $\mathfrak{A}_I$ with $\mathfrak{A}_I^{-1}$ and identifying the curves $\mathfrak{B}_I$ with $\mathfrak{B}_I^{-1}$.

$\mathfrak{A}_I$ is identical to the length of $\mathfrak{A}_I^{-1}$ and similarly the lengths of the segments $\mathfrak{B}_I$ and $\mathfrak{B}_I^{-1}$ are equal to one another.

B.5.1 Compactification of $\Gamma\backslash\mathcal{H}$

We choose the standard topology of $\mathcal{H}$ generated by open sets given by open metric discs. Consider a Fuchsian group $\Gamma \subset \mathrm{SL}(2,\mathbb{R})$ whose cusps form a set $\mathcal{C} \subset \mathbb{R} \cup \{\infty\}$. To Γ we associate the space,

$$\bar{\mathcal{H}} = \mathcal{H} \cup \mathcal{C} \tag{B.5.3}$$

To specify the topology of $\bar{\mathcal{H}}$ we add open sets that contain the cusps in $\mathcal{C}$ to the topology of $\mathcal{H}$. For the cusp $x = \infty$ and the cusps $x \neq \infty$, these open sets may be chosen as follows,

$$
\begin{aligned}
x &= \infty & \{\infty\} \cup \{z \in \mathcal{H} \text{ such that } \mathrm{Im}\,(z) > y > 0 \text{ for } y \in \mathbb{R}^+\} \\
x &\neq \infty & \{x\} \cup \{z \in \mathcal{H} \text{ such that } |z - x - iy| < y \text{ for } y \in \mathbb{R}^+\}
\end{aligned}
$$

$$\tag{B.5.4}$$

The space $\bar{\mathcal{H}}$ equipped with this topology is clearly not compact. In fact, it is not even locally compact.[5] The group Γ consistently acts of $\bar{\mathcal{H}}$ which allows us to define the quotient,

$$X(\Gamma) = \Gamma\backslash\bar{\mathcal{H}} \qquad (B.5.5)$$

The quotient space $\Gamma\backslash\bar{\mathcal{H}}$ is locally compact. Finally, we state the following proposition here without proof.

Proposition B.5.1. *If the space $\Gamma\backslash\bar{\mathcal{H}}$ is compact then the number of Γ-inequivalent cusps (resp. the number of elliptic points) is finite.*

• Bibliographical notes

Classic references on Riemann surfaces are the books by Gunning [Gun15], Farkas and Kra [FK80] and by Arbarello, Cornalba, Griffiths, and Harris [ACGH85]. Comprehensive lecture notes by Bost may be found in the collection [Bos92]. A detailed treatment of modular curves and their compactification may be found in Chapters 2 and 3 of [DS05], where a proof of Proposition B.5.1 is also given.

---

[5] A topological space X is locally compact if every point in X has an open neighborhood which is contained in a compact subset of X.

Appendix C
Line bundles on Riemann surfaces

In this appendix, we shall define and study complex line bundles over a Riemann surface Σ, provide their topological classification in terms of divisors and the Riemann–Roch theorem, and prove various dimension formulas, including for the dimension of the moduli space of complex or conformal structures on a Riemann surface. We then discuss line bundles from a more physics-oriented point of view in terms of spaces of vector fields, differential forms, and spinor fields.

C.1 Holomorphic line bundles on a Riemann surface

A complex line bundle L on a Riemann surface Σ provides an assignment of a one-dimensional complex vector space L_p to each point $p \in \Sigma$. A local description of the line bundle L may be given by using a covering of Σ with open sets $\mathcal{U}_\alpha$ so that $\bigcup_\alpha \mathcal{U}_\alpha = \Sigma$. Introducing local complex coordinates $z_\alpha, \bar{z}_\alpha$ in each open set makes $\mathcal{U}_\alpha$ into a coordinate chart. Locally in each coordinate chart, the bundle is a direct product $\mathcal{U}_\alpha \times \mathbb{C}$ and each section f of L reduces to a smooth $\mathbb{C}$-valued function f_α on $\mathcal{U}_\alpha$. In the intersection $\mathcal{U}_\alpha \cap \mathcal{U}_\beta$, the functions f_α and f_β are related by a transition function $\varphi_{\alpha\beta}$,

$$f_\alpha = \varphi_{\alpha\beta} f_\beta \qquad \text{in} \quad \mathcal{U}_\alpha \cap \mathcal{U}_\beta \qquad\qquad \text{(C.1.1)}$$

The transition functions $\varphi_{\alpha\beta}$ are nowhere vanishing and, for a holomorphic line bundle, must be holomorphic functions. Considering now the intersection of three coordinate charts, we obtain a condition on products of transition functions,

$$\varphi_{\alpha\beta}\, \varphi_{\beta\gamma} = \varphi_{\alpha\gamma} \qquad \text{in} \quad \mathcal{U}_\alpha \cap \mathcal{U}_\beta \cap \mathcal{U}_\gamma \qquad\qquad \text{(C.1.2)}$$

It is important to stress that, while the transition functions for a holomorphic line bundle L are holomorphic, the sections f and the local functions f_α need not be holomorphic.

C.1.1 Topological classification

To obtain the topological classification of line bundles over Σ, we solve the nonvanishing condition on the transition functions $\varphi_{\alpha\beta}$ in terms of exponentials, $\varphi_{\alpha\beta} = \exp(2\pi i \psi_{\alpha\beta})$, where $\psi_{\alpha\beta}$ is defined by $\varphi_{\alpha\beta}$ up to the addition of an arbitrary integer $n_{\alpha\beta}$. The product relation (C.1.2) is then equivalent to the fact that the combinations,

$$c_{\alpha\beta\gamma} = \psi_{\alpha\beta} + \psi_{\beta\gamma} - \psi_{\alpha\gamma} \tag{C.1.3}$$

are integers since their exponential equals 1 by (C.1.2). The transformations $\psi_{\alpha\beta} \to \psi_{\alpha\beta} + n_{\alpha\beta}$ imply transformations of $c_{\alpha\beta\gamma}$ by an *exact cocycle*,

$$c_{\alpha\beta\gamma} \to c_{\alpha\beta\gamma} + n_{\alpha\beta} + n_{\beta\gamma} - n_{\alpha\gamma} \tag{C.1.4}$$

By construction in terms of $\psi_{\alpha\beta}$, the integers $c_{\alpha\beta\gamma}$ satisfy the following *closed cocycle* relation on the intersection of four coordinate charts, $\mathcal{U}_\alpha, \mathcal{U}_\beta, \mathcal{U}_\gamma$, and $\mathcal{U}_\delta$,

$$c_{\alpha\beta\gamma} - c_{\beta\gamma\delta} + c_{\alpha\gamma\delta} - c_{\alpha\beta\delta} = 0 \tag{C.1.5}$$

The space of closed cocycles $c_{\alpha\beta\gamma}$ satisfying (C.1.5) modulo exact cocycles $n_{\alpha\beta} + n_{\beta\gamma} - n_{\alpha\gamma}$ is the second Čech cohomology group of Σ with integer coefficients, denoted by $H^2(\Sigma, \mathbb{Z})$.

One may translate the Čech cohomology formulation of line bundles into the language of differential forms, de Rham cohomology, and gauge fields, which is more familiar to physicists. It will be convenient here to assume that Σ is compact so that one may extract from the covering of Σ by open sets $\mathcal{U}_\alpha$ a covering with only a finite number of open sets, and avoid issues of convergence.

Each line bundle L on Σ corresponds to a cohomology class of $H^2(\Sigma, \mathbb{Z})$, labeled by the first Chern class $c_1(L)$. In turn, the first Chern class $c_1(L)$ may be represented in terms of a de Rham cohomology class $[F]$, where F is the curvature 2-form of a $U(1)$ connection 1-form. In each coordinate chart $\mathcal{U}_\alpha$, a representative F of the class $[F]$ may be expressed in terms of a local connection form A_α by $F = dA_\alpha$. The transition functions for the connection are,[1]

$$A_\alpha - A_\beta = d\lambda_{\alpha\beta} \tag{C.1.6}$$

---

[1] To be precise the de Rham theorem states an isomorphism between the de Rham cohomology group $H^2_{\mathrm{dR}}(\Sigma)$ and the Čech cohomology group with real coefficients $H^2(\Sigma, \mathbb{R})$.

with $c_{\alpha\beta\gamma} = \lambda_{\alpha\beta} + \lambda_{\beta\gamma} - \lambda_{\alpha\gamma}$. The first Chern class, defined by,

$$c_1(L) = \frac{i}{2\pi} \int_\Sigma F \tag{C.1.7}$$

takes integer values, so that $H^2(\Sigma, \mathbb{Z}) = \mathbb{Z}$. For later use, we note that the first Chern class of the tensor product of two line bundles L_1 and L_2 over Σ is given as follows,

$$c_1(L_1 \otimes L_2) = c_1(L_1) + c_1(L_2) \tag{C.1.8}$$

We conclude that the topological classification of line bundles L over a compact Riemann surface Σ is in terms of a single integer d, referred to as the *degree* of the line bundle L, which labels the first Chern class $c_1(L)$ of L.

C.1.2 Holomorphic classification: divisors

Holomorphic line bundles in a given *topological class*, characterized by the degree $c_1(L) = d$, are not necessarily equivalent *holomorphically*. The space of holomorphic line bundles of degree d is referred to as the *Picard variety*,

$$\text{Pic}_d(\Sigma) = \{\text{holomorphic line bundles } L \text{ on } \Sigma \text{ with } c_1(L) = d\} \tag{C.1.9}$$

The *Jacobian variety* $J(\Sigma) = \text{Pic}_0(\Sigma)$ corresponds to the case $d = 0$.

A convenient way to characterize a holomorphic line bundle is by its divisor. A divisor is a complex co-dimension-one subset of Σ given by a formal sum over distinct points $p_\alpha \in \Sigma$,

$$D = \sum_{\alpha=1}^{N} n_\alpha p_\alpha \tag{C.1.10}$$

The integer n_α denotes the order of the point p_α in the divisor D and may take positive or negative values, while points for which $n_\alpha = 0$ are usually omitted from the sum.

To construct the line bundle L associated with a divisor D, we introduce coordinate charts $\mathcal{U}_\alpha$ with $p_\alpha \in \mathcal{U}_\alpha$ and a local complex coordinate z_α in $\mathcal{U}_\alpha$ that vanishes at p_α for each $\alpha = 1, \cdots, N$. By choosing each $\mathcal{U}_\alpha$ small enough, we can make them mutually disjoint $\mathcal{U}_\alpha \cap \mathcal{U}_\beta = \emptyset$ for $\alpha \neq \beta$. We denote the open complement to all the points p_α by $\mathcal{U}_\infty = \Sigma \setminus \{p_1, \cdots, p_N\}$ so that $\mathcal{U}_\infty \bigcup_\alpha \mathcal{U}_\alpha = \Sigma$. The line bundle L corresponding to the divisor D is then constructed by taking transition functions in the intersections $\mathcal{U}_\alpha \cap \mathcal{U}_\infty$ to be $z_\alpha^{n_\alpha}$, the integers n_α corresponding to the order of the point p_α in the divisor D in (C.1.10). A section f of L, described locally by functions f_α on

$\mathcal{U}_\alpha$ and f_∞ in $\mathcal{U}_\infty$, satisfies $f_\alpha = z_\alpha^{n_\alpha} f_\infty$. Conversely, given a line bundle L and a section f, the corresponding divisor may be recovered uniquely from the zeros and poles of f. Since the open sets $\mathcal{U}_\alpha$ are mutually disjoint, all intersections with more than two $\mathcal{U}$ coordinate charts are empty. As a result, the cocycle conditions (C.1.2) and (C.1.5) are automatically satisfied.

Adding the contributions from each transition function to the first Chern class, we find,

$$d = c_1(L) = \sum_{\alpha=1}^{N} n_\alpha \qquad (C.1.11)$$

In other words, the degree d of L is the total number of zeros (counted with their orders n_α when $n_\alpha > 0$) minus the total number of poles (counted with their orders $-n_\alpha$ when $n_\alpha < 0$). A topologically trivial bundle L with $d = 0$ has an equal number of zeros and poles (counted with their orders) and its meromorphic sections are the meromorphic functions on Σ. One may introduce an equivalence relation between divisors such that D_1 and D_2 are equivalent to one another if they differ by a divisor of degree 0. When this is the case, the ratio f_2/f_1 of any two sections f_1 and f_2 of D_1 and D_2 respectively, has degree 0. The equivalence class with representative divisor D is denoted by $[D]$.

C.1.3 Examples

A familiar example in physics is provided by the Dirac magnetic monopole, which may be thought of having a constant field strength F on the Riemann sphere $\hat{\mathbb{C}}$. Denoting the North and South poles by $p_\pm$, we introduce coordinate charts $\mathcal{U}_\pm = \hat{\mathbb{C}} \setminus \{p_\mp\}$ and local coordinates $z_\pm$ which vanish at $p_\pm$, respectively. Choosing $f_- = 1$ and $f_+ = z_+^{n_+}$ gives $\psi_{+-} = n_+ \ln(z_+)/(2\pi i)$ and $A_+ - A_- = n_+ dz_+/(2\pi i z_+)$, so that $c_1(L) = n_+$ for a Dirac monopole of charge n_+.

The tangent bundle $T\Sigma$ over a Riemann surface Σ is a bundle of rank 2 over the reals. Using the complex structure of Σ, it may be decomposed into the direct sum of its holomorphic component $T\Sigma_{(1,0)}$ and its complex conjugate $T\Sigma_{(0,1)}$. The cotangent bundle $T^*\Sigma$ may be decomposed analogously,

$$T\Sigma = T\Sigma_{(1,0)} \oplus T\Sigma_{(0,1)}$$
$$T^*\Sigma = T^*\Sigma_{(1,0)} \oplus T^*\Sigma_{(0,1)} \qquad (C.1.12)$$

The bundles $T\Sigma_{(1,0)}$ and $T^*\Sigma_{(1,0)}$ are holomorphic line bundles over Σ. The holomorphic cotangent bundle $T^*\Sigma_{(1,0)}$ is also referred to as the *canonical bundle* over Σ and denoted K, while $T\Sigma_{(1,0)}$ is isomorphic to K^{-1}.

This decomposition may be rendered explicit on the sections of the bundles. In terms of real local coordinates ξ^m on Σ a section of $T\Sigma$ is a vector field and may be expressed as $v^m \partial_m$ while a section of $T^*\Sigma$ is a differential one-form that takes the form $\omega_m d\xi^m$. In local complex coordinates z and $\bar{z}$, they may be decomposed as follows,

$$v^m \, \partial_m = v^z \, \partial_z + v^{\bar{z}} \, \partial_{\bar{z}}$$
$$\omega_m \, d\xi^m = \omega_z \, dz + \omega_{\bar{z}} \, d\bar{z} \qquad \text{(C.1.13)}$$

where $v^z \, \partial_z$ is a section of $T\Sigma_{(1,0)}$ and $\omega_z \, dz$ is a section of $T^*\Sigma_{(1,0)}$ and similarly for their complex conjugates. Their transition functions are given as follows,

$$(v^z)_\alpha(z_\alpha) = (v^z)_\beta(z_\beta) \, (\partial z_\alpha/\partial z_\beta)$$
$$(\omega_z)_\alpha(z_\alpha) = (\omega_z)_\beta(z_\beta) \, (\partial z_\alpha/\partial z_\beta)^{-1} \qquad \text{(C.1.14)}$$

where $\partial z_\alpha/\partial z_\beta$ is a holomorphic nowhere vanishing transition function in $\mathcal{U}_\alpha \cap \mathcal{U}_\beta$.

Finally, a spin bundle S on Σ is a holomorphic line bundle whose tensor square is isomorphic to the canonical bundle K,

$$S \otimes S \approx K \qquad\qquad c_1(S) = \tfrac{1}{2}c_1(K) \qquad \text{(C.1.15)}$$

The relation between the Chern numbers follows from (C.1.8). By a slight abuse of notation, a spin bundle is sometimes denoted by $K^{\frac{1}{2}}$. For $g = 0$, the spin bundle is unique. However, for $g \geq 1$, the relation to the canonical bundle does not specify the spin bundle S uniquely. The different spin bundles are referred to as *spin structures*. On a surface of genus g, there are 2^{2g} independent spin structures.

C.2 Holomorphic sections and the Riemann–Roch theorem

As emphasized earlier, while the transition functions of a holomorphic line bundle are holomorphic functions, the sections of a holomorphic line bundle are not generally holomorphic. The Cauchy–Riemann operator $\bar{\partial}_L$ for a holomorphic line bundle L acts on the sections of L by the ordinary Cauchy–Riemann operator $\bar{\partial} = d\bar{z}^\alpha \partial_{\bar{z}_\alpha}$ with respect to the local coordinate, z_α in each open set $\mathcal{U}_\alpha$. This action is covariant without the need for a connection, as may be seen by applying $\bar{\partial}$ to local sections in the overlap of open sets $\mathcal{U}_\alpha$

given by (C.1.1),

$$\bar{\partial} f_\alpha = \varphi_{\alpha\beta} \bar{\partial} f_\beta \qquad \text{in} \qquad \mathcal{U}_\alpha \cap \mathcal{U}_\beta \qquad (C.2.1)$$

The subspace of all holomorphic sections of a holomorphic line bundle plays a special role and contains a lot of information on the line bundle. It may be defined as follows. To a holomorphic line bundle L, we associate the vector space of its holomorphic sections,

$$\mathrm{Ker}(\bar{\partial}_L) = \{\text{holomorphic sections of } L\} \qquad (C.2.2)$$

These spaces are finite dimensional, and their dimensions are related by the Riemann–Roch theorem, which will be proven, in a slightly different guise, in Appendix C.5.

Theorem C.2.1. *The Riemann–Roch theorem for a holomorphic line bundle L states*

$$\dim \mathrm{Ker}\left(\bar{\partial}_L\right) - \dim \mathrm{Ker}\left(\bar{\partial}_{K \otimes L^{-1}}\right) = c_1(L) + \frac{1}{2}\chi(\Sigma) \qquad (C.2.3)$$

where K is the canonical bundle, $\chi(\Sigma)$ is the Euler number of Σ, defined in (B.1.1), and $\bar{\partial}_L$ is the Cauchy Riemann operator acting on L.

C.3 Vanishing theorem and dimension formulas

In this section, we shall state a vanishing theorem and combine its implications with the Riemann–Roch theorem to obtain general expressions for the dimensions of the spaces of homomorphic sections of line bundles. We shall specialize to the case of compact Σ and make use of the fact that the Euler number is then given in terms of the genus g of Σ by $\chi(\Sigma) = 2 - 2g$.

We begin by obtaining some immediate consequences of the Riemann–Roch theorem. Since the only holomorphic functions on Σ are the constant functions, the corresponding bundle L is the trivial bundle with $c_1(L) = 0$ and we have $\mathrm{Ker}(\bar{\partial}_L) = \mathbb{C}$ and $\dim \mathrm{Ker}(\bar{\partial}_L) = 1$. Using the Riemann–Roch theorem for the trivial bundle L, we find

$$\dim \mathrm{Ker}(\bar{\partial}_K) = g \qquad (C.3.1)$$

For $g = 0$, there are no holomorphic 1-forms. For $g = 1$, we recover the observation, made long ago, that the space of holomorphic 1-forms is one-dimensional. For $g \geq 1$, there are g linearly independent holomorphic 1-forms on Σ already identified in Appendix B.1.4. Setting $L = K$, and using the fact that $K \otimes K^{-1}$ is the trivial bundle, the Riemann–Roch theorem implies $c_1(K) = 2g - 2$. For $g = 1$, we recover the fact, observed long ago, that the holomorphic 1-form dz is nowhere vanishing, while for $g \geq 2$ we obtain the result that every holomorphic 1-form has $2g - 2$ zeros (counted with multiplicities).

More generally, representing the divisor class $[L]$ of an arbitrary line bundle L by a tensor power of the canonical bundle $[K^n]$ for $n \in \mathbb{Z}/2$, where it is understood that $K^{\frac{1}{2}} = S$ is a spin bundle with specified spin structure, we have the following vanishing theorem.

Theorem C.3.1. *The following vanishing results hold,*

$$\dim \mathrm{Ker}(\bar{\partial}_L) = 0 \qquad \begin{cases} g = 0 & \text{and} \quad n > 0 \\ g \geq 2 & \text{and} \quad n < 0 \end{cases} \tag{C.3.2}$$

The entry for $g = 0$ states the absence of holomorphic forms or spinors of degree greater than 0 on the sphere. The entry for $g \geq 2$ states that there are no holomorphic vector fields on a higher genus Riemann surface. The proof of this theorem will be giving in Appendix C.6, formulated in a slightly different guise.

Combining the results of the vanishing Theorem C.3.1 with the results obtained from the Riemann–Roch Theorem C.2.1 for arbitrary divisor class $[L] = [K^n]$ and the relations $c_1(L) = c_1(K^n) = nc_1(K)$ for $n \in \mathbb{Z}/2$, we obtain the following dimension formulas for $n \neq \frac{1}{2}$,

$$\dim \mathrm{Ker}(\bar{\partial}_L) = \begin{cases} (2n-1)(g-1) & \text{for} \quad g \geq 2 \quad \text{and} \quad n \geq \frac{3}{2} \\ g & g \geq 0 & n = 1 \\ 1 & g \geq 0 & n = 0 \\ 1 & g = 1 & n \neq \frac{1}{2} \\ 1 - 2n & g = 0 & n \leq -\frac{1}{2} \end{cases} \tag{C.3.3}$$

For n a half-odd-integer different from $\frac{1}{2}$, these dimensions are independent of the spin structure of L. But for $n = \frac{1}{2}$, the dimension of $\mathrm{Ker}(\bar{\partial}_L)$ depends on the spin structure of L. Spin structures are partitioned into even and odd depending on whether $\dim \mathrm{Ker}(\bar{\partial}_L)$ is even or odd,

$$\dim \mathrm{Ker}(\bar{\partial}_L) = \begin{cases} 1 \ (\mathrm{mod}\ 2) & g \geq 1 & n = \frac{1}{2} \quad (\text{odd}) \\ 0 \ (\mathrm{mod}\ 2) & g \geq 1 & n = \frac{1}{2} \quad (\text{even}) \end{cases} \tag{C.3.4}$$

The dimensions in (C.3.4) equal 1 and 0, respectively, throughout moduli space for $g = 1, 2$ and at generic points in moduli space for arbitrary $g \geq 3$. For $g \geq 3$, the dimensions may increase by a multiple of 2 at certain subvarieties of moduli space, including at the locus of hyperelliptic Riemann surfaces.

C.4 Tensors and spinors on Σ

In physics, we will deal with scalar, vector, tensor, and spinor fields on a Riemann surface Σ viewed as a two real dimensional surface equipped with a Riemannian metric,

$$\mathfrak{g} = \mathfrak{g}_{mn}(\xi)\, d\xi^m d\xi^n \tag{C.4.1}$$

in a system of real coordinates ξ^m. We shall now analyze complex line bundles, covariant derivatives, and Laplace–Beltrami operators on line bundles in a tensorial formulation and use it to provide a physics-minded proof of the Riemann–Roch and vanishing theorems.

An arbitrary tensor field $\mathfrak{t}_{m_1 \dots m_n}$ of rank n may be decomposed into a direct sum of symmetric traceless tensors and scalars. Indeed, any antisymmetric pair of indices may be contracted with the tensor $\sqrt{\det \mathfrak{g}}\, \varepsilon_{mn}$ to produce a scalar, while the metric $\mathfrak{g}_{mn}$ may be used to eliminate all traces. The resulting symmetric traceless tensor may be further decomposed into the eigenspaces of the complex structure $\mathcal{J}$, already introduced in (B.2.4). The convention $v^z \partial_z = v^z(dz)^{-1}$ may be used to identify a form of negative weight with a vector field of opposite weight. Henceforth, we shall suppress the z and $\bar{z}$ indices on these reduced tensors, so that $\mathfrak{t} = t_{z,\dots z}$ and $\bar{\mathfrak{t}} = t_{\bar{z}\dots,\bar{z}}$. Thus, all tensor fields decompose into a direct sum of the following spaces,

$$T^*_{(n,0)} = \Big\{ \mathfrak{t}(z) : \Sigma \to \mathbb{C} \text{ such that } \mathfrak{t}'(z')(dz')^n = \mathfrak{t}(z)(dz)^n \Big\} \tag{C.4.2}$$

The space $T^*_{(1,0)}$ may be identified with the space of sections of the canonical bundle K, while the spaces $T^*_{(n,0)}$ for $n > 0$ may be identified with the space of sections of the nth tensor power K^n of K. Finally, we include spinors by allowing n to take half-integer values that further requires specifying a spin structure.

C.4.1 Covariant derivatives and Laplace operators

On each space of sections $T^*_{(n,0)}$, we define an $L^2(\Sigma)$ inner product as follows,

$$\langle \mathfrak{t}_1 | \mathfrak{t}_2 \rangle = \int_\Sigma d\mu_{\mathfrak{g}}\, (\mathfrak{g}_{z\bar{z}})^{-n}\, \overline{\mathfrak{t}_1(z)}\, \mathfrak{t}_2(z) \tag{C.4.3}$$

The Cauchy–Riemann operator $\bar{\partial} = (d\bar{z})\partial_{\bar{z}}$ acts covariantly on the spaces $T^*_{(n,0)}$, without the need for a Christoffel connection,

$$\bar{\partial} : T^*_{(n,0)} \to T^*_{(n,1)}$$
$$\mathfrak{t}(z)(dz)^n \to \big(\partial_{\bar{z}}\mathfrak{t}(z)\big)(dz)^n d\bar{z} \tag{C.4.4}$$

It is often more convenient to use a covariant derivative which maps a space into a space of the same type. This may be achieved by the following Cauchy–Riemann operator acting on $T^*_{(n,0)}$,

$$\nabla^z_{(n)} : T^*_{(n,0)} \to T^*_{(n-1,0)}$$
$$\mathfrak{t}(z)(dz)^n \to \nabla^z_{(n)}\mathfrak{t}(z) = \left(g^{z\bar{z}}\partial_{\bar{z}}\mathfrak{t}(z)\right)(dz)^{n-1} \tag{C.4.5}$$

To define the adjoint operator $-\nabla^{(n-1)}_z$ of $\nabla^z_{(n)}$ with respect to the $L^2(\Sigma)$ inner product (C.4.3) does require a Christoffel connection and is given by,

$$\nabla^{(n)}_z : T^*_{(n,0)} \to T^*_{(n+1,0)}$$
$$\mathfrak{t}(z)(dz)^n \to \left(\nabla_z\mathfrak{t}(z)\right)(dz)^{n+1} \tag{C.4.6}$$

The adjoint operators with respect to the inner product (C.4.3), and the associated Laplace–Beltrami operators, are as follows,

$$\left(\nabla^z_{(n)}\right)^\dagger = -\nabla^{(n-1)}_z \qquad\qquad \Delta^+_{(n)} = 2\,\nabla^z_{(n+1)}\nabla^{(n)}_z$$
$$\left(\nabla^{(n)}_z\right)^\dagger = -\nabla^z_{(n+1)} \qquad\qquad \Delta^-_{(n)} = 2\,\nabla^{(n-1)}_z\nabla^z_{(n)} \tag{C.4.7}$$

The Laplace–Beltrami operators both map $T_{(n,0)} \to T_{(n,0)}$. The nonzero eigenvalues and their corresponding eigenfunctions are related. To see this, we observe that,

$$\Delta^+_{(n)}\psi = \lambda\psi \quad\Longrightarrow\quad \Delta^-_{(n+1)}\left(\nabla^{(n)}_z\psi\right) = \lambda\left(\nabla^{(n)}_z\psi\right)$$
$$\Delta^-_{(n)}\varphi = \mu\varphi \quad\Longrightarrow\quad \Delta^+_{(n-1)}\left(\nabla^z_{(n)}\varphi\right) = \mu\left(\nabla^z_{(n)}\varphi\right) \tag{C.4.8}$$

When $\lambda, \mu \neq 0$, these maps are invertible. As a result, the spectra of non-vanishing eigenvalues of $\Delta^+_{(n)}$ and $\Delta^-_{(n+1)}$ coincide and the one-to-one map between their eigenfunctions is provided by the above relations. No such correspondence follows for the kernels of these operators. Since the Laplace–Beltrami operators are negative, their kernels are related to the kernels of the Cauchy–Riemann operators as follows,

$$\text{Ker}\,\Delta^+_{(n)} = \text{Ker}\,\nabla^{(n)}_z$$
$$\text{Ker}\,\Delta^-_{(n)} = \text{Ker}\,\nabla^z_{(n)} \tag{C.4.9}$$

In Sections C.5 and C.6, we shall relate the dimensions of these kernels to one another by the Riemann–Roch and vanishing theorems.

C.5 Proof of the Riemann–Roch theorem

The kernel of the operator $\nabla^z_{(n)}$, introduced in (C.4.5), is the space of holomorphic differentials of rank n for $n \geq 1$, and holomorphic vector fields of rank $-n$ for $n \leq -1$. In this formulation, the Riemann–Roch theorem for a compact Riemann surface Σ may be stated as follows,

$$\dim \operatorname{Ker} \nabla^z_{(n+1)} - \dim \operatorname{Ker} \nabla^{(n)}_z = -\left(n + \tfrac{1}{2}\right)\chi(\Sigma)$$
$$= (2n+1)(g-1) \qquad (C.5.1)$$

where $\chi(\Sigma) = 2 - 2g$ is the Euler number of a surface Σ of genus g. The Riemann–Roch theorem C.2.1 may equivalently be recast in terms of Laplace–Beltrami operators,

$$\dim \operatorname{Ker} \Delta^-_{(n+1)} - \dim \operatorname{Ker} \Delta^+_{(n)} = (2n+1)(g-1) \qquad (C.5.2)$$

To prove this result, we use the fact established earlier that the nonzero eigenvalues of $\Delta^+_{(n)}$ and $\Delta^-_{(n+1)}$ are identical to obtain the following relation,

$$\dim \operatorname{Ker} \Delta^-_{(n+1)} - \dim \operatorname{Ker} \Delta^+_{(n)} = \operatorname{Tr}\left(e^{s\Delta^-_{(n+1)}}\right) - \operatorname{Tr}\left(e^{s\Delta^+_{(n)}}\right) \quad (C.5.3)$$

valid for all $0 < s \in \mathbb{R}$. In particular, the right side may be evaluated in the limit $s \to 0$ where the short-time asymptotics of the heat-kernel may be used to derive an explicit formula. The short-time expansion, or Bott-Seeley expansion, is given as follows,

$$\operatorname{Tr}\left(f\, e^{s\Delta^\pm_{(n)}}\right) = \frac{1}{4\pi s}\int_\Sigma d\mu_{\mathfrak{g}} f + \frac{1 \pm 3n}{12\pi}\int_\Sigma d\mu_{\mathfrak{g}} R_{\mathfrak{g}} f + \mathcal{O}(s) \quad (C.5.4)$$

Setting $f = 1$, substituting the expressions into (C.5.3), and using the integral expression for the Euler number, we establish (C.5.2), which proves the Riemann–Roch theorem.

C.6 Proof of the vanishing theorem

In terms of the differential operators $\nabla^z_{(n)}$ and $\nabla^z_{(n)}$, the vanishing Theorem C.3.1 reads as follows,

$$\dim \operatorname{Ker} \nabla^z_{(n)} = 0 \qquad \begin{cases} g = 0 & \text{and} \quad n > 0 \\ g \geq 2 & \text{and} \quad n < 0 \end{cases} \qquad (C.6.1)$$

The proof uses the following equations from differential geometry,

$$[\nabla_m, \nabla_n]\, t_p = -t_q\, R^q{}_{pmn}$$
$$[\nabla_m, \nabla_n]\, t^p = +t^q\, R^p{}_{qmn} \qquad (C.6.2)$$

where ∇_m is the covariant derivative for the affine connection which satisfies $\nabla \mathfrak{g}_{pq} = 0$, $R^q{}_{pmn}$ is the corresponding Riemann tensor, and t_q is a rank one tensor, that is, a one-form. Adapted to the case of two dimensions, and working in complex coordinates, the formulas reduce to,

$$[\nabla_z, \nabla_{\bar{z}}]\, \mathsf{t}_z = -\mathfrak{g}_{z\bar{z}} R_{\mathfrak{g}}\, \mathsf{t}_z \qquad (C.6.3)$$

where $R_{\mathfrak{g}}$ is the Gaussian curvature (normalized to be one for the round sphere of unit radius). Formulas for tensors and spinors $\mathsf{t}_{z\cdots z} = \mathsf{t} \in T^*_{(n,0)}$ of arbitrary rank $n \in \mathbb{Z}/2$ may be derived by taking the tensor product and the square root, and we find,

$$[\nabla_z, \nabla_{\bar{z}}]\, \mathsf{t} = -n\, \mathfrak{g}_{z\bar{z}}\, R_{\mathfrak{g}}\, \mathsf{t} \qquad (C.6.4)$$

We are now ready to prove (C.6.1) and thereby also the vanishing Theorem C.3.1. We use the following rearrangement formula, obtained by integrating by parts (no boundary terms arise since Σ has no boundary),

$$\int_\Sigma d\mu_{\mathfrak{g}}\, (\mathfrak{g}_{z\bar{z}})^{n-1} |\nabla_{\bar{z}}\, \mathsf{t}|^2 = \int_\Sigma d\mu_{\mathfrak{g}} \left[(\mathfrak{g}_{z\bar{z}})^{n-1} |\nabla_{\bar{z}}\, \bar{\mathsf{t}}|^2 + n(\mathfrak{g}_{z\bar{z}})^n R_{\mathfrak{g}} |\mathsf{t}|^2 \right] \quad (C.6.5)$$

- For the case $g = 0$, we may choose a metric of constant positive curvature $R_{\mathfrak{g}} = 1$ on the sphere. Since we have $n > 0$, the right side of (C.6.5) is positive definite in t so that the left side cannot vanish unless $\mathsf{t} = 0$. As a result, there are no holomorphic differential forms of any positive weight $n > 0$ on the sphere which proves the result on the first line of (C.6.1).

- For the case $g \geq 2$, we may choose a metric of constant negative curvature $R_{\mathfrak{g}} = -1$. Since we have $n < 0$, the right side of (C.6.5) is again positive definite so that the left side cannot vanish unless $\mathsf{t} = 0$. As a result, there are no holomorphic vector fields of any negative weight $n < 0$ on a higher genus Riemann surface. In particular, this implies that there are no conformal Killing vectors or conformal Killing spinors on a Riemann surface of genus $g \geq 2$ and thus no continuous symmetries.

- For the case $g = 1$, and a flat metric $R_{\mathfrak{g}} = 0$ on the torus, the relation (C.6.5) imposes no restrictions.

C.7 The dimension of moduli space

The moduli space $\mathcal{M}_g$ of compact Riemann surfaces of arbitrary genus g is the space of inequivalent complex structures or, equivalently, the space of inequivalent conformal structures, which may be parametrized with the help of the Weyl-invariant tensor $\mathcal{J}_m{}^n = \sqrt{\det \mathfrak{g}}\, \varepsilon_{mp}\, \mathfrak{g}^{pn}$ already introduced in

(B.2.4). Two tensors that are related by a diffeomorphism of Σ are equivalent to one another, so that the moduli space is given by,

$$\mathcal{M}_g = \{\mathcal{J}\}/\mathrm{Diff}(\Sigma) \tag{C.7.1}$$

To compute the dimension of $\mathcal{M}_g$, we evaluate the dimension of its tangent space at a given complex structure with local complex coordinates z and $\bar{z}$ in terms of which the metric is given by $\mathfrak{g} = 2\mathfrak{g}_{z\bar{z}}|dz|^2$, and the complex structure tensor $\mathcal{J}$ has the following components,

$$\mathcal{J}_z{}^{\bar{z}} = -\mathcal{J}_{\bar{z}}{}^z = 0 \qquad\qquad \mathcal{J}_z{}^z = -\mathcal{J}_{\bar{z}}{}^{\bar{z}} = i \tag{C.7.2}$$

Next, we compute the variation of the complex structure tensor, as the metric $\mathfrak{g}$ is varied, *leaving the system of complex coordinates $z, \bar{z}$ unchanged.* The variation of the metric is used to first order,

$$\delta\mathfrak{g} = 2\delta\mathfrak{g}_{z\bar{z}}|dz|^2 + \delta\mathfrak{g}_{zz}\,dz^2 + \delta\mathfrak{g}_{\bar{z}\bar{z}}\,d\bar{z}^2 \tag{C.7.3}$$

Infinitesimal Weyl transformations $\delta\mathfrak{g}_{z\bar{z}}$ leave $\mathcal{J}$ invariant. Thus, we obtain the following variations, to first order in $\delta\mathfrak{g}_{zz}$ and $\delta\mathfrak{g}_{\bar{z}\bar{z}}$,

$$\begin{aligned} \delta\mathcal{J}_z{}^z &= 0 & \delta\mathcal{J}_z{}^{\bar{z}} &= -i\,\mathfrak{g}^{z\bar{z}}\,\delta\mathfrak{g}_{zz} &&\in T^*_{(1,-1)} \\ \delta\mathcal{J}_{\bar{z}}{}^{\bar{z}} &= 0 & \delta\mathcal{J}_{\bar{z}}{}^z &= +i\,\mathfrak{g}^{\bar{z}z}\,\delta\mathfrak{g}_{\bar{z}\bar{z}} &&\in T^*_{(-1,1)} \end{aligned} \tag{C.7.4}$$

Moduli deformations, however, are variations of the complex structure modulo diffeomorphisms. The action of an infinitesimal diffeomorphism $v^m\partial_m$ on the metric is given as follows,

$$\delta_v\mathfrak{g}_{mn} = \nabla_m v_n + \nabla_n v_m \tag{C.7.5}$$

Under this variation, the complex structure transforms as follows,

$$\begin{aligned} \delta_v\,\mathcal{J}_z{}^{\bar{z}} &= -2i\,\nabla_z^{(1)}v^{\bar{z}} \\ \delta_v\,\mathcal{J}_{\bar{z}}{}^z &= +2i\,\nabla_{\bar{z}}^{(-1)}v^z \end{aligned} \tag{C.7.6}$$

Vectors in the tangent space $T_\mathfrak{g}\mathcal{M}_g$ to moduli space $\mathcal{M}_g$ at a metric $\mathfrak{g}$ may be identified with complex structure deformations that are in the orthogonal complement to the ranges of these operators, evaluated at the metric $\mathfrak{g}$,

$$T_\mathfrak{g}\mathcal{M}_g = \left(\mathrm{Range}\,\nabla_z^{(1)}\right)^\dagger \oplus \left(\mathrm{Range}\,\nabla_{\bar{z}}^{(-1)}\right)^\dagger \tag{C.7.7}$$

But the orthogonal complement to the range of any operator is the kernel of the adjoint of that operator. Since we have $(\nabla_z^{(1)})^\dagger = -\nabla_{(2)}^z$ by (C.4.7), we obtain the following fundamental expression for the cotangent space

$$T_\mathfrak{g}\mathcal{M}_g = \mathrm{Ker}\nabla_{(2)}^z \oplus \mathrm{Ker}\nabla_{(-2)}^{\bar{z}} \tag{C.7.8}$$

First of all, we note that the tangent space to $\mathcal{M}_g$ is split according to holomorphic and anti-holomorphic directions, so that $\mathcal{M}_g$ is a complex manifold (actually, more precisely a complex orbifold). The dimension of moduli space may be read off from the table of dimensions given in (C.3.3),

$$\dim_{\mathbb{C}} \mathcal{M}_g = \begin{cases} 0 & g = 0 \\ 1 & g = 1 \\ 3g - 3 & g \geq 2 \end{cases} \qquad (\text{C.7.9})$$

The combination $\delta \mathcal{J}_{\bar{z}}{}^z$ and its complex conjugate are often referred to as *Beltrami differentials*. The elements of the dual space $\mathrm{Ker} \nabla_{(2)}^z$ are the *holomorphic quadratic differentials*.

• Bibliographical notes

A summary of the ingredients of holomorphic line bundles on compact Riemann surfaces, accessible to physicists, was given in the review paper on string perturbation theory [DP88] and in the lecture notes [Bos92]. A fundamental and useful reference is the book by Fay [Fay06]. Helpful discussions of line bundles characterized by divisors may be found in the books by Farkas and Kra [FK80] and by Arbarello, Cornalba, Griffiths, and Harris [ACGH85], while the deformation theory of complex structures for Riemann surfaces is treated quite explicitly in the book by Schiffer and Spencer [SS14]. More general treatments of vector bundles on complex and Kähler manifolds may be found in the books by Gunning [Gun67] and Kodaira [Kod86]. The Bott-Seeley expansion in the context of two-dimensional Riemannian manifolds is discussed and proven in [Alv83] and [DP88].

Appendix D

Riemann ϑ-functions and meromorphic forms

In this appendix, we shall review the modular geometry of the Siegel half-space at higher rank, Riemann ϑ-functions of higher rank, the embedding of higher-genus Riemann surfaces into the Jacobian variety via the Abel map, and use these ingredients to construct the prime form, the Szegö kernel, and other meromorphic functions and differential forms on higher-genus Riemann surfaces.

D.1 The Siegel half-space

The rank g Siegel half-space $\mathcal{H}_g$ may be defined as the space of complex $g \times g$ symmetric matrices with positive definite imaginary part,

$$\mathcal{H}_g = \left\{ \Omega \in \mathbb{C}^{g \times g}, \ \Omega^t = \Omega, \ \mathrm{Im}\,(\Omega) > 0 \right\} \tag{D.1.1}$$

Its dimension is $\dim_{\mathbb{C}} \mathcal{H}_g = \frac{1}{2}g(g+1)$. Alternatively, $\mathcal{H}_g$ is given as the coset,

$$\mathcal{H}_g = \mathrm{Sp}(2g, \mathbb{R})/\mathrm{U}(g) \tag{D.1.2}$$

The group $\mathrm{Sp}(2g, \mathbb{R})$ acts on Ω by

$$\Omega \to \tilde{\Omega} = (A\Omega + B)(C\Omega + D)^{-1} \tag{D.1.3}$$

where the $g \times g$ real matrices $A, B, C,$ and D are given in terms of $M \in \mathrm{Sp}(2g, \mathbb{R})$,

$$M = \begin{pmatrix} A & B \\ C & D \end{pmatrix} \qquad M^t \mathfrak{J} M = \mathfrak{J} \qquad \mathfrak{J} = \begin{pmatrix} 0 & -I \\ I & 0 \end{pmatrix} \tag{D.1.4}$$

The unitary subgroup $\mathrm{U}(g)$ is generated by setting $C = -B$ and $D = A$ subject to the conditions $A^t A + B^t B = I$ and $A^t B - B^t A = 0$. Since the group $\mathrm{U}(g)$ contains a $\mathrm{U}(1)$ factor, $\mathcal{H}_g$ is a Kähler manifold, with the following

$\mathrm{Sp}(2g, \mathbb{R})$-invariant Kähler metric,

$$ds^2 = \sum_{I,J,K,L=1}^{g} (Y^{-1})^{IJ} \, d\Omega_{IK} \, (Y^{-1})^{KL} \, d\bar{\Omega}_{JL} \qquad \text{(D.1.5)}$$

where $Y = \mathrm{Im}\,(\Omega)$ is positive definite on $\mathcal{H}_g$ in view of (D.1.1), and Y^{-1} is its matrix inverse. The $\mathrm{Sp}(2g, \mathbb{R})$-invariant Laplace–Beltrami operator Δ for the metric ds^2 of (D.1.5), acting on scalar functions of $\mathcal{H}_g$, is given by,

$$\Delta = \sum_{I,J,K,L=1}^{g} 4 Y_{IJ} Y_{KL} \, \partial^{IK} \, \bar{\partial}^{JL} \qquad \text{(D.1.6)}$$

where the derivatives with respect to the components of Ω_{IJ} are defined by,

$$\partial^{II} = \frac{\partial}{\partial \Omega_{II}} \qquad\qquad \partial^{IJ} = \frac{1}{2} \frac{\partial}{\partial \Omega_{IJ}} \qquad \text{for } J \neq I \quad \text{(D.1.7)}$$

For genus one, $\mathcal{H}_1 = \mathcal{H}$ coincides with the Poincaré upper half plane, the metric ds^2 coincides with the Poincaré metric on $\mathcal{H}$, and Δ is the Laplace–Beltrami operator on scalars functions on $\mathcal{H}$.

D.2 The Riemann theta function

The *Riemann ϑ-functions* are complex-valued functions on $\mathcal{H}_g \times \mathbb{C}^g$ that generalize the Jacobi ϑ-functions on $\mathcal{H} \times \mathbb{C}$ to arbitrary rank g. Just as for Jacobi ϑ-functions, Riemann ϑ-functions may be considered for arbitrary complex characteristics $\delta = [\delta'|\delta'']$ with $\delta', \delta'' \in \mathbb{C}^g$. In terms of local coordinates Ω on $\mathcal{H}_g$ and $\zeta = (\zeta_1, \cdots, \zeta_g)^t$ on $\mathbb{C}^g$, the Riemann ϑ-function for characteristic δ may be defined by the absolutely convergent series,

$$\vartheta[\delta](\zeta|\Omega) = \sum_{n \in \mathbb{Z}^g} \exp\left(i\pi(n + \delta')^t \Omega (n + \delta') + 2\pi i(n + \delta')^t(\zeta + \delta'') \right) \quad \text{(D.2.1)}$$

Upon shifting by $m, n \in \mathbb{Z}^g$, we have the periodicity relations,

$$\vartheta[\delta](\zeta + m + \Omega n|\Omega) = e^{-i\pi n^t \Omega n - 2\pi i n^t(\zeta + \delta') + 2\pi i m^t \delta''} \, \vartheta[\delta](\zeta|\Omega)$$
$$\vartheta[\delta' + n|\delta'' + m](\zeta|\Omega) = e^{2\pi i m^t \delta'} \, \vartheta[\delta'|\delta''](\zeta|\Omega) \qquad \text{(D.2.2)}$$

The ϑ-function without characteristics is denoted by $\vartheta(\zeta|\Omega) = \vartheta[0](\zeta|\Omega)$.

We shall focus on *half-integer characteristics* δ with values $\delta', \delta'' \in (\mathbb{Z}/2\mathbb{Z})^g$ corresponding to spin structures. The parity $4\,\delta' \cdot \delta''$ (mod 2) of a ϑ-function with half-integer characteristics (we shall use the notation $\delta \cdot \delta'' = (\delta')^t \delta''$) depends on δ and, at a point $(\zeta|\Omega)$ where $\vartheta[\delta](\zeta|\Omega) \neq 0$, is defined by,

$$\vartheta[\delta](-\zeta|\Omega) = (-1)^{4\delta' \cdot \delta''} \vartheta[\delta](\zeta|\Omega) \qquad \text{(D.2.3)}$$

The spin structure δ is referred to as an even or odd spin structure according to whether the integer $4\,\delta' \cdot \delta''$ is even or odd.

D.2.1 Modular transformations

The discrete (or *arithmetic*) subgroup $\mathrm{Sp}(2g, \mathbb{Z})$ of $\mathrm{Sp}(2g, \mathbb{R})$ is the group of modular transformations, whose action on the Siegel upper half-space $\mathcal{H}_g$ is given by (D.1.3), but now for $M \in \mathrm{Sp}(2g, \mathbb{Z})$ with the entries of the matrices $A, B, C,$ and D being integers. Under a modular transformation $M \in \mathrm{Sp}(2g, \mathbb{Z})$, a half-integer characteristic $\delta = [\delta' | \delta'']$ transforms into another half-integer characteristic $\tilde{\delta} = [\tilde{\delta}' | \tilde{\delta}'']$ given as follows,

$$\begin{pmatrix} \tilde{\delta}' \\ \tilde{\delta}'' \end{pmatrix} = \begin{pmatrix} D & -C \\ -B & A \end{pmatrix} \begin{pmatrix} \delta' \\ \delta'' \end{pmatrix} + \frac{1}{2} \operatorname{diag} \begin{pmatrix} CD^t \\ AB^t \end{pmatrix} \qquad M = \begin{pmatrix} A & B \\ C & D \end{pmatrix} \quad \text{(D.2.4)}$$

Under a modular transformation acting jointly on Ω, ζ, and δ, the Riemann ϑ-function with characteristics transforms as follows,

$$\vartheta[\tilde{\delta}]\left(\{(C\Omega + D)^{-1}\}^t \zeta | \tilde{\Omega} \right) = \varepsilon(\delta; M) \exp\left\{ i\pi \zeta^t (C\Omega + D)^{-1} C\zeta \right\}$$
$$\times \det (C\Omega + D)^{\frac{1}{2}}\, \vartheta[\delta](\zeta|\Omega) \qquad \text{(D.2.5)}$$

where $\varepsilon(\delta; M)^8 = 1$ and the precise dependence of $\varepsilon(\delta, M)$ on its arguments is given in [Fay06] and [Igu12]. The above transformation law reduces to the one for Jacobi ϑ-functions given in (3.7.7) with $\varepsilon(\delta; M) \to \varepsilon(\alpha, \beta; \gamma)$.

D.2.2 Riemann relations on ϑ-functions

For each spin structure λ, there exists a Riemann relation which may be expressed as a quadri-linear sum over all spin structures κ,

$$\sum_\kappa \langle \kappa | \lambda \rangle \prod_{i=1}^4 \vartheta[\kappa](\zeta_i|\Omega) = 2^h \prod_{i=1}^4 \vartheta[\lambda](\zeta_i'|\Omega) \qquad \text{(D.2.6)}$$

The signature symbol for the pairing of two spin structures is defined by,

$$\langle \kappa | \lambda \rangle = e^{4\pi i (\kappa' \cdot \lambda'' - \kappa'' \cdot \lambda')} \qquad \text{(D.2.7)}$$

The relation between the vectors ζ and ζ' is given in terms of a matrix T,

$$\begin{pmatrix} \zeta_1' \\ \zeta_2' \\ \zeta_3' \\ \zeta_4' \end{pmatrix} = T \begin{pmatrix} \zeta_1 \\ \zeta_2 \\ \zeta_3 \\ \zeta_4 \end{pmatrix} \qquad T = \frac{1}{2} \begin{pmatrix} 1 & 1 & 1 & 1 \\ 1 & 1 & -1 & -1 \\ 1 & -1 & 1 & -1 \\ 1 & -1 & -1 & 1 \end{pmatrix} \qquad \text{(D.2.8)}$$

which satisfies $T^2 = I$. In the special case where $\zeta = \zeta' = 0$, only even spin structures $\kappa = \delta$ contribute to the sum and we recover a Riemann relation for each odd spin structure $\lambda = \nu$,

$$\sum_\delta \langle \nu | \delta \rangle \vartheta[\delta]^4(0|\Omega) = 0 \tag{D.2.9}$$

More generally, if $\zeta_1 + \zeta_2 + \zeta_3 + \zeta_4 = 0$ then $\zeta'_1 = 0$, the right side of (D.2.6) vanishes for any odd spin structure λ, and the Riemann relations become,

$$\sum_\kappa \langle \kappa | \lambda \rangle \prod_{i=1}^4 \vartheta[\kappa](\zeta_i|\Omega) = 0 \tag{D.2.10}$$

For genus one, there is only a single odd spin structure and the Riemann relation reduces to the famous Jacobi identity $\vartheta_2^4 - \vartheta_3^4 + \vartheta_4^4 = 0$ of (2.6.18).

D.3 Jacobian, Abel map, and Riemann vanishing theorem

We recall from Appendix B that a canonical basis may be chosen for the holomorphic $(1,0)$-forms ω_I with $I = 1, \cdots, g$ on a compact Riemann surface Σ of genus g by normalizing the integrals of ω_I on the $\mathfrak{A}$-cycles of a canonical basis $(\mathfrak{A}, \mathfrak{B})$ of generators for the homology group $H_1(\Sigma, \mathbb{Z})$. The integrals on the $\mathfrak{B}$-cycles then give the period matrix Ω of the Riemann surface Σ in the homology basis $(\mathfrak{A}, \mathfrak{B})$,

$$\oint_{\mathfrak{A}_I} \omega_J = \delta_{IJ} \qquad\qquad \oint_{\mathfrak{B}_I} \omega_J = \Omega_{IJ} \tag{D.3.1}$$

Under a symplectic change of canonical homology basis, the holomorphic $(1,0)$-forms ω and period matrix Ω transform as given in (B.1.13). In view of the Riemann relations, we have $\Omega^t = \Omega$ and $\mathrm{Im}\,(\Omega) > 0$, so that the period matrix of a compact Riemann surface takes values in the Siegel half-space $\mathcal{H}_g$. The set of $\mathfrak{A}$ and $\mathfrak{B}$ periods defines a lattice $\Lambda = \mathbb{Z}^g + \Omega\mathbb{Z}^g$ in $\mathbb{C}^g$, whose quotient is an Abelian variety referred to as the *Jacobian variety*,

$$J(\Sigma) = \mathbb{C}^g/(\mathbb{Z}^g + \Omega\mathbb{Z}^g) \tag{D.3.2}$$

In other words, the Jacobian is a g-dimensional complex torus. For genus $g = 1$, this torus gives an equivalent representation of the Riemann surface itself. For higher genus $g \geq 2$, this simple correspondence no longer holds but is replaced by a more subtle identification produced by the Abel map.

Given a base point $z_0 \in \Sigma$, the Abel map sends a divisor D of n points

$z_i \in \Sigma$ with weights $q_i \in \mathbb{Z}$ for $i = 1, \cdots, n$, which is formally denoted by $D = q_1 z_1 + \cdots q_n z_n$, into $\mathbb{C}^g$ by,

$$D = q_1 z_1 + \cdots + q_n z_n \to \sum_{i=1}^{n} q_i \int_{z_0}^{z_i} (\omega_1, \cdots, \omega_g) \qquad (D.3.3)$$

where the g-tuple $(\omega_1, \cdots, \omega_g)$ stands for the vector of holomorphic $(1,0)$-forms ω_I. The Abel map into $\mathbb{C}^g$ is multiple valued because each integral $\int_{z_0}^{z_i} \omega_I$ is multiple-valued in $\mathbb{C}$, but the Abel map is single-valued as a map into the Jacobian $J(\Sigma)$.

Theorem D.3.1. *The Riemann vanishing theorem states that, for a Riemann surface Σ with period matrix Ω, the equation $\vartheta(\zeta|\Omega) = 0$ holds if and only if there exist $g - 1$ points $p_1, \cdots, p_{g-1} \in \Sigma$ such that,*

$$\zeta = p_1 + \cdots p_{g-1} - \Delta(z_0) \qquad \zeta_I = \sum_{i=1}^{g-1} \int_{z_0}^{p_i} \omega_I - \Delta_I(z_0) \quad (D.3.4)$$

where $\Delta_I(z_0)$ are the components of the Riemann vector given by,

$$\Delta_I(z_0) = -\frac{1}{2} - \frac{1}{2}\Omega_{II} + \sum_{J \neq I} \oint_{\mathfrak{A}_J} \omega_J(z) \int_{z_0}^{z} \omega_I \qquad (D.3.5)$$

for base-point z_0. The combination ζ is independent of the base-point.

The proof of the Riemann vanishing theorem may be found in any book on ϑ-functions, or in the physics literature in [DP88]. For the special case of genus $g = 1$, the set of points p_i is empty, and the Riemann vector reduces to $\Delta = -\frac{1}{2} - \frac{1}{2}\tau$ which merely states that $\vartheta(\frac{1}{2} + \frac{1}{2}\tau|\tau) = 0$, a fact that we have established long ago.

An important application of the Riemann vanishing theorem is to the existence of the following holomorphic $(1,0)$-forms, defined for an arbitrary odd spin structure ν by,

$$\omega_\nu(z) = \sum_{I=1}^{g} \omega_I(z)\partial^I \vartheta[\nu](0|\Omega) \qquad (D.3.6)$$

Its $2(h-1)$ zeros are all double zeros at points p_i such that $\zeta_I = \nu_I$ in (D.3.4). Therefore, $\omega_\nu(z)$ admits a holomorphic square root $h_\nu(z)$. Since the square of h_ν is $\omega_\nu(z)$, which is a section of the canonical bundle, h_ν is a section of a spin bundle with spin structure ν. For each odd spin structure ν, the holomorphic $(\frac{1}{2}, 0)$-form h_ν is unique up to a sign factor. In the special case of genus one, there is a unique odd spin structure ν and we have $\omega_\nu(z) = \vartheta_1'(0|\tau)dz$ so that $h_\nu(z) = \vartheta_1'(0|\tau)^{\frac{1}{2}} dz^{\frac{1}{2}}$.

D.4 The prime form

To construct the *prime form*, we consider an arbitrary odd spin structure ν. Its associated holomorphic section $h_\nu(z)$ of the spin bundle with spin structure ν has $g-1$ simple zeros. The prime form is a $(-\frac{1}{2},0)$-form in both variables z and w, defined by,

$$E(z,w|\Omega) = \frac{\vartheta[\nu](z-w|\Omega)}{h_\nu(z)h_\nu(w)} \qquad (D.4.1)$$

The argument $z-w$ of the ϑ-function stands for the Abel map of (D.3.3) with divisor $D = z-w$. The form $E(z,w|\Omega)$ defined this way is independent of ν. It is holomorphic in z and w and has a unique simple zero at $z=w$. It is double-valued when z is moved around $\mathfrak{A}_I$ cycles and has nontrivial monodromy when $z \to z+\mathfrak{B}_I$ is moved around $\mathfrak{B}_I$ cycles,[1]

$$E(z+\mathfrak{A}_I, w|\Omega) = -E(z,w|\Omega)$$

$$E(z+\mathfrak{B}_I, w|\Omega) = -\exp\left(-i\pi\Omega_{II} - 2\pi i \int_w^z \omega_I\right) E(z,w|\Omega) \qquad (D.4.2)$$

Thus, $E(z,w|\Omega)$ is properly defined on the simply connected covering space of $\Sigma \times \Sigma$, or it may be defined in a fundamental domain. For genus one, the prime form reduces to $E(z,w|\tau) = \vartheta_1(z-w|\tau)/\vartheta_1'(0|\tau)dz^{-\frac{1}{2}}dw^{-\frac{1}{2}}$.

D.5 Holomorphic differentials

Combining the Riemann–Roch Theorem C.2.1 and the vanishing Theorem C.3.1, we obtained the dimension of the space of holomorphic $(n,0)$-forms in (C.3.3) and (C.3.4), which we shall denote by $\Upsilon(n)$. A basis of holomorphic differentials is denoted by $\phi_a^{(n)}$, $a = 1,\cdots,\Upsilon(n)$. They are holomorphic sections of the line bundles K^n, the nth power of the canonical bundle K, for which the number of zeros is given by,

$$\left|\left\{\text{zeros of } \phi_a^{(n)}\right\}\right| = c_1(K^n) = 2n(g-1) \qquad (D.5.1)$$

For a given a, there are $\Upsilon(n) - 1$ zeros fixed to be z_b with $b \neq a$, and another g zeros due to the Riemann vanishing theorem, giving a total of $2n(g-1)$, in agreement with the value of the first Chern class. For $n = 0$, they are constants, for $n = 1/2$ and ν odd they are denoted by $h_\nu(z)$, while for $n = 1$ they are the holomorphic Abelian differentials ω_I, for $I = 1,\cdots,g$.

---

[1] We note that the sign of the second term in the argument of the exponential is opposite to that of formula (6.55a) in [DP88], which was incorrect.

Given any set of $\Upsilon(n)$ points $z_1, \cdots, z_{\Upsilon(n)}$ on the surface, we may choose a basis $\hat{\phi}_a^{(n)}$ for the holomorphic n-differentials normalized at the points z_b,

$$\hat{\phi}_a^{(n)}(z_b) = \delta_a^b \qquad \text{(D.5.2)}$$

The holomorphic differentials with this normalization may be exhibited explicitly in terms of the prime form $E(z, w)$, the $g/2$-differential $\sigma(z)$, and the ϑ-functions. For $n \geq 3/2$, we have,

$$\hat{\phi}_a^{(n)}(z) = \frac{\vartheta[\delta](z - z_a + D_n)}{\vartheta[\delta](D_n)} \frac{\prod_{b \neq a} E(z, z_b)}{\prod_{b \neq a} E(z_a, z_b)} \frac{\sigma(z)^{2n-1}}{\sigma(z_a)^{2n-1}} \qquad \text{(D.5.3)}$$

where $D_n = \sum z_b - (2n - 1)\Delta$ and $\sigma(z)$ is a holomorphic $(\frac{g}{2}, 0)$-form without zeros or poles defined, up to a constant, by the following ratio,

$$\frac{\sigma(z)}{\sigma(w)} = \frac{\vartheta(z - \sum p_i + \Delta)}{\vartheta(w - \sum p_i + \Delta)} \prod_{i=1}^{g} \frac{E(w, p_i)}{E(z, p_i)} \qquad \text{(D.5.4)}$$

where p_i, $i = 1, \cdots, g$ are arbitrary points on the surface. Note that $\sigma(z)$ is single valued around $\mathfrak{A}_I$ cycles but multivalued around $\mathfrak{B}_I$ cycles in the following way,

$$\sigma(z + \mathfrak{B}_I) = \sigma(z) \exp\{-i\pi(g - 1)\Omega_{II} + 2\pi i \Delta_{Iz}\} \qquad \text{(D.5.5)}$$

Besides the $\Upsilon(n) - 1$ zeros z_b, $b \neq a$, the differential $\hat{\phi}_a^{(n)}(z)$ has g additional zeros. The differential $\hat{\phi}_a^{(n)}(z)$ is an $(n, 0)$-form in z, a $(-n, 0)$-form in z_a and a $(0, 0)$-form z_b with $b \neq a$. For $n = 1$, we have,

$$\hat{\phi}_a^{(1)}(z) = \frac{\vartheta(z - z_a + D_1)}{\vartheta(D_1)} \frac{E(z_a, w_0)\sigma(z)}{E(z, w_0)\sigma(z_a)} \prod_{b \neq a} \frac{E(z, z_b)}{E(z_a, z_b)} \qquad \text{(D.5.6)}$$

where $D_1 = \sum z_b - w_0 - \Delta$.

D.6 Meromorphic differentials

Meromorphic differentials play the role of Green functions for the Cauchy–Riemann operators acting on holomorphic line bundles. The Green function $G_n(z, w) = G_n(z, w; z_1, \cdots, z_{\Upsilon(n)})$ for the operator $\partial_{\bar{z}}^{(n)}$ is a meromorphic $(n, 0)$ form in z and $(1 - n, 0)$-form in w. When $n \geq 3/2$ for general spin structure δ, and for $n = 1/2$ and even spin structure δ and generic moduli for $g \geq 3$, the Green function G_n may be defined by the following relations,[2]

$$\partial_{\bar{z}}^{(n)} G_n(z, w) = +2\pi\delta(z, w)$$

$$\partial_{\bar{w}}^{(1-n)} G_n(z, w) = -2\pi\delta(z, w) + 2\pi \sum_{a=1}^{\Upsilon(n)} \hat{\phi}_a^{(n)}(z)\delta(w, z_a) \qquad \text{(D.6.1)}$$

The properly normalized holomorphic n-differentials $\hat{\phi}_a^{(n)}$ are defined in (D.5.2) and (D.5.3). Setting $z = z_a$, we have $\partial_{\bar{w}}^{(1-n)} G_n(z_a, w) = 0$, so that $G_n(z_a, w) = 0$. Explicit expressions for the Green's function are,

$$G_n(z, w) = \frac{\vartheta[\delta]\big(z - w + D_n\big)}{\vartheta[\delta]\big(D_n\big)} \frac{\sigma(z)^{2n-1}}{E(z, w)} \prod_a \frac{E(z, z_a)}{E(w, z_a)} \qquad \text{(D.6.2)}$$

where $D_n = \sum z_b - (2n-1)\Delta$. For $n = 1/2$, and generic moduli when $g \geq 3$, this reduces to the Szegö kernel, usually denoted by,

$$S_\delta(z, w) = \frac{\vartheta[\delta]\big(z - w\big)}{\vartheta[\delta]\big(0\big) E(z, w)} \qquad \text{(D.6.3)}$$

For $n = 1$, the Green's function $G_1(z, w) = G_1(z, w; z_1, \cdots, z_g, w_0)$ is the Abelian differential of the third kind, satisfying,

$$\partial_{\bar{z}}^{(1)} G_1(z, w) = 2\pi\delta(z, w) - 2\pi\delta(z, w_0)$$

$$\partial_{\bar{w}}^{(0)} G_1(z, w) = -2\pi\delta(z, w) + 2\pi \sum_{a=1}^{g} \hat{\phi}_a^{(1)}(z)\,\delta(w, z_a) \qquad \text{(D.6.4)}$$

and explicitly given by the following expression,

$$G_1(z, w) = \frac{\vartheta\big(z - w + D_1\big) E(w, w_0)\sigma(z)}{\vartheta\big(D_1\big) E(z, w)E(z, w_0)\sigma(w)} \prod_a \frac{E(z, z_a)}{E(w, z_a)} \qquad \text{(D.6.5)}$$

where $D_1 = -w_0 + \sum z_b - \Delta$. The combination $\partial_z\partial_w \ln E(z, w)$ is a meromorphic differential (Abelian of the second kind) with a single double pole at $z = w$. Its integrals around homology cycles are given by,

$$\oint_{\mathfrak{A}_I} dz\, \partial_z\partial_w \ln E(z, w) = 0$$

$$\oint_{\mathfrak{B}_I} dz\, \partial_z\partial_w \ln E(z, w) = 2\pi i\omega_I(w) \qquad \text{(D.6.6)}$$

and will be of use throughout.

---

[2] For genus $g \geq 3$, nongeneric moduli correspond to the hyperelliptic divisor, in which case the presence of extra zero modes requires modification of the right side of these relations.

D.7 The bc system

A useful unification of the formulas for holomorphic and meromorphic forms is provided by the conformal field theory correlators of anticommuting fields b and c of weights $(n, 0)$ and $(1 - n, 0)$.[3] This quantum field theory was already introduced in Section 5.2 for a Riemann surface of genus 1, namely, a torus. Here, we shall generalize the correlators to arbitrary genus $g \geq 2$. Since the roles of b and c are swapped by letting $n \to 1 - n$, we shall restrict to considering the cases $n \geq \frac{1}{2}$ without loss of generality.

As in the case of the torus, the fields b and c are locally holomorphic,

$$\partial_{\bar{z}} b = \partial_{\bar{z}} c = 0 \tag{D.7.1}$$

However, as b and c fields approach one another poles may develop. The poles may be represented schematically by the singular contribution to the operator product expansion, given by a simple pole with unit residue,

$$b(z) \, c(w) \sim \frac{1}{z - w} \tag{D.7.2}$$

All poles must arise from these operator coincidences. In view of the anti-commuting nature of the fields, the operator product of two like fields produces a simple zero,

$$b(z_1) b(z_2) \sim (z_1 - z_2) \, \partial b(z_2) \, b(z_2)$$
$$c(w_1) c(w_2) \sim (w_1 - w_2) \, \partial c(w_2) \, c(w_2) \tag{D.7.3}$$

In the case of the torus, we used Fourier analysis to solve for the fields and their correlators. Since higher-genus Riemann surfaces have no continuous isometries, such as translation symmetry, Fourier analysis is not applicable.

Now consider the correlator of an arbitrary number of the b and c fields of weight $(n, 0)$ and $(1 - n, 0)$, respectively,

$$\langle b(z_1) \cdots b(z_N) \, c(w_1) \cdots c(w_M) \rangle \tag{D.7.4}$$

Viewed as a function of z_1, the correlator has simple zeros at $z_2, \cdots, z_N$ and simple poles at $w_1, \cdots, w_M$. But since all poles arise from operator coincidences, and no other poles can appear, the number of poles in z_1 is exactly M, and thus the n-form must have $M + 2n(g-1)$ zeros. Now $N-1$ of these zeros are specified by the positions of the points $z_2, \cdots, z_N$, leaving an

---

[3] The bc system may also be considered for commuting fields, in which case standard physics terminology designates it as the $\beta\gamma$ system [FMS86]. Its mathematical foundation is more involved than for the case of anticommuting fields and was clarified recently in [Wit12].

extra g zeros due to the Riemann vanishing theorem, for a total of $N-1+g$ zeros. Equating the number of zeros computed both ways gives,

$$N = M + (2n-1)(g-1) \qquad (D.7.5)$$

In other words, the correlator must vanish unless the number of b and c fields satisfies this relation.

D.7.1 The case $n = \frac{1}{2}$ with even spin structure

In the case $n = \frac{1}{2}$ with even spin structure δ, and generic moduli, neither the b nor the c field has zero modes, and (D.7.5) shows that we must have $M = N$. By using Wick contractions, the correlator is given by,

$$\langle b(z_1) \cdots b(z_M)\, c(w_M) \cdots c(w_1)\rangle = \det S_\delta(z_i, w_j) \qquad (D.7.6)$$

where S_δ is the Szegö kernel for even spin structure δ, the determinant is taken of the $M \times M$ matrix $S(z_i, w_j)$, and the numbering of the points w_j has been chosen in descending order so that the sign multiplying the determinant on the right side is positive. An alternative formula is obtained by matching poles and zeros in each variable, and we obtain,

$$\langle b(z_1) \cdots b(z_M)\, c(w_M) \cdots c(w_1)\rangle$$
$$= \frac{\vartheta[\delta](D|\Omega) \prod_{i<j} E(z_i, z_j) \prod_{i<j} E(w_i, w_j)}{\vartheta[\delta](0|\Omega) \prod_{i,j} E(z_i, w_j)} \qquad (D.7.7)$$

for the divisor $D = (z_1 - w_1) + \cdots + (z_M - w_M)$. This formula is referred to as the *bosonized correlator* as it may be obtained by representing the spin-$\frac{1}{2}$ fermion fields b and c in terms of the chiral half of a complex scalar field.

D.7.2 The case $n = \frac{1}{2}$ with odd spin structure

In the case $n = \frac{1}{2}$ with odd spin structure ν, and generic moduli, there is a single zero mode h_ν of the Dirac equation. Clearly, the 2-point correlator $\langle b(z)c(w)\rangle$ must be saturated by the zero modes and be proportional to $h_\nu(z)h_\nu(w)$. For higher point functions, we need a Green function which may be defined by an equation that is orthogonal to the Dirac zero mode. Such an equation is not unique, but different definitions will produce the same correlator. Preserving meromorphicity, we choose the following definition,

$$\partial_{\bar{z}} S_\nu(z, w) = 2\pi\delta(z, w) - 2\pi\delta(z, w_0)\frac{h_\nu(w)}{h_\nu(w_0)} \qquad (D.7.8)$$

which is orthogonal to $h_\nu(z)$. The general correlator is then given by,

$$\langle b(z_1) \cdots b(z_M) c(w_M) \cdots c(w_1) \rangle$$
$$= \sum_{i,j=1}^{M} (-)^{i+j} \langle b(z_i)c(w_j) \rangle \det S_\nu(z_k, z_l) \Big|_{\substack{k \neq i, \\ l \neq j}} \qquad \text{(D.7.9)}$$

The constant of proportionality in $\langle b(z)c(w) \rangle \sim h_\nu(z)h_\nu(w)$ is given by the functional determinant of the Dirac operator on $n = \frac{1}{2}$ sections with odd spin structure ν.

D.7.3 The case $n = 1$

In any nonvanishing correlator for $n = 1$, the b field has g zero modes, while the c field has one zero mode, namely, the constants. We shall normalize the correlator so that the simplest nonzero amplitude is given by,

$$\mathcal{G}_0(z_1, \cdots, z_g; w) = \langle b(z_1) \cdots b(z_g)c(w) \rangle = \det \omega_I(z_i) \qquad \text{(D.7.10)}$$

where the determinant is taken of the $g \times g$ matrix with entries $\omega_I(z_i)$ for $i, I = 1, \cdots, g$. The correlator with an arbitrary number of b and c fields subject to the condition (D.7.5),

$$\mathcal{G}_M(z_1, \cdots w_{M+1}) = \langle b(z_1) \cdots b(z_{g+M}) c(w_1) \cdots c(w_{M+1}) \rangle \qquad \text{(D.7.11)}$$

with this normalization is given by,

$$\mathcal{G}_M(z_1, \cdots, w_{M+1}) = \frac{\vartheta(D)}{Z^3} \frac{\prod_{i<j} E(z_i, z_j) \prod_{a<b} E(w_a, w_b) \prod_i \sigma(z_i)}{\prod_{i,a} E(z_i, w_a) \prod_a \sigma(w_a)} \qquad \text{(D.7.12)}$$

where $i, j = 1, \cdots, g + M$ and $a, b = 1, \cdots, M + 1$ and D is defined by

$$D = \sum_{i=1}^{g+M} z_i - \sum_{a=1}^{M+1} w_a - \Delta \qquad \text{(D.7.13)}$$

The factor Z is the chiral boson partition function given by setting $M = 0$,

$$\det \omega_I(p_i) = \frac{\vartheta(p_1 + \cdots p_g - q - \Delta)}{Z^3} \frac{\prod_{i<j} E(p_i, p_j) \prod_i \sigma(p_i)}{\prod_i E(p_i, q)\sigma(q)} \qquad \text{(D.7.14)}$$

for arbitrary points q and p_i with $i = 1, \cdots, g$. With this normalization, the correlator transforms under modular transformations by,

$$\tilde{\mathcal{G}}_M(z_1, \cdots, w_{M+1}) = \det(C\Omega + D)^{-1}\mathcal{G}_M(z_1, \cdots, w_{M+1}) \qquad \text{(D.7.15)}$$

D.7.4 The case $n \geq \frac{3}{2}$

For $n \geq \frac{3}{2}$, the b field has $\Upsilon = (2n - 1)(g - 1)$ zero modes, while the c field has no zero modes. We have the following bosonization formula,

$$\langle b(z_1) \cdots b(z_{M+\Upsilon}) \, c(w_1) \cdots c(w_M) \rangle$$

$$= \frac{\vartheta[\delta](D_n | \Omega) \prod_{a<b} E(z_a, z_b) \prod_{i<j} E(w_i, w_j) \prod_a \sigma(z_a)^{2n-1}}{Z \qquad \prod_{i,a} E(z_a, w_i) \prod_i \sigma(w_i)^{2n-1}} \qquad \text{(D.7.16)}$$

where $a, b = 1, \cdots, M + \Upsilon$, while $i, j = 1, \cdots, M$ and ζ is given by,

$$D_n = \sum_{a=1}^{M+\Upsilon} z_a - \sum_{i=1}^{M} w_i - (2n - 1)\Delta \qquad \text{(D.7.17)}$$

In the special case where $M = 0$, we recover a purely holomorphic correlator, which is entirely saturated by the zero modes of the field b,

$$\langle b(z_1) \cdots b(z_\Upsilon) \rangle = \frac{\vartheta[\delta](D_n | \Omega)}{Z} \prod_{a<b} E(z_a, z_b) \prod_a \sigma(z_a)^{2n-1} \qquad \text{(D.7.18)}$$

which is an unnormalized version of the formula (D.5.3) for holomorphic differentials of weight $(n, 0)$.

• Bibliographical notes

Classic introductions to Siegel modular forms and higher rank modular geometry may be found in the books by Siegel [Sie89a, Sie89b], Igusa [Igu12], and Klingen [Kli90]. Useful lecture notes on higher rank modular forms are those by van der Geer [VDG08]. The connection between higher rank ϑ-functions and Riemann surfaces is spelled out in the classic book by Fay [Fay06]. The connection between higher rank ϑ-functions and nonlinear equations may be found in the survey article by Dubrovin [Dub81]. Early uses of Riemann ϑ-functions in string perturbation theory are discussed in [AGMV86], [Moo86], and [BKMP86]. The solution of the bc quantum system in terms of ϑ-functions was given in [VV87], building on the work of [Fay06].

Appendix E

Solutions to exercises

In this final appendix, we shall present solutions to the exercises listed at the end of each chapter. Equations will remain unnumbered in this final chapter, unless they are being referred to in the text.

Solution to Exercise 2.1

The integral representation for $\Gamma(1+z)$, given in (2.1.17), is absolutely convergent for $\text{Re}(z) > -1$. It's analytic continuation, given by the relation (2.1.18), shows that $\Gamma(1+z)$ has no zeros and that the only singularities are single poles at $-z \in \mathbb{N}$. Therefore poles match on both sides of (2.11.13). Using the Sterling formula for $z \to \infty$, namely, $\ln\Gamma(1+z) = (z+1)\ln(z+1) - z + \mathcal{O}(z^0)$, the left side of (2.11.13) behaves as $1/z$ as $z \to \infty$, as does the right side. Thus, the difference of the two sides is holomorphic in $\mathbb{C}$, vanishing at ∞, and therefore zero by Liouville's theorem. The right side of (2.11.13) has radius of convergence $|z| < 1$. Carrying out the expansion term-by-term in powers of z and then integrating in z twice, we find,

$$\ln\Gamma(1+z) = az + b + \sum_{k=2}^{\infty} \frac{(-z)^k}{k}\zeta(k)$$

The integration constants a and b are determined by using $\Gamma(1) = 1$ and $\Gamma'(1) = -\gamma$ and so that $b = 0$ and $a = -\gamma$, where γ is Euler's constant.

Solution to Exercise 2.2

The functions $\wp(2z|\tau)$, $\wp(z|\frac{\tau}{2})$, and $\wp(2z|\frac{\tau}{2})$ are even in z and have periods $(\frac{1}{2}, \frac{\tau}{2})$, $(1, \frac{\tau}{2})$, and $(\frac{1}{2}, \frac{\tau}{4})$, respectively. They are thus even doubly periodic functions with periods $(1, \tau)$ and may be expressed as rational functions functions of $\wp(z|\tau)$.

The expression for $\wp(2z|\tau)$ is given by the duplication formula, which follows from taking the limit $z_P, z_Q \to z$ in the addition formula (2.9.1), and then using the differential equation for $\wp$ to eliminate its derivatives,

$$\wp(2z|\tau) = \frac{1}{4}\wp(z|\tau) + \frac{12g_2(\tau)\wp(z|\tau)^2 + 36g_3(\tau)\wp(z|\tau) + g_2(\tau)^2}{16(4\wp(z|\tau)^2 - g_2(\tau)\wp(z|\tau) - g_3(\tau))}$$

An alternative derivation uses the fact that $\wp(2z|\tau)$ has double poles at the half-periods, whose normalizations are readily evaluated,

$$\wp(2z|\tau) = \frac{1}{4}\wp(z|\tau) + \frac{1}{16}\sum_{\alpha=1}^{3}\frac{12e_\alpha(\tau)^2 - g_2(\tau)}{\wp(z|\tau) - e_\alpha(\tau)}$$

Using the values $e_\alpha(\tau)$ of $\wp(z|\tau)$ at the half-periods of (2.4.10), the two expressions are found to agree upon combining the three fractions and using the relations of (2.4.12) between g_2, g_3, and e_α.

The function $\wp(z|\frac{\tau}{2})$ has double poles at 0 and $\frac{\tau}{2}$ so that we have,

$$\wp(z|\tfrac{\tau}{2}) = \wp(z|\tau) + \wp(z + \tfrac{\tau}{2}|\tau) - \wp(\tfrac{\tau}{2}|\tau)$$

where the z-independent term is determined by requiring the constant term in the Laurent expansion at $z = 0$ to vanish on both sides. We may now again use the addition theorem (2.9.1) to combine all terms as follows,

$$\wp(z|\tfrac{\tau}{2}) = \frac{(\wp(z|\tau) - e_1(\tau))(\wp(z|\tau) - e_3(\tau))}{\wp(z|\tau) - e_2(\tau)} - 2e_2(\tau) \qquad (E.1)$$

The expression for $\wp(2z|\frac{\tau}{2})$ may be obtained by combining the results for $\wp(2z|\tau)$ and $\wp(z|\frac{\tau}{2})$.

Solution to Exercise 2.3

The function $\wp(z|2\tau)$ has period $(1, 2\tau)$ but is not doubly periodic with periods $(1, \tau)$ since it has a double pole at $z = 0$, while it is regular at $z = \tau$. Therefore it cannot be expressed as a *rational function* of $\wp(z|\tau)$. Instead, $\wp(z|2\tau)$ satisfies a *quadratic equation* whose coefficients are rational functions of $\wp(z|\tau)$, which may be obtained by letting $\tau \to 2\tau$ in (E.1),

$$\wp(z|\tau) = \frac{(\wp(z|2\tau) - e_1(2\tau))(\wp(z|2\tau) - e_3(2\tau))}{\wp(z|2\tau) - e_2(2\tau)} - 2e_2(2\tau)$$

Solution to Exercise 2.4

Expressing the right side of (2.6.12) in terms of the infinite sum representation of the Jacobi ϑ-functions with $\zeta_i' = (T\zeta)_i$, and arranging the summation and the summand in matrix notation, we obtain,

$$\prod_{i=1}^{4} \vartheta[\lambda](\zeta_i'|\tau) = \sum_{N \in \mathbb{Z}^4} \exp \left\{ i\pi\tau (N + \hat{\lambda}')^t (N + \hat{\lambda}') \right.$$

$$\left. + 2\pi i (N + \hat{\lambda}')^t (T\zeta + \hat{\lambda}'') \right\}$$

where we shall use the notations,

$$M = \begin{pmatrix} m_1 \\ m_2 \\ m_3 \\ m_4 \end{pmatrix} \qquad N = \begin{pmatrix} n_1 \\ n_2 \\ n_3 \\ n_4 \end{pmatrix} \qquad \hat{\lambda}' = \begin{pmatrix} \lambda' \\ \lambda' \\ \lambda' \\ \lambda' \end{pmatrix} \qquad \hat{\lambda}'' = \begin{pmatrix} \lambda'' \\ \lambda'' \\ \lambda'' \\ \lambda'' \end{pmatrix}$$

Recall that the matrix T satisfies $T^t = T$ and $T^2 = I$. We now change summation variables, for given λ, as follows,

$$M + \hat{\kappa}' = T(N + \hat{\lambda}') \qquad \Leftrightarrow \qquad T(M + \hat{\kappa}') = N + \hat{\lambda}'$$

where the column matrix $\hat{\kappa}'$ is defined in terms of κ' as $\hat{\lambda}'$ is in terms of λ'. To a given value of N and λ' corresponds a unique pair M and $\kappa' \in \{0, \frac{1}{2}\}$, as may be verified by writing out the left relation in components,

$$m_1 + \kappa' = \tfrac{1}{2}(n_1 + n_2 + n_3 + n_4) + 2\lambda'$$
$$m_2 + \kappa' = \tfrac{1}{2}(n_1 + n_2 - n_3 - n_4)$$
$$m_3 + \kappa' = \tfrac{1}{2}(n_1 - n_2 + n_3 - n_4)$$
$$m_4 + \kappa' = \tfrac{1}{2}(n_1 - n_2 - n_3 + n_4) \qquad \text{(E.2)}$$

Multiplying the first line by 2 and using the fact that $4\lambda'$ is an even integer, we obtain a unique determination for κ' in terms of N,

$$2\kappa' \equiv n_1 + n_2 + n_3 + n_4 \qquad (\text{mod}\, 2)$$

It then follows that $m_1, m_2, m_3,$ and m_4 are integers, which are uniquely defined by the above equations. As a result, for given λ', the map between $N \in \mathbb{Z}^4$ and the pair (M, κ') with $M \in \mathbb{Z}^4$ and $\kappa' \in \{0, \frac{1}{2}\}$ is bijective and we may replace the sum over N by the sum over M and κ',

$$\prod_{i=1}^{4} \vartheta[\lambda](\zeta_i'|\tau) = \sum_{\kappa'} \sum_{M \in \mathbb{Z}^4} \exp \left\{ i\pi\tau (M + \hat{\kappa}')^t (M + \hat{\kappa}') \right.$$

$$\left. + 2\pi i (M + \hat{\kappa}')^t (\zeta + T\hat{\lambda}'') \right\}$$

Since $T\hat{\lambda}''$ has integer entries, we have,

$$e^{2\pi i (M + \hat{\kappa}')^t T \hat{\lambda}''} = e^{4\pi i \kappa' \lambda''} = \langle \kappa | \lambda \rangle e^{4\pi i \kappa'' \lambda'}$$

where $\kappa'' \in \{0, \frac{1}{2}\}$ is arbitrary. Summing also over κ'' and dividing by a factor of 2 leaves the expression unchanged. As a result, we have,

$$e^{2\pi i(M+\hat{\kappa}')^t(\zeta+T\hat{\lambda}'')} = \langle \kappa|\lambda \rangle \, e^{2\pi i(M+\hat{\kappa}')^t(\zeta+\hat{\kappa}'')} \tag{E.3}$$

where we have used the fact that the sum of the four equations in (E.2) is always an even integer. The resulting sum over M, κ', and κ'' gives the promised Riemann relations (2.6.12).

Solution to Exercise 2.5

To prove (2.11.14), we shall begin by applying the Eisenstein summation prescription to the series representation of Weierstrass ζ-function $\zeta(z|\tau)$,

$$\zeta(z|\tau) = \sum_{n\in\mathbb{Z}} \frac{1}{z+n} + \sum_{m=1}^{\infty}\sum_{n\in\mathbb{Z}} \left(\frac{1}{z+m\tau+n} + \frac{1}{z-m\tau-n} \right) + zG_2(\tau)$$

Carrying out the summations over n using (2.2.2) and (2.2.3), we obtain,

$$\zeta(z|\tau) = \pi \frac{\cos \pi z}{\sin \pi z} - 2\pi i \sum_{m=1}^{\infty} \left(\frac{1}{1-q^m e^{2\pi i z}} - \frac{1}{1-q^{-m}e^{-2\pi i z}} \right) + zG_2(\tau)$$

Using the elementary derivatives,

$$\partial_z \ln(1 - q^m e^{\pm 2\pi i z}) = \pm 2\pi i \left(1 - \frac{1}{1 - q^m e^{\pm 2\pi i z}} \right)$$

the second relation in (2.6.21), and the formula in (2.6.22), we obtain,

$$\partial_z \ln \vartheta_1(z|\tau) = \pi \frac{\cos \pi z}{\sin \pi z} + \partial_z \sum_{m=1}^{\infty} \left(\ln(1 - q^m e^{2\pi i z}) + \ln(1 - q^m e^{-2\pi i z}) \right)$$

$$+ \left(G_2(\tau) + \frac{\vartheta_1'''(0|\tau)}{\vartheta_1'(0|\tau)} \right) z \tag{E.4}$$

Since the left side and the first line on the right side are invariant under $z \to z+1$, the second line on the right side must also be invariant. This requires the following relation,

$$G_2(\tau) = -\frac{\vartheta_1'''(0|\tau)}{\vartheta_1'(0|\tau)} \tag{E.5}$$

a formula to which we shall return in Chapter 3 and interpret in terms of quasi-modular forms. Integrating (E.4) in z, subject to (E.5), and matching the behavior as $z \to 0$ gives the promised product formula (2.11.14).

Solution to Exercise 3.1

Since $\mathcal{H} = \mathrm{SL}(2,\mathbb{R})/\mathrm{SO}(2)$ is a Kähler manifold, its affine connection in local complex coordinates $z = x + iy$ has only a single component $\Gamma^z_{zz} = i/y$ and its complex conjugate, so that the geodesic equation is $y\ddot{z} + i\dot{z}^2 = 0$, where $\dot{z} = dz/dt$ for an arbitrary parameter t. The equation is invariant under the isometries of the metric, namely, translations in x and simultaneous dilations in x and y and thereby produces two first integrals.

The real part produces the first integral $\dot{x} = cy^2$. The case $c = 0$ yields vertical geodesics at arbitrary values of x. For $c \neq 0$, the other first integral is given by $(x - x_0)\dot{x} + y\dot{y} = 0$ for an arbitrary value of $x_0 \in \mathbb{R}$. Its integral is a circle of arbitrary radius R, centered at an arbitrary point $x_0 \in \mathbb{R}$,

$$(x - x_0)^2 + y^2 = R^2$$

A point on the geodesic may be parametrized by an angle θ so that $z = x_0 + Re^{i\theta}$, in terms of which the metric along the geodesic takes the form $ds = d\theta/\sin\theta$. The distance between two points z_1 and z_2 on this geodesic is obtained by integrating the metric and given by,

$$\ell(z_1, z_2) = \ln\tan\frac{\theta_2}{2} - \ln\tan\frac{\theta_1}{2} \qquad \theta_1 < \theta_2$$

In terms of the $\mathrm{SL}(2,\mathbb{R})$-invariant *cordal variable* $u(z_1, z_2)$,

$$u(z_1, z_2) = \frac{|z_1 - z_2|^2}{2(\mathrm{Im}\,z_1)(\mathrm{Im}\,z_2)} = -2\frac{(z_1 - z_2)(\bar{z}_1 - \bar{z}_2)}{(z_1 - \bar{z}_1)(z_2 - \bar{z}_2)}$$

one has $\cosh\ell(z_1, z_2) = 1 + u(z_1, z_2)$, To obtain the area of a hyperbolic triangle, we first compute the area inside the semi-infinite triangle D_β of opening angles $\alpha = \frac{\pi}{2}$ and $\beta > 0$ as shown in Figure 3.1, with the left vertical edge translated to $x = x_0 = 0$. The area is then given by the integral,

$$\mathrm{area}(D_\beta) = \int_0^{R\cos\beta} dx \int_{\sqrt{R^2-x^2}}^\infty \frac{dy}{y^2} = \frac{\pi}{2} - \beta$$

Formula (3.2.4) follows by combining two hyperbolic right triangles.

Solution to Exercise 3.2

(a) Recasting the positive definite quadratic form $Q(x, y) = \alpha x^2 + \beta xy + \gamma y^2$ with $\alpha, \gamma > 0$ and discriminant $D = \beta^2 - 4\alpha\gamma < 0$ in the form $Q(x, y) = \alpha|x - \tau y|^2$ with $\mathrm{Im}\,(\tau) > 0$ gives a one-to-one correspondence between quadratic forms for given D and points τ in the Poincaré upper half plane $\tau \in \mathcal{H}$,

$$\tau = -\frac{\beta}{2\alpha} + i\frac{\sqrt{-D}}{2\alpha}$$

where $\sqrt{-D}$ is taken to be the positive square root. The linear transformations on x, y induce the Möbius transformation $\tau \to (a\tau + b)/(c\tau + d)$. Using this correspondence, and Theorem 3.2.1, we see that a unique representative in each equivalence class may be chosen to correspond to a value of τ in the fundamental domain F for $\mathrm{SL}(2, \mathbb{Z})$. To avoid double counting, F contains $\mathrm{Re}\,(\tau) = -\frac{1}{2}$ but not $\mathrm{Re}\,(\tau) = \frac{1}{2}$ and $|\tau| = 1$ with $\mathrm{Re}\,(\tau) \le 0$ but not $|\tau| = 1$ with $\mathrm{Re}\,(\tau) \ge 0$. Thus the unique representatives in each one of the classes with discriminant $D = \beta^2 - 4\alpha\gamma$ satisfy,

$$1 - \alpha \le \beta \le \alpha < \gamma \qquad \text{or} \qquad 0 \le \beta \le \alpha = \gamma$$

(b) Since $\beta^2 \le \alpha^2$ and $4\alpha\gamma = \beta^2 - D$, we have $4\alpha\gamma \le \alpha^2 - D$ and thus $3\alpha^2 \le -D$ so that the number of values taken by α and β for a D is finite and in all cases γ is determined by $\gamma = (\beta^2 - D)/(4\alpha)$. With the help of a computer program, one readily computes $h(D)$ for $-30 < D < 0$,

$$h(D) = 1 \qquad\qquad D = -3, -4, -7, -8, -11, -19$$
$$D = -12, -16, -27, -28$$
$$h(D) = 2 \qquad\qquad D = -15, -20, -24,$$
$$h(D) = 3 \qquad\qquad D = -23$$

while $h(D) = 0$ for all other D. The discriminants $D = -12, -16, -27, -28$ are sometimes referred to as *nonfundamental* since the number of classes when ignoring the restriction to primitive $\gcd(\alpha, \beta, \gamma) = 1$ is larger than 1, while the discriminants $D = -3, -4, -7, -8, -11, -19$ are referred to as *fundamental*. The full list of fundamental discriminants of class number 1 is finite and known to be limited to,

$$D = -3, -4, -7, -8, -11, -19, -43, -67, -163$$

These numbers will arise again when considering complex multiplication and Heegner points in Chapter 11. Gauss considered the problem of evaluating $h(D)$ in his *Disquisitiones Arithmeticae* [Gau01] for the case where β is even, while the present enumeration of a unique representative for each class, and for all β, was given by Zagier in [Zag81b] and [Zag08].

Solution to Exercise 3.3

(a) The sum of the left side is absolutely convergent for $|x| < 1$. Expanding the denominator in a Taylor series gives,

$$\sum_{n=1}^{\infty} \frac{n^s x^n}{1 - x^n} = \sum_{m=1}^{\infty} \sum_{n=1}^{\infty} n^s x^{mn} = \sum_{N=1}^{\infty} x^N \sum_{n|N} n^s$$

In the second equality, we have rearranged the sum by setting $N = mn$. The second sum on the right side is precisely $\sigma_s(N)$, which proves the formula.

(b) The divisor sum for the product of positive integers m and n is given by,

$$\sigma_s(mn) = \sum_{d|mn} d^s$$

where the sum is over all positive integers d dividing mn. If $\gcd(m, n) = 1$ and $d|mn$ then d is of the form $d = d_1 d_2$ where $d_1|m$ and $d_2|n$, so that,

$$\sigma_s(mn) = \sum_{d_1|m} d_1^s \sum_{d_2|n} d_2^s = \sigma_s(m)\,\sigma_s(n)$$

thereby proving that $\sigma_s(m)$ is a multiplicative function in n. The simplest instance that shows that σ_s is not completely multiplicative is $m = 2$ and $n = 6$, in which case we have,

$$\sigma_s(2) = 1^s + 2^s \qquad \sigma_s(6) = 1^s + 2^s + 3^s + 6^s$$
$$\sigma_s(3) = 1^s + 3^s \qquad \sigma_s(12) = 1^s + 2^s + 3^s + 4^s + 6^s + 12^s$$

Taking the product of $\sigma_s(2)$ and $\sigma_s(6)$, we obtain,

$$\sigma_s(2)\sigma_s(6) = 1^s + 2^s + 2^s + 3^s + 4^s + 6^s + 6^s + 12^s$$
$$= \sigma_s(12) + 2^s \sigma_s(3)$$

in accord with the general multiplication formula,

$$\sigma_s(m)\sigma_s(n) = \sum_{d|\gcd(m,n)} d^s\, \sigma_s\left(\frac{mn}{d^2}\right)$$

Solution to Exercise 3.4

The first equation in (3.4.26) is proven by substituting the expressions for $\mathsf{E}_4(\tau)$ and $\mathsf{E}_8(\tau)$ given in (3.4.13) into the relation $\mathsf{E}_4^2 = \mathsf{E}_8$ of (3.4.25) and identifying coefficients of q^n. The second equation in (3.4.26) is proven by substituting the expressions for $\mathsf{E}_4(\tau), \mathsf{E}_6(\tau)$, and $\mathsf{E}_{10}(\tau)$ given in (3.4.13) into the relation $\mathsf{E}_4\mathsf{E}_6 = \mathsf{E}_{10}$ of (3.4.25) and identifying coefficients of q^n.

Solution to Exercise 3.5

The relation may be obtained by expanding the cusp form $\Delta(\tau)$ in powers of q with the help of (3.7.13) and then using (3.4.15) to express the cusp form in terms of $E_{12}(\tau)$ and $E_6(\tau)$ using (3.4.25),

$$12^3 \cdot 21^2 \sum_{N=1}^{\infty} \tau(N)q^N = 691\left(E_{12}(\tau) - E_6(\tau)^2\right)$$

The order q^0 term is absent on both sides. The order q contribution matches in view of $\tau(1) = \sigma_{11}(1) = \sigma_5(1) = 1$ and the identity $12^3 \cdot 21^2 = 65520 + 1008\cdot691$. Using (3.4.13) to expand E_6 and E_{12} in powers of q and identifying the coefficients of the higher powers of q on both sides, we obtain,

$$756\,\tau(N) = 65\,\sigma_{11}(N) + 691\left(\sigma_5(N) - 252\sum_{M=1}^{N-1}\sigma_5(M)\,\sigma_5(N-M)\right)$$

Considering this equation (mod 691) cancels the second and third terms on the right side, while the relation $756 \equiv 65\,(\mathrm{mod}\,691)$ reduces the congruence relation to $65\,\tau(N) \equiv 65\,\sigma_{11}(N)\,(\mathrm{mod}\,691)$. Since $\gcd(65,691) = 1$, this implies Ramanujan's relation $\tau(N) \equiv \sigma_{11}(N)\,(\mathrm{mod}\,691)$.

Solution to Exercise 4.1

Applying the modular transformation γ to $E_2^*(\tau)$ defined by (4.1.10),

$$E_2^*(\gamma\tau) = E_2(\gamma\tau) - \frac{3}{\pi\tau_2}(c\tau + d)(c\bar{\tau} + d)$$

and then using (4.1.3) to evaluate $E_2(\gamma\tau)$ and (4.1.10) to reexpress $E_2(\tau)$ in terms of $E_2^*(\tau)$, we obtain,

$$E_2^*(\gamma\tau) = (c\tau + d)^2\left(E_2^*(\tau) + \frac{12}{2\pi i}\frac{c}{c\tau + d} + \frac{3}{\pi\tau_2} - \frac{3}{\pi\tau_2}\frac{c\bar{\tau} + d}{c\tau + d}\right)$$

The last three terms inside the parentheses cancel to give (4.1.9).

Solution to Exercise 4.2

The operator $\nabla_{\bar{\tau}}$ maps any modular form of weight $(k,\bar{k})$ to a modular form of weight $(k-2,\bar{k})$. Since $f(\tau)$ of (4.7.8) is a modular form of weight $(k,0)$ the modular form $\nabla_{\bar{\tau}}f(\tau)$ has weight $(k-2,0)$. The explicit expression for its successive derivatives,

$$\nabla_{\bar{\tau}}^{\ell} f(\tau) = \left(\frac{i}{2}\right)^{\ell} \sum_{n=0}^{N-\ell} \frac{(n+\ell)!}{n!} \frac{f_{n+\ell}(q)}{(\tau - \bar{\tau})^n}$$

shows that $\nabla_{\bar{\tau}} f(\tau)$ has depth $N-1$ and that $\nabla_{\bar{\tau}}^{\ell} f(\tau) \in \mathcal{A}(k-2\ell, N-\ell)$. Its top component $f_N(q)$ is proportional to $\nabla_{\bar{\tau}}^{N} f$ and is annihilated by $\nabla_{\bar{\tau}}$ so that $f \in \mathrm{Ker} \nabla_{\bar{\tau}}^{N+1}$. Thus, $f_N(q)$ is a holomorphic modular form of weight $(k-2N, 0)$, which vanishes when k is odd, $k = 2N+2$, or $k < 2N$.

Performing a modular transformation γ on (4.7.8), decomposing $c\bar{\tau} + d = c\tau + d - c(\tau - \bar{\tau})$, and identifying terms in powers of $\tau - \bar{\tau}$, we obtain,

$$f_n(q) = \sum_{\ell=n}^{N} \binom{\ell}{n} (-c)^{\ell-n} (c\tau + d)^{n+\ell-k} f_\ell(\gamma q)$$

For example, the highest order terms are as follows,

$$f_N(q) = (c\tau + d)^{2N-k} f_N(\gamma q)$$
$$f_{N-1}(q) = (c\tau + d)^{2N-2-k} f_{N-1}(\gamma q) - cN(c\tau + d)^{2N-1-k} f_N(\gamma q)$$
$$f_{N-2}(q) = (c\tau + d)^{2N-4-k} f_{N-2}(\gamma q) - c(N-1)(c\tau + d)^{2N-3-k} f_{N-1}(\gamma q)$$
$$+ \tfrac{1}{2} c^2 N(N-1)(c\tau + d)^{2N-2-k} f_N(\gamma q)$$

The first line confirms that $f_N(q)$ is a holomorphic modular form of weight $(k-2N, 0)$.

Solution to Exercise 4.3

We begin by proving the first part of Theorem 4.1.1 for an arbitrary almost-holomorphic modular form f of modular weight $(k, 0)$ and depth N, given by the expansion in (4.7.8). In particular, we shall use the result of Exercise 4.2 that the function $f_N(q)$ is a modular form of weight $k - 2N$. We begin by using the relation between $E_2(\tau)$ and $E_2^*(\tau)$ to produce,

$$\frac{1}{(\tau - \bar{\tau})^N} - \left(\frac{i\pi E_2^*}{6}\right)^N = -\sum_{n=0}^{N-1} \binom{N}{n} \left(\frac{i\pi E_2}{6}\right)^{N-n} \frac{1}{(\tau - \bar{\tau})^n}$$

Isolating the term proportional to $f_N(q)$ and using the above formula to re-express its dependence on $\bar{\tau}$, we obtain,

$$f(\tau) = \left(\frac{i\pi}{6}\right)^N E_2^*(\tau)^N f_N(q) + \hat{f}(\tau) \qquad\qquad \hat{f}(\tau) = \sum_{n=0}^{N-1} \frac{\hat{f}_n(q)}{(\tau - \bar{\tau})^n}$$

with the coefficients given by,

$$\hat{f}_n(q) = f_n(q) - \binom{N}{n}\left(\frac{i\pi}{6}\right)^{N-n} E_2(\tau)^{N-n} f_N(q)$$

Since f, $\mathsf{E}_2^*(\tau)$, and $f_N(q)$ are modular forms of weight k, 2, and $k-2N$, respectively, the product $f_N(q)\mathsf{E}_2^*(\tau)^N$ is a modular form of weight k and therefore so is $\hat{f}(\tau)$. Although $\mathsf{E}_2(\tau)$ is expressed as a function of τ, it is actually a function of q only, which justifies the notation $\hat{f}_n(q)$. Recalling that $f_N(q)$ is a holomorphic modular form of weight $k-2N$, it may be expressed as a polynomial in E_4 and E_6. Finally, $\hat{f}(\tau)$ is an almost-holomorphic modular form of depth $N-1$. Thus we have shown that any almost-holomorphic modular form of depth N may be expressed as the sum of an almost-holomorphic modular form of depth $N-1$ plus a term which belongs to the ring generated by $\mathsf{E}_2^*, \mathsf{E}_4$, and E_6. By induction on N we conclude that any almost-holomorphic modular form of arbitrary depth N belongs to the polynomial ring generated by $\mathsf{E}_2^*, \mathsf{E}_4$, and E_6, thereby proving the first part of Theorem 4.1.1. The second part of Theorem 4.1.1 immediately follows by tracing the coefficients $f_0(q)$, $\hat{f}_0(q)$ and so on in the induction procedure, all of which belong to the polynomial ring generated by $\mathsf{E}_2, \mathsf{E}_4$, and E_6. The corresponding grading properties are manifest.

Solution to Exercise 4.4

It will be convenient to recast the differential equations of (4.1.20) for E_2 and (4.1.17) for E_4 as follows,

$$12\,q\frac{d\mathsf{E}_2}{dq} + \mathsf{E}_4 = \mathsf{E}_2^2 \qquad\qquad 3\,q\frac{d\mathsf{E}_4}{dq} + \mathsf{E}_6 = \mathsf{E}_2\mathsf{E}_4$$

and use the expansions for $\mathsf{E}_2, \mathsf{E}_4$ and E_6 in powers of q of (4.1.5) for E_2 and (3.4.13) for E_4 and E_6. Equating the expansions gives the following expressions for $\sigma_3(N)$ in terms of $\sigma_1(N)$, and for $\sigma_5(N)$ in terms of $\sigma_3(N)$ and $\sigma_1(N)$ [BVdGHZ08],

$$5\sigma_3(N) = (6N-1)\sigma_1(N) + 12\sum_{M=1}^{N-1}\sigma_1(N-M)\sigma_1(M)$$

$$21\sigma_5(N) = 10(3N-1)\sigma_3(N) + \sigma_1(N) + 240\sum_{M=1}^{N-1}\sigma_1(N-M)\sigma_3(M)$$

Solution to Exercise 4.5

To show that $G_N(\tau)$ admits a formal expansion in terms of non-holomorphic Eisenstein series, we begin by isolating the $m = n = 0$ part and expanding the part with $(m,n) \neq (0,0)$ in powers of t. Since the polynomial $Q_N(t)$ has degree $2N - 1$ and satisfies $Q_N(t) = t^{2N} Q_N(1/t)$, its constant term must vanish. Therefore, the expansion has the following form,

$$\frac{Q_N(t)}{(t+1)^{2N+1}} = \sum_{p=2}^{\infty} S_N(p) t^{p-1} \tag{E.6}$$

where the constant term is absent. We use the following integrals,

$$\sideset{}{'}\sum_{m,n\in\mathbb{Z}} \int_0^\infty dt\, t^{p-1} e^{-\pi t|m+n\tau|^2/\tau_2} = \Gamma(p) E_p(\tau)$$

As a result, the expansion involves non-holomorphic Eisenstein series of integral weight ≥ 2 only, and we have,

$$G_N(\tau) = \int_0^\infty dt\, \frac{Q_N(t)}{(t+1)^{2N+1}} + \sum_{p=2}^{\infty} S_N(p)\Gamma(p) E_p(\tau)$$

To compute the inner product $g_N(s) = \langle G_N | E^*_{\frac{1}{2}+is} \rangle$, we use the unfolding trick,

$$g_N(s) = 2\zeta^*(1+2is) \int_0^\infty \frac{d\tau_2}{\tau_2^2} \tau_2^{\frac{1}{2}+is} \int_0^1 d\tau_1\, G_N(\tau)$$

compute the τ_1-integral using the Poisson resummation formula (4.2.11), and subsequently carry out the τ_2-integral, to obtain,

$$\int_0^\infty d\tau_2\, \tau_2^{-\frac{3}{2}+is} \int_0^1 d\tau_1 \sum_{m,n\in\mathbb{Z}} e^{-\pi t|m+n\tau|^2/\tau_2} = \frac{2}{\sqrt{t}} \frac{\Gamma(is)\zeta(2is)}{\pi^{is}} \left(t^{is} + t^{-is}\right)$$

Collecting all contributions, we find,

$$g_N(s) = 4\zeta^*(1+2is)\zeta^*(2is) \int_0^\infty dt\, \frac{Q_N(t)}{\sqrt{t}\,(t+1)^{2N+1}} \left(t^{is} + t^{-is}\right)$$

One verifies invariance of the integrand under $t \to 1/t$ using the symmetry properties of $Q_N(t)$, so that one may restrict the integration region to $t \in [0,1]$ upon including a factor of 2. Using the expansion of (E.6), which is convergent for $t \in [0,1]$, we obtain the following formal expansion,

$$g_N(s) = 8\zeta^*(1+2is)\zeta^*(2is) \sum_{p=2}^{\infty} S_N(p) \left(\frac{1}{p+is-\frac{1}{2}} + \frac{1}{p-is-\frac{1}{2}}\right)$$

Solution to Exercise 5.1

We begin by obtaining the singular part of the operator product expansion of the current $j(z)$ with the fields $b(x)$ and $c(x)$ using the OPE of (5.2.12) to define the current as a composite operator as follows,

$$j(z) = \lim_{w \to z} \left(c(z)b(w) - \frac{I}{z - w} \right)$$

The OPEs of interest are then given by,

$$j(z)b(x) = \lim_{w \to z} \left(c(z)b(w)b(x) - \frac{b(x)}{z - w} \right) = -\frac{b(x)}{z - x} + \mathcal{O}\big((z - x)^0\big)$$

$$j(z)c(x) = \lim_{w \to z} \left(c(z)b(w)c(x) - \frac{c(x)}{z - w} \right) = +\frac{c(x)}{z - x} + \mathcal{O}\big((z - x)^0\big)$$

To obtain the rightmost equalities, we proceed as follows. In the first line, $c(z)$ has simple poles at the points w and x of which the first one is cancelled by the subtraction term under the parentheses. In the second line, $b(w)$ has simple poles at z and x of which the first again cancels.

The correlator $A(z; x, y)$ is locally holomorphic in z away from the points $x_1, \cdots, x_n$ and $y_1, \cdots, y_n$ and has simple poles with residue $B(x, y)$ and $-B(x, y)$ at the points x_i and y_i, respectively. In particular, we may write the following Cauchy–Riemann differential equation,

$$\partial_{\bar{z}} A(z; x, y) = \pi B(x, y) \sum_{i=1}^{n} \Big(\delta(z - x_i) - \delta(z - y_i) \Big)$$

known as a *Ward identity* in physics. Since $A(z; x, y)$ is a $(1, 0)$ form in z, and there are no holomorphic 1-forms on the sphere, this equation may be solved uniquely for $A(z; x, y)$ in terms of $B(x, y)$. For the torus, however, there exists a holomorphic 1-form and therefore $A(z; x, y)$ may be partially solved in terms of $B(x, y)$, up to a z-independent part that cannot be determined from this Ward identity alone.

Solution to Exercise 5.2

The transformation rule for the 1-form $j(z)dz$ may be obtained from the conformal transformation rules of the primary fields $b(w)$ and $c(z)$, given by,

$$b'(z')dz' = b(z)dz \qquad\qquad c'(z') = c(z)$$

Using the definitions of $j'(z')dz'$ and the above transformation law, we have,

$$j'(z')dz' = \lim_{w' \to z'} \left(c'(z')b'(w') - \frac{I}{z' - w'} \right) f'(z)dz$$

$$= \lim_{w \to z} \left(\frac{c(z)b(w)}{f'(w)} - \frac{I}{f(z) - f(w)} \right) f'(z)dz$$

Next, using the definition of $j(z)dz$ in terms of the OPE of $c(z)b(w)$,

$$j'(z')dz' = j(z)dz + \lim_{w \to z} \left(\frac{I}{(z - w)f'(w)} - \frac{I}{f(z) - f(w)} \right) f'(z)dz$$

The limit is convergent and may be readily evaluated by expanding both denominators around z, and we find,

$$j'(z')dz' = j(z)dz - I \cdot \frac{f''(z)}{2f'(z)} dz$$

so that the current $j(z)dz$ transforms as a connection rather than as a conformal primary field.

Solution to Exercise 5.3

The constant offset $G_0(\tau)$ in (5.5.15) is determined by requiring the integral of $G(z|\tau)$ to vanish. For the ratio $\vartheta_1(z|\tau)/\vartheta_1'(0|\tau)$, we shall use the infinite product relation of (2.11.14). It will be convenient to use real co-moving coordinates $z = \alpha + \beta\tau$ with $\alpha, \beta \in [-\frac{1}{2}, \frac{1}{2}]$, so that $G_0(\tau)$ is determined by,

$$\int \frac{d^2 z}{\tau_2} G(z|\tau) = \int_{-\frac{1}{2}}^{\frac{1}{2}} d\alpha \int_{-\frac{1}{2}}^{\frac{1}{2}} d\beta \, G(\alpha + \beta\tau|\tau) = 0$$

We begin by showing that,

$$\sum_{n=1}^{\infty} \int_{-\frac{1}{2}}^{\frac{1}{2}} d\alpha \int_{-\frac{1}{2}}^{\frac{1}{2}} d\beta \, \ln(1 - q^{n+\beta} e^{2\pi i\alpha}) = 0$$

Indeed, since $n + \beta \geq \frac{1}{2}$ uniformly in β, we may expand the logarithm in powers of $q^{n+\beta}$. The integral of each term over α vanishes, which proves the assertion. Furthermore, we have,

$$\int_{-\frac{1}{2}}^{\frac{1}{2}} d\alpha \int_{-\frac{1}{2}}^{\frac{1}{2}} d\beta \, \ln \left| \frac{\sin \pi z}{\pi} \right|^2 = \frac{\pi\tau_2}{2} - 2\ln(2\pi)$$

Putting all together gives $G_0(\tau) + 2\ln(2\pi) + \ln|\eta(\tau)|^4 = 0$. Using the relation $\vartheta_1'(0|\tau) = 2\pi\eta(\tau)^3$ then produces (5.5.16).

Solution to Exercise 5.4

By Wick's theorem, we have,

$$\langle \partial_{z_1}\varphi(z_1)\partial_{z_2}\varphi(z_2)\partial_{z_3}\varphi(z_3)\partial_{z_4}\varphi(z_4)\rangle$$
$$= \partial_{z_1}\partial_{z_2}G(z_1 - z_2|\tau)\,\partial_{z_3}\partial_{z_4}G(z_3 - z_4|\tau)$$
$$+\partial_{z_1}\partial_{z_3}G(z_1 - z_3|\tau)\,\partial_{z_2}\partial_{z_4}G(z_2 - z_4|\tau)$$
$$+\partial_{z_1}\partial_{z_4}G(z_1 - z_4|\tau)\,\partial_{z_2}\partial_{z_3}G(z_2 - z_3|\tau)$$

Solution to Exercise 5.5

In case (a) of the torus, the sum of the electric charges q_n must vanish, and the electric potential is then given by,

$$\Phi^{(a)}(z) = \frac{1}{4\pi}\sum_{n=1}^{N} q_n\, G(z - z_n|\tau) \qquad\qquad \sum_{n=1}^{N} q_n = 0$$

where $G(z|\tau)$ is the scalar Green function of (5.5.16). For cases (b) and (c), it will be convenient to use real co-moving coordinates $\alpha, \beta \in [0,1]$ with $z = \alpha + \beta\tau$. The two cases may then be obtained by superposition of mirror charges reflected by $\alpha \to -\alpha$, $\beta \to -\beta$ or both, and we have,

$$\Phi^{(b)}(z) = \frac{1}{4\pi}\sum_{n=1}^{N} q_n\left(G(\alpha + \beta\tau - z_n|2\tau) - G(\alpha - \beta\tau - z_n|2\tau)\right)$$

$$\Phi^{(c)}(z) = \frac{1}{4\pi}\sum_{n=1}^{N} q_n\left(G\big(\tfrac{1}{2}(\alpha + \beta\tau - z_n)|\tau\big) - G\big(\tfrac{1}{2}(-\alpha + \beta\tau - z_n)|\tau\big)\right.$$
$$\left. -G\big(\tfrac{1}{2}(\alpha - \beta\tau - z_n)|\tau\big) + G\big(\tfrac{1}{2}(-\alpha - \beta\tau - z_n)|\tau\big)\right)$$

It is readily verified that $\Phi^{(b)}(z)$ is periodic in α and z with period 1 and that $\Phi^{(b)}(z) = 0$ for $\beta = 0, 1$, namely, for Im $z = 0, \tau_2$. Similarly, $\Phi^{(c)}(z) = 0$ for $\alpha = 0, 1$ as the terms on the first and second lines separately cancel one another, and for $\beta = 0, 1$ as the first terms on the first and second lines cancel one another and separately also the second terms. It is instructive to spell out the points that correspond to the reflections of α and β in terms of the original complex variables $z, \bar{z}$. For case (b) we have,

$$\alpha - \tau\beta = \bar{z} + i\frac{\tau_1}{\tau_2}(z - \bar{z})$$

while the entries for case (c) may be obtained by also using 180° rotations $z \to -z$. For the special case of the rectangle, we have $\tau_1 = 0$ and the images result from complex conjugation and 180° rotations.

Solution to Exercise 6.1

Any element $h \in \Gamma(N)$ may be written as $h = I + N\hat{h}$, where $\hat{h}$ is a 2×2 matrix with integer entries. Conjugating h by an arbitrary $g \in \mathrm{SL}(2, \mathbb{Z})$ gives $ghg^{-1} = I + Ng\hat{h}g^{-1}$. Since $g\hat{h}g^{-1}$ has integer entries, and $\det(ghg^{-1}) = \det h = 1$, we see that $ghg^{-1} \in \Gamma(N)$, so that $\Gamma(N)$ is a normal subgroup of $\mathrm{SL}(2, \mathbb{Z})$. The group $\Gamma(N)$ is the kernel of the homomorphism $\mathbb{Z} \to \mathbb{Z}_N$ of $\mathrm{SL}(2, \mathbb{Z})$, and therefore the quotient $\mathrm{SL}(2, \mathbb{Z})/\Gamma(N)$ is isomorphic to $\mathrm{SL}(2, \mathbb{Z}_N)$.

Solution to Exercise 6.2

The hyperbolic distance $\ell(z_1, z_2)$ between points $z_1, z_2 \in \mathcal{H}$ was computed in Exercise 3.1 and expressed in terms of the cordal variable,

$$u(z_1, z_2) = -2\frac{(z_1 - z_2)(\bar{z}_1 - \bar{z}_2)}{(z_1 - \bar{z}_1)(z_2 - \bar{z}_2)}$$

via the relation $\cosh \ell(z_1, z_2) = 1 + u(z_1, z_2)$. Since the scalar Green function $G_{\mathcal{H}}(z_1, z_2)$ on $\mathcal{H}$ depends only on z_1 and z_2 through the distance $\ell(z_1, z_2)$, we introduce the reduced Green function $g(u)$ as follows,

$$G_{\mathcal{H}}(z_1, z_2) = g\big(u(z_1, z_2)\big)$$

The scalar Laplace operator on $G_{\mathcal{H}}$ then reduces to an operator on g,

$$\Delta G_{\mathcal{H}}(z_1, z_2) = 2(1 + u)g'(u) + u(2 + u)g''(u) = \big(u(2 + u)g'(u)\big)'$$

For $z_1 \neq z_2$, the equation $\Delta G_{\mathcal{H}} = 0$ gives $g(u) = c_1 \ln(1 + 2/u) + c_2$ or for the original Green function $G_{\mathcal{H}}$,

$$G_{\mathcal{H}}(z_1, z_2) = c_1 \ln \frac{|z_1 - z_2|^2}{|z_1 - \bar{z}_2|^2} + c_2$$

where c_1 and c_2 are integration constants. Setting $c_1 = -1$ provides the standard short-distance normalization of the Green function,

$$\Delta G_{\mathcal{H}}(z_1, z_2) = -4\pi\delta(z_1, z_2)$$

Dirichlet boundary conditions on $\mathbb{R}$ require that $G_{\mathcal{H}}(z_1, z_2) = 0$ whenever $z_1 \in \mathbb{R}$ or $z_2 \in \mathbb{R}$, which is achieved by setting $c_2 = 0$. Alternatively, the expression obtained for $G_{\mathcal{H}}$ might have been constructed by using the image charge $\bar{z}_2$ of z_2 to enforce the Dirichlet boundary condition on $\mathbb{R}$.

Solution to Exercise 6.3

(a) To show that the Laplace operator $\Delta = y^2(\partial_x^2 + \partial_y^2)$ on real-valued scalar functions with Dirichlet boundary conditions on $\mathbb{R}$ is self-adjoint, we make use of the Dirichlet boundary conditions to integrate by parts for arbitrary $f, g : \mathcal{H} \to \mathbb{R}$,

$$\langle f | \Delta g \rangle = - \int_{\mathcal{H}} dx\, dy\, \Big((\partial_x f)(\partial_x g) + (\partial_y f)(\partial_y g) \Big) = \langle \Delta f | g \rangle$$

Since Δ is self-adjoint, its eigenvalues are real. From the above formula, it is manifest that Δ is negative so that all its eigenvalues are negative.

(b) Integration by parts in y establishes the equality below [Iwa02],

$$\int_{\mathcal{H}} \frac{dx\, dy}{y^2} f^2 = 2 \int_{\mathcal{H}} \frac{dx\, dy}{y} f \partial_y f \le 4 \int_{\mathcal{H}} dx\, dy\, (\partial_y f)^2 \qquad (E.7)$$

while the inequality is obtained by applying Cauchy's inequality to the integral of the product $f/y \cdot \partial_y f$ against the measure $dx\, dy$. Combining with the first formula for $g = f$, we obtain the inequality,

$$\langle f | \Delta f \rangle \le -\tfrac{1}{4} \langle f | f \rangle$$

so that the spectrum of Δ is bounded from above by $-\tfrac{1}{4}$.

(c) Explicit calculation shows that $\Delta C_s(z, \xi) = s(s-1) C_s(z, \xi)$ so that also f_s is an eigenfunction with eigenvalue $s(s-1)$.

Solution to Exercise 6.4

Since Exercise 6.3 shows that the spectrum of Δ is real and bounded from above by $-\tfrac{1}{4}$, we shall parametrize the eigenvalues of Δ by $-\tfrac{1}{4} - s^2$ with $s \in \mathbb{R}^+$. The simultaneous eigenfunctions $f_{k,s}$ of the operators $-i\partial_x$ and Δ for eigenvalues $k \in \mathbb{R}$ and $\tfrac{1}{4} + s^2$, respectively, satisfy the following equations,

$$-i\partial_x f_{k,s}(x, y) = k \varphi_{k,s}(x, y)$$
$$y^2(\partial_x^2 + \partial_y^2) f_{k,s}(x, y) = - \left(\tfrac{1}{4} + s^2 \right) f_{k,s}(x, y)$$

Solving for the first and substituting the result into the second gives,

$$f_{k,s}(x, y) = e^{ikx} \varphi_{k,s}(y) \qquad\qquad y^2 \varphi_{k,s}'' + (\tfrac{1}{4} + s^2) \varphi_{k,s} = k^2 y^2 \varphi_{k,s}$$

whose solutions are given by modified Bessel functions $y^{\frac{1}{2}} K_{is}(|k|y)$ of imaginary order with $K_{-is} = K_{is}$. Thus, the eigenfunctions are given by,

$$f_{k,s}(x, y) = f_{k,s}^0\, y^{\frac{1}{2}} K_{is}(|k|y)\, e^{ikx} \qquad (E.8)$$

where $f^0_{k,s}$ is an x, y-independent normalization factor. Since $f_{k,s}(x, y) \sim y^{\frac{1}{2} \pm is}$ as $y \to 0$, the overlap integral $\langle f_{k',s'} | f_{k,s} \rangle$ is not generally convergent and should be understood in the sense of distributions. To work this out carefully, we include a regulator factor y^ε with $\varepsilon > 0$ in the product,

$$\langle f_{k',s'} | f_{k,s} \rangle = \lim_{\varepsilon \to 0} \int_{\mathbb{R}} dx \int_0^\infty \frac{dy}{y^2} y^\varepsilon \overline{f_{k',s'}(x,y)} f_{k,s}(x,y)$$

$$= 2\pi \overline{f^0_{k',s'}} f^0_{k,s} \delta(k - k') \int_0^\infty \frac{dy}{y} y^\varepsilon K_{is'}(|k|y) K_{is}(|k|y)$$

The factor of $|k|$ may be scaled out and produces a factor of $|k|^\varepsilon$ which will turn out to be immaterial. Page 693 of the book [GRR88] gives the following formula (setting $a = b = 1$ and $\lambda = 1 - \varepsilon$ in their notation),

$$\int_0^\infty \frac{dy}{y} y^\varepsilon K_\mu(y) K_\nu(y) = \frac{2^\varepsilon}{8\Gamma(\varepsilon)} \prod_{\sigma_1 = \pm 1} \prod_{\sigma_2 = \pm 1} \Gamma\left(\frac{\varepsilon + \sigma_1 \mu + \sigma_2 \nu}{2}\right)$$

In our case, $\mu = is'$ and $\nu = is$ with $s, s' > 0$, the only nonvanishing contribution in the limit $\varepsilon \to 0$ arises from the factors for which $\sigma_1 \sigma_2 = -1$ when $s' - s \to 0$. Making use of the limit,

$$\lim_{\varepsilon \to 0} \left(\frac{1}{\xi - i\varepsilon} - \frac{1}{\xi + i\varepsilon}\right) = 2\pi i \delta(\xi)$$

we obtain,

$$\lim_{\varepsilon \to 0} \int_0^\infty \frac{dy}{y} y^\varepsilon K_{is'}(y) K_{is}(y) = \frac{\pi^2 \delta(s - s')}{2s \sinh(\pi s)}$$

and therefore,

$$\langle f_{k',s'} | f_{k,s} \rangle = \delta(k - k')\delta(s - s') \qquad |f^0_{k,s}|^2 = \frac{s \sinh(\pi s)}{\pi^3} \qquad (E.9)$$

We have obtained an orthonormal set of eigenfunctions of the Laplacian, which corresponds to the principal continuous series of unitary representations of SL$(2, \mathbb{R})$ [GGPS69].

The fact that this set of eigenfunctions is complete and provides a basis is confirmed by the Kontorovitch–Lebedev pair of transforms [Erd81b],

$$g(y) = \int_0^\infty ds f(s) K_{is}(y)$$

$$f(s) = \frac{2s}{\pi^2} \sinh(\pi s) \int_0^\infty \frac{dy}{y} K_{is}(y) g(y)$$

In the next exercise, we shall provide a direct and independent proof of the completeness of this set of eigenfunctions.

Solution to Exercise 6.5

We shall right away evaluate the integral kernel $\mathcal{K}_{\mathcal{H}}^{[h]}(z, z')$ associated with the integral operator $h(-\Delta)$ for an arbitrary function h by using the orthonormal set of eigenfunctions obtained in Exercise 6.4,

$$\mathcal{K}_{\mathcal{H}}^{[h]}(z, z') = \int_{\mathbb{R}} dk \int_0^\infty ds\, h\left(\tfrac{1}{4} + s^2\right) \overline{f_{k,s}(x', y')}\, f_{k,s}(x, y) \qquad \text{(E.10)}$$

Making the eigenfunction $f_{k,s}(x, y)$ explicit using (E.8) and (E.9), we have,

$$\mathcal{K}_{\mathcal{H}}^{[h]}(z, z') = (yy')^{\frac{1}{2}} \int_0^\infty ds\, h\left(\tfrac{1}{4} + s^2\right) \frac{s\, \sinh(\pi s)}{\pi^3}$$
$$\times \int_{\mathbb{R}} dk\, K_{is}(|k|y) K_{is}(|k|y') e^{ik(x-x')}$$

The integral over k may be performed with the help of [GRR88], page 732,

$$\int_0^\infty dx\, K_\nu(ax) K_\nu(bx) \cos(cx) = \frac{\pi^2}{4\sqrt{ab}\, \cos \pi\nu} P_{-\frac{1}{2}+\nu}\left(\frac{a^2 + b^2 + c^2}{2ab}\right)$$

in terms of the Legendre function $P_\sigma(x)$. The result is as follows,

$$\mathcal{K}_{\mathcal{H}}^{[h]}(z, z') = \frac{1}{2\pi} \int_0^\infty ds\, h\left(\tfrac{1}{4} + s^2\right) s \frac{\sinh(\pi s)}{\cosh(\pi s)} P_{-\frac{1}{2}+is}(\cosh \ell)$$

where ℓ is the hyperbolic distance between the points z and z' given by,

$$\cosh \ell = \frac{(x - x')^2 + y^2 + (y')^2}{2yy'}$$

The combination of hyperbolic and Legendre functions may be simplified as follows, using [GRR88], page 1006,

$$-i\pi \frac{\sinh(\pi s)}{\cosh(\pi s)} P_{-\frac{1}{2}+is}(\xi) = Q_{-\frac{1}{2}+is}(\xi) - Q_{-\frac{1}{2}-is}(\xi)$$

so that,

$$\mathcal{K}_{\mathcal{H}}^{[h]}(z, z') = \frac{i}{2\pi^2} \int_0^\infty ds\, h\left(\tfrac{1}{4} + s^2\right) s \left(Q_{-\frac{1}{2}+is}(\cosh \ell) - Q_{-\frac{1}{2}-is}(\cosh \ell)\right)$$

Finally, using the following integral representation for the Legendre function,

$$Q_{-\frac{1}{2}+is}(\cosh \ell) = \frac{1}{\sqrt{2}} \int_\ell^\infty db\, \frac{e^{-isb}}{\sqrt{\cosh b - \cosh \ell}}$$

we find our final formula,

$$\mathcal{K}_{\mathcal{H}}^{[h]}(z, z') = \int_\ell^\infty db\, \frac{g(b)}{\sqrt{\cosh b - \cosh \ell}}$$

where $g(b)$ is given by the following transform of the function h,

$$g(b) = \frac{1}{2\sqrt{2}\pi^2} \int_{-\infty}^{\infty} ds \, h\left(\tfrac{1}{4} + s^2\right) s \sin(bs)$$

For the completeness relation, we set $h = 1$, which gives $\pi g(b) = \delta'(b)$ and $\mathcal{K}_{\mathcal{H}}^{[h=1]}(z, z') = 0$ whenever $\ell \neq 0$, so that its support is $z = z'$, and the completeness of the eigenfunctions $f_{k,s}$ for $k \in \mathbb{R}$ and $s \in \mathbb{R}^+$ is proven. Note that for any polynomial function h, the support of $\mathcal{K}_{\mathcal{H}}^{[h]}(z, z')$ will also be $z = z'$ as expected for polynomial functions of a local differential operator. For the propagator, we set $h = (s^2 + \tfrac{1}{4})^{-1}$, so that $g(b) = e^{-b/2}/(2\sqrt{2}\pi)$.

Finally, for the heat-kernel we have $h_t(s) = e^{-t(s^2+1/4)}$ and we get,

$$\mathcal{K}_{\mathcal{H}}^{t}(z, z') = \mathcal{K}_{\mathcal{H}}^{[h_t]}(z, z') = \frac{e^{-t/4}}{2(2\pi t)^{\frac{3}{2}}} \int_{\ell}^{\infty} db \, \frac{b \, e^{-b^2/4t}}{\sqrt{\cosh b - \cosh \ell}}$$

a result quoted in many instances, such as [Ter85].

Solution to Exercise 7.1

An arbitrary element of $\gamma \in \Gamma_0(N)$ and its mirror $\tilde{\gamma} \in \Gamma^0(N)$ may be parametrized as follows,

$$\gamma = \begin{pmatrix} a & b \\ Nc & d \end{pmatrix} \qquad\qquad \tilde{\gamma} = \begin{pmatrix} a & Nb \\ c & d \end{pmatrix}$$

with $a, b, c, d \in \mathbb{Z}$ and $ad - Nbc = 1$. Using the relation,

$$N\gamma\tau = \frac{a(N\tau) + bN}{c(N\tau) + d} = \tilde{\gamma}(N\tau)$$

and the fact that Δ is a modular form of weight 12, we obtain,

$$\Delta(N\gamma\tau) = \Delta\big(\tilde{\gamma}(N\tau)\big) = \big(c(N\tau) + d\big)^{12}\Delta(N\tau)$$
$$\Delta(\gamma\tau) = (cN\tau + d)^{12}\Delta(\tau)$$

so that the ratio $\Delta(N\tau)/\Delta(\tau)$ is invariant, proving the first part.

To prove the second part, we begin by considering invariance under a translation $\gamma = T$ which maps $T : \tau \to \tau + 1$. Using the transformation law (3.8.3) of η under T, we have,

$$\varphi_N(\tau + 1) = \exp\left\{\tfrac{2\pi i}{24}r(N - 1)\right\} \varphi_N(\tau) \qquad\qquad (\text{E}.11)$$

so that the condition $r(N - 1) \equiv 0 \pmod{24}$ is a necessary one. Next, consider an arbitrary $\gamma \in \Gamma_0(N)$, parametrized as above, and use the transformation law for η of (3.8.4) to obtain,

$$\eta(N\gamma\tau) = \eta(\tilde{\gamma}N\tau) = \varepsilon_\eta(\tilde{\gamma})(cN\tau + d)^{\frac{1}{2}}\eta(N\tau)$$

$$\eta(\gamma\tau) = \varepsilon_\eta(\gamma)(cN\tau + d)^{\frac{1}{2}}\eta(\tau)$$

Therefore we have,

$$\varphi_N(\gamma\tau) = \frac{\varepsilon_\eta(\tilde{\gamma})^r}{\varepsilon_\eta(\gamma)^r}\,\varphi_N(\tau)$$

which may be evaluated using (3.8.6). For odd c we find,

$$\frac{\varepsilon_\eta(\tilde{\gamma})^r}{\varepsilon_\eta(\gamma)^r} = \frac{(d|c)^r}{(d|Nc)^r}\exp\left\{\frac{2\pi i}{24}r(N-1)\left(bd(1 + Nc^2) + (3 - a - d)c\right)\right\}$$

Since a, b, c, d, and N are integers, the exponential evaluates to 1 when $r(N-1) \equiv 0 \pmod{24}$. The Legendre symbols evaluate to ± 1, so that invariance requires r to be even. The case of even c is analogous.

Solution to Exercise 7.2

Formula (7.8.16) is obtained from (2.6.1) and the third line in (2.6.24) by setting $w = e^{2\pi i z}$. To obtain the first formula in (7.8.17), we let $q \to q^3$ and set $w = -q^{\frac{1}{2}}$ to obtain,

$$\prod_{n=1}^{\infty}(1 - q^{3n})(1 - q^{3n-2})(1 - q^{3n-1}) = \sum_{n=-\infty}^{\infty}(-)^n q^{\frac{3}{2}n^2 - \frac{1}{2}n} \qquad (E.12)$$

This relation gives the first line in (7.8.17). To prove the second line in (7.8.17), we let $q \to q^2$ and set $w = -1$ in (7.8.16). Finally, to obtain (7.8.18), we let $w \to -w^2$ in (7.8.16) and multiply both sides by w to obtain,

$$w\prod_{n=1}^{\infty}(1 - q^n)(1 - q^{n-\frac{1}{2}}w^2)(1 - q^{n-\frac{1}{2}}w^{-2}) = \sum_{n=-\infty}^{\infty}(-)^n q^{\frac{1}{2}n^2}w^{2n+1}$$

Note that the third factor on the left side vanishes at $w = q^{\frac{1}{4}}$ for $n = 1$. Taking the derivative with respect to w on both sides, letting $w \to q^{\frac{1}{4}}$, and recasting the sum on the right to positive n, produces (7.8.18).

Solution to Exercise 7.3

The first equality on the first line of (7.8.19) is obtained by multiplying both sides of the first line of (7.8.17) by $q^{\frac{1}{24}}$ and rearranging the exponent of q

in the summand. To obtain the second equality, we note that the Jacobi symbol $(12|m)$ is a primitive Dirichlet character with modulus 12, namely, $\mathbb{Z} \to \mathbb{Z}_{12}^* = \{\pm 1, \pm 5\}$ that satisfies,

$$(12|m) = \begin{cases} +1 & m \equiv \pm 1 \ (\mathrm{mod}\,12) \\ -1 & m \equiv \pm 5 \ (\mathrm{mod}\,12) \\ 0 & \text{otherwise} \end{cases}$$

These congruences are solved by $m = 6n - 1$ when $n > 0$ and $m = -6n + 1$ when $n \leq 0$, thereby producing the equality. The second line of (7.8.19) readily follows from recasting the left side of the second line of (7.8.17) in terms of the Dedekind η function. Finally, the third line of (7.8.19) follows from recasting the left side of the (7.8.18) in terms of η and recognizing $(-1|m)$ as a primitive character $\mathbb{Z} \to \mathbb{Z}_4^* = \{\pm 1\}$ that vanishes for even m and takes the value $(-1)^n$ for $m = 2n + 1$.

Solution to Exercise 7.4

We begin by converting equation (7.8.14) from $2\tau \to \tau$ and express E_2 in terms of the discriminant Δ using (4.1.6), to obtain on the one hand,

$$\vartheta_3(0|\tau)^4 = \frac{1}{3\pi i} \partial_\tau \ln \frac{\Delta(2\tau)}{\Delta(\frac{1}{2}\tau)}$$

or in terms of the infinite product representation of Δ in terms of $q = e^{2\pi i \tau}$,

$$\vartheta_3(0|\tau)^4 = \frac{1}{\pi i} \partial_\tau \ln \left(q^{\frac{1}{2}} \prod_{n=1}^{\infty} \frac{(1 - q^{2n})^8}{(1 - q^{\frac{n}{2}})^8} \right)$$

On the other hand, the ratio of ϑ-constants is given by,

$$\frac{\vartheta_2(0|\tau)^4}{\vartheta_4(0|\tau)^4} = 16 q^{\frac{1}{2}} \prod_{n=1}^{\infty} \frac{(1 + q^n)^8}{(1 - q^{n-\frac{1}{2}})^8} = 16 q^{\frac{1}{2}} \prod_{n=1}^{\infty} \frac{(1 - q^{2n})^8}{(1 - q^{n-\frac{1}{2}})^8 (1 - q^n)^8}$$

Clearly the product in the denominator of the rightmost formula runs over all half positive integers and, upon substitution into the argument of the logarithm gives the announced formula.

Solution to Exercise 7.5

One shows that $\mathsf{E}_4(\tau)$, $\mathsf{E}_4(2\tau)$, and $\mathsf{E}_4(4\tau)$ are modular forms of weight 4 of $\Gamma_0(4)$ by the arguments used in Section 7.8. The dimension formulas of (7.2.7) have already been used in (7.6.2) to shown that $\dim(\mathcal{M}_4(\Gamma_0(4))) = 3$, so that these three holomorphic modular forms of $\Gamma_0(4)$ span the entire space $\mathcal{M}_4(\Gamma_0(4))$. By decomposing their Kronecker–Eisenstein series, one find,

$$E_4(\tau) = \frac{45}{\pi^4} G_4(0\ 0|\tau)$$

$$E_4(2\tau) = \frac{45}{16\pi^4} \Big(G_4(0\ 0|\tau) + G_4(0\ 2|\tau)\Big)$$

$$E_4(4\tau) = \frac{45}{256\pi^4} \Big(G_4(0\ 0|\tau) + G_4(0\ 1|\tau) + G_4(0\ 2|\tau) + G_4(0\ 3|\tau)\Big)$$

verifying that the vector space of Eisenstein series of $\Gamma_0(4)$ obtained in (7.6.2) is indeed the same as the one generated by $E_4(\tau)$, $E_4(2\tau)$, and $E_4(4\tau)$. The coefficients in the relation of the Exercise (7.8.21) are determined from the first few Fourier modes, obtained using the Fourier series of E_4,

$$E_4(\tau) = 1 + 240 \sum_{n=1}^{\infty} \sigma_3(n) q^n$$

which was derived in (3.4.13). Expanding the result, one readily obtains,

$$\vartheta(0|2\tau)^8 = 1 + \sum_{n=1}^{\infty} r_8(n) q^n = 1 + 16 \sum_{n=1}^{\infty} \sigma_3(n) \Big(q^n - 2q^{2n} + 16q^{4n}\Big)$$

In particular, for odd n one has $r_8(n) = 16\sigma_3(n)$.

Solution to Exercise 8.1

Using the relation (4.1.6) between E_2 and Δ, we have,

$$\left(q\frac{d}{dq} - kE_2(q)\right)\Delta^k(q) = 0$$

Substituting the Fourier expansions of $E_2(q)$, given in (4.1.5), and $\Delta^k(q)$ given in terms of $a_n^{(k)}$, one finds after some simplifications,

$$\sum_{n=1}^{\infty} n\, a_n^{(k)}\, q^{n+k} = -24k \sum_{n=1}^{\infty} \sum_{m=1}^{n-1} \sigma(m)\, a_{n-m}^{(k)}\, q^{n+k}$$

from which it follows that

$$a_n^{(k)} = -\frac{24}{n} \times k \sum_{m=1}^{n-1} \sigma(m)\, a_{n-m}^{(k)}$$

The Fourier coefficients $a_n^{(k)}$ must all be integer (because the Fourier coefficients of $\Delta(q)$ itself are), as are the divisor sums $\sigma(m)$, and hence it follows that $a_n^{(k)}$ is divisible by $24/\gcd(24, n)$. Incidentally, this divisibility property is closely related to the theory of "topological modular forms," and to the cancellation of global anomalies in Heterotic string theory, though we will not describe this relationship here. The interested reader may consult the following references: [Hop02, Tac22, TY23, LP23, Lin22, AKL23].

Solution to Exercise 8.2

To begin, we use the definition of the hypergeometric functions to obtain the explicit expression,

$$m_1 = \frac{10\mu^2 + 2\mu - 12}{6 - \mu}$$

from which it follows that,

$$\mu = \frac{1}{20}\left[-(m_1 + 2) \pm \sqrt{(m_1 + 2)^2 + 40(12 + 6m_1)}\right]$$

Because we are restricting to $\mu \in \mathbb{Q}$, at least one (and hence both) of the solutions to this equation must be rational. This means that we can write,

$$(m_1 + 2)^2 + 40(12 + 6m_1) = p^2 \qquad p \in \mathbb{Z} \qquad (E.13)$$

where without loss of generality we may take $p > 0$. Next, defining $\ell = 120 + (m_1 + 2) - p$, we rewrite the above equation as,

$$m_1 + 2 = \frac{1}{2}\ell - 120 + 7200\,\ell^{-1}$$

Requiring that $m_1 \in \mathbb{Z}$, it follows that ℓ must be even and divide 7,200. In every such case, $m_1 + 2 \geq 0$. Combining this with (E.13) and the fact that $p > 0$ tells us that $p \geq m_1 + 2$. Since $\ell = 120 + (m_1 + 2) - p$, this means that $\ell < 120$. We are thus left with a finite set of possible values for ℓ,

$$\ell = 2, 4, 6, 8, 10, 12, 16, 18, 20, 24, 30, 32, 36,$$
$$40, 48, 50, 60, 72, 80, 90, 96, 100$$

Of these, it can be explicitly checked that only the following 10 values give rise to m_i which are integer up to $i = 15$,

$$\ell = 20, 24, 30, 40, 48, 60, 72, 80, 90, 96$$

The corresponding values of $2(\mu - 1)$ are as follows,

$$2(\mu - 1) = 8, \frac{38}{5}, 7, 6, \frac{26}{5}, 4, \frac{14}{5}, 2, 1, \frac{2}{5}$$

In the context of rational conformal field theory, these are the central charges of a well-known series of two-character conformal field theories. For more details, the reader is referred to the original calculation in [MMS88].

Solution to Exercise 8.3

The result of Exercise 7.4, given in (7.8.20), along with its modular images under T and S, may be expressed as follows,

$$\partial_\tau \ln \vartheta_2(0|\tau)^4 - \partial_\tau \ln \vartheta_4(0|\tau)^4 = i\pi\vartheta_3(0|\tau)^4$$
$$\partial_\tau \ln \vartheta_2(0|\tau)^4 - \partial_\tau \ln \vartheta_3(0|\tau)^4 = i\pi\vartheta_4(0|\tau)^4$$
$$\partial_\tau \ln \vartheta_3(0|\tau)^4 - \partial_\tau \ln \vartheta_4(0|\tau)^4 = i\pi\vartheta_2(0|\tau)^4$$

One readily verifies that the sum of the second and third lines gives the first line using the relation $\vartheta_3(0|\tau)^4 = \vartheta_2(0|\tau)^4 + \vartheta_4(0|\tau)^4$ of (2.6.18). Adding the first two lines and using the relation $\vartheta_2(0|\tau)^8\vartheta_3(0|\tau)^8\vartheta_4(0|\tau)^8 = 2^8\eta(\tau)^{24}$ derived by combining (3.7.11), (3.7.12), and (3.8.2), we find,

$$\partial_\tau \ln \frac{\vartheta_2(0|\tau)}{\eta(\tau)} = \frac{i\pi}{12}\left(\vartheta_3(0|\tau)^4 + \vartheta_4(0|\tau)^4\right)$$

$$\partial_\tau \ln \frac{\vartheta_3(0|\tau)}{\eta(\tau)} = \frac{i\pi}{12}\left(\vartheta_2(0|\tau)^4 - \vartheta_4(0|\tau)^4\right)$$

$$\partial_\tau \ln \frac{\vartheta_4(0|\tau)}{\eta(\tau)} = \frac{i\pi}{12}\left(-\vartheta_2(0|\tau)^4 - \vartheta_3(0|\tau)^4\right) \tag{E.14}$$

Solution to Exercise 8.4

The first line in (E.14) provides a first-order differential equation for f_3,

$$\partial_\tau f_3 = \frac{i\pi}{24}\left(\vartheta_2(0|\tau)^4 + \vartheta_4(0|\tau)^4\right)f_3 \tag{E.15}$$

and similarly for $f_1 \pm f_2$, for which we will leave the derivation to the reader. Of course, these first-order equations are not MDEs for $SL(2,\mathbb{Z})$ because the coefficients in the equations are not modular forms under this group. To obtain a linear differential equation whose coefficients are modular forms under $SL(2,\mathbb{Z})$, we need to proceed to higher order. We shall use the following identity for the Serre derivative on a modular form g of weight k,

$$D_k g = \left(\frac{1}{2\pi i}\frac{d}{d\tau} - \frac{k}{12}E_2\right)g = \frac{\eta^{2k}}{2\pi i}\frac{d}{d\tau}\left(\frac{g}{\eta^{2k}}\right)$$

to reduce the evaluation of successive derivatives of f_3 to the use of the equations (E.14), and we find,

$$D_0 f_3 = \frac{1}{48} \left(\vartheta_3^4 + \vartheta_4^4 \right) f_3$$

$$D_2 D_0 f_3 = \frac{1}{48^2} \left(9\vartheta_3^8 + 9\vartheta_4^8 - 30\vartheta_3^4 \vartheta_4^4 \right) f_3$$

$$D_4 D_2 D_0 f_3 = \frac{1}{48^3} \left(\vartheta_3^4 + \vartheta_4^4 \right) \left(153\vartheta_3^8 + 153\vartheta_4^8 - 222\vartheta_3^4 \vartheta_4^4 \right) f_3$$

Using the expression for E_4 given in (3.7.19), we have,

$$D_4 D_2 D_0 f_3 - \frac{107}{48^2} \mathsf{E}_4 D_0 f_3 = \frac{46}{48^3} \left(\vartheta_3^4 + \vartheta_4^4 \right) \left(\vartheta_3^8 + \vartheta_4^8 - \tfrac{5}{2}\vartheta_3^4 \vartheta_4^4 \right) f_3$$

Using the expression in (3.7.22) for E_6 in terms of ϑ-constants, we recover the third-order MDE (8.2.17) for the parameters given in (8.2.21),

$$D_4 D_2 D_0 f_3 - \frac{107}{48^2} \mathsf{E}_4 D_0 f_3 = -\frac{46}{48^3} \mathsf{E}_6 f_3$$

Solution to Exercise 8.5

Recall that modular functions satisfy the valence formula of (3.5.1) with $k = 0$ and, for $\Gamma(N)$, with $\varepsilon_2 = \varepsilon_3 = 0$ in view of (6.7.1), so that we have,

$$\sum_{p \in \Gamma(N) \backslash \mathcal{H}} \mathrm{ord}_{f_i}(p) = 0$$

where $\mathrm{ord}_{f_i}(p)$ denotes the order of a zero (counted positively) or a pole (counted negatively) of $f_i(\tau)$ at $\tau = p$. By the assumption of meromorphicity, our modular function has no poles in the interior of $\Gamma(N) \backslash \mathcal{H}$.

Consider the particular component $f_i(q)$ that is a modular function for $\Gamma(N)$. If $n_0^{(i)} < 0$ then, by definition, $n_0 < 0$ and the proof is complete. Assuming instead that $n_0^{(i)} > 0$, we have $\mathrm{ord}_{f_i}(i\infty) > 0$ and, by the valence formula, $f_i(\tau)$ must have at least one pole. Since we are assuming there are no poles in the interior of $\Gamma(N) \backslash \mathcal{H}$, this pole must be at a cusp, say $\tau = \tau_*$.

Now consider the non-holomorphic modular invariant,

$$F(\tau, \bar{\tau}) = \mathbf{f}(\tau) \cdot \overline{\mathbf{f}(\tau)}$$

Since the component $|f_i(\tau)|^2$ has a pole at τ_*, the full expression does as well. But it is always possible to map a cusp of $\Gamma(N) \backslash \mathcal{H}$ to $i\infty$ via an appropriate $\mathrm{SL}(2, \mathbb{Z})$ transformation. Since $F(\tau, \bar{\tau})$ is $\mathrm{SL}(2, \mathbb{Z})$ invariant, this means that $F(\tau, \bar{\tau})$ must have a pole at $i\infty$ as well. In other words, there must be another component $f_j(\tau)$ which has $n_0^{(j)} < 0$, and hence it follows that $n_0 < 0$.

Solution to Exercise 9.1

To prove the first formula, we start from the definition,

$$
C_{w_1,w_2,0} = \sum_{\substack{(m_r,n_r)\neq(0,0) \\ r=1,2,3}} \frac{\tau_2^{w_1+w_2}\,\delta\left(\sum_{r=1}^3 m_r\right)\delta\left(\sum_{r=1}^3 n_r\right)}{\pi^{w_1+w_2}|m_1+n_1\tau|^{2w_1}|m_2+n_2\tau|^{2w_2}}
$$

Since the denominator is independent of m_3 and n_3, we may reduce the sum to be over m_1, n_1 and m_2, n_2 independently, provided we remove the contribution for which $(m_3, n_3) = (0,0)$, namely, $(m_2, n_2) = (-m_1, -n_1)$. These two contributions readily give the desired formula.

To prove the second formula, we start from its definition,

$$
C_{w_1+1,w_2,-1} = \sum_{\substack{(m_r,n_r)\neq(0,0) \\ r=1,2,3}} \frac{\tau_2^{w_1+w_2}\,\delta\left(\sum_{r=1}^3 m_r\right)\delta\left(\sum_{r=1}^3 n_r\right)|m_3+n_3\tau|^2}{\pi^{w_1+w_2}|m_1+n_1\tau|^{2w_1+2}|m_2+n_2\tau|^{2w_2}}
$$

Since the denominator is independent of m_3 and n_3, we may reduce the sum to be over m_1, n_1 and m_2, n_2 independently. In this case, it is unnecessary to remove the contribution for which $(m_3, n_3) = (0,0)$ since the numerator makes the summand for this contribution vanish,

$$
C_{w_1+1,w_2,-1} = \sum_{\substack{(m_1,n_1)\neq(0,0) \\ (m_2,n_2)\neq(0,0)}} \frac{\tau_2^{w_1+w_2}|m_3+n_3\tau|^2}{\pi^{w_1+w_2}|m_1+n_1\tau|^{2w_1+2}|m_2+n_2\tau|^{2w_2}}
$$

Next, we decompose,

$$
\begin{aligned}
|m_3+n_3\tau|^2 &= |m_1+n_1\tau|^2 + (m_1+n_1\tau)(m_2+n_2\bar\tau) \\
&\quad + |m_2+n_2\tau|^2 + (m_1+n_1\bar\tau)(m_2+n_2\tau)
\end{aligned}
$$

Since the sums over (m_1, n_1) and (m_2, n_2) are now completely independent of one another, the contribution from the cross terms vanishes. The remaining terms give the desired formula.

Solution to Exercise 9.2

We wish to solve the following system of equations for n and γ,

$$
(m_1\ n_1) = (0\ n)\gamma \qquad\qquad \gamma = \begin{pmatrix} a & b \\ c & d \end{pmatrix} \in \mathrm{SL}(2,\mathbb{Z})
$$

or $m_1 = nc$ and $n_1 = nd$. Since $\gcd(c,d) = 1$, we take $n = \pm\gcd(m_1,n_1)$, such that $c = m_1/n$, $d = n_1/n$, and $d > 0$ if $d \neq 0$ or $c > 0$ if $d = 0$. Bézout's lemma guarantees that the equation $ad - bc = 1$ has a solution a_0, b_0, so that the general solution a, b satisfies $(a - a_0)d - (b - b_0)c = 0$ and is given by $a = a_0 + kc$ and $b = b_0 + kd$ which amounts to left multiplication of γ by $\left(\begin{smallmatrix} 1 & k \\ 0 & 1 \end{smallmatrix}\right)$, so that the element γ in the coset $\Gamma_\infty \backslash SL(2, \mathbb{Z})$ is unique.

Solution to Exercise 9.3

It will be convenient to swap the roles of m_r and n_r and start from,

$$C_{a_1,a_2,a_3}(\tau) = \sum_{\substack{m_r, n_r \in \mathbb{Z} \\ (m_r, n_r) \neq (0,0)}} \delta\left(\sum_{r=1}^{3} m_r\right) \delta\left(\sum_{r=1}^{3} n_r\right) \prod_{r=1}^{3} \left(\frac{\tau_2}{\pi |m_r \tau + n_r|^2}\right)^{a_r}$$

Using the result of Exercise 9.2 to parametrize the pair $(m_1\ n_1)$, we have,

$$C_{a_1,a_2,a_3}(\tau) = \frac{(2\pi)^{2w} c_w}{2\zeta(2w)} E_w(\tau) + \sum_{\gamma \in \Gamma_\infty \backslash SL(2,\mathbb{Z})} \lambda_{a_1,a_2,a_3}(\gamma \tau)$$

where $w = a_1 + a_2 + a_3$, c_w is the coefficient of the τ_2^w term in the Laurent expansion of $C_{a_1,a_2,a_3}(\tau)$ given in Theorem 9.4.2, and the seed function $\lambda_{a_1,a_2,a_3}(\tau)$ is given by,

$$\lambda_{a_1,a_2,a_3}(\tau) = \sum_{m,n \neq 0} \sum_{\mu \in \mathbb{Z}} \frac{\tau_2^w}{\pi^w n^{2a_1} |m\tau + \mu|^{2a_2} |m\tau + \mu + n|^{2a_3}}$$

For a detailed derivation, we refer the reader to [DK19a].

Solution to Exercise 9.4

We shall prove this relation by induction on n. The relation clearly holds for $n = 1$. Starting from the rightmost expression, we evaluate Δ_{n+1} via,

$$\Delta_{n+1} = \nabla^{n+1} \tau_2^{-2n-2} \bar{\nabla}^{n+1}$$
$$= \Delta \Delta_n + 2i \nabla [\partial_{\bar\tau}, \nabla^n \tau_2^{-2n}] \bar{\nabla}^n \tag{E.16}$$

Using the auxiliary expressions $[\tau_2^{-1}, \nabla] = 1$, $[\tau_2^{-1}, \nabla^n] = n\nabla^{n-1}$ and,

$$[\partial_{\bar\tau}, \nabla^n] = in\tau_2^{-1}\nabla^n - \tfrac{i}{2}n(n-1)\nabla^{n-1}$$

we evaluate the commutator as follows,

$$[\partial_{\bar\tau}, \nabla^n \tau_2^{-2n}] = [\partial_{\bar\tau}, \nabla^n]\tau_2^{-2n} + \nabla^n [\partial_{\bar\tau}, \tau_2^{-2n}]$$
$$= in [\tau_2^{-1}, \nabla^n]\tau_2^{-2n} - \frac{i}{2}n(n-1)\nabla^{n-1}\tau_2^{-2n}$$
$$= \frac{i}{2}n(n+1)\nabla^{n-1}\tau_2^{-2n}$$

Substituting this result into the second line of (E.16), we obtain,

$$\Delta_{n+1} = (\Delta - n(n+1))\Delta_n$$

which implies (9.7.16). Using $\Delta\tau_2^s = s(s-1)\tau_2^s$ so that,

$$\Delta_n\tau_2^s = \prod_{m=1}^{n}\Big(s(s-1) - m(m-1)\Big)\tau_2^s = \prod_{m=1-n}^{n}(s-m)\tau_2^s$$

Solution to Exercise 9.5

We begin by computing ∇^2 of $C_{1,1,1,1}$,

$$\nabla^2 C_{1,1,1,1} = 10\,\mathcal{C}^+\begin{bmatrix}3 & 1 & 1 & 1 & 1 \\ -1 & 1 & 1 & 1 & 1\end{bmatrix} + 20\,\mathcal{C}^+\begin{bmatrix}2 & 2 & 1 & 1 & 1 \\ 0 & 0 & 1 & 1 & 1\end{bmatrix}$$

Using a momentum conservation identity on the first term and holomorphic subgraph reduction on both gives

$$\nabla^2 C_{1,1,1,1} = 40\,\mathcal{C}^+\begin{bmatrix}4 & 1 & 1 & 1 \\ 0 & 1 & 1 & 1\end{bmatrix} + 60\,\mathcal{C}^+\begin{bmatrix}4 & 0 \\ 0 & 0\end{bmatrix}\mathcal{C}^+\begin{bmatrix}1 & 1 & 1 \\ 1 & 1 & 1\end{bmatrix}$$

The second term can be cancelled by adding $-10\nabla^2(E_2 C_{1,1,1})$, and having eliminated all terms with more than one nonpositive lower index, we proceed to order ∇^3. Upon use of momentum conservation identities and holomorphic subgraph reduction, we obtain

$$\nabla^3(C_{1,1,1,1} - 10E_2 C_{1,1,1}) = -720\,\mathcal{C}^+\begin{bmatrix}3 & 0 \\ 1 & 0\end{bmatrix}\mathcal{C}^+\begin{bmatrix}5 & 0 \\ 1 & 0\end{bmatrix} + 1080\,\mathcal{C}^+\begin{bmatrix}6 & 1 & 1 \\ 0 & 1 & 1\end{bmatrix}$$

$$-720\,\mathcal{C}^+\begin{bmatrix}4 & 0 \\ 0 & 0\end{bmatrix}\mathcal{C}^+\begin{bmatrix}2 & 1 & 1 \\ 0 & 1 & 1\end{bmatrix}$$

In order to cancel the final term, we add $-60\nabla^3 C_{3,1,1}$. Having eliminated all terms with more than one nonpositive lower index, we proceed to ∇^4,

$$\nabla^4(C_{1,1,1,1} - 10E_2 C_{1,1,1} - 60C_{3,1,1}) = -80640\,\mathcal{C}^+\begin{bmatrix}9 & 0 \\ 1 & 0\end{bmatrix} \qquad \text{(E.17)}$$

Finally, we note that the right-hand side can be cancelled by adding $48\nabla^4 E_5$. Lemma 9.6.1 then tells us that $C_{1,1,1,1} - 10E_2 C_{1,1,1} - 60C_{3,1,1} + 48\nabla^4 E_5$, with the constant being fixed to $16\zeta(5)$ by analyzing the asymptotics at the cusp.

Solution to Exercise 10.1

Recall the defining equation of the Möbius function for $n \in \mathbb{N}$,

$$\sum_{d|n} \mu(d) = \delta_{n,1}$$

where $\delta_{n,1}$ is the Kronecker delta function equaling 1 if $n = 1$ and zero otherwise. To prove the proposed formula, substitute (10.6.1) into the right side of equation (10.10.24),

$$\sum_{d|(m,n)} \mu(d) d^{k-1} \mathsf{T}_{m/d} \mathsf{T}_{n/d} = \sum_{d|(m,n)} \mu(d) d^{k-1} \sum_{\delta|(\frac{m}{d},\frac{n}{d})} \delta^{k-1} \mathsf{T}_{mn/(d\delta)^2}$$

The product $D = d\delta$ must divide m and n, while d must divide D, so that,

$$\sum_{d|(m,n)} \mu(d) d^{k-1} \mathsf{T}_{m/d} \mathsf{T}_{n/d} = \sum_{D|(m,n)} D^{k-1} \mathsf{T}_{mn/D^2} \sum_{d|D} \mu(d)$$

The result follows from the fact that the sum over d vanishes whenever $D > 1$ and equals 1 when $D = 1$.

Solution to Exercise 10.2

Using formula (10.10.24) of Exercise 10.1 for $m = p$ and $n = p^\nu$, we find,

$$\mathsf{T}_{p \cdot p^\nu} = \mu(1) \mathsf{T}_p \mathsf{T}_{p^\nu} + \mu(p) p^{k-1} \mathsf{T}_1 \mathsf{T}_{p^{\nu-1}}$$

Using $\mathsf{T}_1 = 1$, the definition of the Möbius function, and the fact that $p \geq 2$ is prime, we have $\mu(1) = 1$ and $\mu(p) = -1$, giving the recursion relation,

$$\mathsf{T}_{p^{\nu+1}} = \mathsf{T}_p \mathsf{T}_{p^\nu} - p^{k-1} \mathsf{T}_{p^{\nu-1}}$$

Solution to Exercise 10.3

To calculate $f_p(x)$, we use the recursion relation of Exercise 10.2 in the last term of the following rearrangement,

$$f_p(x) = 1 + x \mathsf{T}_p + \sum_{\nu=1}^{\infty} x^{\nu+1} \left(\mathsf{T}_p \mathsf{T}_{p^\nu} - p^{k-1} \mathsf{T}_{p^{\nu-1}} \right)$$

Recasting the sums on the right side in terms of $f_p(x)$, we obtain,

$$f_p(x) = \left(1 - x \mathsf{T}_p + x^2 p^{k-1} \right)^{-1}$$

For the derivation of this and further relations involving the Hecke operators, see, for example, the book by Iwaniec [Iwa97].

Solution to Exercise 10.4

One begins by using the definition of the Hecke operator in (10.8.4) to write,

$$\sqrt{n}\,\langle g|\mathsf{T}_n f\rangle = \sum_{\gamma \in \mathrm{SL}(2,\mathbb{Z})\backslash M_n} \int_{\mathrm{SL}(2,\mathbb{Z})\backslash \mathcal{H}} \frac{d\tau_1 d\tau_2}{\tau_2^2} \overline{g(\tau)} f(\gamma\tau)$$

Note that $h(\tau) = f(\gamma\tau)$ is a modular form for $\Gamma(n)$. Indeed, since $\gamma \in M_n$, the matrix $n\gamma^{-1}$ has integer coefficients, and hence for any $\tilde{\gamma} \in \Gamma(n)$, we have $n\gamma\tilde{\gamma}\gamma^{-1} \equiv 0 \pmod{n}$, and consequently $\gamma\tilde{\gamma}\gamma^{-1} \in \mathrm{SL}(2,\mathbb{Z})$. It follows that,

$$h(\tilde{\gamma}\tau) = f(\gamma\tilde{\gamma}\gamma^{-1}\gamma\tau) = f(\gamma\tau) = h(\tau)$$

since $f(\tau)$ is a modular form for $\mathrm{SL}(2,\mathbb{Z})$.

Since $\Gamma(n)\backslash\mathcal{H}$ can be written as $[\Gamma : \Gamma(n)]$ copies of $\Gamma\backslash\mathcal{H}$, we may write,

$$\sqrt{n}\,\langle g|\mathsf{T}_n f\rangle = \sum_{\gamma \in \mathrm{SL}(2,\mathbb{Z})\backslash M_n} \frac{1}{[\Gamma : \Gamma(n)]} \int_{\Gamma(n)\backslash\mathcal{H}} \frac{d\tau_1 d\tau_2}{\tau_2^2} \overline{g(\tau)}\,f(\gamma\tau)$$

Setting $\tilde{\tau} = \gamma\tau = \tilde{\tau}_1 + i\tilde{\tau}_2$ and noting that $[\Gamma : \gamma\Gamma(n)\gamma^{-1}] = [\Gamma : \Gamma(n)]$, this expression may be rewritten as,

$$\sqrt{n}\,\langle g|\mathsf{T}_n f\rangle = \sum_{\gamma \in \mathrm{SL}(2,\mathbb{Z})\backslash M_n} \frac{1}{[\Gamma : \gamma\Gamma(n)\gamma^{-1}]} \int_{\gamma\Gamma(n)\gamma^{-1}\backslash\mathcal{H}} \frac{d\tilde{\tau}_1 d\tilde{\tau}_2}{\tilde{\tau}_2^2} \overline{g(\gamma^{-1}\tilde{\tau})}\,f(\tilde{\tau})$$

Finally, using Lemma 10.2.2, we conclude that the right side equals $\sqrt{n}\,\langle \mathsf{T}_n g|f\rangle$, which is the desired result. This proof can be found in [Pet39] and [Ter85].

Solution to Exercise 10.5

The result follows by Taylor expanding the logarithms for $|p|, |q| < 1$,

$$\ln \prod_{\substack{m>0 \\ n\in\mathbb{Z}}} (1 - p^m q^n)^{-a_{mn}} = \sum_{m>0}\sum_{n\in\mathbb{Z}}\sum_{d>0} a_{mn} \frac{p^{md}q^{md}}{d}$$

where we have made use of the Taylor expansion for the logarithm. Redefining m and n, we arrive at,

$$\ln \prod_{\substack{m>0 \\ n\in\mathbb{Z}}} (1 - p^m q^n)^{-a_{mn}} = \sum_{m>0}\sum_{n\in\mathbb{Z}}\sum_{d|(m,n)} d^{-1}p^m q^n\, a_{mn/d^2} \qquad (\mathrm{E}.18)$$

from which the result follows upon using the definition (10.4.3).

Solution to Exercise 11.1

Beginning with part (a), we first rewrite the curve as $y^2 = 4\tilde{x}^3 - 15 \times 4^{1/3}\tilde{x} + 22$, from which we read off $g_2 = 15 \times 4^{1/3}$ and $g_3 = -22$. This allows one to compute the value of the j-function using (3.6.1), and we find $j(\tau) = 54{,}000$. We now invert $j(\tau)$ to obtain τ by using the relations,

$$\tau = i\,\frac{{}_2F_1\left(\frac{1}{6};\frac{5}{6};1;1-a\right)}{{}_2F_1\left(\frac{1}{6};\frac{5}{6};1;a\right)} \qquad\qquad j(\tau) = \frac{1728}{4a(1-a)} \qquad (\text{E.19})$$

Solving for a gives $a = \frac{1}{50}(25 - 11\sqrt{5})$, and using this value in the expression for τ gives $\tau = i\sqrt{3}$ upon computer implementation (the other solution for a corresponds to $-1/\tau$, since $j(\tau) = j(-1/\tau)$). The elliptic curve thus has complex multiplication and corresponds to the second entry of Table 11.2 with $\tau \in \mathbb{Q}(\sqrt{D})$ for $D = -3$. In the same way one can show that for (b) we have $D = -19$, while for (c) we have $D = -163$. Other explicit examples of elliptic curves with complex multiplication can be found in, for example, [Sil85].

Solution to Exercise 11.2

The double poles match on both sides, leaving a z-independent term. To fix it, we derive the identity from the definition of $\wp$,

$$\sum_{\alpha,\beta \in \mathbb{Z}_N} \wp\left(z + \tfrac{\alpha+\beta\tau}{N}\Big|\tau\right) \qquad\qquad (\text{E.20})$$

$$= \sum_{\alpha,\beta \in \mathbb{Z}_N}\sum_{m,n \in \mathbb{Z}} \left(\frac{1}{(z + \frac{\alpha+\beta\tau}{N} + m + n\tau)^2} - \frac{1 - \delta_{m,0}\delta_{n,0}}{(m+n\tau)^2}\right)$$

Multiplying numerator and denominator in the first sum by N^2, recognizing that $\alpha + Nm$ and $\nu + Nn$ run over all integers, and that the summation over α, β in the last sum gives a factor of N^2 readily proves the formula.

Solution to Exercise 11.3

For $n = 3$, the assignments for (a, d, b) in the formula (11.4.3) are $(3, 1, 0)$, $(1, 3, 0)$, $(1, 3, 1)$, and $(1, 3, 2)$ giving the following expression for $P_3(x|\tau)$,

$$P_3(x|\tau) = (x - j(3\tau))\left(x - j(\tfrac{\tau}{3})\right)\left(x - j(\tfrac{\tau+1}{3})\right)\left(x - j(\tfrac{\tau+2}{3})\right)$$

By construction, $P_3(x|\tau)$ is a polynomial in x of degree 4, whose coefficients are polynomials in $j(\tau)$, and therefore may be expressed as a polynomial $Q_3(x, j)$ in x and $j - j(\tau)$ as follows,

$$Q_3(x, j) = x^4 - A(j)x^3 + B(j)x^2 - C(j)x + D(j) \qquad (\text{E.21})$$

The degree in j of each of the coefficients may be read off from the expression of Q_3 and the asymptotic behavior $j(q) \sim q^{-1}$ as $q \to 0$, so that we have $P_3(x|\tau) \to q^{-4}$ as $q \to 0$, and more specifically, A, B, and C are of degree 3 while D is of degree 4. The remainder of the work consists in expanding $j(\tau)$ on the right side of (E.21), and $j(3\tau)$, and $j(\frac{\tau+b}{3})$ for $b = 0, 1, 2$ on the left side of (E.21) in powers of q using (3.6.3). Needless to say, this requires use of Mathematica or Maple. Matching all the terms on both sides of negative or zero powers in q suffices to produce the identity between modular functions,

$$A(j) = j^3 - 2232j^2 + 1069956j - 36864000$$

$$B(j) = 2232j^3 + 2587918086j^2 + 8900222976000j + 452984832000000$$

$$C(j) = 1069956j^3 - 8900222976000j^2 + 7708459663360000000j$$
$$\quad -18554258718720000000000$$

$$D(j) = j^4 + 36864000j^3 + 452984832000000j^2$$
$$\quad +18554258718720000000000j$$

Evaluating $Q_3(x, x)$ using the above expressions gives,

$$Q_3(x, x) = -x(x - 54000)(x + 32768)^2(x - 8000)^2 \qquad \text{(E.22)}$$

which reproduces the expression in (11.4.12).

Solution to Exercise 11.4

Define the integral,

$$I = \frac{1}{2\pi i} \int_{\partial F_{\text{reg}}} d\tau \frac{j'(\tau)}{j(\tau) - \alpha} \qquad \text{(E.23)}$$

where F_{reg} is the regularized fundamental domain defined in Figure 3.3. If there is no τ such that $j(\tau) = \alpha$, this integral must vanish by Cauchy's argument principle. However, it is easy to prove that this is not the case. Indeed, note that invariance of $j(\tau)$ under $T \in \text{SL}(2, \mathbb{Z})$ implies that the contribution from the left and right contours cancel, while invariance under $S \in \text{SL}(2, \mathbb{Z})$ implies that the left and right sides of the lower arc cancel. The integral I then reduces to an integral over the horizontal portion of the contour at $\tau_2 = L$. Switching variables to $q = e^{2\pi i\tau}$ gives an integral over the circle $|q| = e^{-2\pi L}$, and noting that $j(\tau) = \frac{1}{q} + \mathcal{O}(q^0)$, we have,

$$I = \frac{1}{2\pi i} \int_{|q|=e^{-2\pi L}} \left(-\frac{1}{q} + \mathcal{O}(q^0) \right) = -1 \qquad \text{(E.24)}$$

Thus by contradiction, we conclude that for any $\alpha \in \mathbb{C}$, there must exist a τ such that $j(\tau) = \alpha$.

Solution to Exercise 11.5

One begins by searching for k such that $F(k)$ defined in (11.5.3) satisfies $F(k) = \frac{1}{2}(1 + i)$, or in other words,

$$\frac{1}{2}(1 + i) \, {}_2F_1\left(\tfrac{1}{2}, \tfrac{1}{2}; 1; k^2\right) = {}_2F_1\left(\tfrac{1}{2}, \tfrac{1}{2}; 1; 1 - k^2\right)$$

By the identity (11.5.10), we see that a solution is given by $k = \sqrt{2}$. Plugging this into (11.5.2), we arrive at the identity,

$$\vartheta_3\left(0 \,\middle|\, \tfrac{1}{2}(-1 + i)\right) = \sqrt{{}_2F_1\left(\tfrac{1}{2}, \tfrac{1}{2}; 1; 2\right)}$$

The hypergeometric function appearing on the right evaluates to,

$$ {}_2F_1\left(\tfrac{1}{2}, \tfrac{1}{2}; 1; 2\right) = \frac{\Gamma\left(\tfrac{1}{4}\right)^2}{2\pi^{3/2} e^{-\frac{i\pi}{4}}}$$

from which the result follows.

Solution to Exercise 12.1

In both integrals, we shall use the differential equation $\Delta E_s^* = s(s-1)E_s^*$ for the Eisenstein series $E_s^*(\tau) = \frac{1}{2}\Gamma(s)E_s(\tau)$ defined in (12.11.8), and its Fourier series representation,

$$E_s^*(\tau) = \zeta^*(2s)\tau_2^s + \zeta^*(2s - 1)\tau_2^{1-s}$$

$$+ 4\sqrt{\tau_2} \sum_{N=1}^{\infty} N^{\frac{1}{2}-s} \sigma_{2s-1}(N) K_{s-\frac{1}{2}}(2\pi\tau_2 N) \cos(2\pi N\tau_1)$$

Multiplying G_1 by $s(s-1)$ and using the differential equation, we find,

$$s(s-1)G_1(s; L) = \int_{F_L} \frac{d^2\tau}{\tau_2^2} \Delta E_s^*(\tau) = \int_0^1 d\tau_1 \frac{\partial}{\partial L} E_s^*(\tau_1 + iL)$$

Since only the constant Fourier mode contributes, we find,

$$s(s-1)G_1(s; L) = s\zeta^*(2s)L^{s-1} + (1-s)\zeta^*(2s-1)L^{-s}$$

To evaluate G_2, we make use of the following identity,

$$(s_1 - s_2)(s_1 + s_2 - 1)E_{s_1}^* E_{s_2}^* = E_{s_2}^* \Delta E_{s_1}^* - E_{s_1}^* \Delta E_{s_2}^*$$

$$= 4\tau_2^2 \partial_{\bar{\tau}}\left(E_{s_2}^* \partial_\tau E_{s_1}^* - E_{s_1}^* \partial_\tau E_{s_2}^*\right)$$

and carry out the integral over F_L using Stokes's theorem,

$$(s_1 - s_2)(s_1 + s_2 - 1)G_2(s_1, s_2; L)$$
$$= \int_0^1 d\tau_1 \left(E_{s_2}^*(\tau)\partial_{\tau_2} E_{s_1}^*(\tau) - E_{s_1}^*(\tau)\partial_{\tau_2} E_{s_2}^*(\tau) \right)\Big|_{\tau_2=L}$$

The pairing of the constant Fourier mode decouples from the pairing of the nonconstant Fourier modes, producing the following result,

$$G_2(s_1, s_2; L) = \sum_{\sigma_1 = s_1, 1 - s_1} \sum_{\sigma_2 = s_2, 1 - s_2} \zeta^*(2\sigma_1)\, \zeta^*(2\sigma_2)\, \frac{L^{\sigma_1 + \sigma_2 - 1}}{\sigma_1 + \sigma_2 - 1}$$
$$+ \frac{\mathcal{C}(s_1, s_2; L)}{(s_1 - s_2)(s_1 + s_2 - 1)}$$

The pairing of the nonconstant Fourier modes contributes,

$$\mathcal{C}(s_1, s_2; L) = 8 \sum_{N=1}^{\infty} N^{1 - s_1 - s_2} \sigma_{2s_1 - 1}(N)\sigma_{2s_2 - 1}(N)$$
$$\times \left(L\, K_{s_2 - \frac{1}{2}}(2\pi L N)\frac{\partial}{\partial L}K_{s_1 - \frac{1}{2}}(2\pi L N) - (s_1 \leftrightarrow s_2) \right)$$

Solution to Exercise 12.2

Since the Laurent polynomial of $C_{a,b,c}(\tau)$ has degree $(n, 1-n)$ in τ_2, each one of its terms is in the kernel of Δ_n in view of Exercise 9.4. Since $\Delta_4 C_{2,1,1}(\tau)$ decays exponentially at ∞, the integral is absolutely convergent. Application of the Rankin–Selberg method allows us to unfold the integral over the fundamental domain for $\mathrm{SL}(2, \mathbb{Z})$ to the fundamental domain for Γ_∞,

$$f(s) = c_s \int_0^\infty d\tau_2\, \tau_2^{s-2} \int_0^1 d\tau_1\, \Delta_4 C_{2,1,1,}(\tau)$$

where $c_s = 2\zeta(2s)/\pi^s$. Using $(\Delta - 2)C_{2,1,1} = 9E_4 - E_2^2$, we readily deduce,

$$\Delta_4 C_{2,1,1} = -(\Delta - 12)(\Delta - 6)\Delta E_2^2$$

where we have used the fact that $(\Delta - 12)E_4 = 0$. The expression for E_2^2 is readily obtained from (4.2.17) for $s = 2$. The cross-terms between the Laurent part and the exponential part cancel out in the Fourier zero mode, while the square of the Laurent part is in the kernel of $(\Delta - 12)(\Delta - 6)\Delta$.

Carrying out the integral over τ_1 gives,

$$f(s) = -8c_s \sum_{N=1}^{\infty} \sigma_{-3}(N)^2 \int_0^{\infty} d\tau_2\, \tau_2^{s-2}\, \Delta_4 \left(\frac{e^{-4\pi N \tau_2}}{(4\pi \tau_2)^2} \right)$$

$$= -8c_s \lambda_4(s) \frac{\Gamma(s-3)}{(4\pi)^{s-1}} \sum_{N=1}^{\infty} \frac{\sigma_{-3}(N)^2}{N^{s-3}}$$

where $\lambda_n(s)$ was defined in Exercise 9.4. The sum may be performed with the help of the Ramanujan identity (see page 170 of [Erd81c]),

$$\sum \frac{\sigma_a(N)\sigma_b(N)}{N^s} = \frac{\zeta(s)\zeta(s-a)\zeta(s-b)\zeta(s-a-b)}{\zeta(2s-a-b)}$$

giving the final result,

$$f(s) = -32\lambda_4(s) \frac{\Gamma(s-3)}{(2\pi)^{2s-1}} \zeta(s)^2 \zeta(s+3)\zeta(s-3)$$

An alternative derivation was given in [DK19a].

Solution to Exercise 12.3

The partial amplitude $\mathcal{B}_{(0,2)}(\tau)$ is obtained via straightforward expansion of (12.10.2) and is given by

$$240\,\mathcal{B}_{(3,0)} = C_{1,1,1,1,1,1} + 45\, E_2 C_{1,1,1,1} - 10\, C_{1,1,1}^2 + 120\, C_{3,1,1,1}$$
$$+ 180\, C_{2,2,1,1} - 300\, C_{2,2,2} - 1440\, C_{3,2,1} - 360\, C_{4,1,1}$$
$$- 180\, E_2 E_4 - 360\, E_3^2 + 1200\, E_6 + 90\, E_2^3$$

Using the following algebraic identities [DK16],

$$C_{1,1,1,1,1,1} = 15 E_2 C_{1,1,1,1} - 720 C_{2,2,1,1} + 3460 C_{2,2,2} + 13440 C_{3,2,1}$$
$$+ 720 C_{4,1,1} + 10 C_{1,1,1}^2 - 30 E_2^3 + 1440 E_3^2 + 720 E_2 E_4 - 7600 E_6$$
$$2 C_{3,1,1,1} = -3 C_{2,2,1,1} + 18 C_{4,1,1} + 24 C_{3,2,1} + 2 C_{2,2,2} + 9 E_2 E_4 + 6 E_3^2 - 32 E_6$$

one may eliminate $C_{1,1,1,1,1,1}$ and $C_{3,1,1,1}$ and then using the following differential identities [DG19],

$$24 C_{3,2,1} = \Delta(6 C_{3,2,1} + C_{2,2,2}) + 12 E_3^2 - 84 E_6$$
$$3 C_{2,2,2} = \Delta(3 C_{3,2,1} + C_{2,2,2}) - 36 E_6$$
$$18 C_{2,2,1,1} = 9\Delta(C_{2,2,1,1} - 2 E_3^2 - E_2 E_4) - 14 C_{2,2,2} + 48 C_{3,2,1}$$
$$- 36 C_{4,1,1} - 36 C_{2,1,1} E_2 + 6 E_2^3 + 72 E_2 E_4 + 180 E_3^2 + 284 E_6$$

one may rewrite $\mathcal{B}_{(3,0)}$ in the form of a total derivative, plus terms that are linear and bilinear in Eisenstein series, and a term proportional to $E_2C_{2,1,1}$,

$$6\mathcal{B}_{(3,0)} = \Delta\left(-9C_{2,2,1,1} + 6C_{4,1,1} + 156C_{3,2,1} + 41C_{2,2,2} + 18E_3^2 + 9E_2E_4\right)$$
$$-12E_3^2 - 36E_2E_4 - 2652E_6 + 72E_2C_{2,1,1}$$

Given a modular graph function $\mathcal{C}(\tau)$ of weight w with Laurent polynomial

$$\mathcal{C}(\tau) = \sum_{k=1-w}^{w} c_k\tau_2^k + O(e^{-2\pi\tau_2})$$

the integral of $\Delta\mathcal{C}(\tau)$ over $\mathcal{M}_L$ is given as follows,

$$\int_{\mathcal{M}_L}\frac{d^2\tau}{\tau_2^2}\Delta\mathcal{C}(\tau) = \int_0^1 d\tau_1\partial_{\tau_2}\mathcal{C}(\tau)\big|_{\tau_2=L} = c_1 + O(L^{\pm 1})$$

On the other hand, the integral of products of Eisenstein series may be evaluated using Lemma 12.11.1 or the results of Exercise 12.1. Finally, the integral of $E_2C_{2,1,1}$ may be evaluated using the results of Exercise 12.2 and the following differential identity,

$$2C_{2,1,1} = \Delta C_{2,1,1} + E_2^2 - 9E_4$$

with the final result being,

$$\int_{\mathcal{M}_L}\frac{d^2\tau}{\tau_2^2}E_2(\tau)C_{2,1,1}(\tau) = \frac{\pi\zeta(5)}{108}\left(\ln(2L) + 2\frac{\zeta'(4)}{\zeta(4)} - \frac{\zeta'(5)}{\zeta(5)} - \frac{\zeta'(2)}{\zeta(2)}\right)$$

Assembling the pieces, we arrive at the following result for the amplitude,

$$\mathcal{A}_{(3,0)} = \frac{\pi\zeta(5)}{315}\left[33\log(2L) - 2\frac{\zeta'(6)}{\zeta(6)} - 33\frac{\zeta'(5)}{\zeta(5)} + 70\frac{\zeta'(4)}{\zeta(4)} - 35\frac{\zeta'(2)}{\zeta(2)} + \frac{13}{2}\right]$$

Additional details can be found in [DG19].

Solution to Exercise 12.4

(a) Substituting the Poincaré series of (12.13.1) into the trace of (12.13.2), and isolating the element $\gamma = I$, we have,

$$\text{Tr}(\mathcal{K}_\Sigma^t) = \text{area}(\Sigma)\mathcal{K}_\mathcal{H}^t(z,z) + I(t) \tag{E.25}$$

where the contribution from the elements of Γ different from the identity is,

$$I(t) = \sum_{\substack{\gamma\in\Gamma \\ \gamma\neq I}}\int_{\Gamma\backslash\mathcal{H}}d\mu(z)\mathcal{K}_\mathcal{H}^t(z,\gamma z) \tag{E.26}$$

Each $\gamma \in \Gamma$ with $\gamma \neq I$ may be parametrized in terms of one of the representatives of its coset $\{\gamma\}$ and the elements δ of the centralizer of γ by $\gamma = \delta^{-1}\{\gamma\}\delta$ and $\delta \in \Gamma/\Gamma_\gamma$ so that,

$$I(t) = \sum_{\{\gamma\}\neq I} \sum_{\delta\in\Gamma/\Gamma_\gamma} \int_{\Gamma\backslash\mathcal{H}} d\mu(z)\, \mathcal{K}_\mathcal{H}^t(z, \delta^{-1}\{\gamma\}\delta z)$$

Invariance under $\mathrm{SL}(2,\mathbb{R})$ of the Poincaré measure gives $d\mu(\delta z) = d\mu(z)$, while invariance of $\mathcal{K}_\mathcal{H}^t$ gives $\mathcal{K}_\mathcal{H}^t(z, \delta^{-1}\{\gamma\}\delta z) = \mathcal{K}_\mathcal{H}^t(\delta z, \{\gamma\}\delta z)$. Changing variables $\delta z \to z$ transforms the fundamental domain $\Gamma\backslash\mathcal{H}$ to the fundamental domain $\delta(\Gamma\backslash\mathcal{H})$. Assembling these ingredients, we obtain,

$$I(t) = \sum_{\{\gamma\}\neq I} \sum_{\delta\in\Gamma/\Gamma_\gamma} \int_{\delta(\Gamma\backslash\mathcal{H})} d\mu(z)\mathcal{K}_\mathcal{H}^t(z, \{\gamma\}z)$$

Next, we use the fact that,

$$\bigcup_{\delta\in\Gamma/\Gamma_\gamma} \delta(\Gamma\backslash\mathcal{H}) = \Gamma_\gamma\backslash\mathcal{H} \qquad \Longrightarrow \qquad \sum_{\delta\in\Gamma/\Gamma_\gamma} \int_{\delta(\Gamma\backslash\mathcal{H})} = \int_{\Gamma_\gamma\backslash\mathcal{H}}$$

to recast $I(t)$ as follows,

$$I(t) = \sum_{\{\gamma\}\neq I} \int_{\Gamma_\gamma\backslash\mathcal{H}} d\mu(z)\, \mathcal{K}_\mathcal{H}^t(z, \{\gamma\}z)$$

Choosing the primitive element β of $\{\gamma\}$ in diagonal form,

$$\beta = \begin{pmatrix} e^{\frac{1}{2}\ell(\beta)} & 0 \\ 0 & e^{-\frac{1}{2}\ell(\beta)} \end{pmatrix} \tag{E.27}$$

each element of the coset $\{\gamma\} \neq I$ may be represented by β^p for $p \neq 0$, whose action on z is given by $\{\gamma\}z = e^{p\ell(\beta)}z$. The fundamental domain for Γ_γ may be chosen as follows,

$$\Gamma_\gamma\backslash\mathcal{H} = \{z \in \mathcal{H},\, z = x + iy \text{ with } x \in \mathbb{R}, 1 \leq y \leq e^{\ell(\beta)}\}$$

The function $I(t)$ then becomes,

$$I(t) = \sum_{\beta \text{ primitive}} \sum_{p=1}^{\infty} \int_{\mathbb{R}} dx \int_{1}^{e^{\ell(\beta)}} \frac{dy}{y^2}\, \mathcal{K}_\mathcal{H}^t(z, e^{p\ell(\beta)}z)$$

Combining the result of $I(t)$ with the term area$(\Gamma\backslash\mathcal{H})\mathcal{K}_\mathcal{H}^t(z,z)$ gives the proposed formula and concludes the derivation of part (a) of this exercise.

(b) To derive this part, it suffices to perform the integrals over x and y

in the above expression for $I(t)$. To do so, we begin by calculating the hyperbolic lengths $\ell(z, e^{p\ell(\beta)} z)$ for β given in (E.27),

$$\cosh \ell(z, e^{p\ell(\beta)} z) = \cosh(p\ell(\beta)) + v^2$$

where we have defined,

$$v = \sqrt{2}\, \frac{x}{y}\, \sinh\left(\tfrac{1}{2} p\ell(\beta)\right)$$

Since the integrand $\varphi(v) = \mathcal{K}_{\mathcal{H}}^t(z, e^{p\ell(\beta)} z)$ depends on x and y only through v, we change variables from x, y to v, y and carry out the integration over y,

$$\int_{\mathbb{R}} dx \int_1^{e^{\ell(\beta)}} \frac{dy}{y^2}\, \varphi(v) = \frac{\ell(\beta)}{\sqrt{2}\, \sinh(\tfrac{1}{2} p\ell(\beta))} \int_{\mathbb{R}} dv\, \varphi(v)$$

Invoking now the integral representation for the heat-kernel on the upper half plane, obtained in Exercise 6.5, with the appropriate entries,

$$\int_{\mathbb{R}} dv\, \mathcal{K}_{\mathcal{H}}^t(z, e^{p\ell(\beta)} z) = \frac{e^{-t/4}}{2(2\pi t)^{\frac{3}{2}}} \int_{\mathbb{R}} dv \int_{b_*}^{\infty} db\, \frac{b\, e^{-b^2/4t}}{\sqrt{\cosh b - \cosh(p\ell(\beta)) - v^2}}$$

where $b_* > 0$ satisfies $\cosh(b_*) = \cosh(p\ell(\beta)) + v^2$. Interchanging the integrations in v and b, and carrying out the integration over v we obtain,

$$I(t) = \frac{\pi\, e^{-t/4}}{2(2\pi t)^{\frac{3}{2}}} \sum_{\beta\ \text{primitive}} \sum_{p=1}^{\infty} \frac{\ell(\beta)}{\sqrt{2}\, \sinh\left(\tfrac{1}{2} p\ell(\beta)\right)} \int_{p\ell(\beta)}^{\infty} db\, b\, e^{-b^2/4t}$$

Carrying out also the integration over b gives,

$$I(t) = \frac{e^{-t/4}}{2(2\pi t)^{\frac{1}{2}}} \sum_{\beta\ \text{primitive}} \sum_{p=1}^{\infty} \frac{\ell(\beta)}{\sqrt{2}\, \sinh\left(\tfrac{1}{2} p\ell(\beta)\right)} e^{-p^2\ell(\beta)^2/4t}$$

This remarkable formula gives the trace of the heat-kernel, which is the generating function for all the eigenvalues of the Laplacian on the Riemann surface Σ, in terms of the lengths $\ell(\beta)$ of the primitive geodesics on Σ. The construction may be generalized to an arbitrary invariant integration kernel and is referred to as the *Selberg trace formula*.

Solution to Exercise 12.5

To compute the functional determinant of $\mathcal{O} = -\Delta + s(s-1)$, we introduce the corresponding ζ-function,

$$\zeta_{\mathcal{O}}(\sigma) = \text{Tr}(\mathcal{O}^{-\sigma})$$

The operator $\mathcal{O}^{-\sigma}$ is trace-class for $\mathrm{Re}\,(\sigma) > 1$ since the spectrum of eigenvalues λ_n of $-\Delta$ increases linearly with eigenvalue number n on a compact surface Σ. The function $\zeta_{\mathcal{O}}(\sigma)$ admits an analytic continuation to $\sigma \in \mathbb{C}$ with a pole at $\sigma = 1$. The determinant is then given by,

$$\ln \mathrm{Det}\Big(-\Delta + s(s-1) \Big) = -\zeta'_{\mathcal{O}}(0)$$

We compute $\zeta_{\mathcal{O}}(\sigma)$ using the result of Exercise 12.4 for the heat-kernel,

$$\zeta_{\mathcal{O}}(\sigma) = \frac{1}{\Gamma(\sigma)} \int_0^\infty dt\, t^{\sigma-1}\, e^{-s(s-1)t}\, \mathrm{Tr}\big(\mathcal{K}^t_\Sigma\big)$$

Ignoring the term proportional to the area of Σ, the last factor in the integrand is just $I(t)$, and we have,

$$\zeta_{\mathcal{O}}(\sigma) = \frac{1}{4\sqrt{\pi}\,\Gamma(\sigma)} \sum_{\beta \text{ primitive}} \sum_{p=1}^\infty \frac{\ell(\beta)}{\sinh\big(\frac{1}{2}p\ell(\beta)\big)}$$
$$\times \int_0^\infty dt\, t^{\sigma-\frac{3}{2}}\, e^{-(s-\frac{1}{2})^2 t - p^2 \ell(\beta)^2/4t}$$

The presence of the prefactor $1/\Gamma(\sigma)$ and the fact that the t-integral is absolutely convergent for $\mathrm{Re}\,(2s-1)^2 > 0$ in view of $p\ell(\beta) > 0$ makes it very easy to evaluate the derivative of $\zeta_{\mathcal{O}}(\sigma)$ in σ at $\sigma = 0$, and we find,

$$\zeta'_{\mathcal{O}}(0) = \frac{1}{4\sqrt{\pi}} \sum_{\beta \text{ primitive}} \sum_{p=1}^\infty \frac{\ell(\beta)}{\sinh\big(\frac{1}{2}p\ell(\beta)\big)} \int_0^\infty dt\, t^{-\frac{3}{2}}\, e^{-(s-\frac{1}{2})^2 t - p^2 \ell(\beta)^2/4t}$$
$$= \frac{1}{2} \sum_{\beta \text{ primitive}} \sum_{p=1}^\infty \frac{\ell(\beta)}{\sinh\big(\frac{1}{2}p\ell(\beta)\big)} \frac{e^{-(s-\frac{1}{2})p\ell(\beta)}}{p\ell(\beta)}$$

Simplifying the factor of $\ell(\beta)$, we obtain equivalently,

$$\zeta'_{\mathcal{O}}(0) = \sum_{\beta \text{ primitive}} \sum_{p=1}^\infty \frac{1}{p} \frac{e^{-sp\ell(\beta)}}{1 - e^{p\ell(\beta)}} = \sum_{\beta \text{ primitive}} \sum_{p=1}^\infty \frac{1}{p} \sum_{k=0}^\infty e^{-(s+k)p\ell(\beta)}$$

Carrying out the sum in p, we obtain,

$$\zeta'_{\mathcal{O}}(0) = - \sum_{\beta \text{ primitive}} \sum_{k=0}^\infty \ln\Big(1 - e^{-(s+k)\ell(\beta)}\Big)$$

Inspection of the definition of the Selberg Zeta function then gives,

$$e^{-\zeta'_{\mathcal{O}}(0)} = \mathrm{Det}\big(-\Delta + s(s-1)\big) = Z(s)$$

up to the exponential of the area-dependent factor which is actually proportional to the Euler characteristic since curvature is -1. This result was
obtained in [DP86b] and [Sar87].

Solution to Exercise 13.1

The T-duality group $SO(2,2;\mathbb{Z})$ for $\mathbb{T}^2$ is composed of the following three
types of transformations: (i) the $PSL(2,\mathbb{Z})_\sigma$ duality group of the space-time
torus, (ii) integer shifts of the B-field $B_{12} \to B_{12} + N$, and (iii) individual
T-duality transformations along the two cycles of the torus. By definition,
the transformations in $PSL(2,\mathbb{Z})_\sigma$ act as Möbius transformations on the
complex parameter σ of the space-time torus, while integer shifts of the B-
field are generated by $T_\rho : \rho_1 \to \rho_1 + 1$. As for the T-duality transformations,
begin by considering the simultaneous transformation of both directions of
$\mathbb{T}^2$. The fields G and B then transform as in (13.5.10), which using the
symmetry properties of G and B may be recast as,

$$\widetilde{G} = \tfrac{1}{2}\left[(G+B)^{-1} + (G-B)^{-1}\right]$$
$$\widetilde{B} = \tfrac{1}{2}\left[(G+B)^{-1} - (G-B)^{-1}\right]$$

We thus have $\widetilde{G} + \widetilde{B} = (G+B)^{-1}$. By explicitly summing and inverting the
matrices in (13.6.11), one finds,

$$\widetilde{G} + \widetilde{B} = \frac{1}{|\rho|^2}\frac{\rho_2}{\sigma_2}\begin{pmatrix} 1 & -\sigma_1 \\ -\sigma_1 & |\sigma|^2 \end{pmatrix} + \frac{1}{|\rho|^2}\begin{pmatrix} 0 & -\rho_1 \\ \rho_1 & 0 \end{pmatrix} \qquad \text{(E.28)}$$

Conjugation by $\begin{pmatrix} 0 & -1 \\ 1 & 0 \end{pmatrix}$ relates this to $G + B$ itself, up to $\rho \to -1/\rho$. Thus
simultaneous T-duality corresponds $S_\rho : \rho \to -1/\rho$, which together with in-
teger shifts of B generates $PSL(2,\mathbb{Z})_\rho$. The remaining $\mathbb{Z}_2$ actions correspond
to T-duality along just a single cycle of the torus, that is, $(\sigma, \rho) \to (\rho, \sigma)$,
and the exchange of the two cycles of the torus $(\sigma, \rho) \to (-\bar{\sigma}, -\bar{\rho})$.

Finally, by direct substitution into (13.5.13), one finds the following expressions for the left- and right-moving momenta,

$$4p_L^2 = \frac{1}{\sigma_2 \rho_2}\left|(m_1 - m_2\sigma) + \bar{\rho}\,(n_2 + n_1\sigma)\right|^2$$
$$4p_R^2 = \frac{1}{\sigma_2 \rho_2}\left|(m_1 - m_2\sigma) + \rho\,(n_2 + n_1\sigma)\right|^2 \qquad \text{(E.29)}$$

Solution to Exercise 13.2

First note that worldsheet parity Ω acts as $G \to G$ and $B \to -B$, and so by (13.6.11),

$$\Omega: \quad (\sigma, \rho) \to (\sigma, -\bar{\rho}) \tag{E.30}$$

For the choices of σ and ρ given in the problem statement, we have,

$$E = G + B = \begin{pmatrix} \rho_2 & \rho_1 \\ -\rho_1 & \rho_2 \end{pmatrix} \qquad |\rho|^2 = 1 \tag{E.31}$$

from which it follows that

$$E(\rho_1, \rho_2) = E(-\rho_1, \rho_2)^{-1} \tag{E.32}$$

We thus conclude that the sigma model is left-invariant by the combined operation $\Omega \gamma_1 \gamma_2$, with γ_i defined in (13.5.28).

Solution to Exercise 13.3

The starting point is the action,

$$S = \frac{1}{4\pi\alpha'} \int_\Sigma \left(G_{\mu\nu} \, dX^\mu \wedge \star dX^\nu - i B_{\mu\nu} \, dX^\mu \wedge dX^\nu \right)$$

We begin by defining $f^\mu = dX^\mu$ and imposing the Bianchi identity via Lagrange multipliers $\tilde{X}_\mu$,

$$S = \frac{1}{4\pi\alpha'} \int_\Sigma \left(G_{\mu\nu} \, f^\mu \wedge \star f^\nu - i B_{\mu\nu} \, f^\mu \wedge f^\nu \right) + \frac{i}{2\pi\alpha'} \int_\Sigma f^\mu \wedge d\tilde{X}_\mu$$

The field equation for f^μ is,

$$G_{\mu\nu}(\star f^\nu) - i B_{\mu\nu} f^\nu + i d\tilde{X}_\mu = 0$$

which may be solved for f^μ to give,

$$f^\mu = -\tilde{G}^{\mu\nu} B_{\nu\rho} G^{\rho\sigma} d\tilde{X}_\sigma + i \, \tilde{G}^{\mu\nu} \star d\tilde{X}_\nu$$

Here we have defined $\tilde{G}_{\mu\nu} = G_{\mu\nu} - B_{\mu\rho} G^{\rho\sigma} B_{\sigma\nu}$, while $\tilde{G}^{\mu\nu}$ are the components of the matrix $\tilde{G}^{-1}$. Plugging the result for f^μ back into the action, and noting that $\tilde{G} B G^{-1} = G^{-1} B \tilde{G}$ (which follows straightforwardly from the symmetry properties of $G_{\mu\nu}$ and $B_{\mu\nu}$), we obtain,

$$S = \frac{1}{4\pi\alpha'} \int_\Sigma \left(\tilde{G}^{\mu\nu} d\tilde{X}_\mu \wedge \star d\tilde{X}_\nu - i \tilde{B}^{\mu\nu} d\tilde{X}_\mu \wedge d\tilde{X}_\nu \right)$$

where $\tilde{G} = (G - BG^{-1}B)^{-1}$ and $\tilde{B} = -\tilde{G} B G^{-1}$, thereby reproducing the results of (13.5.10).

Solution to Exercise 13.4

We begin from the partition function,

$$Z(\mathsf{L}|\tau) = \frac{1}{|\eta(\tau)|^{2d}} \sum_{p \in \mathsf{L}} e^{i\pi\tau p_L^2 - i\pi\bar{\tau}p_R^2}$$

Demanding invariance under $\tau \to \tau + 1$ gives immediately,

$$p_L^2 - p_R^2 \in 2\mathbb{Z}$$

which is equivalent to evenness, $p \circ p \in 2\mathbb{Z}$. As for invariance under $\tau \to -1/\tau$, we may first note that,

$$\sum_{p' \in \mathsf{L}} \delta^{(2d)}(p - p') = V_\mathsf{L}^{-1} \sum_{p'' \in \mathsf{L}^*} e^{2\pi i\, p'' \circ p}$$

from which we obtain,

$$Z\left(\mathsf{L}\,\Big|\, -\frac{1}{\tau}\right) = \frac{V_\mathsf{L}^{-1}}{|\eta(-1/\tau)|^{2d}} \sum_{p'' \in \mathsf{L}^*} \int d^{2d}p\; e^{2\pi i\, p'' \circ p - \frac{\pi i}{\tau}p_L^2 + \frac{\pi i}{\bar{\tau}}p_R^2}$$

$$= |\tau|^d \frac{V_\mathsf{L}^{-1}}{|\eta(-1/\tau)|^{2d}} \sum_{p'' \in \mathsf{L}^*} e^{\pi i\tau(p_L'')^2 - \pi i\bar{\tau}(p_R'')^2}$$

$$= V_\mathsf{L}^{-1} Z(\mathsf{L}^*|\tau)$$

To have a modular invariant result for all τ, we must thus have $\mathsf{L} = \mathsf{L}^*$, where we note that since $V_\mathsf{L} = V_{\mathsf{L}^*}^{-1}$, self-dual lattices satisfy $V_\mathsf{L} = 1$.

Solution to Exercise 13.5

Let $\Gamma^{2,2}$ be the lattice of left- and right-moving momenta, in terms of which the partition function is,

$$Z(\tau) = \frac{1}{|\eta(\tau)|^4} \sum_{(p_L, p_R) \in \Gamma^{2,2}} q^{\frac{1}{2}p_L^2}\, \bar{q}^{\frac{1}{2}p_R^2}$$

We further define the sets,

$$\Gamma_0 = \left\{ p_L \;\Big|\; \begin{pmatrix} p_L \\ 0 \end{pmatrix} \in \Gamma^{2,2} \right\} \qquad \Gamma_L = \left\{ p_L \;\Big|\; \begin{pmatrix} p_L \\ * \end{pmatrix} \in \Gamma^{2,2} \right\}$$

The set Γ_0 forms a sublattice of $\Gamma^{2,2}$ of rank no greater than 2, and generically 0. This follows since, by Exercise 13.4, elements of $\Gamma^{2,2}$ must have $p_L^2 - p_R^2 \in 2\mathbb{Z}$, which for $p_R = 0$ becomes the constraint,

$$\frac{1}{4}\left(\frac{\sqrt{\alpha'}}{R}m + \frac{R}{\sqrt{\alpha'}}n\right)^2 \in 2\mathbb{Z}$$

For generic R, this admits no solutions. On the other hand, Γ_L is not in general a sublattice. The statement of rationality of the CFT is equivalent to the statement that Γ_0 is a finite index sublattice of Γ_L, and in this case, both Γ_0 and Γ_L are rank two sublattices of $\Gamma^{2,2}$.

We now prove that Γ_0 is a finite index sublattice of Γ_L if and only if σ and ρ take values in an imaginary quadratic field $\mathbb{Q}(\sqrt{D})$. First, assume that Γ_0 is a finite index sublattice. Then by the explicit expression for p_R given in (E.29), together with the fact that Γ_0 is rank 2, we may write two independent linear relations,

$$m_1 - m_2\,\sigma + n_2\,\rho + n_1\,\rho\sigma = 0$$
$$m_1' - m_2'\,\sigma + n_2'\,\rho + n_1'\,\rho\sigma = 0 \tag{E.33}$$

Solving for ρ from the second and plugging into the former gives an equation of the form $a\sigma^2 + b\sigma + c = 0$, which we recognize as the condition to have $\sigma \in \mathbb{Q}(\sqrt{D})$ for $D = b^4 - 4ac$. Plugging this back into (E.33) shows that $\rho \in \mathbb{Q}(\sqrt{D})$ as well.

Conversely, assume that $\sigma, \rho \in \mathbb{Q}(\sqrt{D})$. In this case, it is possible to write σ and $\sigma\rho$ as linear functions of ρ, since everything involved is an element of $\mathbb{Q}(\sqrt{D})$. Multiplying said relations by suitable integers, we obtain a pair of equations of the form (E.33), with $m_1 = 0$ and $n_2' = 0$. These relations are by construction independent, and hence define a rank-2 lattice Γ_0, which completes the proof. A more thorough discussion can be found in [GV04].

Solution to Exercise 14.1

Poincaré invariance allows for a conformal factor $e^{B(z)}$ multiplying the $\mathbb{R}^8$ part of the metric. Einstein's equations require e^{4B} to be harmonic which would require space-time singularities unless B is constant. By rescaling the x^μ, we may set $B = 1$. By suitably choosing complex coordinates z and $\bar{z}$ adapted to the remaining metric in the x^8 and x^9 part, we may put that part of the metric in conformally flat form. The conformal factor e^A drops out of the field equation for τ which reduces to the following expression,

$$(\tau - \bar{\tau})\partial_z\partial_{\bar{z}}\tau - 2(\partial_z\tau)(\partial_{\bar{z}}\tau) = 0$$

This equation manifestly admits solutions satisfying $\partial_{\bar{z}}\tau = 0$.

Solution to Exercise 14.2

The Einstein equations in the metric G_E read as follows (recall that G_E is the Einstein frame metric),

$$R_{\mu\nu} - \frac{1}{2}G_{E\mu\nu}R = \frac{\partial_\mu\tau\partial_\nu\bar\tau}{2\tau_2^2} - G_{E\mu\nu}\,G_E^{\rho\sigma}\frac{\partial_\rho\tau\partial_\sigma\bar\tau}{4\tau_2^2}$$

where the right side is the stress tensor for the field τ. The only nonvanishing entries of the Ricci tensor are $R_{88} = R_{99} = -2\partial_z\partial_{\bar z}A(z)$, which gives the Ricci scalar $R = -4e^{-A}\partial_{\bar z}\partial_z A$. The equations reduce to a single relation for A which, for the solutions with $\partial_{\bar z}\tau = 0$, becomes,

$$\partial_{\bar z}\partial_z A = -\frac{\partial_z\tau\partial_{\bar z}\bar\tau}{4\tau_2^2} = \partial_{\bar z}\partial_z \ln\tau_2$$

The general solution is given by $A(z,\bar z) = \ln\tau_2 + h(\tau,\bar\tau)$, where $h(\tau,\bar\tau)$ is harmonic. Since τ_2 is not modular invariant, that is, not single-valued in z, the right side is not a single-valued function of z. This is easily remedied with the help of the Dedekind η function,

$$e^{A(z,\bar z)} = \tau_2|\eta(\tau)|^4|f(z)|^2$$

and $|f(z)|^2$ is now a single-valued function of z. One may think of the brane as being localized near a point z_B. When observing the brane from afar, we may approximate the brane profile by a Dirac delta-function at z_B, so that $\partial_{\bar z}\partial_z A = c\delta(z - z_B)$. The normalization c of the delta-function is obtained by integrating both sides over z,

$$c = \int d^2z\,\partial_{\bar z}\partial_z A = -\int d^2z\,\frac{\partial_z\tau\partial_{\bar z}\bar\tau}{4\tau_2^2} = -\frac{1}{4}\int_F \frac{d^2\tau}{\tau_2^2} = -\frac{\pi}{12}$$

where we have used the fact that the map $j : F \to \mathbb{C}$ from the fundamental domain F of $\mathrm{SL}(2,\mathbb{Z})$ to $\mathbb{C}$ is one-to-one. Hence as $z \to \infty$ we have,

$$A(z) \sim -\frac{1}{12}\ln|z|^2$$

Next, we show that the asymptotic behavior of the brane is conical by writing out the above asymptotic form of the metric in the x^8 and x^9 plane in polar coordinates $z = \rho\,e^{i\theta}$,

$$ds_{89}^2 = |z|^{-\frac{1}{6}}|dz|^2 = \rho^{-\frac{1}{6}}(d\rho^2 + \rho^2 d\theta^2)$$

Changing coordinates to $\frac{12}{11}\rho^{\frac{11}{12}} = r$, the metric becomes,

$$ds_{89}^2 = dr^2 + \left(\frac{11}{12}\right)^2 r^2 d\theta^2$$

The deficit angle is $2\pi/12$. For more details, we refer the reader to [GSVY90].

Solution to Exercise 14.3

For the solutions with multiple D7-branes at positions z_i, we may use the following Ansatz,

$$f(z) = \prod_{i=1}^{N} \frac{1}{(z - z_i)^{\frac{1}{12}}}$$

The solution is no longer a cone but, in the approximation where the curvature due to each brane is supported at the positions of the branes z_i, the metric is locally flat, with a curvature singularity located at each z_i. Each D7-brane contributes $\pi/6$ to the total deficit angle, which is thus $N\pi/6$. For total deficit angle 4π, the cone closes up to a sphere, which requires a total 24 different D7-branes. The positions of the branes are subject to further monodromy constraints that we shall not analyze here.

Solution to Exercise 14.4

Using the transformations (14.6.1) and (14.6.2), one readily computes,

$$T_{(q,p)} \rightarrow \frac{1}{2\pi\alpha'} \frac{|c\tau + d|}{\sqrt{\tau_2}} \left| p + q\left(\frac{a\tau + b}{c\tau + d}\right) \right|$$
$$= \frac{1}{2\pi\alpha'} \frac{1}{\sqrt{\tau_2}} |(pd + bq) + (cp + aq)\tau|$$

under an $SL(2, \mathbb{Z})$ transformation. From this we see that (p, q) strings turn into (p', q') strings, with

$$(q', p') = (q, p) \begin{pmatrix} a & b \\ c & d \end{pmatrix}$$

This is what we would expect from the worldsheet coupling

$$\int (q, p) \begin{pmatrix} C_2^2 \\ C_2^1 \end{pmatrix} \rightarrow \int (q, p) \begin{pmatrix} a & b \\ c & d \end{pmatrix} \begin{pmatrix} C_2^2 \\ C_2^1 \end{pmatrix} .$$

Solution to Exercise 14.5

The key is that a (p, q) string cannot experience monodromy from a (p, q) 7-brane. If it did, then we could consider moving a (p, q) string anchored on the brane once around it, in which case it would transform into a new type of string, still anchored on the 7-brane. This is not possible if only (p, q)-strings can end on (p, q) 7-branes. In the case of a D7-brane, it is known that a fundamental, that is, $(1, 0)$, string can end on the brane, and indeed we have,

$$(0, 1) \begin{pmatrix} 1 & 1 \\ 0 & 1 \end{pmatrix} = (0, 1)$$

We now consider monodromy matrices $M_{(p,q)}$ defined such that

$$(q, p) M_{(p,q)} = (q, p) \tag{E.34}$$

We may simplify this by noting that,

$$(q, p) = (0, 1) \begin{pmatrix} r & s \\ q & p \end{pmatrix}$$

for any r and s, including those satisfying $rp - sq = 1$. Recasting (E.34) as,

$$(0, 1) \begin{pmatrix} r & s \\ q & p \end{pmatrix} M_{(p,q)} \begin{pmatrix} r & s \\ q & p \end{pmatrix}^{-1} = (0, 1)$$

it follows that the monodromy matrix is,

$$M_{(p,q)} = \begin{pmatrix} r & s \\ q & p \end{pmatrix}^{-1} \begin{pmatrix} 1 & 1 \\ 0 & 1 \end{pmatrix} \begin{pmatrix} r & s \\ q & p \end{pmatrix} = \begin{pmatrix} 1 + pq & p^2 \\ -q^2 & 1 - pq \end{pmatrix}$$

Note that r and s drop out of the final expression (and, as it turns out, out of any physical quantities).

Solution to Exercise 15.1

We begin by Taylor expanding $\Phi(y)$ in terms of θ and $\bar{\theta}$,

$$\Phi = \phi(y) + \sqrt{2}\theta\psi(y) + \theta\theta F(y)$$
$$= \phi(x) + i\theta\sigma^\mu\bar{\theta}\,\partial_\mu\phi(x) + \frac{1}{4}\theta\theta\bar{\theta}\bar{\theta}\,\partial^2\phi(x)$$
$$-\sqrt{2}\theta\,\psi(x) - \frac{i}{\sqrt{2}}\theta\theta\partial_\mu\psi(x)\sigma^\mu\bar{\theta} + \theta\theta F(x)$$

and likewise,

$$\Phi^\dagger = \phi^*(x) - i\theta\sigma^\mu\bar{\theta}\,\partial_\mu\phi^*(x) + \frac{1}{4}\theta\theta\bar{\theta}\bar{\theta}\,\partial^2\phi^*(x)$$
$$+\sqrt{2}\bar{\theta}\,\bar{\psi}(x) + \frac{i}{\sqrt{2}}\bar{\theta}\bar{\theta}\sigma^\mu\partial_\mu\bar{\psi}(x) + \bar{\theta}\bar{\theta}F^*(x)$$

where ϕ^* and F^* denote the complex conjugates of ϕ and F, respectively. On the other hand, from (15.2.2) one obtains,

$$V = -\theta\sigma^\mu\bar{\theta}A_\mu(x) + i\theta\theta\bar{\theta}\bar{\lambda}(x) - i\bar{\theta}\bar{\theta}\theta\lambda(x) + \frac{1}{2}\theta\theta\bar{\theta}\bar{\theta}\dot{D}(x)$$

Using $\int d^2\theta\,\theta\theta = 1$, it is then straightforward to compute,

$$\int d^2\theta d^2\bar{\theta}\,\Phi^\dagger e^V \Phi = \phi\partial^2\phi^* + i\partial_\mu\bar{\psi}\bar{\sigma}^\mu\psi + FF^*$$

$$+ A^\mu \left(\frac{1}{2}\bar{\psi}\bar{\sigma}^\mu\psi + \frac{i}{2}\phi^*\partial_\mu\phi - \frac{i}{2}\partial_\mu\phi^*\phi\right)$$

$$- \frac{i}{\sqrt{2}}\left(\phi\bar{\lambda}\bar{\psi} - \phi^*\lambda\psi\right) + \frac{1}{2}\left(D - \frac{1}{2}A_\mu A^\mu\right)\phi^*\phi$$

using the usual formulas for Berezin integration. All fields on the right-hand side are functions of the usual space-time coordinates x_μ.

As for the kinetic term for the vector superfield, we first compute,

$$W_\alpha = -i\lambda_\alpha(y) + \theta_\alpha D(y) - \frac{i}{2}(\sigma^\mu\bar{\sigma}^\nu\theta)_\alpha F_{\mu\nu}(y) + \theta\theta\left(\sigma^\mu\partial_\mu\bar{\lambda}(y)\right)_\alpha$$

Plugging in $y_\mu = x_\mu + i\theta\sigma_\mu\bar{\theta}$, Taylor expanding in the Grassmann variables, and evaluating the integral then gives,

$$\mathrm{Re}\int d^2\theta\,W^a W_a = D^2 - \frac{1}{2}F^{\mu\nu}F_{\mu\nu} - 2i\lambda\sigma^\mu\partial_\mu\bar{\lambda}$$

where once again all fields on the right-hand side are functions of the usual space-time coordinates x_μ.

Solution to Exercise 15.2

Denote the path integral by,

$$Z[g^2] = \int DA \exp\left(-\frac{1}{2g^2}\int F \wedge *F\right)$$

We now introduce a field B with integer periods that couples as,

$$Z[g^2] = \int \frac{DA\,DB\,D\hat{A}}{\mathcal{G}} \exp\left(-\frac{1}{2g^2}\int (F - B) \wedge *(F - B) + i\int B \wedge d\hat{A}\right)$$

where we have quotiented by the new gauge symmetry $\mathcal{G}$, and the field $\hat{A}$ is a Lagrange multiplier enforcing $dB = 0$. When $dB = 0$ and B has integer periods, the gauge symmetry can be fixed by setting $B = 0$, from which the equality with the original expression follows. But we can instead fix the gauge symmetry by setting $A = 0$, in which case we are left with a Gaussian functional integral over B producing an effective action for $\hat{A}$,

$$Z[g^2] = \int D\hat{A} \exp\left(-\frac{g^2}{2}\int d\hat{A} \wedge *d\hat{A}\right) = Z[1/g^2]$$

Solution to Exercise 15.3

Identifying maximal sets of points (z_e, z_m) that satisfy the mutual locality constraint $z_e z'_m - z_m z'_e = 0$ mod 3, one finds four such solutions,

$$
\begin{aligned}
SU(3): &\qquad \{(0,0),(1,0),(2,0)\} \\
(SU(3)/\mathbb{Z}_3)_0: &\qquad \{(0,0),(0,1),(0,2)\} \\
(SU(3)/\mathbb{Z}_3)_1: &\qquad \{(0,0),(1,1),(2,2)\} \\
(SU(3)/\mathbb{Z}_3)_2: &\qquad \{(0,0),(1,2),(2,1)\}
\end{aligned}
$$

The map between them under $\mathrm{SL}(2,\mathbb{Z})$ can be obtained by using the rules in (15.5.14), with the results being,

$$
\begin{aligned}
S: \quad & SU(3) \to (SU(3)/\mathbb{Z}_3)_0 & \qquad T: \quad & SU(3) \to SU(3) \\
& (SU(3)/\mathbb{Z}_3)_0 \to SU(3) & & (SU(3)/\mathbb{Z}_3)_0 \to (SU(3)/\mathbb{Z}_3)_1 \\
& (SU(3)/\mathbb{Z}_3)_1 \to (SU(3)/\mathbb{Z}_3)_2 & & (SU(3)/\mathbb{Z}_3)_1 \to (SU(3)/\mathbb{Z}_3)_2 \\
& (SU(3)/\mathbb{Z}_3)_2 \to (SU(3)/\mathbb{Z}_3)_1 & & (SU(3)/\mathbb{Z}_3)_2 \to (SU(3)/\mathbb{Z}_3)_0
\end{aligned}
$$

Solution to Exercise 15.4

Comparing the line operators of $\mathrm{SU}(2)$ to those of $\mathrm{SO}(3)_+$, we see that the former include lines of unit electric charge, while the latter includes lines of unit magnetic charge. This is due to the fact that in the former theory all dynamical particles have even electric charge, while in the latter theory all dynamical particles have even magnetic charge. In going between the two, the spectrum of charges are thus rescaled via

$$\left(q^{SO(3)_+}, m^{SO(3)_+}\right) = \left(\tfrac{1}{2}q^{SU(2)}, 2m^{SU(2)}\right)$$

From the BPS mass formula, we immediately conclude that,

$$a^{SO(3)_+} = 2a^{SU(2)} \qquad\qquad a_D^{SO(3)_+} = \tfrac{1}{2}a_D^{SU(2)}$$

from which is follows that the correspond cycles obey,

$$\mathfrak{A}^{SO(3)_+} = 2\,\mathfrak{A}^{SU(2)} \qquad\qquad \mathfrak{B}^{SO(3)_+} = \tfrac{1}{2}\,\mathfrak{B}^{SU(2)}$$

Since $\tau = \partial a_D / \partial a$, we furthermore find that,

$$\tau_{SO(3)_+} = \tfrac{1}{4} \tau_{SU(2)}$$

In the case of $SO(3)_-$, similar considerations lead to,

$$a^{SO(3)-} = 2 \, a^{SU(2)} \qquad\qquad a_D^{SO(3)-} = \tfrac{1}{2} a_D^{SU(2)} + a^{SU(2)}$$

$$\mathfrak{A}^{SO(3)-} = 2 \, \mathfrak{A}^{SU(2)} \qquad\qquad \mathfrak{B}^{SO(3)-} = \tfrac{1}{2} \mathfrak{B}^{SU(2)} + \mathfrak{A}^{SU(2)}$$

as well as,

$$\tau_{SO(3)_-} = \tfrac{1}{4} \tau_{SU(2)} + \tfrac{1}{2}$$

Solution to Exercise 15.5

On the torus $\mathbb{T}^2 = \mathbb{C}/\Lambda$ with $\Lambda = \mathbb{Z}\omega_1 + \mathbb{Z}\omega_2$, we choose local complex coordinates $z, \bar{z}$ with $z = x\omega_1 + y\omega_2$ and $x, y \in \mathbb{R}/\mathbb{Z}$ such that the holomorphic differential dz is normalized by $\oint_{\mathfrak{A}} dz = \omega_1$ and $\oint_{\mathfrak{B}} dz = \omega_2$. Toroidal compactification on $\mathbb{T}^2$ requires that the field H is invariant under translations by $\mathbb{T}^2$, so that it must be of the form $H = F \wedge dx + \tilde{F} \wedge dy$, where F and $\tilde{F}$ are 2-forms on M_4 that are independent of $\mathbb{T}^2$. Since we have assumed that the metric factorizes on $M_4 \times \mathbb{T}^2$, and that M_4 has Minkowski signature while $\mathbb{T}^2$ has Euclidean signature, we have $\star dx = dy$ and $\star dy = -dx$. Therefore, self-duality of H implies $\tilde{F} = \star F$, so that

$$H = F \wedge dx + \star F \wedge dy \qquad\qquad dH = 0 \Longrightarrow \begin{cases} dF &= 0 \\ d \star F &= 0 \end{cases}$$

The equations on the right for the antisymmetric tensor field F are Maxwell's equations. Modular transformations act as follows,

$$\begin{cases} \omega_1' = a\omega_1 + b\omega_2 \\ \omega_2' = c\omega_1 + d\omega_2 \end{cases} \qquad \begin{cases} x = ax' + cy' \\ y = bx' + dy' \end{cases} \qquad \begin{cases} F' &= aF + b \star F \\ \star F' &= cF + d \star F \end{cases}$$

Under the transformation S with $a = d = 0$ and $b = -c = 1$, we get familiar electric–magnetic duality $F' = - \star F$. For more details, we refer the reader to [Wit95a], [Ver95], and [Vaf98].

Solution to Exercise 16.1

The eigenvalues are all the roots of the characteristic polynomial $\varphi(\lambda) = \det(M - \lambda I)$ so that $\mathbb{Q}(\lambda_1, \cdots, \lambda_n)$ is a splitting field for $\varphi(\lambda)$ and a normal extension of $\mathbb{Q}$. Assuming the eigenvalues are all distinct, the characteristic polynomial is separable and the Galois group is a subgroup of S_n by Theorem 16.3.4. Given the eigenvalues $\lambda_1, \cdots, \lambda_n$ of M, the eigenvectors v of M can be constructed using linear algebra to solve the equation $(M - \lambda_i)v_i = 0$ so that the entries of v_i, up to rescaling, take values in $\mathbb{Q}(\lambda_1, \cdots, \lambda_n)$.

Solution to Exercise 16.2

The matrix M commutes with the translation matrix T whose entries are $T_{ij} = \delta_{i,j-1}$, where $\delta_{i,j}$ is the Kronecker delta symbol, which is periodic with period N. One verifies that $T^N = I$ but $T^n \neq I$ for $n \not\equiv 0 \pmod{N}$. Hence the eigenvalues of T are $\mu_n = e^{2\pi i n / N}$ for $n \in \{0, 1, \cdots, N-1\} = \mathbb{Z}_N$. The extension field $\mathbb{Q}(\lambda_1, \cdots, \lambda_{N-1})$ is the Nth cyclotomic field. The Galois group is Abelian and isomorphic to the multiplicative group $\mathbb{Z}_N^*$ whose order is the Euler totient function $\phi(N)$.

Solution to Exercise 16.3

The three modular functions $h_2(v_\alpha | \tau)$ correspond to,

$$h_2(v_\alpha | \tau) = \frac{g_2}{g_3} \wp(\tfrac{\omega_\alpha}{2} | \tau) = \frac{g_2}{g_3} e_\alpha \tag{E.35}$$

using the notation of (2.4.10) for $\omega_1 = 1$ and $\omega_2 = \tau$. Since the e_α satisfy the equation $4x^3 - g_2 x - g_3 = 0$, the functions $h_\alpha = h_2(v_\alpha | \tau)$ satisfy the cubic,

$$4(j - 1728)h_\alpha^3 - 27j(h_\alpha + 1) = 0 \tag{E.36}$$

The Galois group permutes the three roots and is S_3.

Solution to Exercise 16.4

We shall use the addition formula for the elliptic curve $y^2 = 4x^3 - g_2 x - g_3$, given in (2.9.2) and (2.9.3), to characterize the functions $h_N(v | \tau)$ for which $Nv \equiv 0 \pmod{N}$ but $v \not\equiv 0 \pmod{N}$. Rescaling $(x_n, y_n) = (\wp(nv | \tau), \wp'(nv | \tau))$ to obtain modular functions $X_n, Y_n, \lambda = g_2^3 / g_3^2 = 27 \mathsf{E}_4^3 / \mathsf{E}_6^2$, we have,

$$x_n = \frac{g_3}{g_2} X_n \qquad y_n = \left(\frac{g_3}{g_2}\right)^{\frac{3}{2}} Y_n \qquad \lambda = \frac{27j}{j - 1728}$$

In terms of $X = X_1$ and $Y = Y_1$ the elliptic curve is $Y^2 = 4X^3 - \lambda(X+1)$, the doubling formulas become,

$$X_2 = -2X + \frac{(12X^2 - \lambda)^2}{16Y^2}$$

$$Y_2 = -Y + \frac{3X(12X^2 - \lambda)}{2Y} - \frac{(12X^2 - \lambda)^3}{32Y^3}$$

while the addition formulas provide the following recursion relations,

$$X_{n+1} = -X - X_n + \frac{1}{4}\left(\frac{Y_n - Y}{X_n - X}\right)^2$$

$$Y_{n+1} = -\frac{Y_n - Y}{X_n - X}X_{n+1} - \frac{X_n Y - X Y_n}{X_n - X}$$

The iterations that give (X_n, Y_n) in terms of X, Y, and λ produce X_n and Y_n/Y as rational functions of X and λ,

$$X_n = \frac{P_n(X, \lambda)}{Q_n(X, \lambda)} \qquad\qquad Y_n = Y\frac{\tilde{P}_n(X, \lambda)}{\tilde{Q}_n(X, \lambda)}$$

By construction, $P_n, Q_n, \tilde{P}_n$, and $\tilde{Q}_n$ are polynomials in X and λ with integer coefficients, that is, they are elements of the ring $\mathbb{Z}[X, j]$. Here we shall only need X_n since it determines Y_n/Y. The polynomial Q_n has degree $\deg(Q_n) = n^2 - 1$ in X, as expected from the partial fraction decomposition,

$$N^2 \wp(Nz) = \wp(z) + \sum_{v \in \Lambda'_N} \frac{\wp'(v)^2 + \wp''(v)(\wp(z) - \wp(v))}{(\wp(z) - \wp(v))^2} \qquad (\text{E.37})$$

where Λ'_N is the lattice $\mathbb{Z}_N \times \mathbb{Z}_N$ with v and $1 - v$ identified and the origin 0 deleted. For the half-periods v, we have $\wp'(v) = 0$ so that only simple poles in $\wp(z)$ arise. Requiring $Nv \equiv 0 \pmod{N}$ amounts to $(X_N, Y_N) = (\infty, \infty)$ (the identity on the elliptic curve) so that X must be a zero of the polynomial $Q_N(X, \lambda)$. The lowest degree polynomials are given by,

$$\begin{aligned}
&Q_2(X, \lambda) = 4X^3 - \lambda(X+1) && Q_3(X, \lambda) = F_3(X, \lambda)^2 \\
&Q_4(X, \lambda) = Q_2(X, \lambda)F_4(X, \lambda)^2 && Q_5(X, \lambda) = F_5(X, \lambda)^2 \\
&Q_6(X, \lambda) = Q_2(X, \lambda)Q_3(X, \lambda)F_6(X, \lambda)^2 && (\text{E.38})
\end{aligned}$$

with the polynomials $F_n(X, \lambda)$ given by,

$$F_3(X, \lambda) = 48X^4 - 24\lambda X^2 - 48\lambda X - \lambda^2$$

$$F_4(X, \lambda) = 64X^6 - 80\lambda X^4 - 320\lambda X^3 - 20\lambda^2 X^2 - 16\lambda^2 X - 32\lambda^2 + \lambda^3$$

$$F_5(X, \lambda) = 20480X^{12} - 63488\lambda X^{10} - 389120\lambda X^9 + \cdots + \lambda^6$$

$$F_6(X, \lambda) = 4096X^{12} - 22528\lambda X^{10} - 225280\lambda X^9 + \cdots - 3\lambda^6$$

The partial fraction decomposition of (E.37) also shows that all the zeros of $Q_n(X, \lambda)$, other than those corresponding to the half-periods for which $2 \cdot v \equiv 0 \pmod{N}$, must be double, and this result is indeed borne out in the above expressions for the polynomials $Q_N(X, \lambda)$. The polynomials Q_2, F_3, F_4, F_5, and F_6 are irreducible in $\mathbb{Q}(\lambda)$.

Solution to Exercise 16.5

The zeros of Q_2 in Exercise 16.4 are the three nonzero half-periods of Exercise 16.3 and produce zeros of $Q_N(X, \lambda)$ whenever N is even, since $2 \cdot \frac{N}{2} v \equiv 0 \pmod{N}$. Similarly, for any divisor $d|N$, we have $d \cdot \frac{N}{d} v \equiv 0 \pmod{N}$ so that $Q_N(X, \lambda)$ is divisible by $Q_d(X, \lambda)$. This prediction is indeed borne out in the list of Q_N given in Exercise 16.4 since $Q_4(X, \lambda)$ and $Q_6(X, \lambda)$ are divisible by $Q_2(X, \lambda)$, while $Q_6(X, \lambda)$ is divisible by $Q_3(X, \lambda)$. The polynomials Q_2, F_3, F_4, F_5, and F_5 are all irreducible over $\mathbb{C}(j)$.

The field $K = \mathbb{C}(X(N)) = \mathbb{C}(j, \{h_N(v), v \in \Lambda'_N\})$ is a finite algebraic extension of the field $\mathbb{C}(j)$ and is identical to the field $\mathbb{C}(j, h_N(1\ 0), h_N(0\ 1))$. More precisely, $\mathbb{C}(j, \{h_N(v), v \in \Lambda'_N\})$ is a finite extension of $\mathbb{Q}(j)$. A transformation $\gamma \in \mathrm{SL}(2, \mathbb{Z})$ leaves j and λ and thus $\mathbb{C}(j)$, $\mathbb{Q}(j)$, and the polynomials $Q_N(X, \lambda)$ invariant. The functions $h_N(v|\tau)$ with $v = (\mu\ \nu)$ transforms under $\gamma \in \mathrm{SL}(2, \mathbb{Z})$ as follows,

$$h_N(v|\tau) \to h_N(v\gamma|\gamma\tau) \qquad v\gamma = (a\mu + c\nu\ \ b\mu + d\nu) \qquad \text{(E.39)}$$

For each v, the function $h_N(v|\tau)$ is invariant under $\Gamma(N) \times \mathbb{Z}_2$, the factor of $\mathbb{Z}_2$ being due to our freedom to reverse the sign of v. Therefore, $\mathrm{SL}(2, \mathbb{Z})$ induces a representation θ in the Galois group of the extension $K : \mathbb{C}(j)$, which may be shown to be injective, so that the Galois group is isomorphic to $\mathrm{SL}(2, \mathbb{Z})/(\Gamma(N) \times \mathbb{Z}_2) = \mathrm{SL}(2, \mathbb{Z}_N)/\mathbb{Z}_2$; see (6.2.5). The factorization of $\mathrm{SL}(2, \mathbb{Z}_N)$ of (6.3.1) given by the prime factorization of N is reflected in the factorization of the polynomials $Q_N(X, \lambda)$ into irreducible polynomials (E.38). In particular, when N is an odd prime, $Q_N(X, \lambda)$ is the square of a polynomial $F_N(X, \lambda)$ which is irreducible over $\mathbb{Q}(j)$. Thorough discussions of the Galois group structure of the extension field K are presented in Chapter 7, 6, and 11 of [DS05], [Shi71], and [Cox11b], respectively.

References

[AATM+22] Mohammad Akhond, Guillermo Arias-Tamargo, Alessandro Mininno, Hao-Yu Sun, Zhengdi Sun, Yifan Wang, and Fengjun Xu, *The hitchhiker's guide to 4d $\mathcal{N} = 2$ superconformal field theories*, SciPost Phys. Lect. Notes **64** (2022), 1.

[ABP19] Luis F. Alday, Agnese Bissi, and Eric Perlmutter, *Genus-one string amplitudes from conformal field theory*, J. High Energy Phys. **06** (2019), 010.

[AC18] Vassilis Anagiannis and Miranda C. N. Cheng, *TASI lectures on Moonshine*, PoS **TASI2017** (2018), 010.

[ACGH85] Enrico Arbarello, Maurizio Cornalba, Phillip A. Griffiths, and Joseph Harris, *Geometry of Algebraic Curves, Volume I*, Grundlehren der mathematischen Wissenschaften, vol. 267, Springer-Verlag, 1985.

[ACJS18] Thales Azevedo, Marco Chiodaroli, Henrik Johansson, and Oliver Schlotterer, *Heterotic and bosonic string amplitudes via field theory*, J. High Energy Phys. **10** (2018), 012.

[AD95] Philip C. Argyres and Michael R. Douglas, *New phenomena in SU(3) supersymmetric gauge theory*, Nucl. Phys. B **448** (1995), 93–126.

[AE15] Martin Ammon and Johanna Erdmenger, *Gauge/Gravity Duality, Foundations and Applications*, Cambridge University Press, 2015.

[AFM21] Johannes Aspman, Elias Furrer, and Jan Manschot, *Elliptic loci of SU(3) vacua*, Ann. Henri Poincaré **22** (2021), no. 8, 2775–2830.

[AFM22a] _____, *Cutting and gluing with running couplings in N=2 QCD*, Phys. Rev. D **105** (2022), no. 2, 025021.

[AFM22b] _____, *Four flavors, triality, and bimodular forms*, Phys. Rev. D **105** (2022), no. 2, 025017.

[AFP12a] Carlo Angelantonj, Ioannis Florakis, and Boris Pioline, *A new look at one-loop integrals in string theory*, Commun. Number Theory Phys. **6** (2012), 159–201.

[AFP12b] _____, *One-loop BPS amplitudes as BPS-state sums*, J. High Energy Phys. **6** (2012), 070.

[AGBM+87] Luis Alvarez-Gaume, Jean-Benoît Bost, Gregory W. Moore, Philip C. Nelson, and Cumrun Vafa, *Bosonization on higher genus riemann surfaces*, Commun. Math. Phys. **112** (1987), 503.

[AGFM81] Luis Alvarez-Gaume, Daniel Z. Freedman, and Sunil Mukhi, *The background field method and the ultraviolet structure of the supersymmetric nonlinear sigma model*, Ann. Phys. **134** (1981), 85.

[AGH97] Luis Alvarez-Gaume and S. F. Hassan, *Introduction to S duality in N=2 supersymmetric gauge theories: A Pedagogical review of the work of Seiberg and Witten*, Fortsch. Phys. **45** (1997), 159–236.

[AGM⁺00] Ofer Aharony, Steven S. Gubser, Juan Martin Maldacena, Hirosi Ooguri, and Yaron Oz, *Large N field theories, string theory and gravity*, Phys. Rep. **323** (2000), 183–386.

[AGMV86] Luis Alvarez-Gaume, Gregory W. Moore, and Cumrun Vafa, *Theta functions, modular invariance and strings*, Commun. Math. Phys. **106** (1986), 1–40.

[AGW84] Luis Alvarez-Gaume and Edward Witten, *Gravitational anomalies*, Nucl. Phys. B **234** (1984), 269.

[AHBC⁺16] Nima Arkani-Hamed, Jacob Bourjaily, Freddy Cachazo, Alexander Goncharov, Alexander Postnikov, and Jaroslav Trnka, *Grassmannian Geometry of Scattering Amplitudes*, Cambridge University Press, 2016.

[AHEHM22] Nima Arkani-Hamed, Lorenz Eberhardt, Yu-tin Huang, and Sebastian Mizera, *On unitarity of tree-level string amplitudes*, J. High Energy Phys. **02** (2022), 197.

[AK18] Olof Ahlén and Axel Kleinschmidt, $D^6 R^4$ *curvature corrections, modular graph functions and Poincaré series*, J. High Energy Phys. **05** (2018), 194.

[AKL23] Jan Albert, Justin Kaidi, and Ying-Hsuan Lin, *Topological modularity of supermoonshine*, Prog. Theor. Exp. Phys. **2023** (2023), no. 3, 033B06.

[Alv83] Orlando Alvarez, *Theory of strings with boundaries: fluctuations, topology, and quantum geometry*, Nucl. Phys. B **216** (1983), 125.

[AM98] Emil Artin and Arthur Norton Milgram, *Galois Theory*, vol. 2, Courier Corporation, 1998.

[Apo76a] Tom M. Apostol, *Introduction to Analytic Number Theory*, Undergraduate Texts in Mathematics, Springer Verlag, 1976.

[Apo76b] ———, *Modular Functions and Dirichlet Series in Number Theory*, Graduate Texts in Mathematics, vol. 41, Springer Verlag, 1976.

[ASD71] Henry Peter Francis Swinnerton-Dyer and Arthur Oliver Lonsdale Atkin, *Modular forms on non-congruence subgroups*, Combinatorics **19** (1971), 1–25.

[AST13] Ofer Aharony, Nathan Seiberg, and Yuji Tachikawa, *Reading between the lines of four-dimensional gauge theories*, J. High Energy Phys. **08** (2013), 115.

[AT16] Ofer Aharony and Yuji Tachikawa, *S-folds and 4d N=3 superconformal field theories*, J. High Energy Phys. **06** (2016), 044.

[Ban03] Peter Bantay, *The Kernel of the modular representation and the Galois action in RCFT*, Commun. Math. Phys. **233** (2003), 423–438.

[Bas16a] Anirban Basu, *Poisson equation for the Mercedes diagram in string theory at genus one*, Class. Quant. Grav. **33** (2016), no. 5, 055005.

[Bas16b] ———, *Poisson equation for the three loop ladder diagram in string theory at genus one*, Int. J. Mod. Phys. A **31** (2016), no. 32, 1650169.

[Bas16c] _____, *Proving relations between modular graph functions*, Class. Quant. Grav. **33** (2016), no. 23, 235011.

[Bas19] _____, *Eigenvalue equation for the modular graph $C_{a,b,c,d}$*, J. High Energy Phys. **07** (2019), 126.

[BBS06] Katrin Becker, Melanie Becker, and John H. Schwarz, *String Theory and M-theory: A Modern Introduction*, Cambridge University Press, 2006.

[BCC$^+$22] Zvi Bern, John Joseph Carrasco, Marco Chiodaroli, Henrik Johansson, and Radu Roiban, *The SAGEX review on scattering amplitudes Chapter 2: An invitation to color-kinematics duality and the double copy*, J. Phys. A **55** (2022), no. 44, 443003.

[BCF$^+$21] Nathan Benjamin, Scott Collier, A. Liam Fitzpatrick, Alexander Maloney, and Eric Perlmutter, *Harmonic analysis of 2D CFT partition functions*, J. High Energy Phys. **09** (2021), 174.

[BCJ19] Vincent Bouchard, Thomas Creutzig, and Aniket Joshi, *Hecke operators on vector-valued modular forms*, SIGMA **15** (2019), 041.

[BCZ85] Eric Braaten, Thomas L. Curtright, and Cosmas K. Zachos, *Torsion and geometrostasis in nonlinear sigma models*, Nucl. Phys. B **260** (1985), 630 [Erratum: Nucl. Phys. B 266, 748–748 (1986)].

[BDDT18] Johannes Broedel, Claude Duhr, Falko Dulat, and Lorenzo Tancredi, *Elliptic polylogarithms and iterated integrals on elliptic curves. Part I: general formalism*, J. High Energy Phys. **05** (2018), 093.

[BDG$^+$22] Nathan Berkovits, Eric D'Hoker, Michael B. Green, Henrik Johansson, and Oliver Schlotterer, *Snowmass white paper: String perturbation theory*, 2022 Snowmass Summer Study, arXiv:2203.09099 (2022).

[BDL$^+$21] Jin-Beom Bae, Zhihao Duan, Kimyeong Lee, Sungjay Lee, and Matthieu Sarkis, *Fermionic rational conformal field theories and modular linear differential equations*, Prog. Theor. Exp. Phys. **2021** (2021), no. 8, 08B104.

[BDL$^+$22] _____, *Bootstrapping fermionic rational CFTs with three characters*, J. High Energy Phys. **01** (2022), 089.

[Ber94] Arjun Berera, *Unitary string amplitudes*, Nucl. Phys. B **411** (1994), 157–180.

[Ber00] Nathan Berkovits, *Super Poincare covariant quantization of the superstring*, J. High Energy Phys. **04** (2000), 018.

[Ber04] _____, *Multiloop amplitudes and vanishing theorems using the pure spinor formalism for the superstring*, J. High Energy Phys. **09** (2004), 047.

[Ber06] _____, *Super-Poincare covariant two-loop superstring amplitudes*, J. High Energy Phys. **01** (2006), 005.

[Ber12] Bruce C. Berndt, *Ramanujan's Notebooks: Part III*, Springer Science & Business Media, 2012.

[BF04] Jan Hendrik Bruinier and Jens Funke, *On two geometric theta lifts*, Duke Math. J. **125** (2004), 45–90.

[BG03] Lev A. Borisov and Paul E. Gunnels, *Toric modular forms of higher weight*, J. Reine Angew. Math. **560** (2003), 43–64.

[BG07] Peter Bantay and Terry Gannon, *Vector-valued modular functions for the modular group and the hypergeometric equation*, arXiv:0705.2467 (2007).

[BH86] J. David Brown and Marc Henneaux, *Central charges in the canonical realization of asymptotic symmetries: An example from three-dimensional gravity*, Commun. Math. Phys. **104** (1986), 207–226.

[BJ86] Jean-Benoît Bost and Thierry Jolicoeur, *A holomorphy property and critical dimension in string theory from an index theorem*, Phys. Lett. B **174** (1986), 273–276.

[BK86] Alexander A. Belavin and Vadim G. Knizhnik, *Algebraic geometry and the geometry of quantum strings*, Phys. Lett. B **168** (1986), 201–206.

[BKMP86] Alexander A. Belavin, Vadim Knizhnik, Alexei Morozov, and Askold Perelomov, *Two and three loop amplitudes in the bosonic string theory*, JETP Lett. **43** (1986), 411.

[BKOR18] Nathan Benjamin, Shamit Kachru, Ken Ono, and Larry Rolen, *Black holes and class groups*, arXiv:1807.00797 (2018).

[BKP20] Guillaume Bossard, Axel Kleinschmidt, and Boris Pioline, *1/8-BPS couplings and exceptional automorphic functions*, SciPost Phys. **8** (2020), no. 4, 054.

[BL11] Francis Brown and Andrey Levin, *Multiple elliptic polylogarithms*, arXiv:1110.6917 (2011).

[BLL⁺15] Christopher Beem, Madalena Lemos, Pedro Liendo, Wolfger Peelaers, Leonardo Rastelli, and Balt C. van Rees, *Infinite chiral symmetry in four dimensions*, Commun. Math. Phys. **336** (2015), no. 3, 1359–1433.

[BLT13] Ralph Blumenhagen, Dieter Lüst, and Stefan Theisen, *Basic Concepts of String Theory*, Theoretical and Mathematical Physics, Springer, Heidelberg, Germany, 2013.

[BM06] Nathan Berkovits and Carlos R. Mafra, *Equivalence of two-loop superstring amplitudes in the pure spinor and RNS formalisms*, Phys. Rev. Lett. **96** (2006), 011602.

[Bos92] Jean-Benoît Bost, *Introduction to Compact Riemann Surfaces, Jacobians, and Abelian Varieties*, From Number Theory to Physics, Springer, 1992, pp. 64–211.

[BPZ84] A. A. Belavin, Alexander M. Polyakov, and A. B. Zamolodchikov, *Infinite conformal symmetry in two-dimensional quantum field theory*, Nucl. Phys. B **241** (1984), 333–380.

[BR18] Christopher Beem and Leonardo Rastelli, *Vertex operator algebras, Higgs branches, and modular differential equations*, J. High Energy Phys. **08** (2018), 114.

[BR20] Matthew Buican and Rajath Radhakrishnan, *Galois conjugation and multiboundary entanglement entropy*, J. High Energy Phys. **12** (2020), 045.

[BR22] ———, *Galois orbits of TQFTs: Symmetries and unitarity*, J. High Energy Phys. **01** (2022), 004.

[Bro17a] Francis Brown, *A class of non-holomorphic modular forms I*, arXiv:1707.01230 (2017).

[Bro17b] ———, *A class of non-holomorphic modular forms III: Real analytic cusp forms for $SL_2(\mathbb{Z})$*, arXiv preprint arXiv:1710.07912 (2017).

[Bro20] ———, *A class of nonholomorphic modular forms II: Equivariant iterated*, Forum Math. Sigma **8** (2020), e31.

[BSS13] Johannes Broedel, Oliver Schlotterer, and Stephan Stieberger, *Polylogarithms, multiple zeta values and superstring amplitudes*, Fortsch. Phys. **61** (2013), 812–870.

[BSZ19] Johannes Broedel, Oliver Schlotterer, and Federico Zerbini, *From elliptic multiple zeta values to modular graph functions: Open and closed strings at one loop*, J. High Energy Phys. **01** (2019), 155.

[Bum98] Daniel Bump, *Automorphic Forms and Representations*, Cambridge Studies in Advanced Mathematics, vol. 55, Cambridge University Press, 1998.

[BVdGHZ08] Jan Hendrik Bruinier, Gerard Van der Geer, Günter Harder, and Don Zagier, *The 1-2-3 of Modular Forms: Lectures at a Summer School in Nordfjordeid, Norway*, Springer Science & Business Media, 2008.

[Car92] Pierre Cartier, *Introduction to Zeta Functions*, From Number Theory to Physics, Springer, 1992, pp. 1–63.

[CCPS23] Miranda C. N. Cheng, Ioana Coman, Davide Passaro, and Gabriele Sgroi, *Quantum modular $\hat{Z}^G$-invariants*, arXiv:2304.03934 (2023).

[CDPvG08a] Sergio L. Cacciatori, Francesco Dalla Piazza, and Bert van Geemen, *Genus four superstring measures*, Lett. Math. Phys. **85** (2008), 185.

[CDPvG08b] ———, *Modular forms and three loop superstring amplitudes*, Nucl. Phys. B **800** (2008), 565–590.

[CDT21] Frank Calegari, Vesselin Dimitrov, and Yunqing Tang, *The unbounded denominators conjecture*, arXiv:2109.09040 (2021).

[CF06] Henri Cohen and Gerhard Frey, *Handbook of Elliptic and Hyperelliptic Curve Cryptography*, Chapman & Hall/CRC, 2006.

[CG94] A. Coste and T. Gannon, *Remarks on Galois symmetry in rational conformal field theories*, Phys. Lett. B **323** (1994), 316–321.

[CG99] Antoine Coste and Terry Gannon, *Congruence subgroups and rational conformal field theory*, arXiv:math/9909080 (1999).

[CGP+21] Shai M. Chester, Michael B. Green, Silviu S. Pufu, Yifan Wang, and Congkao Wen, *New modular invariants in $\mathcal{N} = 4$ Super-Yang-Mills theory*, J. High Energy Phys. **04** (2021), 212.

[CHST22] Mirjam Cvetic, James Halverson, Gary Shiu, and Washington Taylor, *Snowmass white paper: String theory and particle physics*, arXiv:2204.01742 (2022).

[CKM+17] Henry Cohn, Abhinav Kumar, Stephen Miller, Danylo Radchenko, and Maryna Viazovska, *The sphere packing problem in dimension 24*, Ann. Math. **185** (2017), no. 3, 1017 – 1033.

[CM19] A. Ramesh Chandra and Sunil Mukhi, *Towards a classification of two-character rational conformal field theories*, J. High Energy Phys. **04** (2019), 153.

[Cox11a] David A. Cox, *Galois Theory*, vol. 61, John Wiley & Sons, 2011.

[Cox11b] ———, *Primes of the Form $x^2 + ny^2$: Fermat, Class Field Theory, and Complex Multiplication*, vol. 34, John Wiley & Sons, 2011.

[CP22] Scott Collier and Eric Perlmutter, *Harnessing S-duality in $\mathcal{N} = 4$ SYM & supergravity as $SL(2,\mathbb{Z})$-averaged strings*, arXiv:2201.05093 (2022).

[CS49] Sarvadaman Chowla and Atle Selberg, *On Epstein's zeta function (I)*, Proc. Natl. Acad. Sci. USA **35** (1949), no. 7, 371–374.

[CT21] Changha Choi and Leon A. Takhtajan, *Supersymmetry and trace formulas I. Compact Lie groups*, arXiv:2112.07942 (2021).

[CT23]   _____, *Supersymmetry and trace formulas II. Selberg trace formula*, arXiv:2306.13636 (2023).

[CY97]   John Coates and Shing-Tung Yau (eds.), *Elliptic Curves, Modular Forms and Fermat's Last Theorem*, International Press, 1997.

[dBCD+06] Jan de Boer, Miranda C. N. Cheng, Robbert Dijkgraaf, Jan Manschot, and Erik Verlinde, *A farey tail for attractor black holes*, J. High Energy Phys. **11** (2006), 024.

[DBG91]   Jan De Boer and Jacob Goeree, *Markov traces and III factors in conformal field theory*, Commun. Math. Phys. **139** (1991), 267–304.

[DD18]   Eric D'Hoker and William Duke, *Fourier series of modular graph functions*, J. Number Theory **192** (2018), 1–36.

[DDD+22]   Daniele Dorigoni, Mehregan Doroudiani, Joshua Drewitt, Martijn Hidding, Axel Kleinschmidt, Nils Matthes, Oliver Schlotterer, and Bram Verbeek, *Modular graph forms from equivariant iterated Eisenstein integrals*, J. High Energy Phys. **12** (2022), 162.

[DDHM21]   Arun Debray, Markus Dierigl, Jonathan J. Heckman, and Miguel Montero, *The anomaly that was not meant IIB*, arXiv:2107.14227 (2021).

[Devrg]   Sage Developers, *Sagemath, the Sage Mathematics Software System*, www.sagemath.org.

[DF02]   Eric D'Hoker and Daniel Z. Freedman, *Supersymmetric gauge theories and the AdS / CFT correspondence*, Theoretical Advanced Study Institute in Elementary Particle Physics (TASI 2001): Strings, Branes and EXTRA Dimensions, 2002, pp. 3–158.

[DFMS97]   Philippe Di Francesco, Pierre Mathieu, and David Sénéchal, *Conformal Field Theory*, Springer, 1997.

[DFSZ87]   Philippe Di Francesco, Hubert Saleur, and Jean-Bernard Zuber, *Modular invariance in nonminimal two-dimensional conformal theories*, Nucl. Phys. B **285** (1987), 454.

[DFSZ88]   _____, *Generalized coulomb gas formalism for two-dimensional critical models based on SU(2) Coset Construction*, Nucl. Phys. B **300** (1988), 393–432.

[DG14]   Eric D'Hoker and Michael B. Green, *Zhang–Kawazumi invariants and superstring amplitudes*, J. Number Theory **144** (2014), 111–150.

[DG18]   _____, *Identities between modular graph forms*, J. Number Theory **189** (2018), 25–80.

[DG19]   _____, *Exploring transcendentality in superstring amplitudes*, J. High Energy Phys. **07** (2019), 149.

[DG22]   Eric D'Hoker and Nicholas Geiser, *Integrating three-loop modular graph functions and transcendentality of string amplitudes*, J. High Energy Phys. **02** (2022), 019.

[DGGV17]   Eric D'Hoker, Michael B. Green, Ömer Gürdogan, and Pierre Vanhove, *Modular graph functions*, Commun. Number Theory Phys. **11** (2017), 165–218.

[DGP05]   Eric D'Hoker, Michael Gutperle, and Duong H. Phong, *Two-loop superstrings and S-duality*, Nucl. Phys. B **722** (2005), 81–118.

[DGP19a]   Eric D'Hoker, Michael B. Green, and Boris Pioline, *Asymptotics of the $D^8 R^4$ genus-two string invariant*, Commun. Number Theory Phys. **13** (2019), 351–462.

[DGP19b] _____, *Higher genus modular graph functions, string invariants, and their exact asymptotics*, Commun. Math. Phys. **366** (2019), no. 3, 927–979.

[DGPR15] Eric D'Hoker, Michael B. Green, Boris Pioline, and Rodolfo Russo, *Matching the D^6R^4 interaction at two-loops*, J. High Energy Phys. **01** (2015), 031.

[DGV15] Eric D'Hoker, Michael B. Green, and Pierre Vanhove, *On the modular structure of the genus-one Type II superstring low energy expansion*, J. High Energy Phys. **08** (2015), 041.

[DGW21a] Daniele Dorigoni, Michael B. Green, and Congkao Wen, *Exact expressions for n-point maximal $U(1)_Y$-violating integrated correlators in $SU(N)$ $\mathcal{N} = 4$ SYM*, J. High Energy Phys. **11** (2021), 132.

[DGW21b] _____, *Exact properties of an integrated correlator in $\mathcal{N} = 4$ SU(N) SYM*, J. High Energy Phys. **05** (2021), 089.

[DGW22] _____, *The SAGEX review on scattering amplitudes. Chapter 10: Selected topics on modular covariance of type IIB string amplitudes and their supersymmetric Yang–Mills duals*, J. Phys. A **55** (2022), no. 44, 443011.

[DGWX23] Daniele Dorigoni, Michael B. Green, Congkao Wen, and Haitian Xie, *Modular-invariant large-N completion of an integrated correlator in $\mathcal{N} = 4$ supersymmetric Yang-Mills theory*, J. High Energy Phys. **04** (2023), 114.

[D'H99] Eric D'Hoker, *String theory*, Quantum Fields and Strings: A Course for Mathematicians (P. Deligne and et al., eds.), vol. 2, American Mathematical Society and Institute for Advanced Study, 1999, pp. 807–1012.

[DHVW85] Lance J. Dixon, Jeffrey A. Harvey, C. Vafa, and Edward Witten, *Strings on orbifolds*, Nucl. Phys. B **261** (1985), 678–686.

[DHVW86] _____, *Strings on orbifolds. 2.*, Nucl. Phys. B **274** (1986), 285–314.

[DK16] Eric D'Hoker and Justin Kaidi, *Hierarchy of modular graph identities*, J. High Energy Phys. **11** (2016), 051.

[DK19a] _____, *Modular graph functions and odd cuspidal functions. Fourier and Poincaré series*, J. High Energy Phys. **04** (2019), 136.

[DK19b] Daniele Dorigoni and Axel Kleinschmidt, *Modular graph functions and asymptotic expansions of Poincaré series*, Commun. Number Theory Phys. **13** (2019), no. 3, 569–617.

[DKS21] Eric D'Hoker, Axel Kleinschmidt, and Oliver Schlotterer, *Elliptic modular graph forms. Part I. Identities and generating series*, J. High Energy Phys. **03** (2021), 151.

[DKST17] Frederik Denef, Shamit Kachru, Zimo Sun, and Arnav Tripathy, *Higher genus Siegel forms and multi-center black holes in N=4 supersymmetric string theory*, arXiv:1712.01985 (2017).

[DLN15] Chongying Dong, Xingjun Lin, and Siu-Hung Ng, *Congruence property in conformal field theory*, Algebra Number Theory **9** (2015), no. 9, 2121–2166.

[DM96] Michael R. Douglas and Gregory W. Moore, *D-branes, quivers, and ALE instantons*, arXiv:9603167 (1996).

[DMMV00] Robbert Dijkgraaf, Juan Martin Maldacena, Gregory W. Moore, and Erik P. Verlinde, *A Black hole Farey tail*, arXiv:hep-th/0005003 (2000).

462   *References*

[DMPS20] Eric D'Hoker, Carlos R. Mafra, Boris Pioline, and Oliver Schlotterer, *Two-loop superstring five-point amplitudes. Part I. Construction via chiral splitting and pure spinors*, J. High Energy Phys. **08** (2020), 135.

[DMPS21] _____, *Two-loop superstring five-point amplitudes. Part II. Low energy expansion and S-duality*, J. High Energy Phys. **02** (2021), 139.

[DMVV97] Robbert Dijkgraaf, Gregory W. Moore, Erik P. Verlinde, and Herman L. Verlinde, *Elliptic genera of symmetric products and second quantized strings*, Commun. Math. Phys. **185** (1997), 197–209.

[DMZ12] Atish Dabholkar, Sameer Murthy, and Don Zagier, *Quantum black holes, wall crossing, and mock modular forms*, arXiv:1208.4074 (2012).

[DP86a] Eric D'Hoker and Duong H. Phong, *Multiloop Amplitudes for the Bosonic Polyakov String*, Nucl. Phys. B **269** (1986), 205–234.

[DP86b] _____, *On determinants of laplacians on riemann surfaces*, Commun. Math. Phys. **104** (1986), 537.

[DP88] Eric D'Hoker and Duong H. Phong, *The geometry of string perturbation theory*, Rev. Mod. Phys. **60** (1988), no. 4, 917.

[DP93] Eric D'Hoker and Duong H. Phong, *Momentum analyticity and finiteness of the one loop superstring amplitude*, Phys. Rev. Lett. **70** (1993), 3692–3695.

[DP95] _____, *The Box graph in superstring theory*, Nucl. Phys. B **440** (1995), 24–94.

[DP98a] _____, *Calogero-Moser Lax pairs with spectral parameter for general Lie algebras*, Nucl. Phys. B **530** (1998), 537–610.

[DP98b] _____, *Calogero-Moser systems in SU(N) Seiberg-Witten theory*, Nucl. Phys. B **513** (1998), 405–444.

[DP98c] _____, *Spectral curves for superYang-Mills with adjoint hypermultiplet for general Lie algebras*, Nucl. Phys. B **534** (1998), 697–719.

[DP99] _____, *Lectures on supersymmetric Yang-Mills theory and integrable systems*, 9th CRM Summer School: Theoretical Physics at the End of the 20th Century, 1999, pp. 1–125.

[DP02a] _____, *Two loop superstrings. 1. Main formulas*, Phys. Lett. B **529** (2002), 241–255.

[DP02b] _____, *Two loop superstrings 4: The Cosmological constant and modular forms*, Nucl. Phys. B **639** (2002), 129–181.

[DP02c] Eric D'Hoker and D. H. Phong, *Lectures on two-loop superstrings*, Superstring Theory (S.-T. Yau, K. Liu and C. Zhu, eds.), Advanced Lectures in Mathematics, Vol 1, International Press, 2002, pp. 85–123.

[DP05] Eric D'Hoker and D. H. Phong, *Two-loop superstrings VI: Non-renormalization theorems and the 4-point function*, Nucl. Phys. B **715** (2005), 3–90.

[DR73] Pierre Deligne and Michael Rapoport, *Les schémas de modules de courbes elliptiques*, Lecture Notes in Mathematics, vol. 349, Springer-Verlag, 1973.

[DS05] Fred Diamond and Jerry Michael Shurman, *A First Course in Modular Forms*, Graduate Texts in Mathematics, vol. 228, Springer, 2005.

[DS21] Eric D'Hoker and Oliver Schlotterer, *Two-loop superstring five-point amplitudes. Part III. Construction via the RNS formulation: Even spin structures*, J. High Energy Phys. **12** (2021), 063.

[DS22] ———, *Identities among higher genus modular graph tensors*, Commun. Number Theory Phys. **16** (2022), no. 1, 35–74.

[Dub81] Boris Anatol'evich Dubrovin, *Theta functions and non-linear equations*, Russ. Math. Surv. **36** (1981), no. 2, 11.

[DVV88] Robbert Dijkgraaf, Erik P. Verlinde, and Herman L. Verlinde, *C = 1 conformal field theories on Riemann surfaces*, Commun. Math. Phys. **115** (1988), 649–690.

[DVV97] ———, *Counting dyons in N=4 string theory*, Nucl. Phys. B **484** (1997), 543–561.

[DVVV89] Robbert Dijkgraaf, Cumrun Vafa, Erik P. Verlinde, and Herman L. Verlinde, *The operator algebra of orbifold models*, Commun. Math. Phys. **123** (1989), 485.

[DW96] Ron Donagi and Edward Witten, *Supersymmetric Yang-Mills theory and integrable systems*, Nucl. Phys. B **460** (1996), 299–334.

[EH09] Tohru Eguchi and Kazuhiro Hikami, *Superconformal algebras and mock theta functions 2. Rademacher expansion for K3 surface*, Commun. Number Theory Phys. **3** (2009), 531–554.

[EH15] Henrietta Elvang and Yu-Tin Huang, *Scattering Amplitudes in Gauge Theory and Gravity*, Cambridge University press, 2015.

[EM23] Lorenz Eberhardt and Sebastian Mizera, *Evaluating one-loop string amplitudes*, SciPost Phys. **15** (2023), no. 3, 119.

[Enr16] Benjamin Enriquez, *Analogues elliptiques des nombres multizétas*, Bull. Soc. Math. France **144** (2016), no. 3, 395–427. MR 3558428.

[EOT11] Tohru Eguchi, Hirosi Ooguri, and Yuji Tachikawa, *Notes on the K3 surface and the Mathieu group M_{24}*, Exp. Math. **20** (2011), 91–96.

[EOTY89] Tohru Eguchi, Hirosi Ooguri, Anne Taormina, and Sung-Kil Yang, *Superconformal algebras and string compactification on manifolds with SU(N) Holonomy*, Nucl. Phys. B **315** (1989), 193–221.

[Erd81a] A. Erdèlyi (ed.), *Higher Transcendental Functions*, The Bateman Manuscript Project, vol. 2, Krieger Publishing Company, 1981.

[Erd81b] A. Erdèlyi (ed.), *Higher Transcendental Functions*, The Bateman Manuscript Project, vol. 1, Krieger Publishing Company, 1981.

[Erd81c] A. Erdèlyi (ed.), *Higher Transcendental Functions*, The Bateman Manuscript Project, vol. 3, Krieger Publishing Company, 1981.

[Esc01] Jean-Pierre Escofier, *Galois Theory*, Graduate Texts in Mathematics, vol. 204, Springer, 2001.

[Evt20] Mikhail Evtikhiev, *N = 3 SCFTs in 4 dimensions and non-simply laced groups*, J. High Energy Phys. **06** (2020), 125.

[EZ85] Martin Eichler and Don Zagier, *The theory of jacobi forms, progress in Math. Vol 55 (1985), Birkhäuser-Verlag*, Progress in Math. **55** (1985).

[EZ23] Benjamin Enriquez and Federico Zerbini, *Analogues of hyperlogarithm functions on affine complex curves*, arXiv:2212.03119 (2023).

[Fay06] John D. Fay, *Theta Functions on Riemann Surfaces*, Graduate Texts in Mathematics, vol. 352, Springer-Verlag, 2006.

[FGKP18] Philipp Fleig, Henrik P. A. Gustafsson, Axel Kleinschmidt, and Daniel Persson, *Eisenstein Series and Automorphic Representations*, Cambridge University Press, 2018.

[FHJ73] Sergio Fubini, Andrew J. Hanson, and Roman Jackiw, *New approach to field theory*, Phys. Rev. D **7** (1973), 1732–1760.

[FK80] Hershel M. Farkas and Irwin Kra, *Riemann Surfaces*, Graduate Texts in Mathematics, vol. 71, Springer-Verlag, 1980.

[FM14] Cameron Franc and Geoffrey Mason, *Fourier coefficients of vector-valued modular forms of dimension 2*, Canadian Mathematical Bulletin **57** (2014), no. 3, 485–494.

[FM16a] _____, *Hypergeometric series, modular linear differential equations and vector-valued modular forms*, The Ramanujan Journal **41** (2016), no. 1, 233–267.

[FM16b] _____, *Three-dimensional imprimitive representations of the modular group and their associated modular forms*, J. Number Theory **160** (2016), 186–214.

[FMS86] Daniel Friedan, Emil J. Martinec, and Stephen H. Shenker, *Conformal Invariance, Supersymmetry and String Theory*, Nucl. Phys. B **271** (1986), 93–165.

[FP17] Ioannis Florakis and Boris Pioline, *On the Rankin–Selberg method for higher genus string amplitudes*, Commun. Number Theory Phys. **11** (2017), 337–404.

[Fri85] Daniel Harry Friedan, *Nonlinear models in Two + epsilon dimensions*, Annals Phys. **163** (1985), 318.

[Fri86] David Fried, *Analytic torsion and closed geodesics on hyperbolic manifolds*, Invent. Math. **84** (1986), no. 3, 523–540.

[FVP12] Daniel Z. Freedman and Antoine Van Proeyen, *Supergravity*, Cambridge University Press, Cambridge, UK, 2012.

[Gai12] Davide Gaiotto, *N=2 dualities*, J. High Energy Phys. **08** (2012), 034.

[Gan04] Terry Gannon, *Monstrous moonshine: The First twenty five years*, arXiv:math/0402345 (2004).

[Gan06] _____, *Moonshine beyond the Monster: The Bridge Connecting Algebra, Modular Forms and Physics*, Cambridge University Press, 2006.

[Gan13] _____, *The theory of vector-modular forms for the modular group*, arXiv preprint arXiv:1310.4458 (2013).

[Gau01] Carl Friedrich Gauss, *Disquisitiones Arithmeticae*, Lipsiae, 1801.

[Gaw99] Krzysztof Gawedzki, *Conformal field theory and strings*, Quantum Fields and Strings: A Course for Mathematicians (P. Deligne et al., eds.), vol. 2, American Mathematical Society and Institute for Advanced Study, 1999, pp. 727–806.

[GER16] Iñaki García-Etxebarria and Diego Regalado, $\mathcal{N} = 3$ *four dimensional field theories*, J. High Energy Phys. **03** (2016), 083.

[GER17] _____, *Exceptional* $\mathcal{N} = 3$ *theories*, J. High Energy Phys. **12** (2017), 042.

[Ger20] Jan Erik Gerken, *Modular graph forms and scattering amplitudes in String theory*, Ph.D. thesis, Humboldt U., Berlin, Humboldt U., Berlin, 2020.

[Ger21] Jan E. Gerken, *Basis decompositions and a mathematica package for modular graph forms*, J. Phys. A **54** (2021), no. 19, 195401.

[GG97] Michael B. Green and Michael Gutperle, *Effects of D instantons*, Nucl. Phys. B **498** (1997), 195–227.

[GG98] Matthias R. Gaberdiel and Michael B. Green, *An SL(2, Z) anomaly in IIB supergravity and its F theory interpretation*, J. High Energy Phys. **11** (1998), 026.

[GGPS69] I. M. Gel'land, M. I. Graev, and II. Pyatetskii-Shapiro, *Representation theory and automorphic functions*, (Transl. from Russian) Philadelphia: Saunders, 1969.

[GGV97] Michael B. Green, Michael Gutperle, and Pierre Vanhove, *One loop in eleven-dimensions*, Phys. Lett. B **409** (1997), 177–184.

[GHMR85] David J. Gross, Jeffrey A. Harvey, Emil J. Martinec, and Ryan Rohm, *Heterotic String theory. 1. The free heterotic string*, Nucl. Phys. B **256** (1985), 253.

[GHMR86] _____, *Heterotic String theory. 2. The interacting heterotic string*, Nucl. Phys. B **267** (1986), 75–124.

[Gin88] Paul H. Ginsparg, *Applied conformal field theory*, Les Houches Summer School in Theoretical Physics: Fields, Strings, Critical Phenomena, 1988.

[GK19] Jan E. Gerken and Justin Kaidi, *Holomorphic subgraph reduction of higher-point modular graph forms*, J. High Energy Phys. **01** (2019), 131.

[GKM$^+$95] Alexander Gorsky, Igor Krichever, Andrei Marshakov, Andrei Mironov, and Alexei Morozov, *Integrability and Seiberg-Witten exact solution*, Phys. Lett. B **355** (1995), 466–474.

[GKS19] Jan E. Gerken, Axel Kleinschmidt, and Oliver Schlotterer, *Heterotic-string amplitudes at one loop: Modular graph forms and relations to open strings*, J. High Energy Phys. **01** (2019), 052.

[GKS20a] _____, *All-order differential equations for one-loop closed-string integrals and modular graph forms*, J. High Energy Phys. **01** (2020), 064.

[GKS20b] _____, *Generating series of all modular graph forms from iterated Eisenstein integrals*, J. High Energy Phys. **07** (2020), no. 07, 190.

[GM10] Humberto Gomez and Carlos R. Mafra, *The overall coefficient of the two-loop superstring amplitude using pure spinors*, J. High Energy Phys. **05** (2010), 017.

[GM13] _____, *The closed-string 3-loop amplitude and S-duality*, J. High Energy Phys. **10** (2013), 217.

[GMMM98] Alexander Gorsky, Andrei Marshakov, Andrei Mironov, and Alexei Morozov, *RG equations from Whitham hierarchy*, Nucl. Phys. B **527** (1998), 690–716.

[GMRV10] Michael B. Green, Stephen D. Miller, Jorge G. Russo, and Pierre Vanhove, *Eisenstein series for higher-rank groups and string theory amplitudes*, Commun. Number Theory Phys. **4** (2010), 551–596.

[GMSMa21] Yvonne Geyer, Ricardo Monteiro, and Ricardo Stark-Muchão, *Superstring loop amplitudes from the field theory limit*, Phys. Rev. Lett. **127** (2021), no. 21, 211603.

[GMV15a] Michael B. Green, Stephen D. Miller, and Pierre Vanhove, $SL(2, \mathbb{Z})$-*invariance and D-instanton contributions to the $D^6 R^4$ interaction*, Commun. Number Theory Phys. **09** (2015), 307–344.

[GMV15b] _____ , *Small representations, string instantons, and Fourier modes of Eisenstein series*, J. Number Theory **146** (2015), 187–309.

[GNO77] Peter Goddard, Jean Nuyts, and David I. Olive, *Gauge theories and magnetic charge*, Nucl. Phys. B **125** (1977), 1–28.

[Gol06] Dorian Goldfeld, *Automorphic forms and L-functions for the group $GL(n, \mathrm{R})$*, Cambridge Studies in Advanced Mathematics, vol. 99, Cambridge University Press, 2006.

[Got61a] Erhard Gottschling, *Über die fixpunkte der Siegelschen modulgruppe*, Math. Ann. **143** (1961), no. 2, 111–149.

[Got61b] _____ , *Über die fixpunktuntergruppen der siegelschen modulgruppe*, Math. Ann. **143** (1961), no. 5, 399–430.

[Got67] _____ , *Die Uniformisierbarkeit der Fixpunkte eigentlich diskontinuierlicher Gruppen von biholomorphen Abbildungen*, Math. Ann. **169** (1967), no. 1, 26–54.

[GP88] David J. Gross and Vipul Periwal, *String perturbation theory diverges*, Phys. Rev. Lett. **60** (1988), 2105.

[GPR94] Amit Giveon, Massimo Porrati, and Eliezer Rabinovici, *Target space duality in string theory*, Phys. Rep. **244** (1994), 77–202.

[GRR88] I. S. Gradshteyn, I. M. Ryzhik, and Robert H. Romer, *Tables of integrals, series, and products*, 1988.

[Gru09] Samuel Grushevsky, *Superstring scattering amplitudes in higher genus*, Commun. Math. Phys. **287** (2009), 749–767.

[GRV08] Michael B. Green, Jorge G. Russo, and Pierre Vanhove, *Low energy expansion of the four-particle genus-one amplitude in type II superstring theory*, J. High Energy Phys. **02** (2008), 020.

[GRV10] _____ , *Automorphic properties of low energy string amplitudes in various dimensions*, Phys. Rev. D **81** (2010), 086008.

[GS82] Michael B. Green and John H. Schwarz, *Supersymmetrical String theories*, Phys. Lett. B **109** (1982), 444–448.

[GS84] _____ , *Anomaly cancellation in supersymmetric D=10 Gauge theory and superstring theory*, Phys. Lett. B **149** (1984), 117–122.

[GS87] David J. Gross and John H. Sloan, *The quartic effective action for the heterotic string*, Nucl. Phys. B **291** (1987), 41–89.

[GS99] Michael B. Green and Savdeep Sethi, *Supersymmetry constraints on type IIB supergravity*, Phys. Rev. D **59** (1999), 046006.

[GSB82] Michael B. Green, John H. Schwarz, and Lars Brink, *N=4 Yang-Mills and N=8 supergravity as limits of string theories*, Nucl. Phys. B **198** (1982), 474–492.

[GSO77] Ferdinando Gliozzi, Joel Scherk, and David Olive, *Supersymmetry, supergravity theories and the dual spinor model*, Nucl. Phys. B **122** (1977), no. 2, 253–290.

[GSVY90] Brian R. Greene, Alfred D. Shapere, Cumrun Vafa, and Shing-Tung Yau, *Stringy cosmic strings and noncompact Calabi–Yau manifolds*, Nucl. Phys. B **337** (1990), 1–36.

[GSW12a] Michael B. Green, John H. Schwarz, and Edward Witten, *Superstring theory Vol. 1: 25th anniversary edition*, Cambridge Monographs on Mathematical Physics, Cambridge University Press, 2012.

[GSW12b] _____ , *Superstring theory Vol. 2: 25th anniversary edition*, Cambridge Monographs on Mathematical Physics, Cambridge University Press, 2012.

[Gun61] R. C. Gunning, *The Eichler cohomology groups and automorphic forms*, Trans. Am. Math. Soc. **100** (1961), no. 1, 44–62.

[Gun67] Robert C. Gunning, *Lectures on Vector Bundles over Riemann Surfaces*, vol. 6, Princeton University Press, 1967.

[Gun15] _____, *Lectures on Riemann Surfaces*, Lectures on Riemann Surfaces, Princeton University Press, 2015.

[GV00] Michael B. Green and Pierre Vanhove, *The low-energy expansion of the one loop type II superstring amplitude*, Phys. Rev. D **61** (2000), 104011.

[GV04] Sergei Gukov and Cumrun Vafa, *Rational conformal field theories and complex multiplication*, Commun. Math. Phys. **246** (2004), 181–210.

[GV06] Michael B. Green and Pierre Vanhove, *Duality and higher derivative terms in M theory*, J. High Energy Phys. **01** (2006), 093.

[GvdVZ86] Marcus T. Grisaru, A. E. M. van de Ven, and Daniela Zanon, *Four loop beta function for the N=1 and N=2 supersymmetric nonlinear sigma model in two-dimensions*, Phys. Lett. B **173** (1986), 423–428.

[GVW00] Sergei Gukov, Cumrun Vafa, and Edward Witten, *CFT's from Calabi–Yau four folds*, Nucl. Phys. B **584** (2000), 69–108 [Erratum: Nucl. Phys. B 608, 477–478 (2001)].

[GW86] David J. Gross and Edward Witten, *Superstring modifications of Einstein's equations*, Nucl. Phys. B **277** (1986), 1.

[GW06] Sergei Gukov and Edward Witten, *Gauge theory, ramification, and the geometric Langlands program*, arXiv:0612073 (2006).

[GW19] Michael B. Green and Congkao Wen, *Modular Forms and $SL(2,\mathbb{Z})$-covariance of type IIB superstring theory*, J. High Energy Phys. **06** (2019), 087.

[GZ85] Benedict Gross and Don Zagier, *Singular moduli*, J. Reine Angew. Math. **355** (1985), 191–220.

[Ham82] S. Hamilton, Richard, *Three-Manifolds with Positive Ricci Curvature*, J. Diff. Geom. **17** (1982), 255–306.

[Har05] Jeffrey A. Harvey, *TASI 2003 lectures on anomalies*, TASI 2003, 2005.

[Hec37a] Erich Hecke, *Über Modulfunktionen und die Dirichletschen Reihen mit Eulerscher Produktentwicklung. I*, Mathematische Annalen **114** (1937), no. 1, 1–28.

[Hec37b] _____, *Über Modulfunktionen und die Dirichletschen Reihen mit Eulerscher Produktentwicklung. II*, Mathematische Annalen **114** (1937), no. 1, 316–351.

[Hec81] _____, *Lectures on the Theory of Algebraic Numbers*, Graduate Texts in Mathematics, vol. 77, Springer-Verlag, 1981.

[Hej83] Dennis A. Hejhal, *The Selberg Trace Formula for SL (2, R): Volumes 1 and 2*, vol. 1001, Springer, 1983.

[HHP22] Sarah M. Harrison, Jeffrey A. Harvey, and Natalie M. Paquette, *Snowmass White Paper: Moonshine*, arXiv:2201.13321 (2022).

[HHW20] Jeffrey A. Harvey, Yichen Hu, and Yuxiao Wu, *Galois symmetry induced by Hecke relations in rational conformal field theory and associated modular tensor categories*, J. Phys. A **53** (2020), no. 33, 334003.

[HM16a] Harsha R. Hampapura and Sunil Mukhi, *On 2d conformal field theories with two characters*, J. High Energy Phys. **01** (2016), 005.

[HM16b] ——, *Two-dimensional RCFT's without Kac–Moody symmetry*, J. High Energy Phys. **07** (2016), 138.

[Hop02] Michael J. Hopkins, *Algebraic topology and modular forms*, arXiv preprint math/0212397 (2002).

[HSV22] Martijn Hidding, Oliver Schlotterer and Bram Verbeek, *Elliptic modular graph forms II: Iterated integrals*, arXiv:2208.11116 (2022).

[HT95] Chris M. Hull and Paul K. Townsend, *Unity of superstring dualities*, Nucl. Phys. B **438** (1995), 109–137.

[Hun12] Thomas W. Hungerford, *Abstract Algebra: An Introduction*, Cengage Learning, 2012.

[HW18] Jeffrey A. Harvey and Yuxiao Wu, *Hecke relations in rational conformal field theory*, J. High Energy Phys. **09** (2018), 032.

[Igu12] Jun-ichi Igusa, *Theta Functions*, Grundlehren der mathematischen Wissenschaften, vol. 194, Springer Science & Business Media, 2012.

[IK21] Henryk Iwaniec and Emmanuel Kowalski, *Analytic Number Theory*, vol. 53, American Mathematical Soc., 2021.

[IS96] Kenneth A. Intriligator and Nathan Seiberg, *Lectures on supersymmetric gauge theories and electric–magnetic duality*, Nucl. Phys. B Proc. Suppl. **45BC** (1996), 1–28.

[Iwa97] Henryk Iwaniec, *Topics in Classical Automorphic Forms*, Graduate Studies in Mathematics, vol. 17, American Mathematical Society, 1997.

[Iwa02] ——, *Spectral Methods of Automorphic Forms*, Graduate Studies in Mathematics, vol. 53, American Mathematical Society, 2002.

[JL93] Jay Jorgenson and Serge Lang, *Basic Analysis of Regularized Series and Products*, Lecture Notes in Mathematics **1564** (1993).

[Joh06] Clifford V. Johnson, *D-branes*, Cambridge University Press, 2006.

[Kaw08] Nariya Kawazumi, *Johnson's homomorphisms and the Arakelov-Green function*, arXiv:0801.4218 (2008).

[Kaw16] ——, *Some tensor field on the Teichmüller space*, Lecture at MCM2016 (2016).

[Kir89] Elias B. Kiritsis, *Fuchsian differential equations for characters on the torus: A classification*, Nucl. Phys. B **324** (1989), 475–494.

[Kir19] Elias Kiritsis, *String Theory in a Nutshell*, vol. 21, Princeton University Press, 2019.

[KK03] Masanobu Kaneko and Masao Koike, *On modular forms arising from a differential equation of hypergeometric type*, Ramanujan J. **7** (2003), no. 1, 145–164.

[KKO+22] Justin Kaidi, Zohar Komargodski, Kantaro Ohmori, Sahand Seifnashri, and Shu-Heng Shao, *Higher central charges and topological boundaries in 2+1-dimensional TQFTs*, SciPost Phys. **13** (2022), no. 3, 067.

[Kli90] Helmut Klingen, *Introductory lectures on Siegel modular forms*, Camb. Stud. Adv. Math. **20** (1990).

[KLPM21] Justin Kaidi, Ying-Hsuan Lin, and Julio Parra-Martinez, *Holomorphic modular bootstrap revisited*, J. High Energy Phys. **12** (2021), 151.

[KM11] Kamal Khuri-Makdisi, *Moduli interpretation of Eisenstein series*, arXiv:0903.1439 (2011).

[KMRW22] Justin Kaidi, Mario Martone, Leonardo Rastelli, and Mitch Weaver, *Needles in a haystack. An algorithmic approach to the classification of 4d $\mathcal{N} = 2$ SCFTs*, J. High Energy Phys. **03** (2022), 210.

[KMZ22] Justin Kaidi, Mario Martone, and Gabi Zafrir, *Exceptional moduli spaces for exceptional $\mathcal{N} = 3$ theories*, J. High Energy Phys. **08** (2022), 264.

[Kna92] Anthony W. Knapp, *Elliptic Curves*, Mathematical Notes, Princeton University Press, 1992.

[Kob94] Neal I Koblitz, *A Course in Number Theory and Cryptography*, Graduate Texts in Mathematics, vol. 114, Springer Science & Business Media, 1994.

[Kob12] ———, *Introduction to Elliptic Curves and Modular Forms*, Graduate Texts in Mathematics, vol. 97, Springer Science & Business Media, 2012.

[Kod86] Kunihiko Kodaira, *Complex Manifolds and Deformation of Complex Structures*, Grundlehren der mathematischen Wissenschaften, vol. 283, Springer, 1986.

[Köh11] Günter Köhler, *Eta Products and Theta Series Identities*, Springer Monographs in Mathematics, Springer Science & Business Media, 2011.

[KT17] Shamit Kachru and Arnav Tripathy, *Black holes and Hurwitz class numbers*, Int. J. Mod. Phys. D **26** (2017), no. 12, 1742003.

[Kub73] Tomio Kubota, *Elementary Theory of Eisenstein Series*, Halsted Press, 1973.

[KV17] Axel Kleinschmidt and Valentin Verschinin, *Tetrahedral modular graph functions*, J. High Energy Phys. **09** (2017), 155.

[KW07] Anton Kapustin and Edward Witten, *Electric–magnetic duality and the geometric Langlands program*, Commun. Number Theory Phys. **1** (2007), 1–236.

[Lan75] Serge Lang, *Sl(2,r)*, Addison-Wesley Publishing Company, 1975.

[Lan76] ———, *Introduction to Modular Forms*, Grundlehren der mathematischen Wissenschaften, vol. 222, Springer-Verlag, 1976.

[Lan87] ———, *Elliptic Functions*, Graduate Texts in Mathematics, vol. 112, Springer-Verlag, 1987.

[Lin22] Ying-Hsuan Lin, *Topological modularity of Monstrous Moonshine*, arXiv:2207.14076 (2022).

[LP23] Ying-Hsuan Lin and Du Pei, *Holomorphic CFTs and topological modular forms*, Commun. Math. Phys. **401** (2023), no. 1, 325–332.

[LZ99] Ruth Lawrence and Don Zagier, *Modular forms and quantum invariants of 3-manifolds*, Asian J. Math. **3** (1999), no. 1, 93–108.

[Mar77] Daniel A. Marcus, *Number Fields*, Springer Universitext, 1977.

[Mar86] Emil J. Martinec, *Nonrenormalization theorems and Fermionic string finiteness*, Phys. Lett. B **171** (1986), 189.

[Mar96] ———, *Integrable structures in supersymmetric gauge and string theory*, Phys. Lett. B **367** (1996), 91–96.

[Mar12] Christopher Marks, *Fourier coefficients of three-dimensional vector-valued modular forms*, arXiv preprint arXiv:1201.5165 (2012).

[Mar20] Mario Martone, *The constraining power of Coulomb Branch Geometry: Lectures on Seiberg-Witten theory*, Young Researchers Integrability School and Workshop 2020: A modern primer for superconformal field theories, 6 2020.

[Mas12] Geoffrey Mason, *On the Fourier coefficients of 2-dimensional vector-valued modular forms*, Proc. Am. Math. Soc. **140** (2012), no. 6, 1921–1930.

[McK72] H. McKean, *Selberg's trace formula for PSL(2,R)*, Commun. Pure Appl. Math. **25** (1972), 223.

[Miy06] Toshitsune Miyake, *Modular Forms*, Springer Science & Business Media, 2006.

[MM10] Jan Manschot and Gregory W. Moore, *A Modern Farey Tail*, Commun. Number Theory Phys. **4** (2010), 103–159.

[MM21] _____, *Topological correlators of SU(2), $\mathcal{N} = 2^*$ SYM on four-manifolds*, arXiv:2104.06492 (2021).

[MMS88] Samir D. Mathur, Sunil Mukhi, and Ashoke Sen, *On the classification of rational conformal field theories*, Phys. Lett. B **213** (1988), 303–308.

[MMS89] _____, *Reconstruction of conformal field theories from modular geometry on the torus*, Nucl. Phys. B **318** (1989), 483–540.

[MMZ20] Jan Manschot, Gregory W. Moore, and Xinyu Zhang, *Effective gravitational couplings of four-dimensional $\mathcal{N} = 2$ supersymmetric gauge theories*, J. High Energy Phys. **06** (2020), 150.

[MN96] Joseph A. Minahan and Dennis Nemeschansky, *An N=2 superconformal fixed point with E(6) global symmetry*, Nucl. Phys. B **482** (1996), 142–152.

[MN97] _____, *Superconformal fixed points with E(n) global symmetry*, Nucl. Phys. B **489** (1997), 24–46.

[MNW98] Joseph A. Minahan, Dennis Nemeschansky, and Nicholas P. Warner, *Instanton expansions for mass deformed N=4 super Yang-Mills theories*, Nucl. Phys. B **528** (1998), 109–132.

[MO77] Claus Montonen and David I. Olive, *Magnetic monopoles as Gauge particles?*, Phys. Lett. B **72** (1977), 117–120.

[Moo86] Gregory W. Moore, *Modular forms and two loop string physics*, Phys. Lett. B **176** (1986), 369–379.

[Mor17] Louis Joel. Mordell, *On Mr. Ramanujan's empirical expansion of modular functions*, Proc. Camb. Philos. Soc., vol. 19, 1917, pp. 117–124.

[MPS20] Sunil Mukhi, Rahul Poddar, and Palash Singh, *Rational CFT with three characters: The quasi-character approach*, J. High Energy Phys. **05** (2020), 003.

[MS88] Gregory W. Moore and Nathan Seiberg, *Polynomial equations for rational conformal field theories*, Phys. Lett. B **212** (1988), 451–460.

[MS89a] _____, *Classical and quantum conformal field theory*, Commun. Math. Phys. **123** (1989), 177.

[MS89b] _____, *Lectures on RCFT*, 1989 Banff NATO ASI: Physics, Geometry and Topology, 1989.

[MS89c] _____, *Taming the conformal zoo*, Phys. Lett. B **220** (1989), 422–430.

[MS19a] Carlos R. Mafra and Oliver Schlotterer, *Towards the n-point one-loop superstring amplitude. Part I. Pure spinors and superfield kinematics*, J. High Energy Phys. **08** (2019), 090.

[MS19b] _____, *Towards the n-point one-loop superstring amplitude. Part II. Worldsheet functions and their duality to kinematics*, J. High Energy Phys. **08** (2019), 091.

[MS19c] _____, *Towards the n-point one-loop superstring amplitude. Part III. One-loop correlators and their double-copy structure*, J. High Energy Phys. **08** (2019), 092.

[MS23] _____, *Tree-level amplitudes from the pure spinor superstring*, Phys. Rep. **1020** (2023), 1–162.

[MSS13] Carlos R. Mafra, Oliver Schlotterer, and Stephan Stieberger, *Complete N-Point superstring disk amplitude I. Pure spinor computation*, Nucl. Phys. B **873** (2013), 419–460.

[MSS17] Ruben Minasian, Soumya Sasmal, and Raffaele Savelli, *Discrete anomalies in supergravity and consistency of string backgrounds*, J. High Energy Phys. **02** (2017), 025.

[Muk98] Sunil Mukhi, *Dualities and the SL(2,Z) anomaly*, J. High Energy Phys. **12** (1998), 006.

[Muk19] _____, *Classification of RCFT from Holomorphic Modular Bootstrap: A Status Report*, Pollica Summer Workshop 2019: Mathematical and Geometric Tools for Conformal Field Theories, 10 2019.

[Mum07] David Mumford, *Tata Lectures on Theta. I (Modern Birkhäuser Classics)*, Birkhäuser Boston Incorporated, 2007.

[Nac89] Stephen G. Naculich, *Differential equations for rational conformal characters*, Nucl. Phys. B **323** (1989), 423–440.

[Nar86] Kumar S. Narain, *New heterotic string theories in uncompactified dimensions < 10*, Phys. Lett. B **169** (1986), 41–46.

[Nek03] Nikita A. Nekrasov, *Seiberg-Witten prepotential from instanton counting*, Adv. Theory Math. Phys. **7** (2003), no. 5, 831–864.

[NRSTV21] Hans Peter Nilles, Saul Ramos-Sanchez, Andreas Trautner, and Patrick K. S. Vaudrevange, *Orbifolds from Sp(4,Z) and their modular symmetries*, Nucl. Phys. B **971** (2021), 115534.

[NS71] André Neveu and John H Schwarz, *Factorizable dual model of pions*, Nucl. Phys. B **31** (1971), no. 1, 86–112.

[NSW87] Kumar S. Narain, M. H. Sarmadi, and Edward Witten, *A note on toroidal compactification of heterotic string theory*, Nucl. Phys. B **279** (1987), 369–379.

[OM13] Fritz Oberhettinger and Wilhelm Magnus, *Anwendung der elliptischen Funktionen in Physik und Technik*, Die Grundlehren der mathematischen Wissenschaften in Einzeldarstellungen, vol. 55, Springer-Verlag, 2013.

[Ono04] Ken Ono, *The Web of Modularity: Arithmetic of the Coefficients of Modular Forms and q-Series*, CBMS, vol. 102, American Mathematical Society, 2004.

[Ono13] _____, *Harmonic Maass forms, mock modular forms, and quantum modular forms*, Notes (2013), 347–454.

[OP99] Niels A. Obers and Boris Pioline, *U duality and M theory*, Phys. Rep. **318** (1999), 113–225.

[OP00] _____, *Eisenstein series and string thresholds*, Commun. Math. Phys. **209** (2000), 275–324.

[Osb79] Hugh Osborn, *Topological charges for N=4 supersymmetric Gauge theories and monopoles of Spin 1*, Phys. Lett. B **83** (1979), 321–326.

[Osb91] Hugh Osborn, *Weyl consistency conditions and a local renormalization group equation for general renormalizable field theories*, Nucl. Phys. B **363** (1991), 486–526.

[P+17] Vasily Pestun et al., *Localization techniques in quantum field theories*, J. Phys. A **50** (2017), no. 44, 440301.

[Per02] Giorgi Perelman, *The entropy formula for the Ricci flow and its geometric applications*, arXiv:math.DG 0211159 (2002).

[Per03] _____, *Ricci flow with surgery on three-manifolds*, arXiv:math.DG 0303109 (2003).

[Pes12] Vasily Pestun, *Localization of gauge theory on a four-sphere and supersymmetric Wilson loops*, Commun. Math. Phys. **313** (2012), 71–129.

[Pet32] Hans Petersson, *Über die Entwicklungskoeffizienten der automorphen Formen*, Acta Math. **58** (1932), 169–215.

[Pet39] _____, *Über eine Metrisierung der ganzen Modulformen.*, Jahresbericht der Deutschen Mathematiker-Vereinigung **49** (1939), 49–75.

[Pio98] Boris Pioline, *A Note on nonperturbative R**4 couplings*, Phys. Lett. B **431** (1998), 73–76.

[Pio06] Boris Pioline, *Lectures on black holes, topological strings and quantum attractors*, Class. Quantum Grav. **23** (2006), S981.

[Pio15] _____, *D^6R^4 amplitudes in various dimensions*, J. High Energy Phys. **04** (2015), 057.

[Pio16] _____, *A Theta lift representation for the Kawazumi–Zhang and Faltings invariants of genus-two Riemann surfaces*, J. Number Theory **163** (2016), 520–541.

[Pol86] Joseph Polchinski, *Evaluation of the one loop string path integral*, Commun. Math. Phys. **104** (1986), 37.

[Pol87] Aleksandr Michajlovič Polyakov, *Gauge Fields and Strings*, Taylor & Francis, 1987.

[Pol07a] Joseph Polchinski, *String Theory. Vol. 1: An Introduction to the Bosonic String*, Cambridge Monographs on Mathematical Physics, Cambridge University Press, 2007.

[Pol07b] _____, *String theory. Vol. 2: Superstring theory and beyond*, Cambridge Monographs on Mathematical Physics, Cambridge University Press, 2007.

[PPR23] Hynek Paul, Eric Perlmutter, and Himanshu Raj, *Integrated correlators in $\mathcal{N}=4$ SYM via SL(2, $\mathbb{Z}$) spectral theory*, J. High Energy Phys. **01** (2023), 149.

[PW96] Joseph Polchinski and Edward Witten, *Evidence for heterotic - type I string duality*, Nucl. Phys. B **460** (1996), 525–540.

[Rad38] Hans Rademacher, *The Fourier coefficients of the modular invariant J (τ)*, Am. J. Math. **60** (1938), no. 2, 501–512.

[Rad70] _____, *Topics in Analytic Number Theory*, Die Grundlehren der mathematischen Wissenschaften, vol. 169, Springer, 1970.

[Ram71] Pierre Ramond, *Dual theory for free fermions*, Phys. Rev. D **3** (1971), no. 10, 2415.

[Ran39]   R. Rankin, *Contributions to the theory of Ramanujan's function* $\tau(n)$ *and similar arithmetic functions*, Proc. Camb. Philol. Soc. **35** (1939), 351.

[Ran77]   Robert A. Rankin, *Modular Forms and Functions*, Cambridge University Press, 1977.

[RS71]   Daniel B. Ray and Isadore M. Singer, *R-torsion and the Laplacian on Riemannian manifolds*, Advances in Mathematics **7** (1971), no. 2, 145–210.

[RSSZ22]   Shlomo Razamat, Evyatar Sabag, Orr Sela, and Gabi Zafrir, *Aspects of 4d supersymmetric dynamics and geometry*, arXiv:2203.06880 (2022).

[RV20]   Daniel Robbins and Thomas Vandermeulen, *Modular orbits at higher genus*, J. High Energy Phys. **02** (2020), 113.

[Sar87]   Peter Sarnak, *Determinants of Laplacians*, Comm. Math. Phys. **110** (1987), no. 1, 113–120.

[SC67]   Atle Selberg and Sarvadaman Chowla, *On Epstein's Zeta-function.*, Walter de Gruyter, 1967.

[Sch95]   John H. Schwarz, *An SL(2,Z) multiplet of type IIB superstrings*, Phys. Lett. B **360** (1995), 13–18 [Erratum: Phys. Lett. B 364, 252 (1995)].

[Sch97]   _____, *Lectures on superstring and M theory dualities: Given at ICTP Spring School and at TASI Summer School*, Nucl. Phys. B Proc. Suppl. **55** (1997), 1–32.

[Sel40]   Atle Selberg, *Bemerkungen über eine Dirichletsche Reihe, die mit der Theorie der Modulformen nahe verbunden ist*, Arch. Math. Naturvid. **43** (1940), 47.

[Sel56]   _____, *Harmonic analysis and discontinuous groups in weakly symmetric Riemannian spaces with applications to Dirichlet series*, J. of the Indian Mathematical Society **20** (1956), no. 7, 47–87.

[Sel65]   _____, *On the estimation of Fourier coefficients of modular forms*, Proc. Symp. Pure Math., vol. 8, AMS, 1965, pp. 1–15.

[Ser73]   Jean-Pierre Serre, *A Course in Arithmetic*, Graduate Texts in Mathematics, vol. 7, Springer-Verlag, 1973.

[Shi71]   Goro Shimura, *Introduction to the Arithmetic Theory of Automorphic Functions*, vol. 1, Princeton University Press, 1971.

[Shi22]   Mikhail Shifman, *Advanced Topics in Quantum Field Theory*, Cambridge University Press, 2022.

[Sie61]   Carl Ludwig Siegel, *On Advanced Analytic Number Theory*, Tata Institute of Fundamental Research, 1961.

[Sie89a]   _____, *Topics in Complex Function Theory, Volume 2: Automorphic Functions and Abelian Integrals*, vol. 16, John Wiley & Sons, 1989.

[Sie89b]   _____, *Topics in Complex Function Theory, Volume 3: Abelian Functions and Modular Functions of Several Variables*, vol. 16, John Wiley & Sons, 1989.

[Sil85]   Joseph H. Silverman, *The Arithmetic of Elliptic Curves*, Graduate Texts in Mathematics, vol. 106, Springer-Verlag, 1985.

[Sin02]   Aninda Sinha, *The G(hat)**4 lambda**16 term in IIB supergravity*, J. High Energy Phys. **08** (2002), 017.

[Sre07] Marc Srednicki, *Quantum Field Theory*, Cambridge University Press, 2007.

[SS79] Joel Scherk and John H. Schwarz, *How to get masses from extra dimensions*, Nucl. Phys. B **153** (1979), 61–88.

[SS13] Oliver Schlotterer and Stephan Stieberger, *Motivic multiple zeta values and superstring amplitudes*, J. Phys. A **46** (2013), 475401.

[SS14] Menahem Schiffer and Donald C. Spencer, *Functionals of Finite Riemann Surfaces*, Courier Corporation, 2014.

[ST92] Joseph H. Silverman and John Tate, *Rational Points on Elliptic Curves*, vol. Undergraduate Texts in Mathematics, Springer-Verlag, 1992.

[Sti14] Stephan Stieberger, *Closed superstring amplitudes, single-valued multiple zeta values and the Deligne associator*, J. Phys. A **47** (2014), 155401.

[SW86] N. Seiberg and Edward Witten, *Spin structures in String theory*, Nucl. Phys. B **276** (1986), 272.

[SW94a] _____, *Electric - magnetic duality, monopole condensation, and confinement in N=2 supersymmetric Yang-Mills theory*, Nucl. Phys. B **426** (1994), 19–52 [Erratum: Nucl. Phys. B 430, 485–486 (1994)].

[SW94b] _____, *Monopoles, duality and chiral symmetry breaking in N=2 supersymmetric QCD*, Nucl. Phys. B **431** (1994), 484–550.

[SW15] Ashoke Sen and Edward Witten, *Filling the gaps with PCOs*, J. High Energy Phys. **09** (2015), 004.

[SZ06] J. Stienstra and Don Zagier, *Bimodular forms and holomorphic anomaly equation*, Workshop on Modular Forms and String Duality, Banff International Research Centre, 2006 (2006).

[T$^+$22] Gabriele Travaglini et al., *The SAGEX review on scattering amplitudes*, J. Phys. A **55** (2022), no. 44, 443001.

[Tac13] Yuji Tachikawa, *N=2 Supersymmetric Dynamics for Pedestrians*, Springer Hindustan Book Agency, 2013.

[Tac22] _____, *Topological modular forms and the absence of a heterotic global anomaly*, Prog. Theor. Exp. Phys. **2022** (2022), no. 4, 04A107.

[Ter85] Audrey Terras, *Harmonic Analysis on Symmetric Spaces and Applications I*, Springer Science & Business Media, 1985.

[Ter88] _____, *Harmonic Analysis on Symmetric Spaces and Applications II*, Springer-Verlag, 1988.

[TY19] Yuji Tachikawa and Kazuya Yonekura, *Why are fractional charges of orientifolds compatible with Dirac quantization?*, SciPost Phys. **7** (2019), no. 5, 058.

[TY23] Yuji Tachikawa and Mayuko Yamashita, *Topological modular forms and the absence of all heterotic global anomalies*, Commun. Math. Phys. **402** (2023), no. 2, 1585–1620 [Erratum: Commun. Math. Phys. 402, 2131 (2023)].

[Vaf96] Cumrun Vafa, *Evidence for F theory*, Nucl. Phys. B **469** (1996), 403.

[Vaf98] _____, *Geometric origin of Montonen-Olive duality*, Adv. Theor. Math. Phys. **1** (1998), 158–166.

[VDG08] Gerard Van Der Geer, *Siegel modular forms and their applications*, The 1-2-3 of modular forms, Springer, 2008, pp. 181–245.

[VdW91] Bartel Leendert Van der Waerden, *Algebra*, vol. 2, Springer, 1991.

[Ver95] Erik P. Verlinde, *Global aspects of electric – magnetic duality*, Nucl. Phys. B **455** (1995), 211–228.

[Via17] Maryna Viazovska, *The sphere packing problem in dimension 8*, Annals of Mathematics **185** (2017), no. 3, 991 – 1015.

[VV87] Erik P. Verlinde and Herman L. Verlinde, *Chiral bosonization, determinants and the string partition function*, Nucl. Phys. B **288** (1987), 357.

[VW94] Cumrun Vafa and Edward Witten, *A Strong coupling test of S duality*, Nucl. Phys. B **431** (1994), 3–77.

[Was03] Lawrence C. Washington, *Elliptic Curves, Number Theory, and Cryptography*, Chapman & Hall/CRC, 2003.

[WB92] Julius Wess and Jonathan Bagger, *Supersymmetry and Supergravity*, Princeton University Press, 1992.

[Wei76] André Weil, *Elliptic Functions According to Eisenstein and Kronecker*, Ergebnisse der Mathematik und ihrer Grenzgebiete, vol. 88, Springer-Verlag, 1976.

[Wei95] Steven Weinberg, *Quantum Theory of Fields, Volume I*, Cambridge University Press, 1995.

[Wei96] ———, *Quantum Theory of Fields, Volume II*, Cambridge University Press, 1996.

[Wei13] ———, *The Quantum Theory of Fields. Vol. 3: Supersymmetry*, Cambridge University Press, 2013.

[Wit79] Edward Witten, *Dyons of charge e theta/2 pi*, Phys. Lett. B **86** (1979), 283–287.

[Wit95a] ———, *On S duality in Abelian gauge theory*, Selecta Math. **1** (1995), 383.

[Wit95b] ———, *String theory dynamics in various dimensions*, Nucl. Phys. B **443** (1995), 85–126.

[Wit97] ———, *Solutions of four-dimensional field theories via M theory*, Nucl. Phys. B **500** (1997), 3–42.

[Wit99] ———, *Dynamics of quantum field theory*, Quantum Fields and Strings: A Course for Mathematicians (P. Deligne et al., eds.), vol. 2, American Mathematical Society and Institute for Advanced Study, 1999, pp. 1119–1424.

[Wit07] ———, *Three-dimensional gravity revisited*, arXiv:0706.3359 (2007).

[Wit12] ———, *Superstring perturbation theory revisited*, arXiv:1209.5461 (2012).

[Wit13] ———, *Notes on holomorphic string and superstring theory measures of low genus*, arXiv:1306.3621 (2013).

[Wit15a] ———, *The Feynman iε in String theory*, J. High Energy Phys. **04** (2015), 055.

[Wit15b] ———, *The super period matrix With Ramond punctures*, J. Geom. Phys. **92** (2015), 210–239.

[Wit19] ———, *Notes on super Riemann surfaces and their moduli*, Pure Appl. Math. Q. **15** (2019), no. 1, 57–211.

[WO78] Edward Witten and David I. Olive, *Supersymmetry algebras that include topological charges*, Phys. Lett. B **78** (1978), 97–101.

[WW69] Edmund Taylor Whittaker and George Neville Watson, *A Course of Modern Analysis*, Cambridge University Press, 1969.

[Yi04]   Jinhee Yi, *Theta-function identities and the explicit formulas for theta-function and their applications*, J. Math. Anal. Appl. **292** (2004), no. 2, 381–400.

[Zag81a] Don Zagier, *Eisenstein Series and the Selberg Trace Formula*, Automorphic Forms, Representation Theory and Arithmetic, Springer-Verlag, 1981, pp. 303–355.

[Zag81b] _____, *Zetafunktionen und quadratische Körper: eine Einfürhrung in die höhere Zahlentheorie*, Springer, 1981.

[Zag82] _____, *The Rankin–Selberg method for automorphic functions which are not of rapid decay*, J. Fac. Sci., Univ. Tokyo, Sect. 1A, Math. **28** (1982), 415.

[Zag90] Don Zagier, *The Bloch-Wigner-Ramakrishnan polylogarithm function*, Math. Ann. **286** (1990), 613–624.

[Zag92] Don Zagier, *Introduction to Modular Forms*, From Number Theory to Physics, Springer, 1992, pp. 238–291.

[Zag94] Don Zagier, *Modular forms and differential operators*, Proc. Ind. Acad. Sci. **104** (1994), 57–75.

[Zag01] Don Zagier, *Vassiliev invariants and a strange identity related to the Dedekind eta-function*, Topology, Elsevier **40** (2001), no. 5, 945–960.

[Zag07] _____, *Ramanujan's mock theta functions and their applications*, Séminaire BOURBAKI (2007).

[Zag08] _____, *Elliptic Modular Forms and Their Applications*, The 1-2-3 of modular forms, Springer, 2008, pp. 1–103.

[Zag10] _____, *Quantum modular forms*, Quanta of maths, American Mathematical Society Providence **11** (2010), 659–675.

[Zer16] Federico Zerbini, *Single-valued multiple zeta values in genus 1 superstring amplitudes*, Commun. Number Theory Phys. **10** (2016), 703–737.

[Zha08] Shou-Wu Zhang, *Gross–Schoen cycles and dualizing sheaves*, arXiv preprint arXiv:0812.0371 (2008).

[ZJ89]   J. Zinn-Justin, *Quantum Field Theory and Critical Phenomena*, Oxford University Press, 1989.

[Zwi04] Barton Zwiebach, *A First Course in String Theory*, Cambridge University Press, 2004.

Index

Abelian
  extension, 341
  group, 16, 32, 55, 190, 260, 340, 355, 357
Abelian differential, 8, 9, 35, 37–39
  holomorphic, 39, 237, 311, 366, 396
  meromorphic, 37, 38, 40, 398
Abelian integrals, 8, 9, 36, 238
  of the first kind, 39, 40
  of the second kind, 40
  of the third kind, 40
algebraic integer, 188, 212, 214, 217, 355
almost-holomorphic modular forms, xvi, 71,
    74, 76, 87–89, 411, 412
anomaly, 73, 119, 278, 281
anti-symmetric tensor field, 233, 255, 259, 276,
    277, 292, 331, 451
automorphisms, 91, 295
  of a field, 335–338
  of a lattice, 43, 44, 208, 260, 261
  of the supersymmetry algebra, 297
axion field, 275, 284
axion-dilaton field, 2, 3, 221, 275, 277, 284, 292

bc system, 95, 113, 120, 346, 402
  on higher-genus surfaces, 98, 399
  on the annulus, 98
  on the torus, 98, 101, 105, 119, 120
Bézout's lemma, 126, 191, 192
Bernoulli numbers, 15, 54, 80
bi-modular forms, 315
BPS, 304
  bound, 296
  mass formula, 304
  states, 292, 296, 303–305, 320

central charge, 96, 97, 202, 204, 230, 295, 296,
    303, 323, 330
character, 202, 342, 355, 360
  CFT, 163, 166, 201, 203, 270, 343
  conductor, 357

Dirichlet, 147, 149, 346, 347, 356, 359
  primitive, 147, 148, 357, 423
  principal, 148, 357
  Virasoro, 201
characteristic, 332
Chern
  class, 379–381, 396
  numbers, 382
Chern–Simons theory, 203
Chowla–Selberg, 78, 221
closed strings, 2, 223, 224, 265, 274
complex multiplication
  and ϑ-functions, 217, 221
  and Eisenstein series, 220
  and elliptic functions, 209
  and RCFT, 270, 272
  and the j-invariant, 211, 212, 214, 216, 221
  conditions for, 209
complex structure, 229, 261, 271, 302, 367,
    369, 381, 385, 388–390
conformal
  b, c field theory, 98
  algebra, 323
  class, 369
  descendant, 202
  family, 202
  field, 96
  primary field, 96, 97, 100
  spinor field, 119
  structure, 229
  transformation, 31, 96, 98, 232, 234
  weight, 96–98, 113
conformal field theory, 4, 8, 119, 188, 202, 207,
    221, 230, 238, 270, 322, 332, 346, 399
  free, 97
  rational, 97, 167, 188, 342, 344
  superconformal, 188, 294, 315, 323
congruence subgroups, 91, 121, 122, 130, 310
  and Hecke operators, 200, 201
  and vector-valued modular forms, 161, 163

congruence subgroups (cont.)
  Eisenstein series for, 143, 147, 150
  fundamental domain for, 125, 127, 327
  modular curves for, 128, 133
  modular forms for, 137, 138
  orders of, 123
  principal, 121
critical dimension, 230
cusp form, 85, 197, 198, 256
  holomorphic, 49, 50, 52, 55, 60, 61, 66, 128,
      130, 138, 145, 189, 194, 197, 410
  non-holomorphic, 82–86, 93, 198
cusps, 121
cyclotomic
  field, 341, 342, 452
  polynomial, 341

Dedekind
  η-function, 8, 68, 69, 92, 111, 121, 154, 423
degree of a line bundle, 380
diffeomorphism, 202, 226, 368, 370, 389
differential forms, 1, 50, 138, 140, 168, 378,
      379, 388, 391
dihedral graphs, 177
dilaton field, 233, 255, 259, 275, 284
discriminant, 91, 209, 210, 212
  cusp form Δ, 21, 23, 55, 57, 65, 66, 68, 72,
      194, 197
divisor
  class, 384
  of a line bundle, 378, 380, 381, 384, 390,
      394, 396, 400
  sum, 58, 66, 70, 94, 145, 148, 152, 197, 213,
      409, 424

effective interaction, 188, 240, 241, 283–285,
      290, 292
  local, 224, 283
  low energy, 185, 188, 240, 252, 256, 257,
      274, 282, 298
Eisenstein series
  holomorphic, 20, 52–54, 56, 142, 147, 150,
      196, 210, 318
  integrals of, 249
  non-holomorphic, 76, 77, 80, 86, 110, 170,
      177, 198, 248, 285, 288
Eisenstein summation convention, 72, 143,
      145, 406
elliptic curve, 8, 9, 32–35, 37–39, 272
elliptic points, 43, 48, 49, 58, 121, 125, 127,
      129, 132, 139
  absence of, 131, 139
  number of, 130, 132, 135, 377
Euler, 1, 11, 154
  constant, 111, 252, 403
  number, 259, 362, 363, 370, 383
  product formula, 14
  totient function, 122, 348, 357

Faddeev–Popov
  gauge fixing, 228
  ghosts, 229
Fermat's little theorem, 351
Fuchsian group, 50, 346, 362, 373, 374, 376
functional determinant, 8, 95, 97, 110, 120,
      258, 401, 440
fundamental domain
  $SL(2, \mathbb{Z})$, 45, 47, 49, 58, 83, 85, 86, 240, 248,
      256
  congruence subgroup, 125–127, 327
  higher genus, 375, 396
  truncated, 81
fundamental parallelogram, 17, 18, 22, 26, 29,
      32, 47, 190, 210

Galois
  automorphism, 213, 337, 343
  correspondence, 338, 339
  extension, 339
  group, 35, 336–338, 340
  symmetry, 343, 344
  theory, xvi, 332, 339, 340, 344
  theory in physics, 342
Gauss, 1, 154, 211, 355
  curvature, 370, 388
  law, 232
  sum, 354
Grassmann variables, 298, 449
Green function, 8, 95, 97, 107, 108, 397, 400
  Arakelov, 237, 238, 244, 246
  torus, 105, 107, 109, 114–117, 120, 169, 171,
      172, 237, 238, 242, 246, 416
  upper half plane, 136, 417

Hecke, 1, 197, 206, 361
  eigenforms, 196–198
  eigenvalues, 196, 199
  operators, xvi, 4, 67, 133, 165, 188–190, 192,
      194, 195, 198–201, 203, 205, 206, 344,
      432
  theory, 201
  transform, 193, 194, 196, 203, 206
Heegner numbers, 211, 212
Heterotic string, 2, 185, 235–237, 255, 264,
      270, 272, 274, 291, 292, 425
hyperbolic geometry, 43, 46
hypergeometric function, 158, 165, 166, 200,
      217–220, 425, 435

integrality theorem, 163, 166, 200
isometry, 369, 371, 375
  group, 77

j-invariant, 23, 24, 61, 158, 205
  at complex multiplication points,
      211, 212
  at Heegner points, 211
  derivative of, 157

Fourier expansion, 62
  in terms of ϑ, 67
Jacobi, 1
  ϑ-function, 9, 11, 25, 28, 41, 64, 114, 121,
    160, 169, 217, 392, 404
  elliptic function, 9, 23, 25, 31
  form, 257, 292
  identity on ϑ-constants, 394
  symbol, 69, 154, 347, 351, 423
  theorem on sums of squares, 150
Jacobian variety, 394

Kähler
  form, 237
  manifold, 33, 302, 390, 391, 407
  metric, 303, 392
  potential, 303
Kronecker limit formulas, 120
  first, 111
  second, 119

L-function, 357–359
Lagrange, 22
  sum of squares, 151
Laplace
  Beltrami operator, 77, 87, 106, 136, 186,
    189, 372, 385–387, 392
  eigenvalue, 83
  eigenvalue equation, 79, 80, 83, 171, 198,
    252, 288, 289, 291
  operator or Laplacian, 77, 79, 103, 105, 106,
    108, 117, 120, 136, 171, 175, 178, 199,
    253, 417–419, 440
Legendre, 1
  function, 420
  relation, 22, 40
  symbol, 65, 69, 347, 350–352, 354, 422
Liouville
  equation, 372
  theorem, 13, 14, 17, 20, 26, 34, 108

M-theory, 291, 329, 330
Maass
  forms, 80–82, 87–89, 91, 93, 198, 199
  operators, 171, 179, 186
method of images, 8–10, 13, 14, 18, 19, 55
metric, 109, 370, 388
  Einstein-frame, 275, 276
  flat, 259
  hyperbolic, 85, 375
  induced, 226
  Kähler, 303, 392
  Minkowski, 294
  Poincaré, 46, 47, 70, 136, 253, 392
  Riemannian, 46, 106, 226, 367, 369, 371,
    372, 385
  Siegel, 252, 253
  space-time, 2, 3, 225, 227–229, 233, 234,
    259, 267, 272, 275, 284, 285

string-frame, 275
  worldsheet, 230, 260, 262
mock modular forms, 88
modular connection, 73
modular curve, 8, 121, 125, 127, 128, 133, 135,
  136, 346, 377
modular graph
  form, 71, 80, 168, 169, 172, 176–178
  function, xvi, 80, 168, 170, 172–175,
    177–179, 181, 185
moduli space, 229, 238, 261, 321, 326, 378,
  384, 389
  dimension of, 388, 390
  genus-one, 237, 242
  genus-two, 237, 244
  integral over, 246, 247, 252
  of four-punctured sphere, 327–329
  supermoduli space, 256, 257
momentum
  conservation, 178
moonshine, 4, 62
  Mathieu, 90, 93
multiple zeta values, 241
multiplicative function, 54, 66, 70, 409
  completely, 66, 70, 147, 190, 351, 409

Nebentypus, 149
non-critical strings, 229
non-linear sigma model, 97, 221, 227, 233, 270,
  272, 275, 281

open strings, 223

(p, q)
  five-branes, 282
  seven-branes, 293
  strings, 282, 293
periodic functions, 1, 9, 18, 50, 53, 86, 88
  doubly periodic, 16, 22, 107, 108, 403
  of a complex variable, 13, 16
  of a real variable, 9, 10
Petersson, 68, 206, 207
  inner product, 86, 197–199
Planck length, 283
Poincaré, 1
  algebra, 294
  metric, 46, 47, 70, 136
  series, 55, 77, 86, 136
  supersymmetry, 259, 294, 296
  upper half plane, 45, 46, 50, 136, 261, 392
Poisson resummation, 8, 9, 11, 12, 68, 77–79,
  117, 253, 263, 266, 267, 413
polynomial
  characteristic, 313, 342, 451
  cubic, 21
  equation, 340, 341, 346–348
  irreducible, 333, 335, 454
  Laurent, 81–83, 86, 87, 174, 289,
    436, 438

polynomial (cont.)
  minimal, 333, 334, 336–338
  monic, 212, 215, 314
  quartic, 35
  ring, 43, 52, 56, 60, 74, 333, 412
  separable, 337, 452
  symmetric, 241, 242, 245, 312

quadratic reciprocity, 346, 347, 353, 355, 361
quadratic residue, 350, 352, 353, 360, 361
quantum field, 95, 98, 103, 259
  free, 97
  theory, 2, 3, 13, 119, 175, 202, 223, 224, 226,
      241, 242, 259, 278, 294, 298, 329
  theory in 2 dimensions, 29, 207, 226, 399
  theory in 4 dimensions, 296
quasi-modular forms, xvi, 2, 71, 74, 88–90, 93,
      177, 220, 317, 406

Rademacher, 109, 206, 207, 361
  sums, 292
Ramanujan, 66, 88, 212, 221, 410
  τ-function, 66, 70, 194, 197, 206
  identity, 437
Rankin–Selberg, 256
Riemann, 1
  ζ-function, 9, 11, 15, 76–78
  relations, 28, 42
  sphere, 33
  surfaces, xvi, 2, 32, 35, 95, 120, 125, 127
Riemann–Roch theorem, 378, 382–387, 396
Roelcke–Selberg, 86

SageMath, 150
Seiberg–Witten
  curve, 311, 312, 314, 330
  differential, 311, 312
  theory, 4, 221, 313, 330
Selberg, 166
  Chowla-, 78, 221
  Rankin-, 256
  Roelcke-, 86
  trace formula, 253, 258, 438, 440
  Zeta function, 120, 258, 441
Serre derivative, 75, 155, 426
Shapiro, 2, 239
short-distance
  divergences, 2
  normalization, 417
spin structure, 113, 114, 255, 382, 384, 385,
      392, 393, 397
  even, 114–117, 257, 393, 400
  odd, 114, 115, 393, 395, 396, 400
stability subgroup, 55
stress tensor, 105
string
  coupling, 2, 95, 224, 225, 234, 240, 259, 274,
      282
  tension, 274

superconformal
  algebra, 323
  multiplet, 324
  phase, 330
  primary field, 323
  Yang–Mills theory, 330
superfields, 297, 298
  chiral, 298–302
  vector, 298–300, 302
supergravity, 2, 224, 233, 240, 255, 274, 282,
      283
  $\mathcal{N} = 8$, 256
  action, 278
  amplitude, 252
  field equations, 277, 278
  Type IIB, 2, 188, 241, 274–278, 292
Szegö kernel, 117, 398, 400

T-duality, 225, 259, 262, 265, 272
topological modular forms, 425
Type I string theory, 235, 237, 255, 264, 291
Type IIA string theory, 264, 274
Type IIB
  string theory, 264, 274, 282, 284–286, 291,
      330
  supergravity, 274–277

unfolding trick, 9, 10, 86, 413
uniformization, 31, 32
  theorem, 370, 371, 375

Verlinde formula, 342, 343
Verma module, 202
Virasoro
  algebra, 96, 97, 104, 120, 202, 204, 264
  descendants, 270
  generators, 96, 105, 111, 232
Virasoro–Shapiro amplitude, 250

Weierstrass, 1
  σ-function, 22, 41
  $\wp$-function, 9, 23, 31, 34, 36, 56, 209, 319
  ζ-function, 22, 29, 30, 37, 40, 102, 115
  elliptic functions, 19, 21, 25, 29, 31, 72
Witt algebra, 96
worldsheet, 119, 223, 225, 231, 233, 234,
      259–261, 263, 271
  action, 226, 233
  fermion, 255
  indices, 226
  instanton, 262
  metric, 230, 233, 262

Yang–Mills theory, 4, 294–296, 298, 301, 324
  $\mathcal{N} = 1$, 324
  $\mathcal{N} = 2$, 311, 330
  $\mathcal{N} = 3$, 296
  $\mathcal{N} = 4$, 94, 242, 305, 312, 330

Printed in the United States
by Baker & Taylor Publisher Services